AF539375

Farm Tools and Equipment for Agriculture

Farm Tools and Equipment for Agriculture

Prof. Surendra Singh, Ph.D.
Ex. Project Coordinator
AICRP on Farm Implements & Machinery (ICAR)

NEW INDIA PUBLISHING AGENCY
New Delhi – 110 034

NEW INDIA PUBLISHING AGENCY
101, Vikas Surya Plaza, CU Block, LSC Market
Pitam Pura, New Delhi 110 034, India
Phone: + 91 (11)27 34 17 17 Fax: + 91(11) 27 34 16 16
Email: info@nipabooks.com
Web: www.nipabooks.com

Feedback at feedbacks@nipabooks.com

ISBN : 978-93-85516-22-1

Composed and Designed by NIPA

Preface

The relevance of farm equipment and machinery to achieve timeliness of farm operation, efficient use of precious inputs such as HYV seeds, fertilizers, chemicals and irrigation water for sustainable agriculture and enhanced productivity of land and labour does not need further emphasis. They also reduce drudgery on the farm. The Indian farmer needs farm machinery which is site-specific and blending with the socio-economic and agro-climatic conditions that are constantly changing. Indian agriculture must continuously evolve to remain ever responsive to manage the changes and meet the growing and diversified needs of different stakeholders in the entire production to consumption chain. International and national experience has clearly established the benefits of engineering inputs in terms of enhanced productivity by about 15% and reduction in cost of production by 20%. However, engineering inputs in Indian agriculture so far have been limited to only a few crops, farm operations and post harvest activities. Mechanization in livestock and fisheries sectors has been minimal. The Vision of Agricultural Engineering has been to develop and demonstrate appropriate, efficient, safe and gender-friendly equipment and technologies for different farming systems and agro-climatic conditions. Mechanization of hill agriculture, horticultural crops, and dryland farming is the priority. Appropriate machinery for cost effective custom hiring and contract farming is becoming a necessity to obviate individual ownership of implements and yet achieving the desired level of farm mechanization. It is estimated that the energy input to agriculture would have to be increased from the present level of 2.0 to 2.5 kW/ha by 2020 to meet the production and productivity level. About 65% of this power will be through tractors and self-propelled machines. Vagaries of nature and a large tract of cultivable land (60%) remaining rainfed, it is necessary to carry out farming operations in a timely manner to achieve the desired level of production of field and horticultural crops.

The book on *'Farm Tools and Equipment for Agriculture'* is an attempt to put together all relevant information regarding the availability of various tools and equipment for land development, seed bed preparation, seeding, planting & transplanting, weeding & interculture, plant protection, harvesting and threshing, straw management, horticulture and forage crops. Information on each item contains a brief description, its uses and power source required. In this a list of manufacturers who have supported the book writing through photographs and drawings are appended in the last to facilitate the readers for easy location. The information provided is quite exhaustive but by no means complete.

I would like to place on record my gratitude for the inspiration and approval given by Er. S.S. Kohli, Director & Scientist-F, SERC, Department of Science and Technology, Ministry of Science and Technology, Govt. of India, New Delhi for writing this book. Financial support by Department of Science and Technology, Ministry of Science and Technology, Govt. of India, New Delhi is heartfully acknowledged. I am also grateful to Dr. P.A. Turbatmath, Assoc. Dean, College of Agricultural Engineering, Mahatma Phule Krishi Vidyapeeth, Rahuri, Er. S.V. Rane, Head, Department of Agricultural Engineering, College of Agriculture, Pune and Associate Dean, College of Agriculture, Pune for their full hearted support during the execution of project.

Pune, 2015 **Surendra Singh**

Contents

1

Introduction

The application of machines to agricultural production has been one of the outstanding developments in world agriculture during last fifty years. The results of this development can be seen in many aspects such as reduction in burden and drudgery of farm work and worker, increase in production per worker, increase in cropping intensity due to timeliness of operations, reduction in grain losses and increase in farm employment. Mechanization is particularly advantageous when it can minimize a high peak labour demand that occurs over a relatively short period of time each year. For example, sowing of wheat and harvesting of paddy in the months of October and November, transplanting of paddy in the month of June and harvesting of wheat in the months of April and May. Severe labour shortages and high rate of wages during peak period, together with simultaneous demands for increased agricultural production had a marked influence on the mechanization of certain farm operations such as seedbed preparation, sowing/planting/transplanting, harvesting and threshing. Mechanization encourages better management of farms, improvement in working conditions and performance of jobs that would otherwise be difficult by hand. It also helps in reducing the cost of production.

Agriculture is greatly affected by climatic and weather conditions. Power must be taken to work rather than taking work to power. Most field operations are seasonal in nature and very short period is available to perform the task. With the result, most of the farm machines have low annual use i.e. 150-200 hours/year for tillage equipment and 75-125 hours/year for seeding equipment (Singh, 2007; Singh and Verma, 2009). Also farm machines have to work under wide range of agro-climatic conditions. The machines have to be designed to work under different soil and field conditions and to be operated by relatively unskilled worker. In addition to this, manufacturing costs of farm machines have to be kept minimum, so that limited amount of use will not put the cost per hour into a prohibitive range.

In 1951, when the country was in its formative years, there were only 8635 tractors in use and all of them were imported (Singh *et al.*, 2009; Singh *et al.*, 2010; Singh *et al.*, 2011). Production of tractors commenced during 1961-62, turning out 880 of them. This figure peaked to over 262,000 in 1999-2000. The sale of tractors in 2003-2004 was 1,72,000. The quantum of power available for the farming sector rose from 45.29 million kW in 1971-72 to over 282.749 million kW in 2013-14 (Singh *et al.*, 2014). Correspondingly, power intensity on the Indian farm increased from 0.2 kW/ha to 2.02 kW/ha on the basis of net-cropped area. The state of Punjab has the highest average farm-power intensity of 3.5 kW/ha and also has the highest productivity levels. During the same period, contribution of animate power reduced from 60% of the total farm power to less than 12% and mechanical and electrical power sources increased from 40% to over 88%. It is also seen that the adoption of mechanical and electrical power was higher for traction applications than for stationary field operations. Power for traction (tractors and power tillers) increased from 8.46% to 49.3%, indicating that more and more power-operated equipment are coming into use. Human power continues to be a significant component for digging, clod breaking, sowing, interculture, harvesting, threshing, cleaning, and grading for which traditional tools and implements have evolved over time in different parts of the country. The small and marginal farmers rely on draught animals for field operations, transport and agro-processing or go far custom hiring.

The need to achieve timeliness of field operations and effective utilization of inputs has resulted in the development of appropriate machinery, which also reduces drudgery. Traditional tools and implements such as bullock-operated country plough, and *bakhar* for tillage, *dufan* or *tifan, enatigoru* and funnel and tube-attachment on country ploughs as sowing devices; sickles, *khurpi*, spades and *olpad* thresher for harvesting, digging and threshing; and swing basket, Persian wheel (*rahat*) and cradle pump for irrigation*, etc* have been very popular in India. Use of electric or diesel engine-operated irrigation pumps, animal and tractor-operated cultivator and disc harrow for seed-bed preparation; seed drill or seed-cum-fertilizer drill and planter for line sowing with fertilizer application; and mechanical power thresher and combine harvesters has also increased. Farmers have also adopted sprinkler- and drip-irrigation systems in commercial crops.

The first requirement of a farm machine is that it should be able to perform its intended function satisfactorily. At the same time, the management and economic aspects of machine application are also equally important. To achieve this one should have a thorough understanding of the factors affecting field capacities and the cost of owning and operating the field machines. The farm machines can be grouped into different categories depending upon their method of hitching viz. manually operated, pull type or trailed type, mounted type, semi-mounted type and self-propelled machine. Manually operated machines are basically hand tools used in small plots and kitchen gardens. A trailed type implement is pulled from a single hitch point either by draught animal(s), power tiller or tractor. It is never completely supported by the power source. A mounted implement is attached to the tractor through a 3-point linkage system and is completely supported when in the raised position. A semi-mounted implement is attached to the tractor through a horizontal hinge axis and is partially supported by tractor. The rear of the implement is supported by its own wheel(s). A self-propelled machine has power source as an integral part of it.

The manufacture of agricultural machinery in the country is carried out by village artisans, tiny units, small-scale industries in unorganized sector and the large industries in organized sector. Production of tractors, motors, engines and process equipment is the domain of the organized sector. The traditional artisans and small-scale industries rely upon own experience; user's feedback and government owned research and development institutions for technological support and operate from their backyards or on road side establishments without regular utility services. Medium and large-scale industries operate in their own premises with sound infrastructure, usually forming a part of an industrial estate, well-established manufacturing and marketing facilities and employ skilled manpower. Diesel engines, electric motors, irrigation pumps, sprayers and dusters, land development machinery, tractors, spare parts, power tillers, post-harvest and processing machinery and dairy equipment are produced in this sector. They have professional marketing network of dealers and provide effective after sales service. They also have in-house research and development facilities or have joint ventures with advanced countries for technology up-gradation. India is recognized, the world over, as a leader in the manufacture of tractor, agricultural equipment and machinery such as combine harvesters, plant protection equipment, drip irrigation and micro-sprinkler. Sizeable quantities of farm implements are exported to Africa, Middle East, Asia, South America and other countries. A list of agricultural machinery developed in the country is given in Table 1.1.

The book on 'Farm Tools and Equipment for Agriculture' is an attempt to put together all relevant information regarding the availability of various tools and equipment for land development, seed bed preparation, seeding, planting & transplanting, weeding & interculture, plant protection, harvesting and threshing, straw management, horticulture and forage. Information on each item contains a brief description, its uses and power source required. In this appendix, a state-wise list of manufacturers particularly who supported in writing of this book is also appended to facilitate easy location of local manufacturers. The information provided is quite exhaustive but by no means complete. Additions, alterations, suggestions and additional information may be provided by the users for issuing supplements and to incorporate them in future editions. Manuscript has been designed as a textbook for the students of Agricultural Engineering in particular to Farm Machinery. The book includes chapters on land development, tillage equipment, sowing, planting & transplanting equipment, weeding & intercultural tools, fertilizer application equipment, plant protection equipment, harvesting and threshing equipment, forage harvesting equipment, horticultural & plantation crop equipment and on straw management equipment.

Table 1.1 : Farm tools and equipment developed/adapted in India

S. No.	Name of Equipment	Power source	Work capacity, ha/h
1.	Manual tool carrier	One person	0.04
2.	Animal drawn light ridger plough	Bullock pair	0.03
3.	Birsa animal drawn ridger plough	Bullock pair	0.022
4.	Animal drawn Bose plough	Bullock pair	0.01
5.	Animal drawn Chisel plough	Bullock pair	0.2
6.	TNAU Animal drawn tool carrier	Bullock pair	0.10
7.	CIAE Animal drawn multi-purpose tool frame	Bullock pair	0.12
8.	Animal drawn Naveen Bakhar blade	Bullock pair	0.07
9.	Animal drawn channel-cum-bund former	Bullock pair	1.00
10.	Animal drawn disc harrow	Bullock pair	0.13

11.	Animal drawn disc harrow-cum-puddler	Bullock pair	0.15
12.	Animal drawn puddler	Bullock pair	0.07
13.	Animal drawn helical blade puddler	Bullock pair	0.12
14.	Hydrotiller	5 hp diesel engine	0.15
15.	Tractor drawn basin lister	30 hp tractor	0.60
16.	Tractor drawn bed-furrow former	35 hp tractor	0.75
17.	Tractor drawn channel-cum-bund former	20 hp tractor	0.4
18.	Tractor drawn pulverizing roller attachment	35 hp tractor	0.60
19.	Tractor drawn spiked clod crusher	35 hp tractor	0.5
20.	Tractor mounted rotary puddler	35 hp tractor	0.45
21.	Tractor drawn combine tillage tool	35 hp tractor	0.22
22.	Wetland leveler	10-12 hp diesel engine	0.16
23.	Hydraulically controlled levler	35 hp tractor	0.40
24.	Patella harrow	Bullock pair	0.30
25.	Power tiller with Rotary attachment	Power tiller	0.30
26.	Pulverizing roller attachment to tiller	35 hp tractor	0.40
27.	Rotary tiller/ rotavator	35 hp tractor	0.25
28.	Tractor operated laser land leveller	45 hp tractor	0.25
29.	Tractor mounted furrower	35 hp tractor	0.25
30.	Manually operated mustard drill	Two persons	0.05
31.	Naveen dibbler	One person	0.028
32.	Rotary dibbler	One person	0.042
33.	Low land rice seeder	Two person	0.12
34.	Adjustable row marker	One person	0.08
35.	Seeding attachment to animal drawn country plough	Bullock pair	0.05
36.	Birsa animal drawn seed drill	Bullock pair	0.035
37.	Seeding attachment to CIAE animal drawn tool frame	Bullock pair	0.10
38.	HAU Animal drawn seed-cum-fertilizer drill	Bullock pair	0.15
39.	HAU Animal drawn mustard drill	Bullock pair	0.2
40.	Camel operated seed-cum-fertilizer drill	One camel	0.22

41.	APAU Animal drawn seed-cum-fertilizer drill	Bullock pair	0.20
42.	Animal drawn Jyoti multicrop planter	Bullock pair	0.10
43.	TNAU Animal drawn planter	Bullock pair	0.19
44.	CIAE animal drawn planter	Bullock pair	0.12
45.	IISR Animal drawn automatic sugarcane planter	Bullock pair	0.10
46.	Animal drawn potato planter	Bullock pair	0.10
47.	TNAU Tractor mounted cultivator seed planter	35 hp tractor	0.63
48.	APAU Tractor mounted cultivator seed planter	35 hp tractor	0.40
49.	Tractor mounted seed-fertilizer drill-cum-planter	25 hp tractor	0.40
50.	Tractor mounted seed-cum-fertilizer drill for oilseeds	30 hp tractor	0.30
51.	Tractor mounted ridge planter for winter maize	35 hp tractor	0.20
52.	Tractor mounted broad bed former-cum-seed planter	35 hp tractor	0.43
53.	Tractor mounted ridger seeder	35 hp tractor	0.20
54.	Tractor mounted direct rice seeder	35 hp tractor	0.68
55.	Tractor drawn mustard drill attachment to semi-automatic sugarcane planter	35 hp tractor	0.30
56.	Tractor mounted planter	35 hp tractor	0.30
57.	Tractor mounted inclined plate planter	35 hp tractor	0.50
58.	Tractor mounted no-till drill	35 hp tractor	0.46
59.	Power operated sugarcane sett-cutting machine	Tractor PTO & 5 hp diesel engine	2000 kg/h
60.	Tractor drawn semi-automatic sugarcane planter	35 hp tractor	0.18
61.	Tractor drawn sugarcane cutter planter with discs	35 hp tractor	0.20
62.	Tractor operated multipurpose implement for sugarcane	35-45 hp tractor	0.20-0.80
63.	Raised bed seeder-cum-cane planter	35-45 hp tractor	0.25-0.40
64.	Tractor mounted ridger type sugarcane cutter planter	35 hp tractor	0.20

65.	Tractor drawn seed drill for rapeseed / mustard	30 hp tractor	0.60
66.	Manual Garlic planter/multicrop planter	Manual, 2-3 persons	0.04
67.	Animal drawn three row seed-cum-fertilizer drill	Bullock pair	0.18
68.	Power tiller operated till plant machine	Power tiller 10-12 hp	
69.	Tractor operated zero-till seed-cum-fertilizer drill	35 hp tractor	0.30
70.	Tractor operated strip till drill	35 hp tractor	0.25
71.	Tractor mounted raised bed planter	35 hp tractor	0.25
72.	Tractor operated inclined plate planter	35 hp tractor	0.42
73.	Tractor mounted inclined plate planter for intercrop sowing on raised beds	35 hp tractor	0.40
74.	Tractor operated inclined plate planter (PAU)	35 hp tractor	0.45
75.	Tractor operated sett cutter planter	35 hp tractor	0.20
76.	Power tiller mounted air assisted seed drill	Power tiller 10-12 hp	0.25
77.	Tractor operated roto till drill	35 hp tractor	0.25
78.	CIAE animal drawn inclined plate planter	Bullock pair	0.12
79.	NEH paddy seeder	manual	0.05
80.	Tractor operated hill drop planter	35 hp tractor	0.25
81.	Manually operated 5-row rice transplanter	One person	0.012
82.	Manually operated 6-row rice transplanter	One person	0.04
83.	Power tiller mounted rice transplanter	8 hp power tiller	0.13
84.	Self propelled riding type rice transplanter	5 hp diesel engine	0.15
85.	Self propelled rice transplanter	6 hp engine	0.12
86.	Tractor operated vegetable transplanter (Plug type)	45 hp tractor	0.14
87.	Tractor operated vegetable transplanter three row (Plug type)	45 hp tractor	0.15
88.	Tractor operated vegetable transplanter (Picker wheel type)	35 hp tractor	0.10
89.	Tractor operated inclined plate planter with raised bed farming attachment for intercrop	35 hp tractor	0.45

90.	Tractor operated pneumatic planter	35 hp tractor	0.30
91.	Tractor operated raised bed planter	35 hp tractor	0.25
92.	Tractor mounted garlic planter	35 hp tractor	0.40
93.	Tractor mounted check row planter	35 hp tractor	0.61
94.	Twine auger digger sugarcane planter	45hp tractor	0.15
95.	Manually operated fertilizer broadcaster	One person	0.8
96.	RPS Marker cum USG Dispencer	One person	-
97.	TNAU Urea Super Granuler Applicator	One person	0.07
98.	Sugarcane stubble shaver cum fertilizer applicator	35 hp tractor	0.53
99.	Rotary nozzle for tractor operated sprayer for mango orchard	35 hp tractor	-
100.	Power tiller operated manure spreader	8-10 hp power tiller	0.90
101.	Tractor operated farm yard manure spreader	35 hp tractor	0.25
102.	Grubber	One person	0.005
103.	TNAU Peg type dry land weeder	One person	0.007
104.	CIAE Peg type dry land weeder	One person	0.009
105.	Twin wheel hoe	One person	0.01
106.	CIAE Single Wheel hoe	One person	0.009
107.	PAU Wheel hand hoe	One person	0.03
108.	TNAU Star weeder	One person	0.007
109.	APAU Star Weeder	One person	0.0125
110.	Animal drawn Multipurpose hoe	Bullock pair	0.15-0.25
111.	Animal drawn sugarcane earthing hoe	Bullock pair	0.14
112.	Animal drawn sweeps	Bullock pair	0.20
113.	Weeding attachment to CIAE Animal drawn tool frame	Bullock pair	0.15
114.	Self-propelled weeder	2 hp engine	0.12
115.	Self-propelled weeder	3 hp engine	0.06
116.	Rotary weeder	1 kW engine	0.07
117.	Tractor mounted earthing-cum-interculture equipment	35 hp tractor	0.40
118.	Tractor mounted 2-row sugarcane stubble shaver	35 hp tractor	0.35
119.	Small hp tractor mounted sugarcane	25 hp tractor	0.30

	inter-culturing implement		
120.	Power tiller drawn earthing up-cum-fertilizer applicator for wide row sugarcane crop	35 hp tractor	0.25
121.	Ratoon management device (RMD)	35 -45 hp tractor	0.30
122.	Cono weeder	manual	0.18
123.	Self propelled power weeder	5 hp diesel engine	0.10
124.	Tractor operated three row rotary weeder	35 hp tractor	0.24
125.	PAU light weight power tiller	5 hp diesel engine	0.05
126.	Engine operated weeder for low land paddy	2.0 hp engine	0.05
127.	Self-propelled cono weeder for SRI	1.5 hp diesel engine	0.05
128.	Battery powered low volume knapsack spinning disc sprayer	6V rechargeable battery	0.20
129.	Tractor mounted CDA crop sprayer	25-30 hp tractor	0.76
130.	Tractor mounted orchard sprayer	35 hp tractor	0.25
131.	Self-propelled high clearance sprayer	20 hp diesel engine	1.6
132.	Self-propelled boom sprayer	5 hp diesel engine	0.80
133.	Multi orchard sprayer	25-30 hp tractor	0.50
134.	Tractor mounted sprayer	25-30 hp tractor	1.50
135.	Aero blast sprayer	25-30 hp tractor	1.70
136.	Power tiller mounted earthing up-cum-fertilizer applicator	8-10 hp power tiller	0.20
137.	Naveen sickle	One person	0.018
138.	Punjab sickle	One person	0.01
139.	Vaibhav sickle	One person	0.01
140.	Sugarcane knives	One person	0.011 ton/day
141.	TNAU animal drawn groundnut digger	Bullock pair	0.10
142.	Udaipur animal drawn groundnut digger	Bullock pair	0.16
143.	Birsa animal drawn potato digger	Bullock pair	0.03
144.	PAU animal drawn single row potato digger	Bullock pair	0.12
145.	Self-propelled vertical conveyor reaper	5 hp diesel engine	0.20
146.	Self-propelled rice harvester	6 hp diesel engine	0.12
147.	Power tiller operated vertical conveyor reaper windrower	8-10 hp diesel engine	0.25

148.	Self-propelled flail-type bush cutter	6 hp diesel engine	0.20
149.	CIAE tractor front mounted vertical conveyor reaper	25 hp tractor	0.31
150.	PAU tractor front mounted vertical conveyor reaper	25 hp tractor	0.35
151.	Riding type vertical conveyor reaper	6 hp engine	0.28
152.	Forage harvester	25/35 hp tractor	0.15
153.	Tractor mounted groundnut digger shaker windrower	35 hp tractor	0.25
154.	Tractor mounted groundnut digger	35 hp tractor	0.266
155.	Tractor mounted 2-row potato digger	35 hp tractor	0.3
156.	Tractor mounted potato digger elevator	35 hp tractor	0.14
157.	Tractor mounted onion harvester-cum-elevator	35 hp tractor	0.21
158.	Tractor front mounted grass seed collector	35 hp tractor	0.30
159.	Self-propelled riding type reaper	6 hp diesel engine	0.20
160.	Tractor mounted fodder harvester	55 hp tractor	0.20
161.	Self-propelled fodder harvester	5 hp diesel engine	0.18
162.	Power tiller mounted groundnut digger	10 hp power tiller	0.05
163.	Tubular maize sheller	One person	20 kg/h
164.	Pedal operated thresher	Two person	44 kg/h
165.	Groundnut stripper (drum type)	One person	18 kg/h
166.	Groundnut stripper (comb type)	One person	40 kg/h
167.	Phule sunflower thresher	One person	40 kg/h
168.	TNAU Groundnut decorticator	One person	100 kg/h
169.	CIAE Groundnut-cum-castor decorticator	One person	50 kg/h
170.	Lac sheller (Peg type)	One person	4.78 kg/h
171.	Single earhead thresher	0.5 hp motor samples (ears)h	300
172.	Multicrop plot thresher	1 hp motor	8-10 samples of 2 kg size/h
173.	CIAE multicrop thresher	5 hp motor	200-1635 kg/h
174.	Semi-axial flow multicrop thresher	7.5 hp motor	350-1350 kg/h
175.	High capacity multicrop thresher	20 hp motor/ 35 hp tractor	2000 kg/h

176.	APAU multicrop thresher	5 hp motor	33.5 kg/h (Safflower)
177.	High capacity pigeon pea thresher	7.5-10 hp motor/ tractor PTO	250 kg/h
178.	Portable rice thresher	5 hp engine or electric motor	100 kg/h
179.	PAU axial flow rice thresher	35 hp tractor	1300 kg/h
180.	Groundnut pod stripper	2 hp motor	120 kg/h
181.	PAU axial flow groundnut thresher	25 hp tractor	220 kg/h
182.	TNAU groundnut thresher	5 hp motor	150 kg/h
183.	APAU sunflower thresher	5 hp motor	200 kg/h
184.	PAU axial flow sunflower thresher	7.5 hp motor	700 kg/h
185.	APAU power operated castor thresher	3 hp motor	250 kg/h
186.	TNAU power operated castor sheller	0.5 hp motor	163 kg/h
187.	Sunflower seed sheller	3 hp motor	100 kg/h
188.	Pantnagar axial flow multicrop thresher	10-15 hp motor/ tractor	312-1200 kg/h
189.	Flow through rice thresher	8 hp engine or 7.5 hp motor	1620 kg/h
190.	Power operated maize dehusker-cum-sheller	35 hp tractor tractor	20 q/h
191.	Tractor operated straw combine	45 hp tractor	0.40
192.	Tractor mounted banana stem shredder	35-45 hp tractor	52 stems/h
193.	Tractor operated banana clump remover	35-45 hp tractor	0.50
194.	Sunflower thresher	7 hp motor/tractor	940 kg/h
195.	Tractor operated groundnut combine	45 hp tractor	0.13-0.15
196.	Whole crop maize thresher	7.5 hp motor/ tractor	350-400 kg/h
197.	APAU seed treating drum	One person	100 kg/h
198.	Tree climber	One person	1.5 min/tree
199.	Plastic mulch laying machine	35-45 hp tractor	0.16
200.	Power tiller operated shredder-cum-in situ incorporator	8-10 hp power tiller	0.08
201.	Tractor mounted rotary field shredder for sugarcane	35-45 hp tractor	0.37
202.	Sugarcane leaf stripper	5 hp engine	-
203.	Tractor mounted turmeric digger	35-45 hp tractor	0.17
204.	Flial type chopper-cum-spreader	35-45 hp tractor	0.40

205.	Cutter bar type chopper-cum-spreader	35-45 hp tractor	0.35
206.	Self-propelled flial type forage harnesser	10 hp diesel engine	0.50
207.	Self-propelled cutter bar type forage harvester	10.2 hp diesel engine	0.28
208.	Tractor operated cutter bar type forage harvester	35-45 hp tractor	0.25
209.	Tractor operated chaffer-cum-loader	35-45 hp tractor	0.35
210.	Self-propelled lucern harvester	5 hp engine	0.11
211.	Tractor operated sugarcane harvester	35-45 hp tractor	0.20
212.	Sugarcane detrasher	One person	400 kg/h
213.	Sugarcane leaf stripper	One person	400 kg/h
214.	Sugarcane trash shredder with chopper knives	35-45 hp tractor	0.35
215.	Self-propelled billet type sugarcane harvester	150 hp engine	0.20
216.	HAU acid delinter Model-I	One person	3.8-7.5 kg/h
217.	HAU acid delinter Model-II	4 Persons	38-100 kg/h
218.	HAU acid delinter Model-III	4 Persons	125 kg/h

(*Source* : Anonymous, 2008, 2008a, 2010, 2013; Garg and Singh, 2002; Pandey *et al.*, 1997; Singh and Pandey, 2008)

References

Anonymous. 2008. NAAS Report on the evaluation of Plan Scheme Central Institute of Agricultural Engineering, Bhopal for the X Five Year Plan (2002-2007). NAAS New Delhi

Anonymous. 2008a. Research Highlight. AICRP on Farm Implements and Machinery, CIAE Bhopal. Technical Bulletin No.: CIAE/2008/141.

Anonymous. 2010. Research Highlight. AICRP on Farm Implements and Machinery, CIAE Bhopal. Technical Bulletin No.: CIAE/2010/151.

Anonymous. 2013. Research Highlight. AICRP on Farm Implements and Machinery, CIAE Bhopal. Technical Bulletin No.: CIAE/2013/158.

Garg I K; Singh Surendra. 2002. Farm equipment for Punjab agriculture. Department of Farm Power & Machinery, Punjab Agricultural University, Ludhiana.

Pandey M M; Majumdar K L; Singh Gyanendra; Singh Gajendra. 1997. Farm Machinery Research Digest. Technical Bulletin No. CIAE/97/69, Central Institute of Agricultural Engineering, Bhopal, 328 p.

Singh R S; Singh S P; Singh Surendra. 2009. Sales of tractors of different makes in India. *Agricultural Engineering Today*, 33(3): 20-37.

Singh S P; Singh R S; Singh Surendra. 2011. Sales trend of tractors and farm power availability in India. *Agricultural Engineering Today*, 35(2): 25-35.

Singh Surendra. 2007. Farm Machinery – Principles and Applications. Directorate of Information & Publication of Agriculture, Indian Council of Agricultural Research, Krishi Anusandhan Bhawan-I, Pusa Campus, New Delhi.

Singh Surendra; Pandey M M. 2008. X Plan Achievements (2002-2007). AICRP on Farm Implements and Machinery, CIAE Bhopal. Technical Bulletin No.: CIAE/2008/137.

Singh Surendra; Singh R S; Singh S P. 2010. Farm power availability and agricultural production scenario in India. *Agricultural Engineering Today*, 34(1): 9-20.

Singh Surendra; Singh R S; Singh S P. 2014. Farm power availability on Indian farms. *Agricultural Engineering Today*, 38(4).

Singh Surendra; Verma S R. 2009. Farm Machinery Maintenance & Management. Directorate of Information & Publication of Agriculture, Indian Council of Agricultural Research, Krishi Anusandhan Bhawan-I, Pusa Campus, New Delhi.

2

Land Levelling Equipment

Declining water table and degrading soil health are the major concerns for the current growth rate and sustainability of Indian Agriculture. Thus proper emphasis is being given on the management of irrigation water usage for adequate growth of agriculture. Keeping in view, the need for judicious use of our natural resources, concerted efforts are being made to enlighten the farmers for efficient use of irrigation water at farm level. Generally, in rice-wheat rotation farmers believed that their fields are levelled and needed no further levelling, but this is not true. Most of the fields are not adequately levelled and requires further precision land levelling. The enhancement of water use efficiency and farm productivity at field level is one of the best options to redress the problem of declining water level.

Levelling, smoothing and shaping the field surface is as important to the surface irrigation system and as the design of laterals, manifolds, risers and outlets for sprinkler drip irrigation systems. It is a process for ensuring that the depths and discharge variations over the field are relatively uniform and, as a result, that water distributions in the root zone are also uniform. These field operations are required nearly every cropping season, particularly where substantial cultivation following harvest disrupts the field surface. There are basically two types of land levelling viz. (i) to provide a slope which fits a water supply; and (ii) to level the field to its best possible condition with minimal earth movement. The second one is generally the most feasible one, because land levelling is expensive and large earth movements may leave significant areas of the field without fertile topsoil, and is the most economic approach also.

Land levelling always improves the efficiency of water, labour and energy resources utilization. Land levelling operations requires for an area depends upon the topography, soil type, soil depth, prevailing land slope, rainfall characteristics, crops to be grown, source of water supply, method of irrigation and other special features of the site including the preference of farmer and power source to be used for cultivation. Usually the field is not graded to a truly level surface, but a gentle uniform slope is maintained to meet the requirements of irrigation and drainage. Land levelling operations may be grouped into three parts viz. rough grading, land levelling and land smoothing (Singh, 2007). Rough grading is the removal of mounds, dunes and other irregularities on the land surface. It also includes filling of pits, depressions and gullies. Land levelling reshapes the land surface to a planned grade. Land levelling requires movement of large quantities of soil over a considerable distance. Land soothing is done prior to seeding as a regular land preparation practice. Planning and survey is to be done prior to land-levelling operation. The entire area should be taken into consideration for the purpose of planning and land levelling operation. Information about type of soil, existing field boundaries, source of irrigation etc may be obtained. A topographical plan of area with levels marked may be prepared, which helps in earthwork calculations. Major topographical changes will nearly reduce crop production in the cut areas until fertility is replaced. Similarly, equipment traffic can so compact or pulverize the soil that water penetration becomes major problem for some time.

The term "Land Levelling" generally applies to mechanize grading of agricultural land based on a detailed engineering survey, design, and layout. In only a few special situations does the final product of land levelling result in a level field. Normally final slopes are up to three percent for furrow irrigation and up to two percent for border irrigation. Most small-scale farming operations rely on animal power or small mechanized equipment which an individual can own and operate. Over a period of several years individual fields are smoothed enough to be watered fairly well. The one feature common to most small-scale land levelling is the trial and error nature of the practices and the long-term incorporation of land levelling with seed bed preparation. Another feature is that no technical or engineering inputs are needed.

New equipment is continually being introduced which provides the capability for more precise land levelling operations. One of the most significant advances has been the adaptation of laser control in land levelling equipment. The equipment has made level basin irrigation particularly attractive since the final field grade can be very precise. Comparisons with less precise techniques have clearly shown that laser-levelled fields achieve better irrigation and production performance. Nevertheless, for most irrigated agriculture, laser-controlled precision is unfeasible because of the high cost of such equipment unless a large number of farmers form a cooperative or a government programme is started with subsidized land levelling as one component in an effort to improve farm production or having a custom hiring centres. Most land leveling is done using a laser-controlled scraper pulled by a tractor. The laser is set to pre-determined cross and run slopes, and the scraper automatically adjusts the cut or filled land over the plane of the field as the tractor moves.

All levelling work should be designed based on measurement of land elevations (topography). If more than one irrigation method or more than one kind of crop is planned, the land must be levelled to meet the requirements of the most restrictive irrigation method and crop. The levelling work must adhere to the slope limits of the water application method, provide for removal of excess surface water, and control erosion caused by rainfall. Land levelling is typically used on mildly sloping land, whereas contour farming is used to farm on modest slopes and terrace farming is used for steeply sloping land. Land levelling is primarily used by agricultural producers using surface irrigation methods (furrow, border, basin, or flood) or by those wishing to improve surface drainage of their non-irrigated field. Land levelling work falls into two general categories: Large scale land shaping prior to cultivating newly irrigated land or land that has never been graded; or floating of a field prior to preparation of seed beds or borders. The time required to grade a field depends on the size and type of land grading equipment, the quantity of soil to be moved, and the complexity of the existing field surface. Typically, the time required to "touchup" a field prior to planting is measured. Effective land levelling reduces the work in crop establishment & crop management, and increases the yield and quality. Level land improves water coverage that improves crop establishment, reduces weed problems, improves uniformity of crop maturity, decreases the time to complete tasks and reduces the amount of water required for land preparation.

The basic methods for land levelling are plane method, profile method, plan inspection method and contour adjustment method (Singh *et al.*, 2015). Each method has advantages and disadvantages and is best adapted to specific site conditions. In the plane method, the centroid of area is found and plane is passed through this point at an elevation equal to average elevation of field. Under this condition, irrespective of slope of plane, the volume of excavation equals the volume of fill. In profile method, ground profiles are plotted along the grid lines using grid point elevation. Final land profiles are chosen by trial and error method such that cuts and fills balance. This method is suitable for levelling flat lands or lands with undulating topography. In this method, profiles are commonly plotted in one direction and datum lines are kept at proposed profiles. Two-way profiles are used when slopes are given in both directions. In plan inspection method, a suitable down slope and cross slope is assumed and cuts and fills are calculated. Assumed slopes are altered until a balance of cuts and fills are obtained. In contour adjustment method, a contour map of area is drawn using the levels. The ground surface expected after grading is shown on the same map with new contour lines such that a uniform desired slope is obtained. The cuts and fills are estimated at the grid points by interpolating between contour lines and by taking the difference in elevations between original and new surface.

Equipment for Land Levelling and Grading

Land levelling and grading is done by animal-drawn, tractor-drawn and self-propelled equipment. Scoops, bulldozers, scrapers, and levellers drawn by a pair of bullocks or power tillers or tractors commonly do rough grading (Singh *et al.*, 2015). Land smoothing is done by tractor-drawn land planes and animal-drawn wooden floats. Both wheel type and crawler type tractors are used in land levelling, grading and smoothening operations.

Soil scoop

It is used in excavating ditches, cleaning drains and moving soil over short distances. It is available both animal-drawn and tractor operated. It consists of blade, soil trough, and hitching-loop and a handle (Fig. 2.1). Blade is made of high carbon steel with carbon content 0.5-0.6%. The angle of cutting blade varies from 12-15 degrees. The blade is bolted to the soil trough made of mild steel sheet.

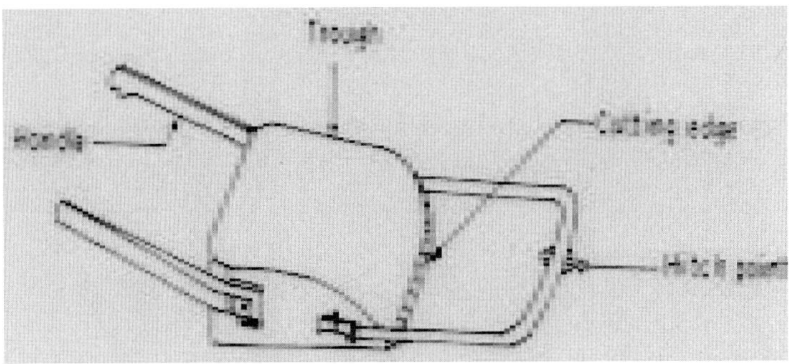

Fig. 2.1 : A view of soil scoop

Buck Scraper

It is a simple machine. It consists of a front board, tailboard, joints, handle and hitch. The machine is most efficient for land grading when animal power is to be used for small and medium fields (Fig. 2.2). It may be used to move the soil loosened by plough. It works well if haulage distance does not exceed 50 m. For long distances tractor operated pull type scraper can be used (Fig. 2.3).

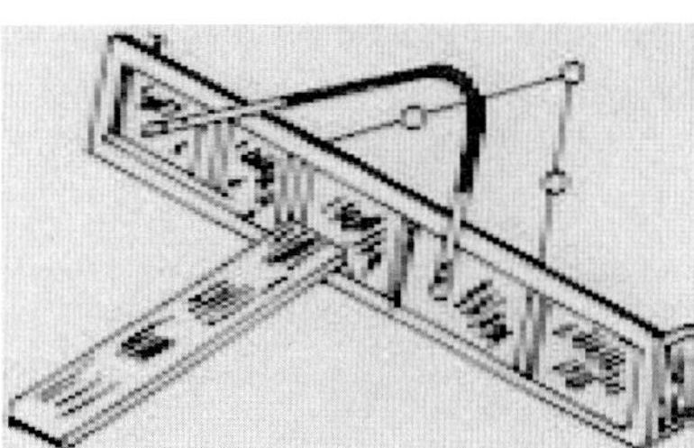

Fig. 2.2 : Animal-operated buck scraper

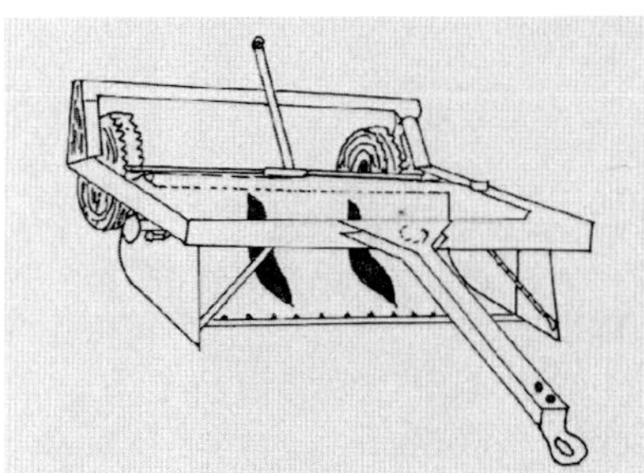

Fig. 2.3 : Tractor-operated pull-type scraper

Leveller

Land levelling is an important farm operation. It is used to level the field to ease not only field operation by machines but receives uniform level of irrigation water and reduces water requirements. The possibility of water logging and soil erosion is reduced considerably. The farmers commonly use wooden logs or metallic planks as levellers (Fig. 2.4a). They may be animal drawn or tractor operated. The other improved type of land leveller generally used on large farms is called levelling 'Karaha' or land leveller. It is suitable for land preparation operations such as scraping, grading, levelling and back filling. It is also used for irrigation, terrace work and general cleaning of field. Leveling of fields and pulling or pushing loosened soil from one place to other. It is available in animal-drawn and tractor-drawn (Singh, 2007; Singh *et al.*, 2015). It consists of hitch system, replaceable cutting blade with sharp edge, and a curved plate with side wings, which form a bucket (Fig. 2.4b). The blade is made from medium carbon steel or low alloy steel, hardened and tempered to suitable hardness. During operation, the blade digs into the soil and extra soil is collected in the bucket, which is released in the depressions of the field. The angle and pitch of leveller is adjustable. The leveller can also be angled left or right, or reversed for back filling. The amount of work done depends on the various factors such as hardness of soil, transportation distance and volume of soil cut each time. The mounted type levellers are usually used for fine grading of small and medium sized fields.

a) Levelling by animal drawn wooden plank in dry condition and tractor operated plank in wet condition

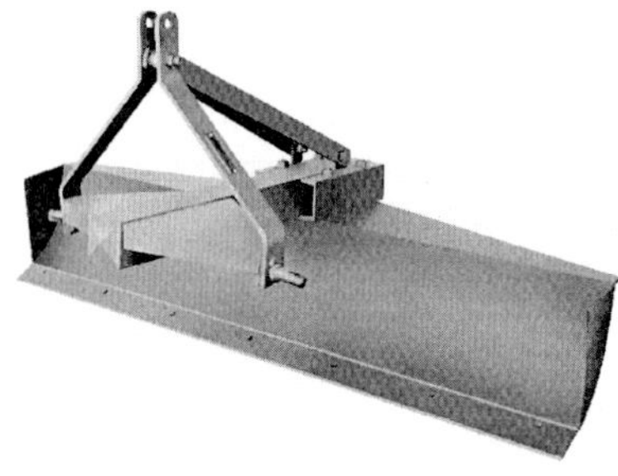

b) Tractor operated land leveller

Fig. 2.4 : Plank and land leveller

The tractor operated blade terracer has an adjustable blade, which is so constructed that it can drag a considerable amount of earth (Pandey *et al.*, 1997). To tilt the blade for ditching or terracing it is tilted to the desired angle by moving the index pin (Fig. 2.5). The blade can be pitched forward and backward or tilted at 15^0 to 30^0 left or right. It can be reversed for back filling. For increasing the length of blade, extensions are provided. The depth of cutting is controlled by hydraulic system of the tractor. It is a tractor-mounted implement controlled by tractor hydraulics and three-point linkage. Tractor mounted leveler and blade terracer can be operated by 35 – 45 hp tractor. Working capacity of the blade is 0.3 to 0.4 ha/h. Some of the manufacturers have started manufacturing tractor operated bund former with leveller to smoothen the soil in between bunds. It has two mould boards at the ends for formation of bunds and leveller in between two to level and smoothen the land (Fig. 2.6).

a) Tractor operated blade terracer

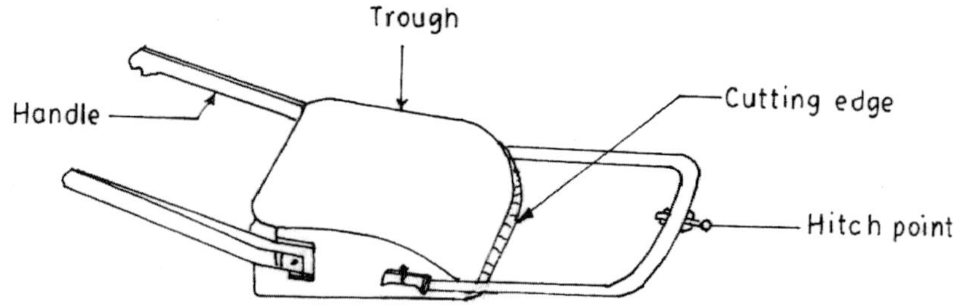

b) Details of tractor-operated blade terracer

Fig. 2.5 : Tractor operated blade terracer

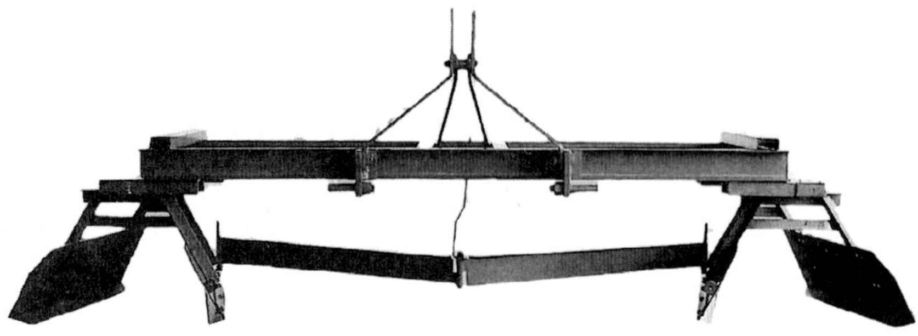

Fig. 2.6 : Tractor operated bund former with leveller
Courtesy: M/s. R J Sekar Industries, Dindigul (Tamil Nadu)

Power tiller operated terrace-cum-leveller

For land levelling, terracing and bund forming 8 to 10 hp power tiller operated terrace-cum-leveller has been developed. The unit consists of 1.0 m wide curved mild steel blade with a steel cutting edge at the bottom (Fig. 2.7). It is attached to the front of the power tiller with the help of a mounting plate. Two solid side support arms made of 25 x 12.5 mm mild steel flat holds the unit rigidly during the operation. The lifting of the

Fig. 2.7 : Power tiller operated terrace-cum-leveller

blade can be made by tail wheel adjustment of the rotary tiller while keeping the tilt angle constant. Two side guards are provided to avoid spilling of soil on both sides of the blade. Bottom skids made of 2 mm mild steel sheet are provided below the blade for maintaining uniform load. This is efficient machine for land levelling with a transportation efficiency of 86.6 per cent.

Bulldozer

Crawler tractors (Fig. 2.8) are employed when heavy earth movement is required (Pandey *et al.*, 1997; Singh, 2007; Singh *et al.*, 2015). Self-propelled bulldozers are generally used where earth cutting and movement is required for short distances (Fig. 2.9). Self-propelled bulldozers equipped with dozer shovel and bucket is more suitable for cutting and moving the earth for slightly longer distances (Fig. 2.10). Since its speed is limited, it cannot be used for long distance haulage. Tractor operated wheel-type bulldozers are used for land grading job where haulage is for long distances (Fig. 2.11). It spreads the soil and compacts. It is very versatile equipment and can climb to very stiff slope. It produces tractive power so it can work under adverse condition of soil. The main disadvantage of this equipment is that it is very expensive and breaks the top surface during transportation.

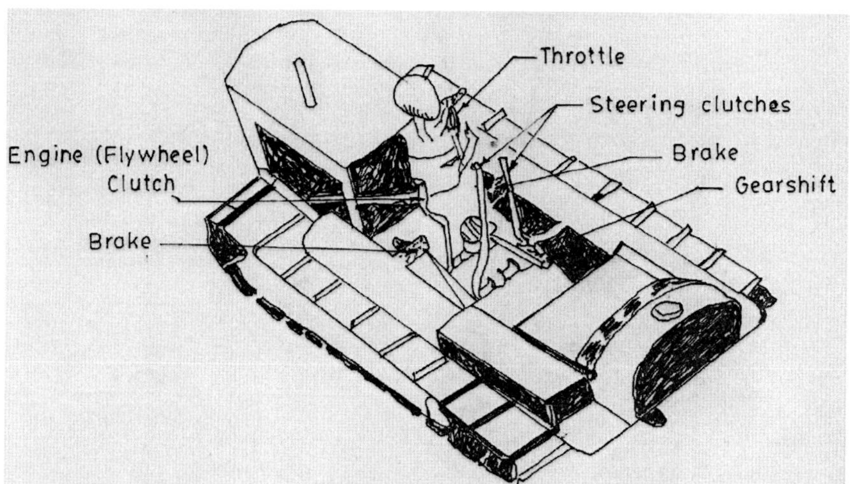

Fig. 2.8 : Crawler tractor

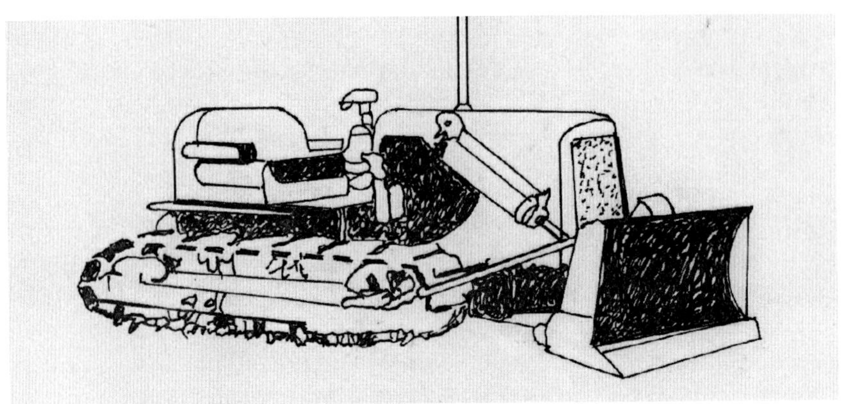

Fig. 2.9 : Self-propelled crawler type bulldozer with straight cutting blade

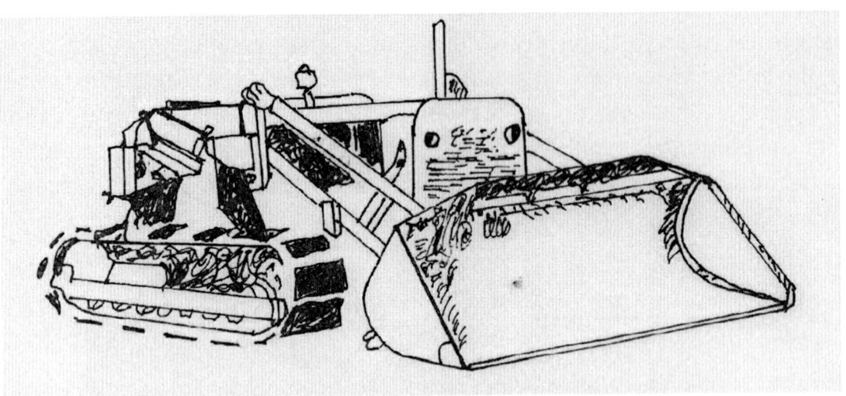

Fig. 2.10 : Self-propelled bulldozer with dozer shovel and bucket

Fig. 2.11 : Tractor mounted wheel type bulldozer with straight cutting blade
Courtesy: Greenfield Equipments India Pvt. Ltd., Coimbatore

Scraper

The scraper is a large piece of equipment used in mining, construction, agriculture and other earthmoving applications. The rear part has a vertically moveable hopper (also known as the bowl) with a sharp horizontal front edge. The hopper can be hydraulically lowered and raised. When the hopper is lowered, the front edge cuts into the soil and fills the hopper. When the hopper is full it is raised, and closed with a vertical blade (known as the apron). The scraper can transport its load to the fill area where the blade is raised and the back panel of the hopper is hydraulically pushed forward and unloads the soil. Then the empty scraper returns to the cut site and repeats the cycle. Scrapers are very efficient on short hauls where the cut and fill areas are close together and have sufficient length to fill the hopper. Some self-propelled scraper have two engines ("tandem powered"), one driving the front wheels and other driving the rear wheels, with engines up to 400 kW or so. Multiple scrapers can work together in a push-pull fashion but this requires a long cut area. Open bowl usually requires a push-cat (bulldozer or similar) to assist in loading. Pull type scraper uses agricultural tractor to pull. Pull type scrapers can be utilized individually or two or three units can be pulled behind a single tractor.

It is used for collecting the soil from one place and unloading at the other. It is used for rough leveling, cutting of high spots and filling of depressions. It is operated by 50 hp tractor and above. The hydraulic scraper is towed behind the tractor (Fig. 2.12a). Self-propelled scrapers are also available (Fig. 2.12b). The scraper consists of cutting blade, hydraulic system, hitch point, hitch bar, apron, bowl, wheels, apron cylinder, side frame, bucket cylinder, spring, and side arm. The scraper working is controlled by the hydraulic arrangement. The blade is made of alloy steel and has self-sharpening tungsten carbide cutting edge. For operation, the scraper is attached to the tractor, hydraulic system connected and apron is raised. With the forward movement of the tractor, the blade penetrates into the soil and bucket bowl gets filled. The apron is closed after the bucket is filled and the scraper is moved to the point of unloading. For unloading, the bucket is tilted hydraulically. The working capacity of scraper is 2.5 to 3.5 m^3/h. The scrapers are cheaper for moving large amount of earth. But it cannot be used for long distances. It takes wide turning circle and cannot be operated on steep slopes. It can climb only on gentle slopes. It cannot cut vertical or near vertical land or rock. Cutting blade width is about 1.5 to 2.1 m.

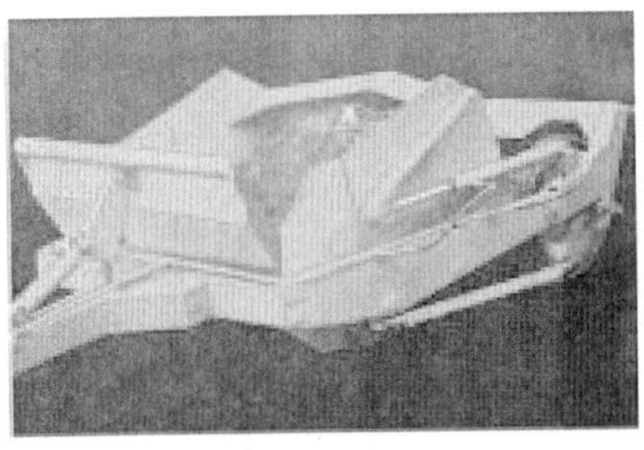

a) Tractor operated hydraulic scraper

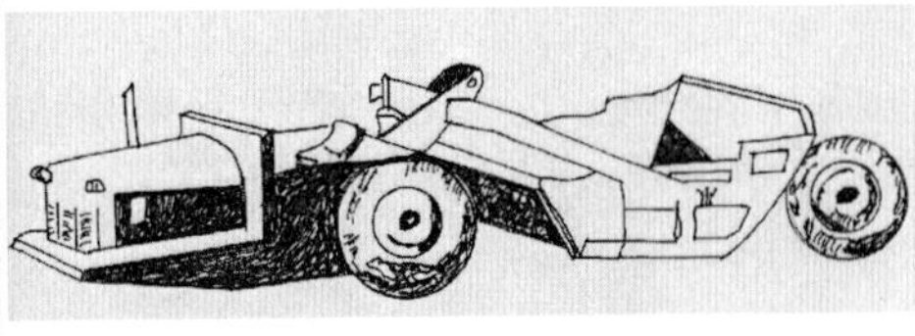

b) Self-powered scraper

Fig. 2.12 : A view of hydraulic scraper

Drainage excavator

Proper field channels and bunds are provided at appropriate spacing depending upon the soil, crop and method of surface irrigation. These channels and bunds have to be made afresh as they get leveled off during land preparation. Therefore, considerable proportion of farm energy has to be diverted for this purpose. The conventional method of making field channels and bunds by use of spade and hand scraper is labourious and time consuming. The other implements such as mould board plough and border plough are being used for channel making. The major disadvantage with these implements is that they have to make a number of trips to make a channel (Pandey *et al.*, 1997). Excavator buckets are of various types for different purposes such as general purpose, heavy duty; ditch cleaning, trapezoidal, tile drain, deep digging, bucket etc (Fig. 2.13).

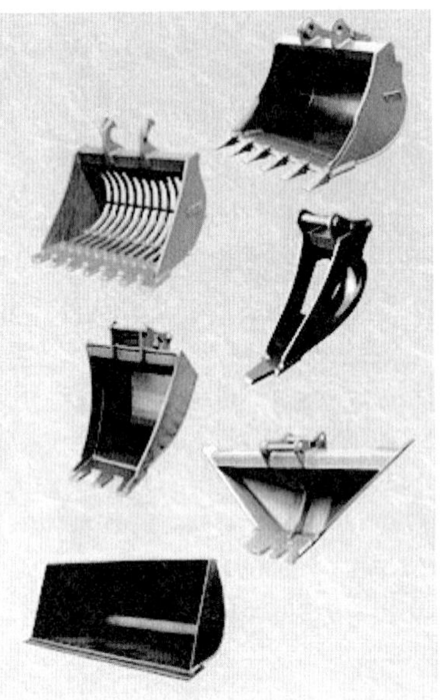

Fig. 2.13 : Various types of excavator buckets used in land development

A bullock-drawn and tractor-drawn channel-cum-bund former can do this job in one run (Singh, 2007). The implement consists of scrapers, intermediate flexible divider, adjusting unit, 3-point linkage hitch and depth adjusting lever. As the implement is pulled forward the scrapers scrap and simultaneously divert the soil inwards towards the divider where it is divided into two halves channeled to form two parallel bunds forming a channel. When the implement is used for making bunds, the intermediate divider is detached from the adjusting unit and scrapers are brought nearer to each other so as to form a bund of required width. As the implement moves forward, the scrapers continue to work as before and scraped soil gets squeezed in between the two scrapers and forms the bund of required width.

Ditching Equipment

Construction of irrigation ditches is one of the most important farm operations under irrigated crop cultivation. These open channels are usually constructed manually, which involves considerable time and hard labour. The bullock-drawn ditcher helps in making trapezoidal ditches in 2-3 successive runs. It requires about 100-150 kg load during first run but goes down during second and third run. Many types of tractor-mounted ditchers are available range in complexity from simple scoop that is connected to 3-point linkage of a tractor to complex self-propelled crawler type ditchers (Fig. 2.14). The frame of the digger is supported by a pair of widely spaced sprang that have independent hydraulic adjustment and carry the whole weight of rear end of tractor and equipment while digging. The machine has a provision for easy lateral adjustment of the king post, which allows good trenching work to be done on an offset line and to increase the reach to one side. There is also a considerable advantage in a large turn angle that enables the outfit to do more work at each move and improves its versatility.

a) Tractor operated ditcher

b) Crawler type self-propelled ditcher

Fig. 2.14 : A view of the tractor operated and self-propelled crawler type ditcher

It is operated by 50 hp tractor and above. It consists of two curved wings with cutting blades, front cutting point, tie bars for adjusting wingspan, and hitch assembly with 3-point linkages (Fig. 2.14a). The cutting blades and cutting point are made of medium carbon or alloy steel, hardened and sharpened. The ditcher penetrates in the soil due to its own weight and suction of the cutting point. Upon drawing the ditcher in the field, it opens the soil in the shape of ditch with either 'V' bottom or flat bottom. The wings enable the ditcher to slice and roll the tough sod, brush and root sets. The depth and width of the ditch is adjusted from the operator's seat. The front cutting point and wings cutting edges are replaceable. The working capacity is 0.2 ha/h.

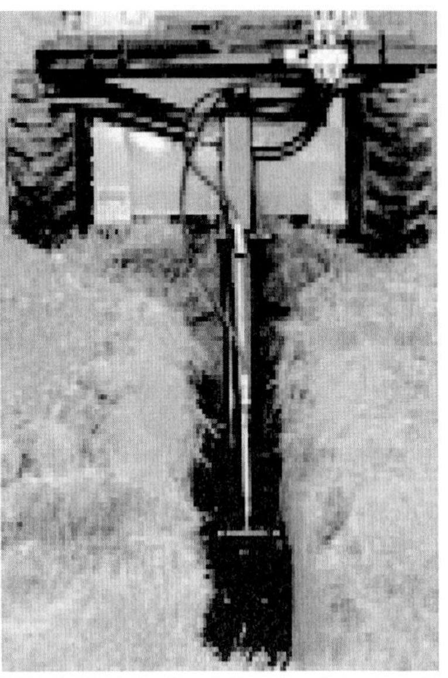

Fig. 2.15 : A view of swing hoe trench digger

There are many different types of ditchers available but swing hoe trench digger and disc ditchers are commonly used (Anonymous, 2014). The swing hoe trench digger features strong, hydraulically-controlled boom swings a full 110 degrees (Fig. 2.15). This allows operators to dump and pick up material on either side of the trench without maneuvering the skid-steer from side to side as is required with rigid, mini-backhoe attachments. It is simple, efficient and saves time and money. Hydraulics is directly connected to the skid steer controls for low maintenance, and versatility. It has power to dig deep trenches, reach high and reposition easily, all from the comfortable operator's position. It digs down up to 2 m, making it ideal for most landscaping or light construction trenching. Buckets available in 30 cm, 37 cm and 50 cm widths suit a variety of digging needs. Disc ditcher is designed to cut down irrigation levees or to dig a 50 cm wide by 30 cm deep ditch, promoting water control and increasing yields for all types of crops (Fig. 2.16).

Fig. 2.16 : A view of disc ditcher

Land planes

Land levelling jobs are finished by land planes, levellers, and float. Tractor-drawn as well as self-propelled land planes (Fig. 2.17) are commonly used for levelling irrigated fields (Anonymous, 2014a). It consists of a long frame supported by two wheels and adjustable levelling blade at some intermediate point. The position of blade is so adjusted that a high spot cut must fill in a depression before the leveller reaches the next high spot. This machine is available in various sizes with effective width of cut varying from 1.5-4.5 m. The machine is provided with wide steel wheels or pneumatic wheels so that blade may not penetrate much into loose soil. The machine is operated in diagonal directions and down fields. The machine is suitable for grading large fields. Land planer ensures even distribution of rainfall and irrigation water. It eliminates high and low spots, sending water all the way to the end of the row. Less watering means less time and labour, and more money saving. Soil smoothing equipment improves surface drainage, controls erosion and helps to handle sub-surface moisture.

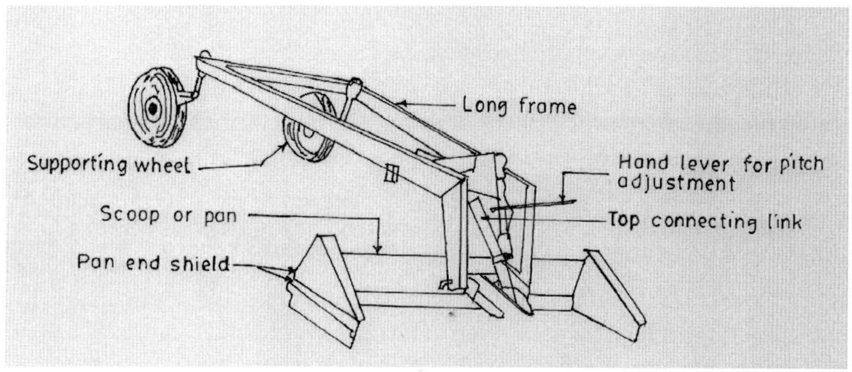

a) Details of tractor operated land planer

b) Tractor operated land planer

Fig. 2.17 : Tractor operated land planer

Backhoe dozer/loader

Backhoe dozer (Fig. 2.18) is used for agricultural land levelling, making bunds on the farms and terracing of farm, road making and site clearance, for trench filling at dam project, after laying cable or pipe etc (Pandey *et al.*, 1997; Singh, 2007). Loader is used for removal of mud and loose soil at canal worksite, for loading of crushed stones into the dumpers and trailers at stone crusher unit, for loading of salt into trucks and trailers, for handling of clay and soil and loading it into the trucks and trailers etc. Backhoe is used for excavating soil, making foundation for building, making trenches for pipe and cable laying, garbage handling, widening of rural roads and removal of bushes and trees etc. The dozer is mounted in front of the tractor and backhoe in the rear. The dozer and backhoe can be easily be removed and joined to the tractor. The dozer consists of a thick curved plate and hardened strip. The strip has sharp cutting edge and is joined to the curved plate of the dozer with fasteners. Therefore, the strip can be replaced on wearing or becoming blunt. The dozer plate is joined to the tractor with sturdy arms and can be raised or lowered with hydraulic system of the tractor. Backhoe consists of a bucket with digging fingers, hydraulic cylinder, arms and base for attaching to the rear of the tractor. The bucket position is manipulated by hydraulic system. The digging fingers are hardened and can be replaced on wearing or becoming blunt. Working capacity of dozer is working depth 255 mm, loader bucket payload capacity 750 kg and loader bucket capacity 0.50 m^3. A simple front end loader (Fig. 2.19) is also available and used frequently for handling of clay, soil and even crop residue/produce at farm level or grain 'mandis' and loading it into the trucks and trailers. It is low cost and can be easily installed even on second hand tractors. It can be operated even on hard and tough soils. Diesel consumption is low. Machines can be mounted at customer site without any problem.

Fig. 2.18 : Tractor operated backhoe dozer/loader

Courtesy: Bull Machines Pvt Ltd., Coimbatore (Tamil Nadu)

Fig. 2.19 : Tractor operated front end loader

Courtesy: Essey Engineering Co., Nashik (Maharashtra)

Power tiller operated bench terracer-cum-leveller for hilly region

In hilly areas, the farming is practiced in terraces and sloppy land. Bench terracing is done only in mountainsides with slopes of up to 40% (Anonymous, 2010). The width of the bench terraces which ranges from 2-5 m depends on the slope of the land. Lands with sharp slopes require narrower bench terraces while those with gentle slopes require wider ones. Soil depth also dictates the width of bench terraces. A narrower width is recommended for shallow soils so that digging and earth moving will not be too deep. A depth of 40 cm with decomposed or weathered parent material is adequate for terracing. For bench terracing, the site is first cleared of brush cover, and then the earth is dug and moved by pick and shovel. Work starts from the upper slope and proceeds downward. All these work, presently done by manual labour in hilly areas which enhances the cost of operation and drudgery. To mechanize the bench terracing work, a mechanical device (power tiller operated) has been developed which will make leveled terrace with proper slope for raising agricultural crops. This increases the annual usage of the power tiller with matching implement. The unit consists of 1.0 m wide curved mild steel blade with a steel cutting edge at the bottom (Fig. 2.20 and Fig. 2.21). It is attached to the rear of the power tiller with the help of two solid side support arms made of 25 x 12.5 mm mild steel flat that holds the unit rigidly during the operation. Two side guards are provided to avoid spilling of soil on both sides of the blade. Bottom skids made of 3 mm mild steel sheet are provided below the blade for maintaining uniform load. The capacity of the terracer is 0.12 ha/h for terracing the

Fig. 2.20 : Bench terrace-cum-leveller attached with power tiller.

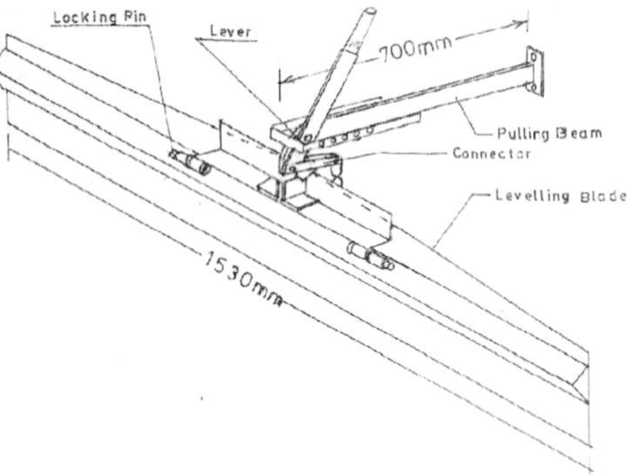

Fig. 2.21 : Details of power tiller operated levelling blade.

Fig. 2.22 : A view of wetland leveller attached to power tiller.

field. The maximum volume of soil handled at a time by the terracer is 0.6 m^3. Wetland leveller attached with power tiller is also used for leveling the field before transplanting to keep the water level uniform throughout the field (Fig. 2.22).

Tractor operated rotary trencher

In order to maintain sustained production in irrigated agriculture, removal of excess water from agricultural land is essential. At present drainage trenches are being formed manually. In certain crops such as banana, inter running of trenches between rows is a prevalent practice for draining excessive water when in abundance and also for water retention during semi dry spells. Trenching machines are available in foreign countries for digging the utility trenches for oil, gas and water pipelines, drainage ditches, sewers, cables and foundations as well as for the construction and road-building jobs. Some of the trenching machines which are used for drainage purpose are, backhoe, endless-chain trencher, slanted boom and drain-tube plow. A trenching machine should operate efficiently in a variety of soil conditions.

A rotary trencher (Fig. 2.23) has been developed which is capable of making a 30 cm × 30 cm trench (Anonymous, 2008). The developed machine uses a rotating cutter disc with radial soil cutting blades. The disc consists of 16 blades of 10 cm width and 30 cm length. The blades are mounted on appropriately positioned frog plates and are made replaceable. When the cutting action is in progress, the disc along with the blades cut to a depth of 30 cm and hence makes a 20 × 30 cm. trench. The rotating cutter disc is mounted on a shaft supported on roller bearings on either side. Rotary power is transferred from the tractor PTO to the rotating cutter shaft through a gear box and a chain drive at a speed ratio of 2.4:1, thus providing a 225 rpm rotational speed of the cutter disc at a PTO speed of 540 rpm and further the rotary cutter's speed is reduced from 225 to 110 rpm by providing a countershaft in between. The details of machine are given in Fig. 2.24. A chisel tool cutting to a depth of 30 cm is mounted at the front of the cutter disc's path to counter the negative draft and also to make a pre-cut. The unit works well with 60 hp tractor and above at 1 kmph (Fig. 2.25). The cutting effort is very low and the formation of cut trench is also superior. Soil evacuation is about 70% and the trench edges are also well formed. Field efficiency is 80% and covers 750-800 m/h.

Fig. 2.23 : A view of tractor operated rotary trencher

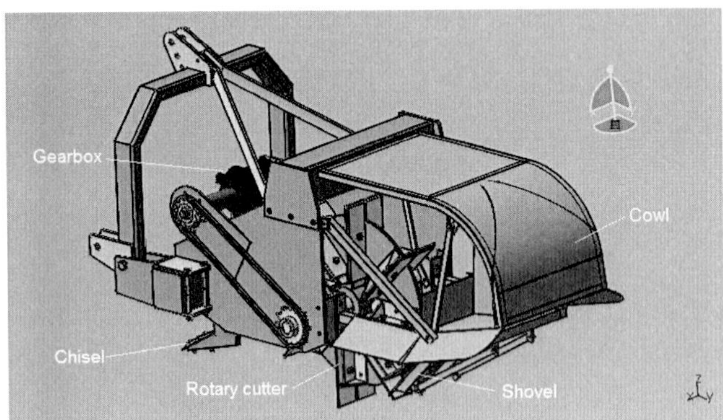

Fig. 2.24 : CAD view of tractor operated rotary trencher

Fig. 2.25 : Tractor mounted rotary ditcher for making trenches

Courtesy : Punjab Engineers, Meerut (Uttar Pradesh)

Tractor operated rectangular trencher

To form rectangular trench of 30 x 30 cm 35 to 45 hp tractor is required. The unit consists of two mould board bottoms placed in line one behind the other (Fig. 2.26). The front and rear bottoms operate at a depth of 0-15 cm and 15-30 cm respectively. The two bottoms throw the removed soil in opposite directions and form vertical walls one on each side of the trench. The mould board shape is formed for easy lifting and throwing of soil away from the trench opened. A safety pin is provided to protect the unit from over loading. An adjustable bar point share is provided in addition to the trench bottom cutting share. It can also be used for laying drip irrigation pipes by opening trenches, application of manure in coconut fields and making drainage around the sugarcane. The cost, time and energy saved by machine trenching in comparison to manual trenching are 95, 99 and 53 per cent respectively.

Fig. 2.26 : Tractor operated trencher

Tractor operated spiked clod crusher

It is used as a combination tillage tool with tractor drawn harrow or cultivator. It is suitable for breaking and segregation of clods for seedbed preparation after paddy harvest. The clod crusher consists of a mild steel sheet drum and pegs are welded on its surface, a rectangular frame made from mild steel angle section, hitch frame and a shaft for carrying the drum (Fig. 2.27), Pandey *et al.*, 1997. The shaft is mounted on bearing pedestals. The length of roller is about 150 cm and diameter 35 cm. The soil clods are pierced and broken by the pegs of the crusher. It can be operated by 25 hp tractor or more. Field capacity is 0.4 - 0.6 ha/h. Sometimes it is used as a combination tillage tool with tractor drawn harrow or cultivator.

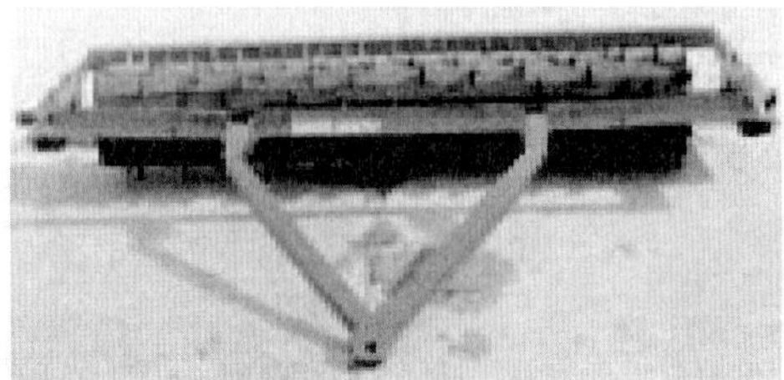

Fig. 2.27 : Tractor operated spiked clod crusher

Tractor operated channel former

It is used for forming alternate beds and channels. The beds are suitable for planting crops like sorghum, maize, and cotton. This bed and furrow system is ideal for efficient irrigation management. The channel former consists of two inner blades, two outer blades, hitch frame, mainframe and shovel (Fig. 2.28), Pandey *et al.,* 1997. The front portions of the two inner blades are joined together and form an angle of 30° in between them. At the junction of these two inner blades a cultivator shovel is fixed to penetrate into the soil. The inner blades can be mounted 50 to 100 mm lower than the outer blades and form a furrow at a lower depth than the surface of the bed for the flow of irrigation water. The two outer blades are placed one on each side of the inner blades and at an angle of 60° to the direction of the travel. The soil collected from the furrow is formed as bund on both the sides of the irrigation furrow. It can be operated by 35 hp tractor and above. It can cover about 1.2 - 1.5 ha/day.

Fig. 2.28 : Tractor operated channel former

Tractor operated laser land leveller

Unevenness of the soil surface has a major impact on the germination, stand and yield of crops through nutrient water interaction and salt and soil moisture distribution pattern. Land levelling is a precursor to good agronomic, soil and crop management practices. Resource conserving technologies perform better on well levelled and laid-out fields. Farmers recognize this and therefore devote considerable attention and resources in levelling their fields properly. However, traditional methods of levelling land are not only more cumbersome and time consuming but more expensive as well. Very often most rice farmers level their fields under ponded water conditions. The others dry level their fields and check level by ponding water. Thus in the process of having good levelling in fields, a considerable amount of water is wasted.

It is a common knowledge that most of the farmers apply irrigation water until all the parts of field are fully wetted and covered with a thin sheet of water. Studies have indicated that a significant (20-25%) amount of irrigation water is lost during its application at the farm due to poor farm designing and unevenness of the fields. This problem is more pronounced in the case of rice fields. Unevenness of fields leads to inefficient use of irrigation water and also delays tillage and crop establishment. Fields that are not levelled have uneven crop stands, increased weed burdens and uneven maturing of crops. All these factors tend to contribute to reduced yield and grain quality which reduce the potential farm income. Effective land levelling is meant to optimize water-use efficiency, improve crop establishment, reduce the irrigation time and effort required to manage crop.

Laser land levelling is one such important technology for using water efficiently as it reduces irrigation time and enhances productivity not only of water but also of other non-water farm inputs. Results in technologically advanced countries have indicated that it saves water to the tune of 25-30% and time by 30%. It has also been observed that with Laser land levelling 2-3% effective cropped area in case of flat fields and even more in ridge sown fields become available for cultivation of crops, as the number of bunds and irrigation channels get reduced considerably. Because of increasing the flatness, it decreases the dose of fertilizer and increases the rate of utilization of fertilizer to 20%, ensuring better seedling emergence efficiency. Every unit area increases the production by 20 to 30%. Using the technology decreases the production cost by 6.3 to 15.4% (Sidhu *et al.*,

2007). Better crop stand due to even application of fertilizers and other inputs resulting improvement in crop yield by 10 to 15%, improves weed control efficiency and reduces labor requirement.

Laser levelling is a user guided precision levelling technique used for achieving very fine levelling with desired grade on the agricultural field (Sidhu *et al.*, 2007; Singh *et al.*, 2015). The complete laser land leveler equipment includes laser emitter, laser receiver, two way hydraulic valve, laser eye, grade rod, tripod stand, control box on tractor and scraper unit (Fig. 2.29). The laser emitter unit sends continuous self levelled laser beam signal with 360° laser reference up to a command radius of 300-400 m (depending upon its range) for auto-guidance of the receiving unit. The laser receiver emitter is mounted on a tripod stand placed just outside the field to be laser levelled and high enough to have unobstructed laser beam travel. Different working components & controls on the laser emitter unit includes laser emission indicator, low battery indicator, off/on power button, manual grade buttons, charge jack, battery assembly and manual mode indicator for setting of desired grade. The trouble free usage of these components should be made by following the relevant instructions mentioned in the operator's manual.

Laser levelling uses a laser transmitter unit that constantly emits 360° rotating beams parallel to the required field plane. This beam is received by a laser receiver (receiving unit) fitted on a mast on the scraper unit. The signal received is converted into cut and fill level adjustments and the corresponding changes in scrapper level are carried out automatically by a two way hydraulic control valve. Laser levelling maintains the grade by automatically performing the cutting and levelling operations. Both level grade and slope grade (one way or two ways) can be achieved with the help of this precision equipment. The field is cultivated and planked before using the Laser Land Leveller. A grid survey is per-formed using grade rod to identify highs and lows in the field and mean grade is found. A grid spacing of 10m x 10m is maintained for accurate land survey; however this spacing can be varied depending upon the size of the field. For practical purposes and with experience, grid survey can be done by pacing off the distances rather than (measuring). A map is then drawn to indicate which areas are high; require soil to be cut and the lows which require soil to be added. It is operated by 45 hp tractor and above. The working capacity is 0.2-0.3 ha/h depending upon the land and soil conditions. A view of machine working under dry condition and wet condition are given in Fig. 2.30 and 2.31.

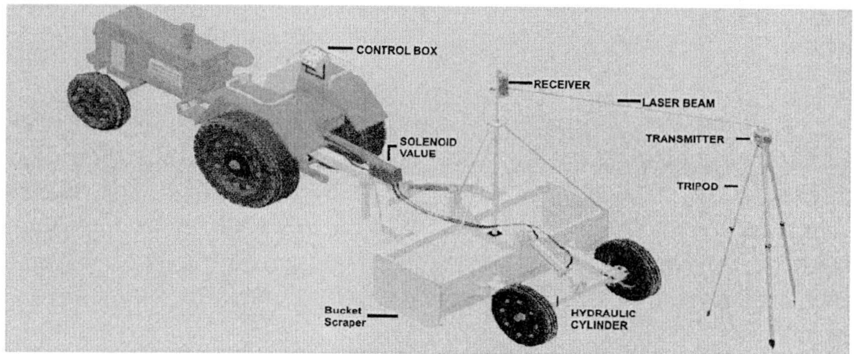

Fig. 2.29 : Details of components of tractor operated laser land leveller
Courtesy: A J Precision & Automation Pvt. Ltd., NOIDA (Uttar Pradesh)

Fig. 2.30 : A view of land levelling by laser land levelling under dry conditions

Fig. 2.31 : A view of land levelling by laser land levelling under puddle conditions

References

Anonymous. 2008. Research Highlight. AICRP on Farm Implements and Machinery, CIAE Bhopal. Technical Bulletin No.: CIAE/2008/141.

Anonymous. 2010. Research Highlight. AICRP on Farm Implements and Machinery, CIAE Bhopal. Technical Bulletin No.: CIAE/2010/151.

Anonymous. 2014. Visited on 22.12.2014. www.equipmentland.com/categories/Tractor-Implements/Ditchers/.

Anonymous. 2014a. www.artsway-mfg.com/products/land-planes/.

Pandey M M; Majumdar K L; Singh Gyanendra; Singh Gajendra. 1997. Farm Machinery Research Digest. Technical Bulletin No. CIAE/97/69, Central Institute of Agricultural Engineering, Bhopal, 328 p.

Sidhu H S; Mahal J S; Dhaliwal I S; Bector Vishal; Singh Manjit; Sharda Ajay; Singh Thakur. 2007. Laser Land Leveling: A Boon for Sustaining Punjab Agriculture. Department of Farm Power and Machinery, Punjab Agricultural University, Ludhiana.

Singh Surendra. 2007. Farm Machinery – Principles and Applications. Directorate of Information & Publication of Agriculture, Indian Council of Agricultural Research, Krishi Anusandhan Bhawan-I, Pusa Campus, New Delhi.

Singh Surendra; Manes G S; Dixit Anoop. 2015. Mechanization of cultivated crops. New India Publishing Agency, Pitam Pura, New Delhi.

3

Tillage Equipment

Tillage is a mechanical manipulation of soil to provide favourable condition for crop production. Soil tillage consists of breaking the compact surface of earth to a certain depth and to loosen the soil mass, so as to enable the roots of the crops to penetrate and spread into the soil. A good seedbed is generally considered to imply finer particles and greater firmness in the vicinity of seeds. The depth up to which tillage operations disturb the soil can classify the operation as shallow, medium or deep. The depth of tillage depends on the crop and soil characteristics and also on the source of power or energy available. The optimum seedbed preparation for raising upland crops, involves the many unit operations. Strength of soil is affected by soil moisture content and can be represented in terms of cone index. Compacted soil has more cone index than the loose soils. Movement of machine in the field causes soil compaction. Sometimes it forms hard pan which is to be broken by the tillage operations to enhance root growth and penetration and infiltration of excess water.

Ploughing is the first operation performed for preparation of the seedbed. It contributes materially in obtaining good tilth. The seedbed is considered the ploughed layer of soil that has been so prepared that sowed seeds will germinate and plants will have proper condition for root development and growth. The crop spends about 95% of its life cycle in its root bed and only about 5% of the time in its seedbed. Therefore, i) the plant roots must be able to extend easily through the soil; ii) the soil must be able to supply a large amount of moisture to growing plants; iii) the root-bed should be well ventilated; iv) the root-bed must supply nutrients to the plants; and v) roots must be in contact with active soil. Thus, the preparation of the seedbed is accomplished with the following objectives:

a) To create a deep seedbed physically, chemically and bio-logically suited to the growth of crops,

b) To add humus and fertility to the soil by covering and burying crop residues and manure so that they are incorporated in the soil,

c) To prevent and destroy weeds or other unwanted vegetation,

d) To aerate the soil for proper growth of crops or to leave the soil in such condition that air will circulate freely,

e) To leave the soil in such conditions as to retain moisture from rain, and to increase water-absorbing capacity of the soil,

f) To destroy insects as well as their eggs, and breeding places, and

g) To leave the surface in a condition to prevent erosion by winds.

In general, the purpose of ploughing is to bring about desirable physical changes in the soil that improve crop growth conditions. The physical condition of the soil in its relationship to plant growth is called soil tilth. The results of ploughing are good if the farmer knows the soil conditions he wants to provide for crop growth. A good job of ploughing reduces the amount of secondary tillage required to prepare the soil for crop growth.

Loosening of soil is done to achieve a desired granular soil structure for a seedbed and to allow rapid infiltration and good retention of moisture, to provide adequate air exchange capacity within the soil and to minimize resistance to root penetration and shoot growth. Local plough *(Hal)* and blade harrow (*Bakhar)* are traditional implements used for loosening of soil. These are simplest tools designed to break the topsoil and multi-passes are carried out to prepare seedbed. Mould board plough, disc plough, soil stirring plough, ridger plough, tool frames/carriers with mouldboard plough or tillage sweeps, etc are improved implements designed for breaking soil. Ploughs are used to break soil and invert furrow slice to control weeds, etc.

Clod breaking operation is required to produce a granular soil structure in the final seedbed. Tine cultivator and disc harrow are used for breaking of clods. Generally these are operated after one pass of mouldboard plough or ridger plough. Direct harrowing or cultivator operation is also performed when the fields are clean and free from plant residues of previous crop. Clod crushers, *patela* harrow etc are very effective for clod crushing under favourable soil moisture conditions but their effect is confined to soil surface only. Power driven implements like rotavator disintegrate the clods over a wide range of soil moisture and provide uniform and fine size clods or aggregates in seedbed. Operation of tools with narrow tines such as comb harrow and spike tooth harrow, in loosened soil, produces a sorting effect, bringing larger clods and aggregates on surface. The sorting effect increases with increasing forward inclination of tines and share width and decreasing speed and soil moisture. Large size clods on the surface are recommended because of their stability under rainfall, which helps in reducing soil erosion. Wide, backward inclined implements compact soil as well as break clods in top surface of soil. Direct compaction at seed depth can best be achieved using narrow press wheels/discs. Planking is widely used to compact the soil at the surface. Smoothening of seedbed is required for proper operation of seeding machines, better distribution of irrigation water and quick disposal of excess rainwater. Smoothening can be best achieved by using wide backward inclined blades, such as levelling boards, floats and *patela* harrow with closely spaced shallow working narrow tines. Wooden plank, *patela* harrow, is recommended for smoothening operation.

Puddling of soil generally refers to breaking down soil aggregates at near saturation into ultimate soil particles and it is one on the common operations in low land rice fields. It is normally done after initial ploughing and allowing about 5 to 10 cm of standing water in the field. In low land condition the farmers often flood the field prior to ploughing and puddling is done to weaken the mechanical strength of the soil. By retaining standing water on the rice field the farmer receives the benefits of weed control and oxidation-reduction conditions that favor nutrient balance and a soft soil into which the rice seedlings are transplanted. Puddling helps to retain standing water in the rice field by producing fine soil particles that reduce soil porosity, thus reducing percolation losses of nutrients. Puddling is also beneficial because it controls weeds, levels the soil surface and provides a homogenized puddle tilth. Puddling is done when there is standing water in the field.

For a farmer puddling is mixing soil with water to make it soft for transplanting and impervious to water. The simplest method of puddling is done by the farmer's feet. Puddling is also done by animal's feet when they are driven around and around in the field so that their hooves smear and compresses the soil and breakdown the soil aggregates. Most puddling is done with animal or tractor drawn implement (puddler) such as ploughs, comb harrow, patela puddler, ladder puddler and rotary puddlers. Power tillers are also used for puddling. The degree of puddling is however depended on the type of puddler and on intensity of puddling. Rotary puddlers generally are better than ploughs because their rotary motion continually changes the direction of the shear stress and therefore matches the weakest fracture plane within a clod. Further the rotary puddlers tend to compact the sub soil, chop and press down organic matter and require relatively low draft compared to ploughs. Puddling index is the combination of the mechanical and puddling performances of puddlers to quantify the state of puddle for softness of soil (for ease of transplanting) and reduces percolation rate (for water and nutrient economy). A combined evaluation of bulk density (specific volume), percolation rate, amount of dispersion of silt and clay and size of water stable aggregates corresponding to specific operational energy and effective field capacity of the puddlers is considered to be more appropriate to quantity the degree of puddling in terms of a puddling performance number (PPN).

Types of tillage equipment

Tillage implements are broadly categorized into several groups depending on the purpose for which they are used (Singh, 2007; Singh and Verma, 2009; Pandey *et al*., 1997; Pandey and Ganesan, 2005; Pandey *et al*., 2006; Rautaray, 2002; Singh *et al*., 2015; Bhardwaj *et al.*, 2004). Implements used for opening and loosening of the soil are known as ploughs. Ploughs are used for primary tillage. Ploughs are of three types: wooden ploughs, iron or inversion ploughs and special purpose ploughs. In ploughing, top layer of the soil is separated into furrow slices, turned sideways and inverted to a varying degree depending upon the type of plough being used. There are two types of tillage operations viz. primary and secondary. The operation performed to open up any cultivable land with a view to prepare a seedbed for growing crops is known as 'Primary Tillage Operations'. Implement used for this purpose is called 'Primary Tillage Implement'. Lighter and finer operations performed on the soil after primary tillage operations are known as 'Secondary Tillage Operations'. The implement used is called 'Secondary Tillage Implement'. There are three types of tillage implements commonly used: manually operated tillage equipment, animal-drawn tillage equipment and tractor/power operated tillage equipment. Sometimes spade is used as manually operated tillage tools for primary as well as secondary tillage operations.

Primary tillage constitutes the initial major soil working operation. It is normally designed to reduce soil strength, cover plant materials and rearrange aggregates. Animal drawn implements mostly include indigenous plough and mould-board plough. Tractor drawn implements include mould-board plough, disc plough, subsoil plough, chisel plough and other similar implements. Secondary tillage operations following primary tillage are performed to create proper soil tilth for seeding and planting called secondary tillage. These are lighter and finer operations, performed on the soil after primary tillage operations. Secondary tillage consists of conditioning the soil to meet the different tillage objectives of the farm. The implements include different types of harrow, cultivators, levellers, clod crushers etc.

There are different types of tillage used in agriculture depending upon requirements. **Minimum tillage** is the minimum soil manipulation necessary to meet tillage requirements for crop production. **Strip tillage** is a tillage system in which only isolated bands of soil are tilled. **Rotary tillage** is the tillage operations employing rotary action to cut, break and mix the soil. **Mulch tillage** is the preparations of soil in such a way that plant residues or other mulching materials are specially left on or near the surface. **Combined tillage** operations simultaneously utilizing two or more different types of tillage tools or implements to simplify, control or reduce the number of operations over a field. Sometimes it uses seed drills/planters to complete the job of tillage and seeding. Different names are used for equipment used in agriculture such as tools/implements/machines. **Tool** is an individual working element such as disc or shovel. **Implement** is equipment generally having no driven moving parts, such as harrow or having only simple mechanism such as plough. **Machine** is a combination of rigid or resistant bodies having definite motions and capable of performing useful work.

Animal-drawn tillage equipment

Animal-drawn plough

Animal-drawn ploughs used for primary tillage operations as well as secondary tillage operations are indigenous ploughs and mould board plough. The traditional seedbed preparation equipment of India is *desi hal* (country plough) which continues to be used in many parts of the country (Fig. 3.1a). Indigenous plough is an implement which is made of wood with an iron share point. It consists of body, shoe, share, beam and handle. The body is the main part of plough to which shoe, beam and handle are generally attached. The share is working part of plough and is attached to shoe with which it penetrates into the soil. The shoe also supports and stabilizes the plough at a required depth. The beam is generally a long wooden piece that connects main body of plough to the yoke. A wooden piece called handle is attached vertically to the body to control the plough. It cuts the soil to a depth of 5-10 cm but does not invert the soil. It cuts a V shaped furrow and opens the soil but there is no inversion. Ploughing operation is also not perfect because some unploughed strip is always left between furrows. This is reduced by cross ploughing, but even then small squares remain unploughed. The field capacity of this plough is 0.2-0.3 ha per day and its draft requirement is about 50-70 kg. It is drawn mainly with a pair of bullocks or a buffalo; but other animals locally available can also be used. To plough one hectare of land a man has to walk about 68 km. Their designs vary from region to region and sometimes craftsman to craftsman. This is used for both dry seed bed preparation (Fig. 3.1b) and puddling (Fig. 3.1c).

a) A view of animal drawn indigenous plough

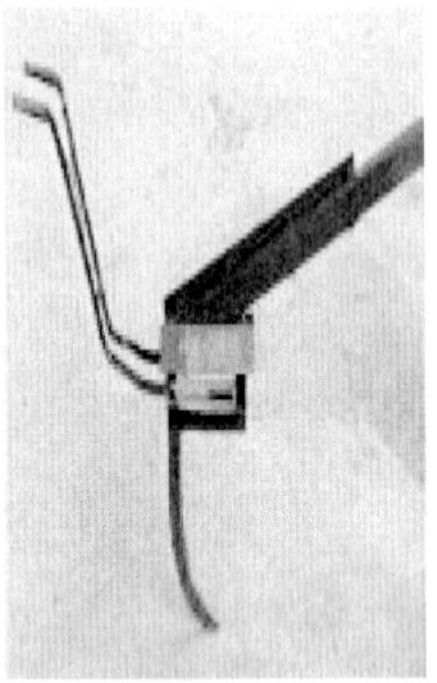

b) Indigenous plough used for dry seed bed preparation

Courtesy : Bhagwati Krushi Udhyog Pvt. Ltd., Ahmedabad (Gujarat)

c) Indigenous plough used for puddling

Fig. 3.1 : Animal drawn indigenous (*Desi*) plough

Soil turning plough: Soil turning ploughs are made of iron and drawn by a pair of bullocks depending on the type of soil (Fig. 3.2). The important feature of this type of plough is that a slice of soil is cut from the ground and inverted, so that weeds and surface trash are buried and the soil is exposed to weathering agents and further cultivation. This type of plough leaves no un-ploughed land as the furrow slices are cut clean and inverted to one side resulting in better pulverization. The animal drawn mould board plough ploughs to a depth of 15 cm. Mould board ploughs are used where soil inversion is necessary. Victory plough is an animal drawn mould board plough with a short shaft. In turn-wrest or reversible or one-way plough the plough bottom is hinged to the beam such that the mould board and the share can be reversed to the left or to the right side of the beam. This adjustment saves the trouble of turning the plough in hilly tracts, but yet facilitates inversion of the furrow slice to one side only.

Many localized animal drawn soil stirring ploughs are used in different part of country depending upon the type of soil and terrain available. Maco soil stirring ploughs are pulled by a pair of animals, this plough opens the soil to aid aeration; its working width is 175 mm and depth of cut is 110 mm. Animal drawn Sudan type ploughs are made from high grade steel. The handle and beam are adjustable and the landside is fitted with a heel for better control. Animal drawn Bose plough is a mould board plough suitable for upland paddy and dry land cultivation. This implement was developed by a Karnataka Farmer Mr. Bose. The effective field capacity is 0.01 ha/h, field efficiency 58% and labour requirement 98 man-h/ha. The draft of the plough is about 410 N.

There are two types of animal-drawn ploughs; viz. one or two way ploughs and left hand or right hand ploughs. Most of the walking type mould board ploughs are one way ploughs, that is they are designed to throw the furrow slice to only one side in the direction of motion. Two ways ploughs or Turn Wrest ploughs are suitable for terraced land of hilly tracts and have the advantage that they do not upset the slope of the land nor leave dead or back furrows in the middle of narrow fields. This is because the two bottoms are used alternatively and the furrow slices are thrown on the same side. Some of the two-way ploughs have a single bottom, but the provision is made to change the direction of throw of furrow slice at the end of the plot where the bullocks turn.

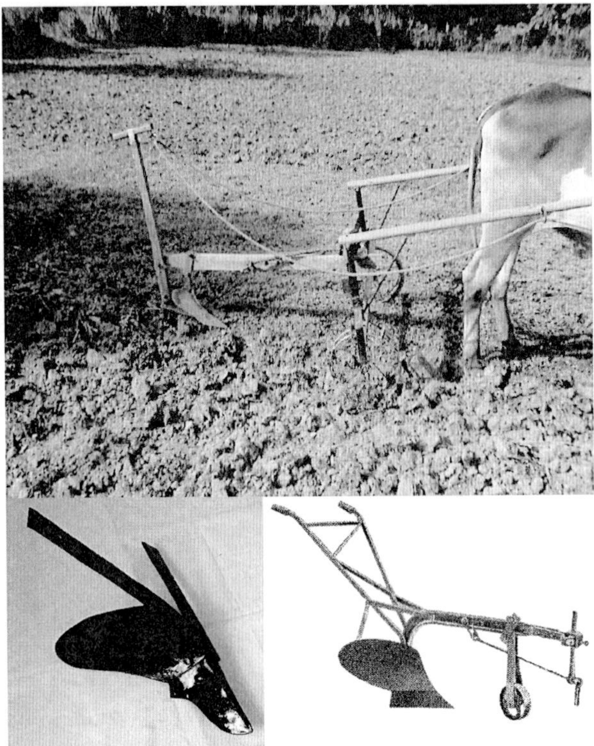

Fig. 3.2 : Different views of animal drawn soil turning plow

Most of the mould board ploughs are right hand ploughs, throwing furrow slice to right. Left-hand type ploughs are rare in India because bullocks are trained to take turn on their left while ploughing. Mouldboards are generally made of high carbon steel. The main parts of mouldboard plough are share, mouldboard, landside, frog, beam and handle (Fig. 3.3a and 3.3b). The share is the part of the plough that penetrates into soil, cuts the furrow slice and passes it onto the mouldboard. It takes the greatest amount of wear. It is fastened to the frog with the help of countersunk bolt to keep the top surface smooth. There are five main parts of the share that play important role in stabilizing the plough bottom on the ground and in cutting the furrow slice. *Share point* enters first in the soil and supports the plough bottom. *Cutting edge* also called throat of the share cuts the furrow slice from main soil body. *Wing of the share* supports the plough bottom. *Gunnel* of the share supports the plough bottom against the furrow wall. *Cleavage* forms joint between mould board and share on the frog. There are different types of share viz. Slip share, slip nose share, shin share and bar share (Fig. 3.3c). The share of plough is commonly made of high carbon steel or chilled cast iron. The steel mainly contains 0.7-0.8% carbon and 0.5-0.8% manganese besides other minor elements. There are self-sharpening shares with a hard alloy surface along the share blade that always remain sharp despite wear. The blade sharpens itself in all soils except sand and stone. These shares do not require periodic forging and have much longer life as compared to conventional shares. When share point is blunt by 3-4 mm, the plough draft increases by about 25 per cent depending upon type and moisture content of the soil. Fuel consumption too increases by 6-8% and productivity of machine drops down. The working side of share is sharpened at an angle of 25-40 degrees.

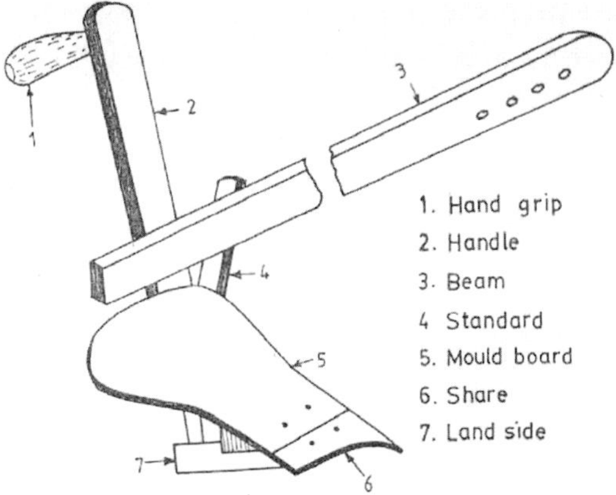

a) Components of animal drawn mould board plough

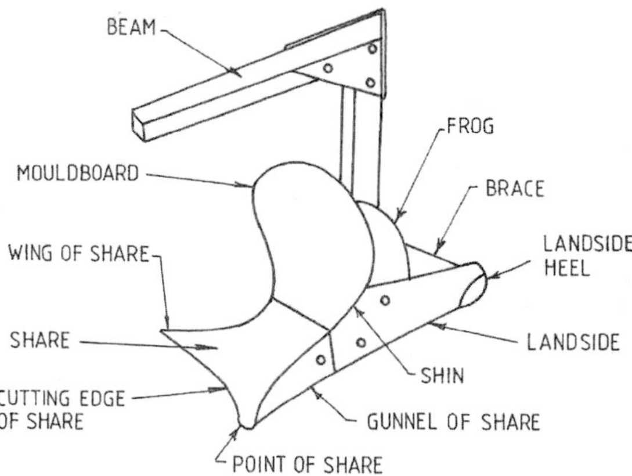

b) Components of animal drawn mould board plough

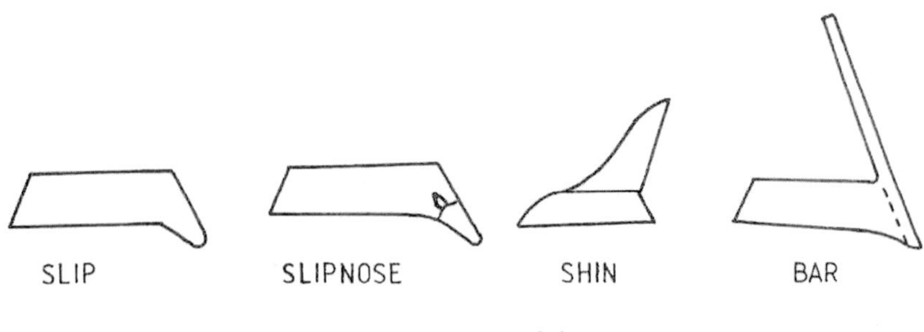

c) Different types of shares

Fig. 3.3 : An animal-drawn mould board plough.

Slip share is a common type of share used by the manufacturers. It is one piece share with curved cutting edge and entire share has to be replaced if it is worn out. In a slip nose, share point of share is provided with a small detachable piece and can be replaced as and when required. In a shin share, shin is an additional part. In a bar share, point of share is provided with an adjustable bar. Full-cut shares are used for high speed and slat bottom where soil is having heavy root crop. Narrow-cut shares are used for soils where roots are not the problem, penetrate better and pull lighter than full-cut shares. Heavy-duty deep suck shares are used in abrasive, rocky and hard soil. Hard surfaced shares are recommended for extremely abrasive soils where regular shares wear out quickly. Chilled cast shares are used in light, abrasive, sandy and gravely soils where good scouring is not a problem. Landside is the part of the plough that slides along the face of furrow wall. It helps to resist the side pressure exerted by furrow slice on the mould board. It helps in stabilizing the plough while it is in operation. Frog is the part of plough bottom to which the share, mould board and landslide are attached rigidly. It is an irregular piece of metal and made of cast iron and to which the different parts of the plough are attached.

Animal drawn chisel plough

It is a pointed curved bar type implement useful for breaking hard layer of soil below the normal ploughing depth to facilitate infiltration of rainwater. The plough is used by a pair of bullocks (Fig. 3.4). Use of this implement leads to increase in yield by up to 15 per cent compared to conventional method of field preparation by country plough without any deep ploughing. Working capacity of chisel plough is 0.02 ha/h. Radius of curvature of tyne is 370 mm and width of cut 40 mm.

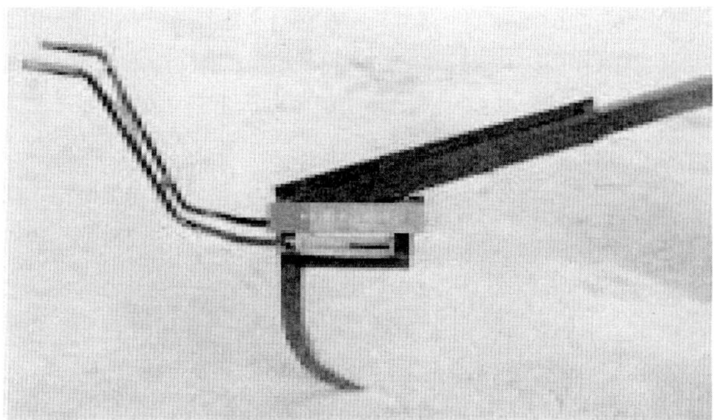

Fig. 3.4 : A view of animal drawn chisel plough

Animal drawn Palam plough

It is used for ploughing. It is operated by pair of animal/ bullocks. It is made of MS plate of 8 mm thick in V-shape embedded with long shovel attached with handle and beam (Fig. 3.5). It is very useful equipment for ploughing and opening the soil effectively with minimum clod formation. Due to light in weight and low cost, it is most suitable for ploughing in low moisture content also. The working width of plough is 180 mm. The equipment has field capacity of 0.036 ha/h at field efficiency of 61%.

Fig. 3.5 : A view of animal drawn Palam plough

Animal drawn melur plough

This type of plough is used in Madurai, Ramnad and Tanjore districts of Tamil Nadu. This plough is suitable for ploughing in wetlands for raising rice crop using a pair of bullocks (Fig. 3.6). This type of plough is used for seed bed preparation and puddling

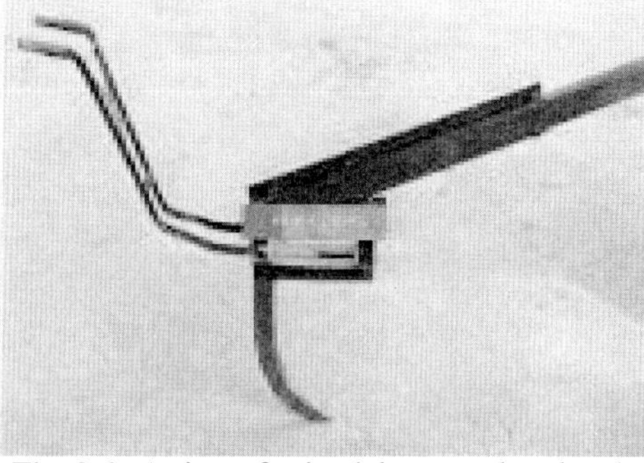

Fig. 3.6 : A view of animal drawn melur plough

in wetlands. The plough is used by a pair of bullocks. The depth of cut in black soil is 3.5 cm and the width is 100-130 mm. The shoe is made of single casting with ribbed surface. The pole shaft is made of babul or vengai wood. Working capacity is 0.025 ha/h.

Animal drawn bose plough

It is a mould board type plough suitable for upland paddy and dryland cultivation (Fig. 3.7). The plough is used by a pair of bullocks. It consists of wooden beam, Wooden handle with a grip, wooden body and a share cum mould board made of mild steel. The share cum mould board has sharp point and cutting edge, which cut and invert the furrow slice. Upon becoming dull it can be replaced. Working capacity is 0.01 ha/h.

Fig. 3.7 : A view of animal drawn bose plough

Animal drawn disc harrow

Animal drawn disc harrow pulverizes soil, provides surface mulch and compacts sub surface. Notched discs provide more effective pulverization of soil than plain discs (Singh, 2007; Singh *et al.*, 2015). It is a single acting double gang type disc harrow suitable for secondary tillage operation. Harrows are used to break the clods, cut weeds, pulverize soil, cover seeds and smoothen the surface. The disc harrow is used by a pair of bullocks. Animal-drawn disc harrow (Pandey *et al.*, 1997, Pandey and Ganesan, 2005) has usually six or eight discs fixed on two gangs; each gang has 3 or 4 discs. It consists of a beam, adjusting lever, frame, spool, transport wheels, centre shovel and plane or notched discs (Fig. 3.8). The diameter of disc is about 40 cm. The width of cut is about 75 cm. The frame is made of mild steel on which gangs with discs are mounted. It is a single acting double gang type disc harrow suitable for secondary tillage operation. The harrow is provided with an operator's seat and a transport wheel which aids in easy transportation. The operator's seat enables the operator to ride instead of walking, which helps in deeper penetration and reduces drudgery. Also the disc angle can be adjusted easily with the help of the slotted bracket and the depth and width of operation can be controlled. It saves 90 per cent labour and operating time and also results in 80 per cent saving in cost of operation compared to conventional method of ploughing by bullock drawn country plough. Working capacity of disc harrow is 0.16 ha/h.

Fig. 3.8 : A view of animal drawn disc harrow
Courtesy: Govind Industries Pvt. Ltd., Barabanki (Uttar Pradesh)

Animal drawn harrow-cum-puddler

It is a disc type modified harrow used for puddling as well as dry seedbed preparation for raising wheat, cotton, bajra, raya and paddy etc. It is operated by a pair of bullock. Animal-drawn disc harrow has usually six or eight discs fixed on two gangs; each gang has 3 or 4 discs of 45 cm diameter developed at Punjab Agricultural University, Ludhiana (Garg and Singh, 2002; Anonymous, 2008, 2010, 2012). In this discs are spaced at 155 mm with the help of circular drum of 220-mm diameter instead of spools (Fig. 3.9a). This can be used for both dry seedbed preparation (Fig. 3.9b) and puddling for rice transplantation (Fig. 3.9c). It consists of a frame made of mild steel angle sections on which two gangs of discs are mounted with their concave surfaces in the opposing direction. Thus the side draft is considerably eliminated. It has provision for seating the operator due to which it can be operated for extended periods without drudgery. The discs are able to cut the trash in the soil and scrapers are also provided for self-cleaning of the discs when used for puddling or when the soil moisture is high enough to cause sticking. It consists of a beam, adjusting lever, frame, spool, transport wheels, centre shovel and plane or notched discs (Fig. 3.9d). The frame is made of mild steel on which gangs with discs are mounted. The discs are spaced at 155 mm with the help of circular drum of 220-mm diameter instead of spools. This leaves only 115 mm of exposed edge of disc to penetrate into the soil, which does not allow excessive sinkage of discs. At maximum depth of 115 mm the disc starts floating and reduces draft requirements. A shovel has been provided in between two gangs to loosen the soil left. Two gangs are mounted on a mild steel frame with help of brackets and wooden bushes. Provision is made in the frame to adjust gang angle up to 25 degrees for increasing depth of cut. It saves 50-65 per cent labour and operating time and 50-60 per cent on cost of operation compared to conventional method of using bullock drawn country plough. Working capacity for intercultural operation is 0.2 ha/h, harrowing 0.16 ha/h and puddling 0.14 ha/h.

a) A view of bullock operated harrow puddler

b) Harrow-cum-puddler used for dry seedbed preparation

c) Harrow-cum-puddler used for puddling

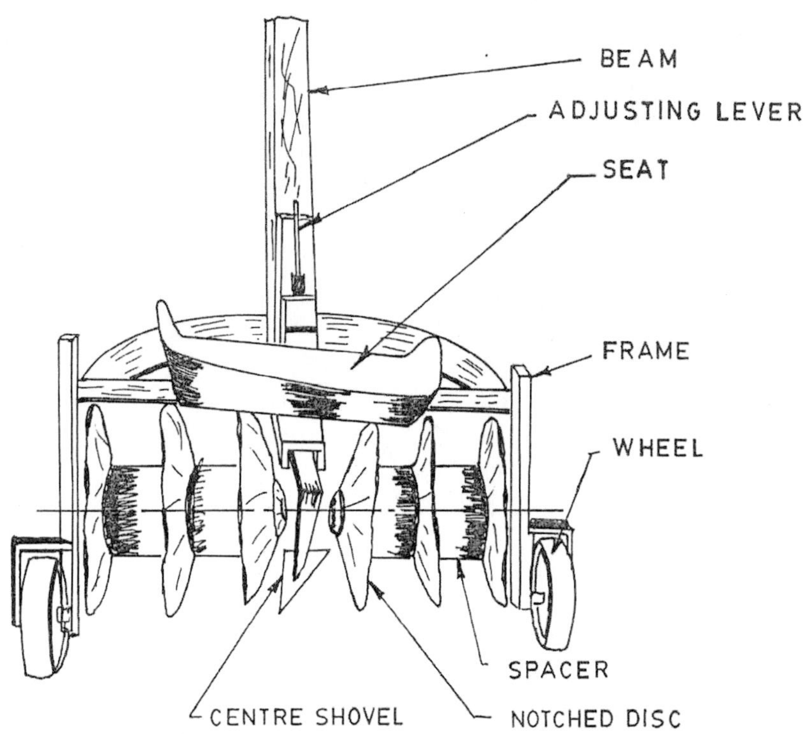

d) Details of harrow-cum-puddler

Fig. 3.9 : Animal drawn harrow-cum-puddler

There are many types of harrows viz. disc harrow, spring tooth harrow, spike tooth harrow, blade harrow, triangular harrow, guntaka harrow and acme or knife harrow. It is used for seedbed preparation, cutting and burying of grass and weeds. Some of them are discussed here.

Animal drawn spike tooth harrow: It breaks clods, stirs the soil, uproots the early weeds, levels the ground, and breaks soil crust. This harrow also covers the seeds. Usually it operates at shallow depths up to 5 cm. It is a harrow with peg shaped teeth of diamond cross-section attached to either rectangular or a triangular frame. The triangular frame is made of wood and is pulled by a pair of animals. The rectangular type spike tooth harrow frame is made of steel and is pulled by tractor. The spike tooth harrow is also known as peg-tooth harrow, bar harrow, drag harrow, section harrow or smoothing harrow. Its principal use is to smooth and level the soil directly after ploughing. It is used to break clods, stir the soil, uproot the weeds, level the ground, break the soil crust and cover the seed. It can be used to cultivate corn and cotton and other row crops in early stages of growth. It is of two types rigid and flexible. Animal-drawn spike tooth harrows are usually of rigid type. The teeth of spike tooth harrows are made of hardened steel. The teeth may be square, triangular or circular in section. It can stir the soil up to the depth of 5 cm. Animal drawn triangular harrow is made of all steel parts and has thirty five tines made from 16 mm square section steel and held in place with standard fasteners. Animal drawn triangular tine harrow consists of a triangular fabricated frame from which seven tines (200 mm) protrude on each side. Its width of coverage is 1.5 m.

Animal drawn spring tooth harrow: It is a harrow with tough flexible teeth, suitable to work in hard and stony soils. Spring-tine harrows are versatile secondary tillage tools.It is fitted with springs, having loops of elliptical shape. It gives springing action in working conditions. It pulverizes soil and helps in killing weeds. It can also be used to loosen previously ploughed soil ahead of grain drill. In this machine, teeth penetrate deeper than spike-tooth harrow and tears out the roots and bring them to surface. This harrow consists of teeth, tooth bars, clamps, frame, clevis, lever and links (Fig. 3.10). The teeth consists of wide, flat, curved, oil-tempered bars of spring steel, one end of which is fastened rigidly to a bar; the other end is pointed to give good penetration. Adjusting angle of teeth by means of levers controls the depth to which teeth will penetrate the soil. The teeth are available with points of various widths and shapes with detachable points for different types of work. By adjusting the depth of work the curved spring tines present a different angle of attack to the soil. Thus at shallow settings the tines are almost vertical while at full depth the tips are nearly horizontal. It works up to a depth of 150 mm. It requires three to five times the draft of a spike tooth harrow of similar size.

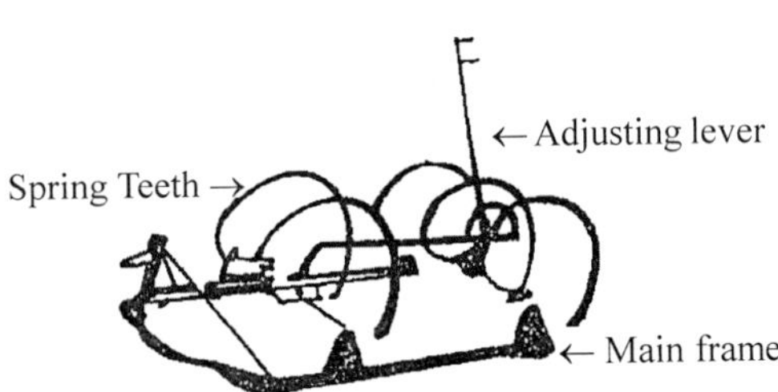

Fig. 3.10 : Animal-drawn spring tooth harrow.

Animal drawn blade harrow: Animal drawn blade harrow popularly known as Bakhar, is an implement of Indian origin (Singh, 2007; Singh et al., 2015). It is one of the agricultural implement of the Buddhist period. It is used mainly for secondary tillage operations like breaking clods and land smoothening in seed bed preparation. It is also used for breaking soil surface crust, controlling weeds, removing old stubble from the field and for soil mulching after the rains (Fig. 3.11a). It works like a sweep skimming under the soil surface without inverting it. In some parts of the country it is used for covering the seeds broadcasted and for harvesting groundnut crop. In the deep black vertisol regions of the country, the implement is used even for primary tillage. The implement consists of a blade, blade holding prongs, wooden body and a wooden beam (Fig. 3.11b). The blade is attached to the bakhar frame prongs with the help of rings, which are hammered into place. It cuts a thin slice of soil and removes the weeds etc and requires a pair of bullocks to pull the implement. The V-shape of the blade has a self-cleaning advantage, which prevents adherence of weeds and its roots, thus enabling speedier operation and improving the uniformity of cut as compared to the straight bladed version. Working capacity is 0.062 to 0.075 ha/h and cutting width varies from 40 to 100 cm. **Guntaka Harrow is** an improved type of blade harrow. The functions of Guntaka are same as that of Bakhar. **Acme Harrow (Knife Harrow)** is a special type of harrow having curved knives. Front part of knife compacts soil and crushes the clods. It is good for mulching also. The knife is made of high carbon steel. Putting additional weight on the frame may increase depth of penetration.

(a) Blade harrow removing weeds and breaking soil surface crust

b) A view of different types of blade harrow

Fig. 3.11 : A view of animal drawn blade harrow

Improved bakhar: It is used for seedbed preparation in black soil. The improved bakhar consists of a frame made of mild steel angle section in which a V-shaped blade is attached by means of fasteners (Fig. 3.12). Behind the blade a roller made of 150 mm diameter Mild Steel pipe is provided to crush soil clods. The frame also supports a handle by which the operator controls the implement. The unit is suitable for seedbed preparation in black soil. This has a modified V-blade replacing conventional curved blade for reduced draft. A pair of bullocks is used for pulling the implement. The working width is about 50 cm and it can work up to 6 cm deep. Field capacity varies from 0.06 to 0.07 ha/h.

Fig. 3.12 : Animal drawn improved bakhar

Animal drawn patela harrow: It is secondary tillage equipment and used for breaking of clods, collection of stubble or trash and leveling of the field surface before the application of seedlings. The patella harrow is used by a pair of bullocks. The animal drawn patela harrow consists of a heavy batten made of Sal wood having 1.5 m length and 10 cm thickness on which a mild steel angle frame is fixed by means of screws (Fig. 3.13). The frame carries a bar to which curved and pointed hooks are attached. The bar can be raised or lowered by means of the lever having a slotted sector to lock its position. Some models may be up to 2.0 m in length. Working capacity is 0.037 ha/h.

Fig. 3.13 : A view of bullock drawn patella harrows in field

Animal drawn diamond harrow has diamond pattern frame of twenty tines. This harrow can be used for pulling out weeds and grass from light ploughed land. It is available in light, medium and heavy sections. **Animal drawn zig-zag harrow** has zig-zag pattern of tines allows each tine to cut a separate track. There are twenty tines in total. The harrow is available in light, medium and heavy sections.

Animal drawn cultivator: It has three or five tines steel cultivator and can be adjusted for width from 30 to 62.5 cm by means of a lever. Animal drawn screw adjustable seven-tine cultivator is designed for preparing seedbeds or inter-row cultivation of crops such as sugarcane, cotton, maize, tobacco and potato (Singh, 2007; Pandey *et al.*, 1997). The seven tines made in steel are reversible. Animal drawn three-tine cultivator has reversible carbon steel tines and the width can be adjusted across the beam. The depth of work can also be altered by changing the angle of the drawbar. The animal drawn cultivators are available in fixed or expandable frame designs (Fig. 3.14a). It is used for shallow ploughing, weeding and intercultural operations. The unit consists of reversible tines, frame, handle and a wooden beam. For operation the implement is attached to a pair of bullock by beam and moved in the soil. The beam is attached to the unit with the help of U-bolts and can be adjusted as per the height of the animal pair. The operator's handle can also be adjusted by varying the position of the supporting bracket. The tines dig into the soil and cut a small furrow slice.

During weeding operation, the tines also uproot weeds and create soil mulch. Seeding attachment can also be provided on the frame, increasing the versatility of the unit. The working capacity is about 0.6-0.7 ha/day. Three tine cultivator mostly used in Gird and Bundelkhand region of Madhya Pradesh has entirely steel body (Fig. 3.14b). The field capacity of this equipment is 0.06 to 0.08 ha/h. Bullock-drawn 4-tine cultivator mostly found in Chhattisgarh and Madhya Pradesh is also entirely made of mild steel suitable for seedbed preparation for rice crop (Fig. 3.14b). This has field capacity of 0.08 to 0.10 ha/h.

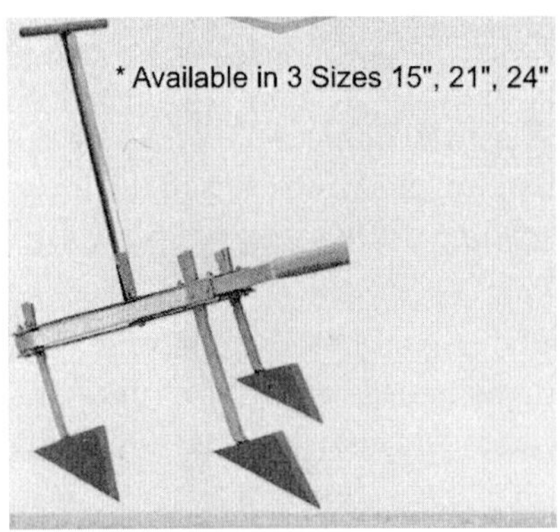

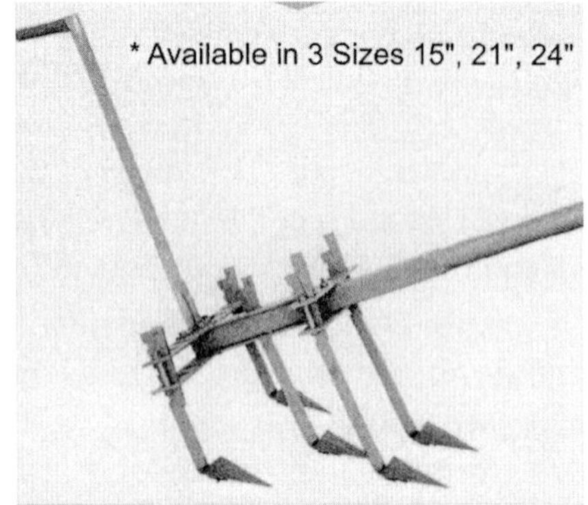

a) A view of 3 and 5 tine bullock-drawn cultivator

Courtesy: Bhagvati Krushi Udhyog, Ahmedabad (Gujarat)

b) Bullock-drawn 3-tine and 4-tine cultivator

Source : Saha *et al.* (2015)

Fig. 3.14 : Bullock-drawn cultivator

a) Animal drawn ridger working in field

Animal drawn ridger: It is used for making ridges and furrows for potato and sugarcane planting and making field channels for irrigation (Fig. 3.15a). The ridger is used by a pair of bullocks (Fig. 3.15). The ridger plough consists of single share and double mould board attached to a short (steel) or long wooden beams (Fig. 3.15b). Ridger has two mould boards, one for turning the soil to the right and another to the left. The share is common for both the mould boards i.e. double winged. These mould boards are mounted on a common body. The, ridger is used to split the field into ridges and furrows and for earthing up of crops. Ridgers are also used to make broad bed and furrows by attaching two ridge ploughs on a frame at 150 cm spacing between them. Sometimes two handles are provided to balance the ridger during operation. During operation the share cuts furrow slice and pulverizes the soil, which moves on to the mould board and is inverted. During the onward movement, a furrow is formed and ridges get formed in its return pass. The point of the ridger can be replaced after it is worn out. In certain designs, the width of ridges can also be adjusted. Working capacity of ridger is 0.02 ha/h.

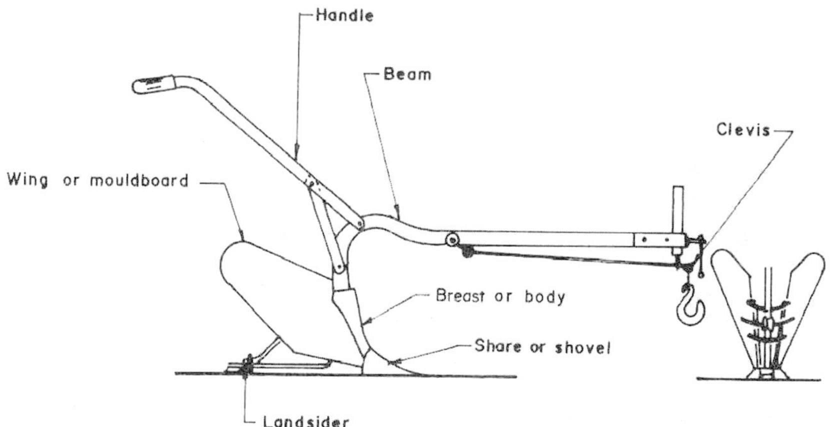

b) Details of animal drawn ridger

Fig. 3.15 : Animal drawn ridger

Animal drawn bund former: Country plough and ridge plough are used for laying out the field into ridges and furrows or to layout irrigation channels. Ridge ploughs, when attached to a frame can be used for making broad-bed furrows. Bunds for irrigation in the garden lands are made usually by manual labour using spades. Bunds are also formed across the contours in the low rainfall regions to conserve soil moisture. The bund farmer is designed to form these bunds replacing manual labour. This implement consists of a pair of iron mould boards fixed in opposite direction facing each other with the front end opening outwards and rear and closing in to form bunds. The implement consisted of two blades, flat iron frame bent at an angle, a handle attached to the frame with tie bars and wooden beam (Fig. 3.16). The operator's handle is made of wood for providing better grip and convenience in operation and attached to the frame with the help of suitable brackets. The frame is bent at an angle and has holes for adjusting the space between the blades. The profile of blades is made to a shape so that bund formed is trapezoidal and remains stable. The blades are attached to the frame with fasteners. For operation, a pair of bullock pulls the implement; the blades gather the loose soil and accumulate it in the form of bund. Working capacity is 0.375 to 1.0 ha/h. It works under well prepared and pulverised soil condition and can be used under all types of soil and crop conditions. The size of bund and soil-collecting capacity can be increased or decreased by adjusting the mould boards.

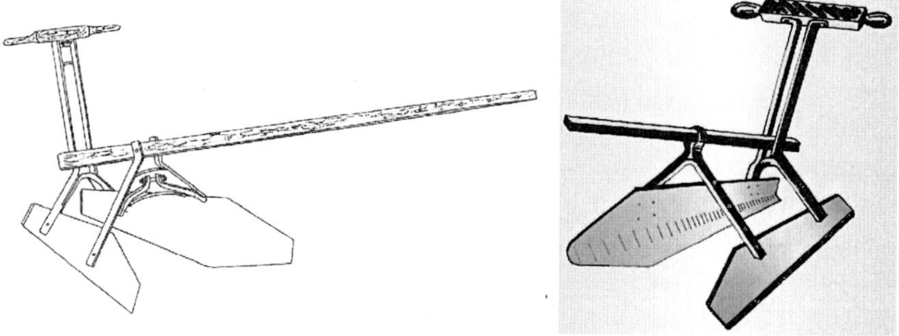

Fig. 3.16 : Different views of animal drawn bund former

Animal drawn clod crushers: Clod crushers are used after ploughing or harrowing for a fine tilth. They can work on an unevenly ploughed soil to produce a well packed seedbed and they also help in minimizing evaporation losses from the land surface. Of the several types in use, clod crushers of the harrow type have been found to be quite efficient in crushing clods. They consist of a rectangular frame mounted on wheels. On the rear side of the frame, two square axles are fixed one behind the other. At each axle, cast iron stars with tapering spikes are fixed. The number of stars varies from 12 to 20 in each gang. The stars are fixed in such a way that there is continuous rolling action without jerks.

Animal drawn chisel ploughs and sub-soilers: The chisel plough is a tool having a rigid curved or straight shank with a relatively narrow shovel point. It may also be termed as a heavy duty deep cultivator. Chisel plough and sub-soiler are similar in their actions, but differ mainly in their working depths. Because the chisel plough does not work as deep as a sub soiler, its draft is lower. The chisel plough works primarily in the top soil and hard pan at shallow depths and is therefore suitable to work with animal power, with a heavy pair or with two pairs of bullocks. The subsoiler is usually operated with a tractor, as the power required to open the hard pan at deeper depths is high. The effective field capacity and labour requirement are 0.02 ha/h and 5 man-h/ha, respectively. The draft of the implement is 1200 N.

Animal drawn scraper: The animal drawn scraper is suitable for use in tropical conditions and can be pulled by two draught animals. The capacity is between 0.1 and 0.15 cubic meters.

Animal drawn wooden leveler: The animal drawn wooden leveler consists of two wooden planks hinged together. A steel handle is provided to control scraping and dumping. The working width is 1.25 m.

Animal drawn patela puddler: The unit consists of a frame made of mild steel, a wooden plank, a gang consisting of pegs for turning the soil, and sliding type pegs (10 nos.), handle and hitch system (Fig. 3.17). The operator can stand on the plank while in operation to provide additional depth. A pair of bullocks is used for the operation of the implement to prepare homogenized puddled tilth for mechanised paddy transplanting. It breaks the soil clods to obtain a smooth puddle. Soil moisture is maintained near the saturation level at the time of operation. It is suitable for shallow puddling with high mechanical dispersion of soil particles. Its working width is 1.5 m and depth 10 cm. Draft requirements is about 100 N. It can cover 0.15 ha/h at a field efficiency of 71%.

Fig. 3.17 : Animal drawn patela puddler

Animal drawn puddler: It is used for preparation of paddy fields with standing water (5-10 cm depth) after initial ploughing. It breaks up the clods and churns the soil. The main purpose of puddling is to reduce leaching of water and to kill weeds. Puddling facilitates transplanting of paddy seedlings. Puddler consists of puddling units each having four paddles mounted on an axle, frame, beam, metal-cross and handle (Fig. 3.18a). Paddles are made of mild steel sheet having thickness of 3.15 mm. While moving, blades (paddles) churn the soil and mix it properly (Fig. 3.18b). The weeds are also chopped and mixed with soil for decomposition. Two to three operations are good enough to get desired puddled soil. According to the power used puddler can be classified as hand operated; animal-drawn and tractor-drawn puddler. Animal-drawn puddlers are commonly used in

India. There are three types of puddler viz. rectangular, angular and helical blade type (Fig. 3.18). Rectangular blade type puddler is suitable for puddling operation under wetland conditions. It has a wooden frame on which bushes are mounted using mild steel flats. The shaft carrying the blades fixed in a staggered fashion rotates in the bushes. The blades are fixed at an angle to the direction of motion. The blades on the implement rotate and impart a lateral and turning action on the soil particles, thus achieving a good puddle. It has an operator's seat, which helps in continuous operation and reduces drudgery. It saves 66 to 88 per cent labour and operating time and 66 to 82 per cent on cost of operation compared to conventional method by using bullock drawn country plough. Working capacity is 0.07 to 0.11 ha/h. Other features are width of cut 700 mm, depth of cut 100 mm, operating speed 1.5 km/h, field efficiency 60 per cent, draft 480 N and labour requirement 9 man-h/ha. Animal drawn helical blade puddler is used for shallow puddling and levelling of puddled soil and also to uproot the weeds (Fig. 3.18c). The puddler is used by a pair of bullocks. The implement consists of tow angle iron brackets carrying a bush. A shaft carrying slots accommodates the helical blades. These blades are imparted a twist along their length to form a helix, thus its nomenclature. The blades on the implement rotate and impart a lateral and turning action on the soil particles, thus achieving a good puddle. The handle for the operator is made of wood and helps in comfortable operation. It saves 30 per cent labour, 46 per cent operating time and 30 per cent on cost of operation compared to conventional method of using country plough or tractor with cage wheels for puddling. Working capacity is 0.12 ha/h.

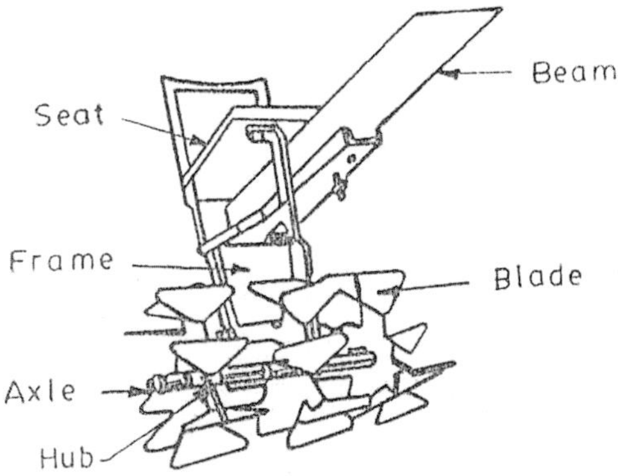

a) Details of angular blade puddler

b) A view of animal drawn flat blade puddler

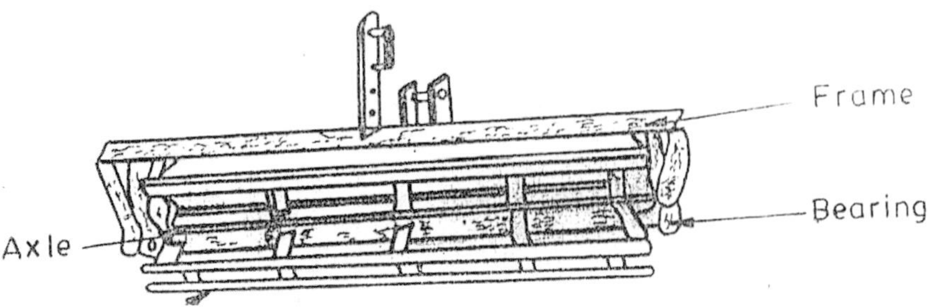

c) Components of animal drawn helical blade puddler

Fig. 3.18 : Animal drawn puddler

Animal drawn improved puddler: Animal drawn improved puddler (Fig. 3.19) developed at CIAE, Bhopal; IGKVV, Raipur, OUAT, Bhubaneswar; GBPUA&T, Pantnagar; and AAU, Jorhat has advantages in terms of better puddling with reduced number of passes (2 no) compared to the traditional method of puddling by use of wooden comb harrow (4 passes), Pandey *et al.*, 1997. Due to rolling type of improved puddlers the draft load on the animals is although at par (60 kg) to the traditional puddler but the work rate is higher. Field capacity is 0.08 ha/h.

Fig. 3.19 : Animal drawn improved puddler

Cossul puddler: The Cossul puddler has three axle mounted cast iron hubs. The each hub is provided with four blades which act on the soil to break up clods and stir the surface layer. The draught power of one animal is sufficient to pull the implement. Its work rate is 0.14 ha/h.

Animal drawn cono puddlers: The design uses the concept of conical shaped rotors for wetland preparation. The blades on the rotors puddle the top 10 cm by a differential soil displacing motion which also buries weeds and trash with a rolling action. There are six rotors individually clamped to a toolbar for easy removal and spacing adjustment. Two wooden beams are rigidly fixed to the tool bar for hitching to the yoke. A removable seat is provided for the driver, whose additional weight will assist in penetration on harder soils. Its draft requirement is 30-80 kg. For 1.6 m working width work rate of the puddler is 0.8 ha/day.

Animal drawn lugged wheel puddler: It is an animal drawn equipment to break the soil clods near saturation level into soil particles in order to prepare homogenized puddled tilth for mechanized paddy transplanting. The equipment is suitable for shallow puddling with higher mechanical dispersion of soil. The unit consists of main frame, hitch, operator seat, angle lugs, and roller wheel. The effective field capacity, field efficiency and labour requirement are 0.10 ha/h, 65% and 10 man-h/ha, respectively.

Animal drawn paddy disc harrow: Disc is a circular, concave revolving plates used for cutting and inverting the soil. Usually two types of discs are used on disc harrows viz. plain and cut away or notched type disc. Plain discs have plain edge and used for normal work. Notched discs have serrated edge and cut stalk, grasses and vegetative matters better than plain discs. Cut away discs are not effective for pulverization of soil but it is very useful for puddling the field especially for paddy cultivation (Fig. 3.20). Higher penetration on disc harrow can be obtained by increasing disc angle, by adding additional weight, by lowering hitch point, by using sharp edge discs of small diameter and lesser concavity and regulating the optimum speed. Bearings of disc harrows must be thoroughly greased at regular interval and blunt edges of discs should be sharpened regularly. Scraper is provided on each disc to prevent soil clogging. It removes the soil that may stick to the concave side of the disc.

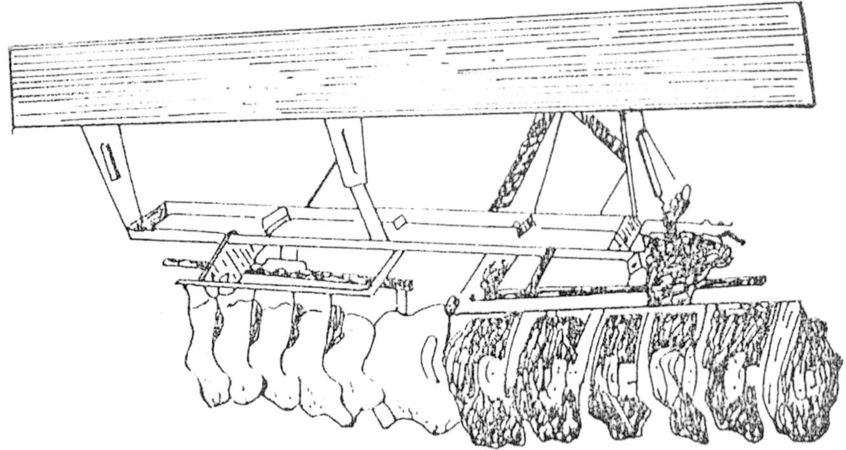

Fig. 3.20 : Paddy disc harrow

Animal drawn engine operated rotary tiller: A bullock drawn, engine operated rotary tiller has been developed at JNKVV, Jabalpur during seventies (Pandey *et al.*, 1997). It is fitted with 10 hp diesel engine which powers only the soil working tool namely, tiller, while the forward motion for the unit is pulled by a pair of bullocks. As most of the two wheel tractors (power tillers) possessed low drawbar pull, much of the engine power being used up in providing traction. This arrangement of bullocks hauling the machine and the engine power being used for operating the soil working tool is an effective way of utilizing power with the minimum loss. The animals gradually experienced to the engine noise and perform their work normally.

Tractor operated tillage equipment

Different types of ploughs used all over the world may be grouped as Indigenous ploughs and Soil turning or stirring ploughs. Soil stirring ploughs are further classified as mouldboard ploughs, disc ploughs, sub-surface ploughs, chisel ploughs and other ploughs. There are four types of tractor-drawn implements viz. trailed or pull type, semi-mounted type, mounted type and self-propelled. A pull or trailed type implement is one that is pulled and guided from a single hitch point and is never completely supported by the tractor. A semi-mounted type implement is attached to the tractor through a horizontal hinge axes and is partially supported by tractor during transportation. It is controlled through a 3-point hitch linkage. A mounted type implement is one that is attached through a 3-point hitch linkage and is completely supported by the tractor when in raised position. A self-propelled machine is one in which the propelling power unit is an integral part of the implement. Tractor operated implements are used for primary as well as secondary tillage operations. Implements used for primary tillage are mouldboard plough, disc plough, sub-soil plough and chisel plough. Secondary tillage implements are disc harrow, spike tooth harrow, spring tooth harrow, rigid tine cultivator and spring tine cultivator.

Power operated mould board ploughs

Mould board ploughs are normally of two types: fixed or reversible bottom plough and mounted or trailed type ploughs. Fixed bottom ploughs throw the soil in one direction, usually to the right (Fig. 3.21). Reversible bottom ploughs have the bottoms so arranged that the right turning bottoms can be quickly and readily replaced with a set that turns the soil to the left (Fig. 3.22 and 3.23). Thus, when end of the furrow is reached, the plough is raised, turned and returned across the field ploughing into the furrows just made. With reversible ploughs there are no dead furrows, no uneven and sunken spots. The mounted ploughs are also called integral ploughs because the tool and power are used as one unit. Hydraulically operated mounted ploughs are easy to handle. The trailing ploughs, which have standard hydraulic cylinder for raising and lowering, are also easy to operate. However, it is more difficult to back trailing equipment than a mounted one. Land wheel is used in raising and lowering the trail-type plough by means of hydraulic or mechanical lift. The plough bottom acts as a 3-sided wedge as it passes through the soil. The main parts of the plough are beam, top link connection, vertical stut, cross shaft, coulter stop, coulter arm, coulter disc, scraper, share point, share wing, mould board and vertical beam (Fig. 3.24). Tractor operated two bottom hydraulic reversible mould board plough has unique turnover mechanism with double acting arm that provides precise, safe and quick turnover action (Fig. 3.23). It has been provided with strong turnover axle with special hardened steel that can withstand hard shock loads. Lateral forces working on the plough are eliminated easily with optiquick adjusment setting.

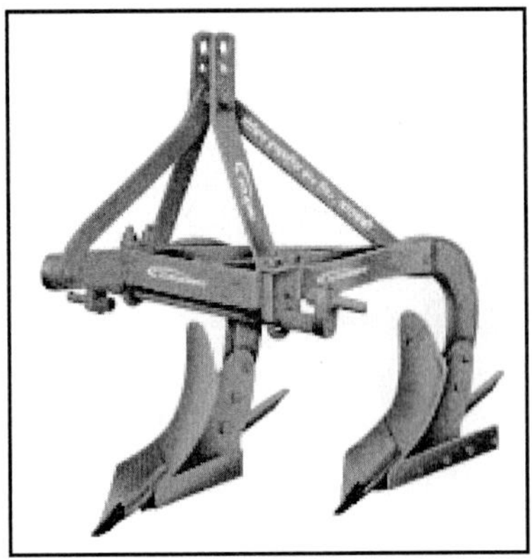

a) A view of tractor operated two bottom mould board plough

Courtesy: Govind Industries Pvt. Ltd., Barabanki (Uttar Pradesh)

b) Tractor operated two bottom mould board plough working in field

c) Tractor operated four bottom mould board plough
Courtesy: Govind Industries Pvt. Ltd., Barabanki (Uttar Pradesh)

Fig. 3.21 : Tractor operated mould board ploughs

Fig. 3.22 : Tractor operated mechanically reversible plow
Courtesy: Kisan Engineering Works, Dhule (Maharashtra)

Courtesy: Lemken India Agro Equipment Pvt. Ltd. Nagpur (Maharashtra)

Courtesy: Deccan Farm Equipment Pvt. Ltd. Kolhapur (Maharashtra)

Fig. 3.23 : Tractor operated two bottom hydraulic reversible mould board plough

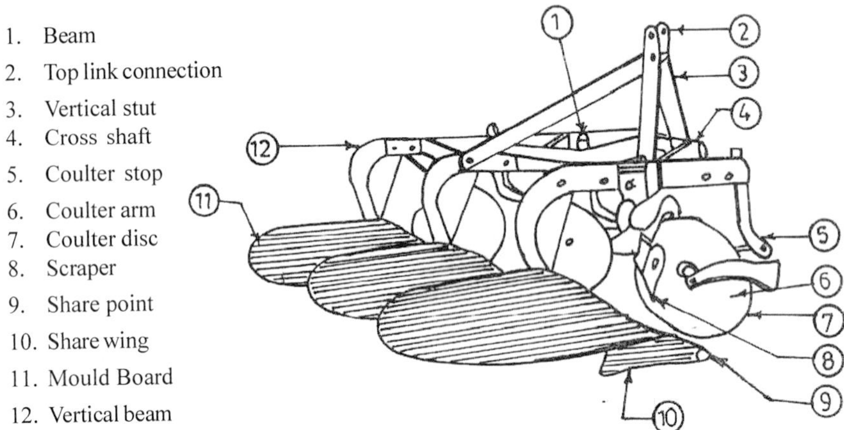

Fig. 3.24 : Tractor mounted 3-bottom mould board plough

There are different types of bottoms viz. stubble bottom, general purpose bottom, high speed general purpose bottom, slatted general purpose bottom, and breaker or sod bottom (Fig. 3.25). The stubble bottom mould board has an abrupt curvature. It turns the furrow slice quickly and provides maximum granulation. This bottom scours better in sticky soil. It is used for ploughing stubble land and not suitable for high speed ploughing. The general purpose bottom has a longer mould board with less curvature than stubble bottom. It has a fairly long mould board with a gradual twist, the surface being slightly convex. It turns the soil less abruptly and may be used in heavy soil, sod or stubble ground. It may be operated at higher speed than stubble bottom. However, speed should be less when ploughing sod than stubble ground. The high speed general purpose bottom has slightly less curvature in the upper part of the mould board as compared to general purpose bottom. It has a less twisting action and can be operated at a higher speed without excessive throwing of the furrow slice. The slatted general purpose bottom is made of slots placed along the length of the mould board so that there are gaps between the slots. This increases the soil pressure against the remaining portions of the mould board that aids scouring in sticky soils. The breaker or sod bottom mould boards are long with gentle curvature that lifts and overturns the furrow slice. They turn the furrow slice more slowly. It is mainly used in heavy sod or clay soil. It is used in tough soil with full of grasses. This is very useful where total inversion of soil is required.

There are mainly three accessories used on plough bottoms viz. jointer, coulter and gauge wheel. Jointer is a small irregular piece of metal having a shape similar to ordinary plough bottom. It looks like a miniature plough. Its main function is to cut and turn over a small ribbon like furrow slice directly in front of main plough bottom. It is set to cut 4-5 cm deep. This furrow slice is inverted so that all trash on the top of the soil is completely turned down. Coulter is used to cut furrow slice vertically from the land ahead of bottom and leaves a clear wall. It also cuts trashes covered under the soil. In general, coulter should be set about 5 cm shallower than the depth of ploughing. To obtain a neat furrow wall, the coulter is kept 2 cm outside the landside of the plough. It may be rolling or sliding type. Rolling coulter is a round steel disc suspended on a shank and yoke from the beam. Mostly plain rolling coulters of 30-45 cm diameter are used. The sharpened edge of coulter may be smooth or notched. It can be adjusted up-down and sideways. The up-down adjustment controls depth and sideways width of cut. Sliding coulter does not roll over the ground but slides. The knife may be of different shape and size. Gauge wheel helps in maintaining depth of cut under different soil conditions. It is usually placed in hanging position.

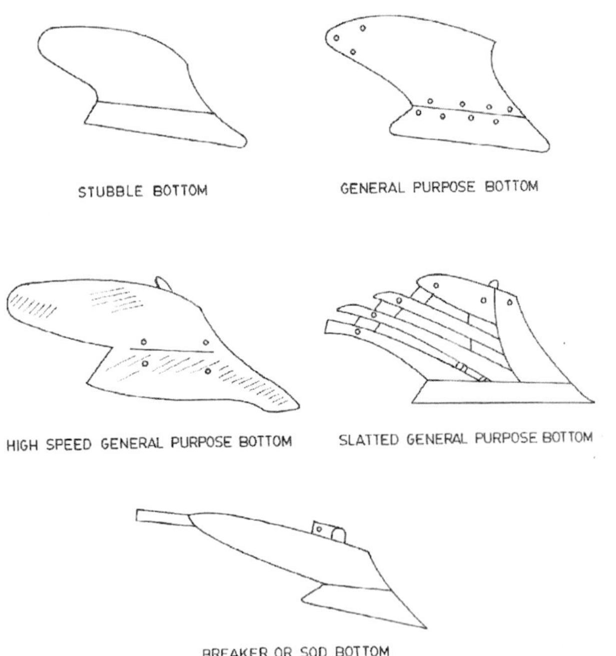

Fig. 3.25 : Different types of plough bottoms

For proper penetration and efficient operation some clearance is provided in the mould board plough and that is called "suction". The suction is of two types viz. vertical and horizontal suction (Fig. 3.26). Vertical clearance is the maximum clearance under the landside and the horizontal surface. Vertical suction is measured at the joining point of share and landside. It helps in maintaining depth of cut. Horizontal suction is the clearance between the landside and the furrow wall. It helps in maintaining width of cut. Mould board plough needs 3-5 mm vertical clearance (vertical suction) and about 5 mm side clearance (horizontal suction). Vertical clevis is a vertical plate with a number of holes, controls the depth of cut and adjusts the line of pull. Horizontal clevis affects lateral adjustment of plough relative to line of draught.

Share points are easily replaceable. Landsides are extra broad with large surface area that guides plough properly. The plough body is tempered and very strong. Each share is adjustable for pitch that allows better entry of the plough in the soil. Furrow cutting width can be adjusted as per soil type and moisture condition with the help of angle plate. It has a strong over load safety device. It can be operated by 50 hp tractors and above depending upon the working width. Power tiller operated single bottom and two bottom mould board ploughs are used on small farms and hilly areas (Fig. 3.27).

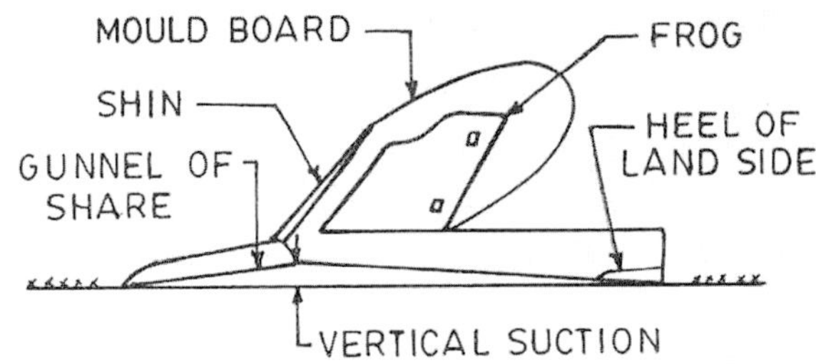

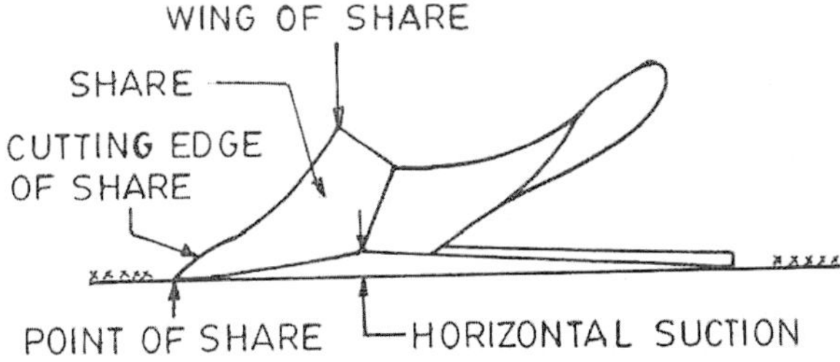

Fig. 3.26 : Horizontal and vertical suction in mould board plough

a) Power tiller operated two bottom mould board plough

b) Power tiller operated single bottom heavy duty mould board plough

Fig. 3.27 : Power tiller operated mould

Courtesy: VST Tillers Tractors Ltd., Bangalore (Karnataka)

Tractor operated disc plough

It is plough that cuts, turns and in some cases breaks the furrow slice by separately mounted steel discs (Fig. 3.28a, 3.28b). It consists of discs, clamp stand, disc bracket, scrappers, rear furrow wheel, furrow wheel spring, standard, beam (frame), and top link connections (Fig. 3.28c). Disc is a circular, concave revolving steel plate used for cutting and inverting the soil. A disc plough is designed with a view to reduce friction by making a rolling plough bottom instead of a sliding plough bottom. A disc plough works well in the conditions where mouldboard plough does not work satisfactorily. The main advantages of a disc plough are; i) works well in too hard and dry soil, ii) works well in sticky soil, iii) more useful for deep ploughing, iv) can be used safely in stony and stumpy soil, and v) works in loose soil also without much clogging. The main disadvantages are i) not suitable for covering surface trash and weeds, ii) leaves the soil in rough and cloddy conditions, and iii) much heavier as compared to mouldboard ploughs. There are two types of disc ploughs viz. Standard disc plough and Vertical disc plough. The disc plough bears little resemblance to the common mould board plough. A large, revolving, concave steel disc replaces the share and the mould board. The disc turns the furrow slice to one side with a scooping action. The usual size of the disc is 60 cm in diameter and this turns a 30 to 35 cm furrow slice. The disc plough is more suitable for land in which there is much fibrous growth of weeds as the disc cuts and incorporates the weeds. No harrowing is necessary to break the clods of the upturned soil as in a mould board plough. The disc plough is lighter in draft than the mould board plough, turning same volume of soil in similar conditions. In very hard soil, some extra weight is added to the wheel which increases the draft.

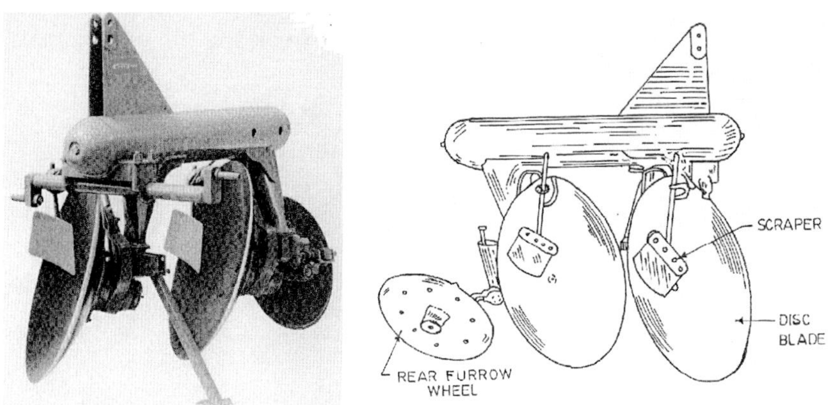

a) Tractor operated two bottom disc plough

Courtesy: Govind Industries Pvt. Ltd., Barabanki (Uttar Pradesh)

b) Tractor operated 3 bottom disc plough

Courtesy: Dasmesh Mechanical Works, Amargarh (Punjab)

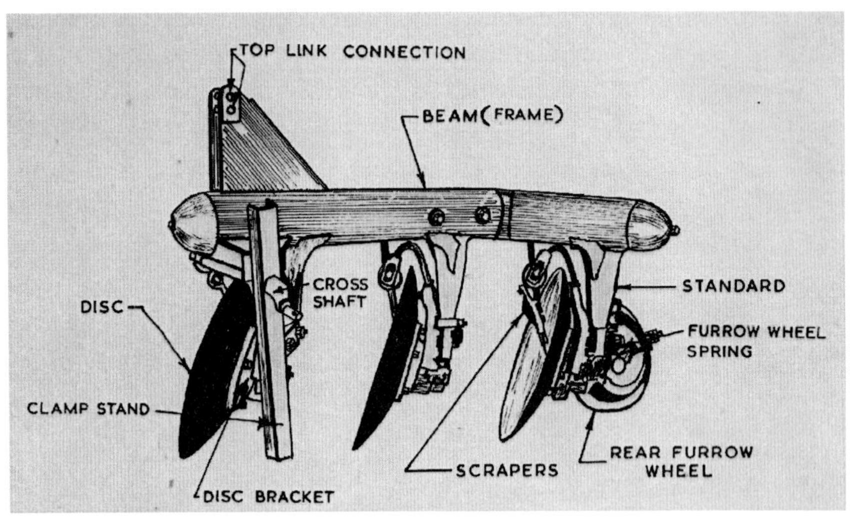

c) Components of disc plough

Fig. 3.28 : Tractor operated three bottom disc plough

Standard disc plough: It consists of steel disc of 60-90 cm diameter set at a certain angle to the direction of travel (Fig. 3.29). Each disc revolves on a stub axle in a thrust bearing. The angle of disc to the vertical and to the furrow wall is adjustable. In action, the disc cuts the soil, breaks it and pushes it sideways. There is a little inversion of furrow slice as well as little burying of weeds and trashes. Disc plough may be mounted or trailed type. In mounted type disc plough, side thrust is taken by wheel of the tractor. In trailed type, side thrust is taken by furrow wheel of plough. Disc is made of heat treated steel of 5-10 mm thickness. The edge of the disc is well sharpened to cut the soil. The amount of concavity varies with the diameter of the disc and it is approximately 80 mm for 60 cm and 160 mm for 95cm diameter.

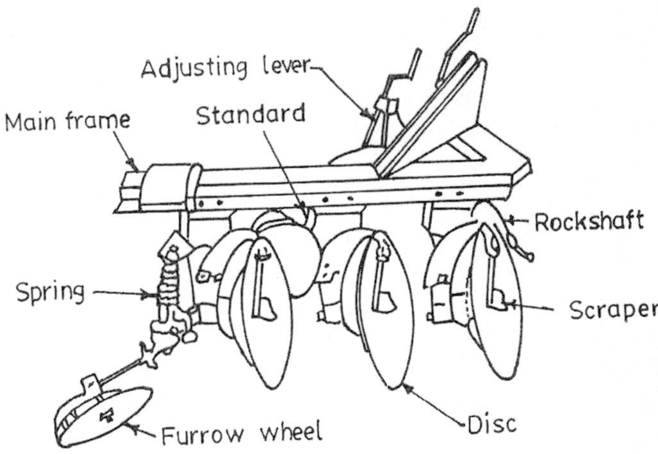

Fig. 3.29 : Tractor operated standard disc plough

Vertical disc plough: This is also called Harrow plough or one way plough (Fig. 3.30). It is the plough which combines the principle of regular disc plough and disc harrow and is used for shallow working in the soil. It has a frame, wheel arrangement and depth adjusting devices same as regular disc plough, but the discs are fitted on a single shaft and turn as one unit like a gang of disc harrow. The spacing between discs may be 20-25 cm. Size of the disc varies between 50-65 cm. Disc angle varies from 40-45^0 and tilt angle is zero.

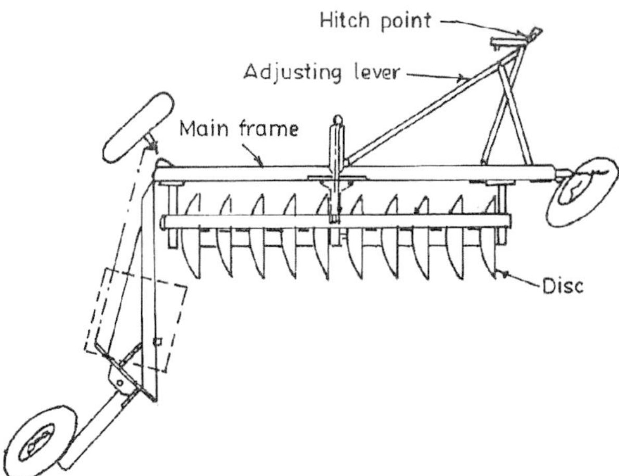

Fig. 3.30 : Tractor operated vertical disc plough

Disc angle is the angle at which the plane of the cutting edge of the disc is inclined to the direction of travel (Fig. 3.31). Usually the disc angle of good plough varies between 42-45 degree. **Tilt angle** is the angle at which the plane of cutting edge of disc is inclined to a vertical line (Fig. 3.31). Tilt angle varies from 15-25 degree for a good plough. **Scraper** is a device to remove soil that tends to stick to the surface of a disc. **Concavity** is the depth measured at the centre of the disc by placing its concave side on a flat surface.

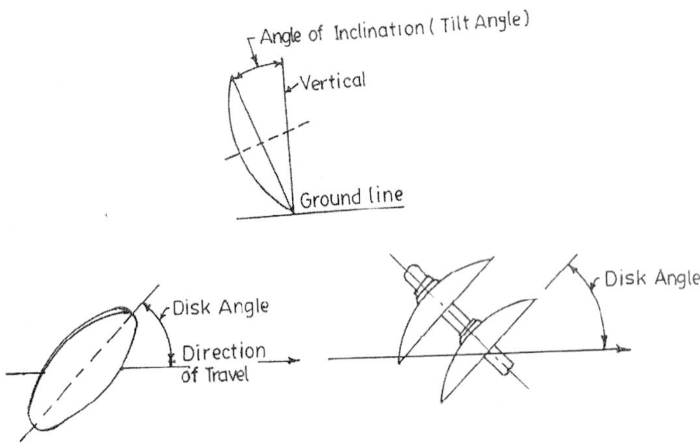

Fig. 3.31 : Disc and tilt angle on disc plough

Various adjustments can be done on the disc plough to control depth, width and pulverization. By increasing tilt angle, penetration can be improved. However, on standard disc plough penetration is improved by decreasing the tilt angle. Increasing disc angle improves penetration but reduces width of cut. Penetration can be improved by adding weight to the plough. Width of cut on the plough can be adjusted by adjusting angle between the frame and land wheel axle. The plough wheels may be properly adjusted to keep the plough running level. It works well in hard soil, rocky, stony and rooted soil. It cuts and breaks the soil, raise the soil, turn and mix the soil well. It works well where scouring is the major problem. It also has extra heavy duty tubular/rectangular frame to make disc plough sturdier. It can be operated by 50 hp tractors and above.

Tractor operated disc harrow

Harrows are used for shallow cultivation in operations such as preparation of seedbed, covering seeds and destroying weed seedlings. Harrows are used to break the clods, cut weeds, pulverize soil, cover seeds and smoothen the surface (Singh, 2007; Singh and Verma, 2009). Harrows are of two types: disc harrow and blade harrow. The disc harrow consists of a number of concave discs of 45 to 55 cm in diameter. These discs are smaller in size than disc plough, but more number of discs are arranged on a frame. These discs are fitted 15 cm apart on axles. Two sets of discs are mounted on two axles. All the discs revolve together with axles. The discs cut through the soil and effectively pulverize the clods.

The disc harrow is used for primary and secondary tillage. It is ideal for field disking, especially in orchards, plantations and vineyard. It is suitable for working under trees close to bunds and fence posts. Disc harrows are divided into two classes depending upon the arrangement of disc viz. single action and double action disc harrows (Fig. 3.32). Double action disc harrows are classified as tandem and offset disc harrows. Single action disc harrow with two gangs placed end to end arranged in such a way that right side gang throws soil towards right and left side gang throws soil towards left. Double disc harrow consisting of two or more gangs in which a set of one or two gangs follows the set of another one or two gangs, arranged in such a way that front and rear gangs throw soil in opposite directions. Thus entire field is worked twice in each trip. Tandem disc harrow consists of four gangs in which each gang is angled in opposite direction. The discs on the front gang throw soil outward and the rear gang inward. Therefore, no soil remains uncut by the offset disc harrow. Discs are important component of the harrow and are made from high carbon steel or alloy steel; the cutting edges are hardened and tempered to suitable hardness.

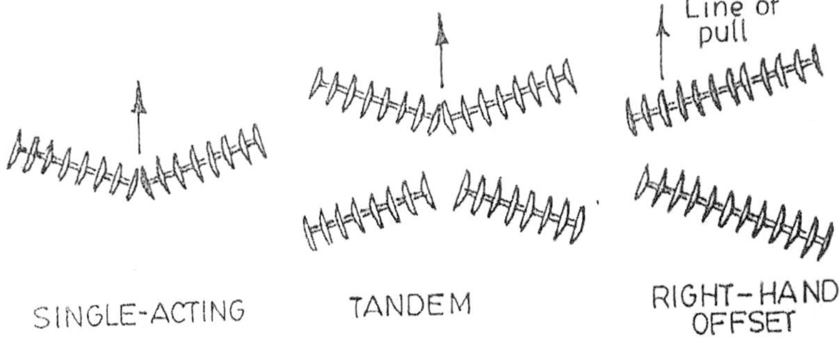

Fig. 3.32 : A view of single action and double action disc harrows

Offset disc harrow mounted type has two gangs in tandem and most commonly used in India (Fig. 3.33). It consists of tool bar, front and rear disc gangs, gang angle adjustment, spacing spool and steel discs (Fig. 3.34). It is capable of being offset to either side of the centre line of tractor. It travels to the left or right of the tractor. The line of pull is not in the middle and that is why it is called offset disc harrow. It is generally used for cultivating under the trees. If the discs are not in offset position, then difference between lateral force components becomes equal to side draft. If the implement is hitched such that it moves to right or left from the no side draft position, side draft is introduced and operating condition of harrow changes. The gangs can be moved in either direction on the hitch frame. The rear gang can be moved the same amount as the front gang. When operating in orchards or plantations, the harrow can be offset to the right or left, thus enabling soil to be thrown towards or away from the trees. The offset feature makes it possible to work under low-hanging branches. Discs with notches on the outer rim are also available for operation in weed-infested fields.

Fig. 3.33 : Tractor-operated offset disc harrow plain and notched discs in front

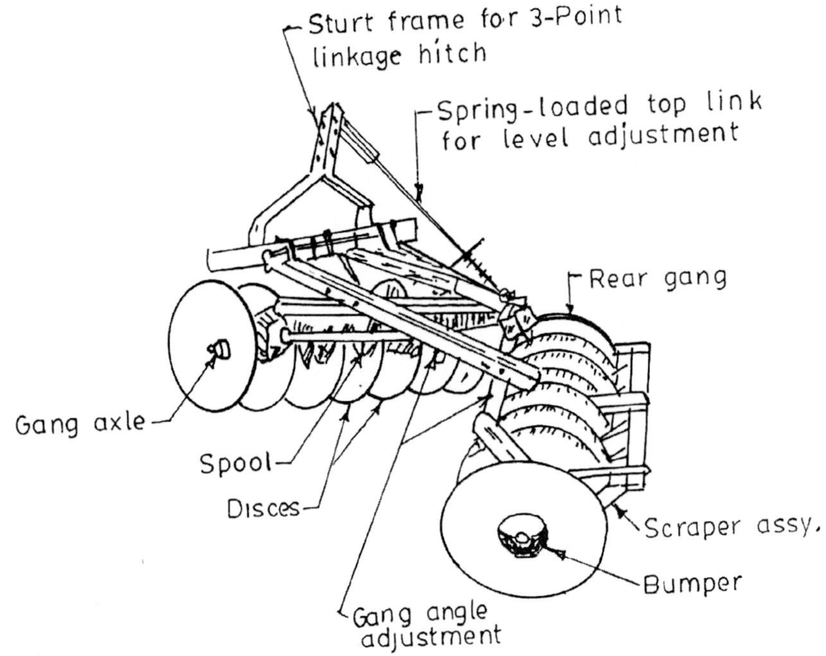

Fig. 3.34 : Details of tractor-operated offset disc harrow

Three point hydraulic linkage and hydraulic control makes it highly maneuverable. Number of discs may vary from 10 to 16. Two operations with these implements are sufficient for good seedbed preparation in light and medium soils. The average capacity of this implement is about 2.5-3.0 ha/day. Notched discs in front is used when organic substance and remains are there on the soil surface. Discs could be of different diameters i.e. 560 mm or 610 mm. Width of cut depends on number of discs used. Depth of cut may vary up to 15 cm. The spherical discs are ground on the convex side to an angle of 50 degree. Discs are made of high-grade heat-treated hardened steel. Tractor-drawn disc harrows have concave discs of size varying from 35 to 70 cm diameter. Concavity of disc affects penetration, inversion and pulverization of soil. All nuts and bolts must be checked daily before transporting the implement to the field. Blunt edges of the discs should be sharpened regularly. Disc harrows are also of trailed type and not very common on Indian forms (Fig. 3.35 and Fig. 3.36).

Fig. 3.35 : A view of tractor operated pull type disc harrow

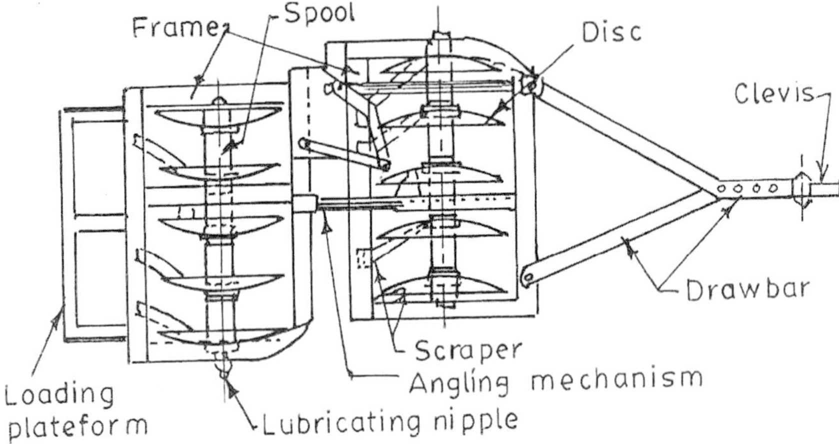

Fig. 3.36 : Detailed components of tractor operated pull type disc harrow

Tractor operated paddy disc harrow

Disc is a circular, concave revolving plates used for cutting and inverting the soil. Usually two types of discs are used on disc harrows viz. plain and cut away or notched type disc. Plain discs have plain edge and used for normal work. Notched discs have serrated edge and cut stalk, grasses and vegetative matters better than plain discs. Cut away discs are not effective for pulverization of soil but it is very useful for puddling the field especially for paddy cultivation (Fig. 3.37). Higher penetration on disc harrow can be obtained by increasing disc angle, by adding additional weight, by lowering hitch point, by using sharp edge discs of small diameter and lesser concavity and regulating the optimum speed. Bearings of disc harrows must be thoroughly greased at regular interval and blunt edges of discs should be sharpened regularly. Scraper is provided on each disc to prevent soil clogging. It removes the soil that may stick to the concave side of the disc.

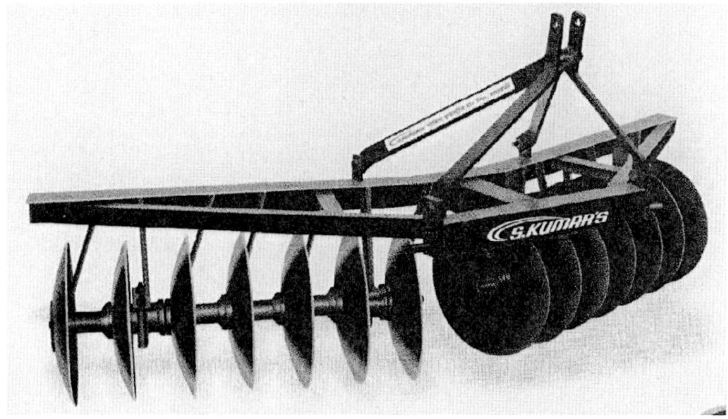

Fig. 3.37 : Tractor operated paddy disc harrow

Courtesy: Govind Industries Pvt. Ltd., Barabanki (Uttar Pradesh)

There are many other types of harrows besides disc harrow such as spring tooth harrow (Fig. 3.38), spike tooth harrow (Fig. 3.39), triangular harrow (Fig. 3.40) and blade harrow (Fig. 3.41).

Fig. 3.38 : Tractor operated spring tooth harrow

Source: https://www.google.co.in/search?q=spring+tooth+harrow

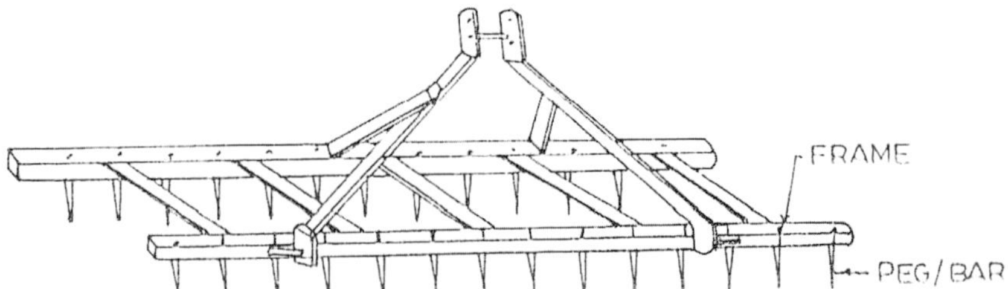

Fig. 3.39 : Tractor mounted spike tooth harrow

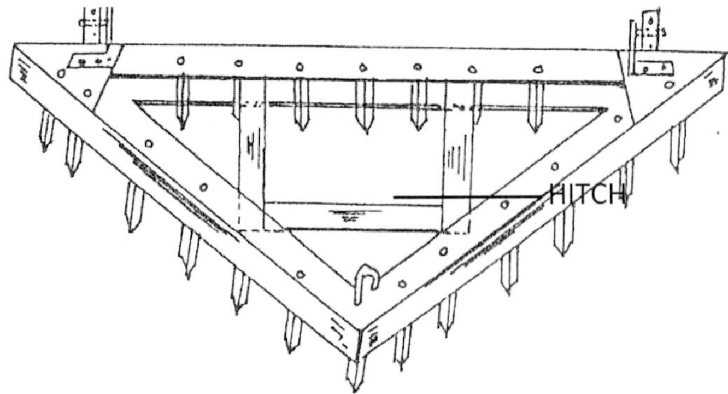

Fig. 3.40 : Tractor mounted triangular harrow

Fig. 3.41 : Tractor mounted blade harrow

Tractor operated cultivator

Cultivator is an implement used for finer operations like breaking clods and working the soil to a fine tilth in the preparation of seedbed. Cultivator is also known as tiller or tooth harrow. It is used to further loosen the previously ploughed land before sowing. It is also used to destroy weeds that germinate after ploughing. Cultivator has two rows of tynes attached to its frame in staggered form. The main object of providing two rows and staggering the position of tynes is to provide clearance between tynes so that clods and plant residues can freely pass through without blocking. Provision is also made in the frame by drilling holes so that tynes can be set close or apart as desired. The number of tynes ranges from 7 to 15. The shares of the tynes can be replaced when they are worn out.

A cultivator performs functions intermediate between those of plough and the harrow. Destruction of weeds is the primary function of a cultivator. The important functions performed by a cultivator are to destroy the weeds in the field, to aerate the soil for proper growth of crops, to conserve moisture by preparing mulch on the surface, to sow seeds when provided with sowing attachments and to prevent surface evaporation and encourage rapid infiltration of rainwater into the soil. According to the source of power used, cultivators may be classified as hand operated, animal-drawn and tractor-drawn. Cultivators are of two types viz. cultivator with spring loaded tines (Fig. 3.42) and with rigid tines (Fig. 3.43). Depending upon the type of soil and crop grown, shovels of different types are used such as straight shovel, reversible shovel, hoop or spear head shovel, sweep, furrower, and half sweep (Fig. 3.44).

Spring tine cultivators: Cultivators are used for seedbed preparation both in dry and wet soils. It is also used for interculture purpose by adjusting the tynes in wider row crops. It is also used for puddling purposes. Cultivator consists of a frame, tynes with reversible shovels, land wheel, hitch system and heavy-duty springs (Fig. 3.42). The function of springs is to save the cultivator tynes from breaking when some hard object comes in contact with the shovel or under the tyne. The shovels are made of heat-treated steel for longer life. The implement is mounted type and is controlled by the hydraulic system of the tractor. Cultivators having tines loaded with springs so that it swings back when an obstacle is encountered are called spring tine cultivators. Each tine in this cultivator is provided with two heavy coil springs. The springs operate when shovel point encounters with crop roots or stones during the field operation. On passing over the obstruction, the tines automatically trips and work continuously without interruption. Tines are made of high carbon or spring steel. These types of cultivators are generally used in soils having stones and stumps and are pulled by tractors. The cultivators are available with 7, 9, 11, 13, 15 tines or more depending upon the requirements. Capacity varies from 0.35-0.50 ha/h.

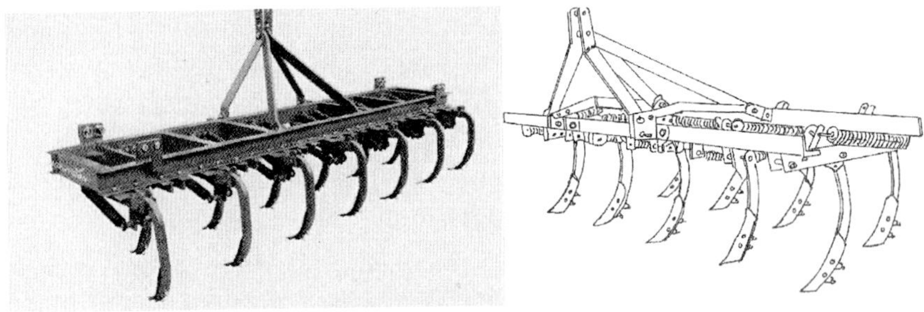

Fig. 3.42 : Tractor mounted spring tine cultivator

Courtesy: India Agrovision Implements Pvt. Ltd., Jaipur (Rajasthan)

Rigid tine cultivators: It is a versatile implement used for loosening and aerating the soil and preparing seed beds quickly and economically. It is used for loosening and aerating of the soil, preparing seedbed, subsoil cultivation, and weeding/interculture. These are one of the most selling cultivators used in agricultural operation. It consists of a rectangular frame made of mild steel angle or channel section, heavy-duty rigid tynes made of mild steel flat or plate section, U-clamps, reversible shovels joined to tynes with fasteners, and hitch assembly (Fig. 3.43). The clamping of tines makes possible to adjust the distance between them according to crop rows. The shovels are made of medium carbon steel, old leaf spring steel or low alloy steel and hardened to 40-45 HRC. Number of tynes varies from 7-15. The shovels are made from medium carbon steel or low alloy steel, hardened and tempered to suitable hardness. The shovels are mounted on the tynes with fasteners and can be replaced easily on wearing or becoming dull. Different types of shovels are used on cultivators (Fig. 3.44). The depth of operation is controlled by the hydraulic system of the tractor. The shovels can be replaced by duck foot sweeps for shallow tillage. In these types of cultivator the tines are fixed to the frame and tines do not deflect during the field operation. Spacing between tines may be achieved simply by sliding the braces to the desired position or with the help of nuts and bolts.

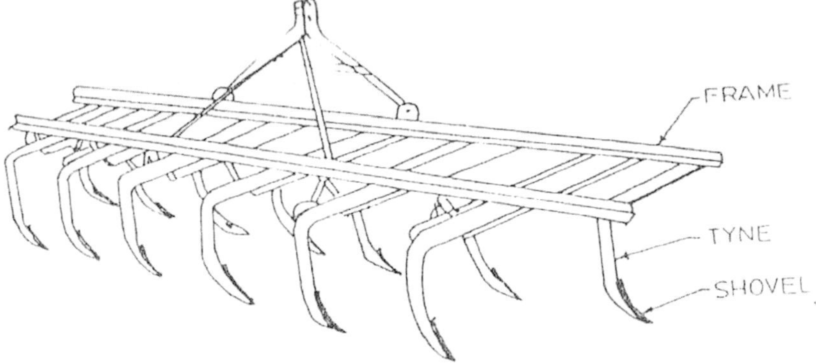

Fig. 3.43 : Tractor mounted rigid tine cultivator (Clock-wise Stationery view, working in field and isometric view)

Courtesy: Kisan Engineering Works, Dhule (Maharashtra)

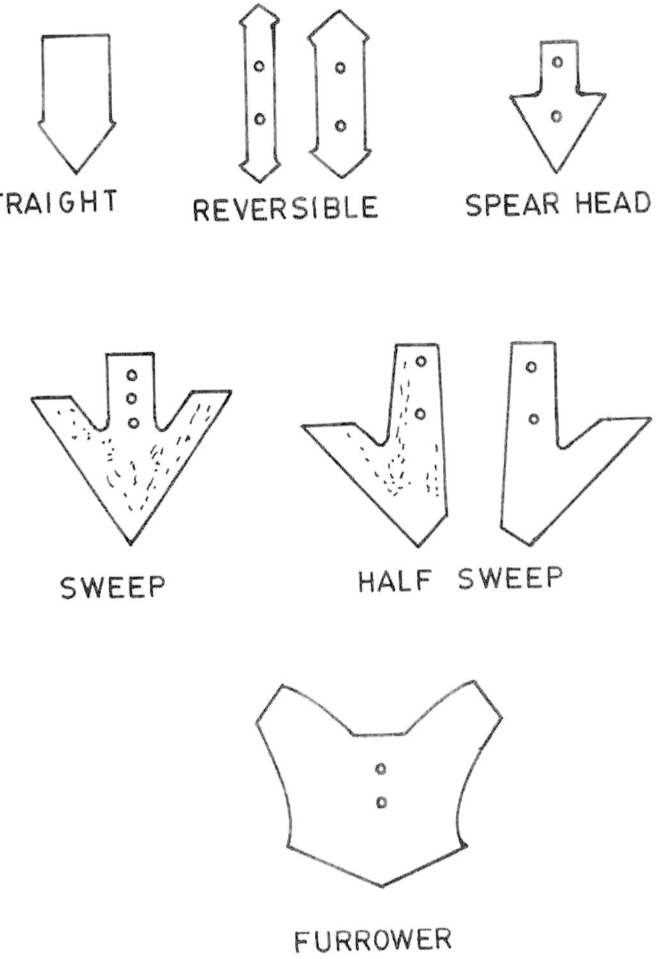

Fig. 3.44 : Different types of shovels used on cultivators

Tractor operated plank: Tractor operated plank is a very simple implement and consists *of* a heavy wooden or metallic beam of 2 m in length (Fig. 3.45). In addition, shafts and handle are fixed to the beams. When it works most of the clods are crushed due to its weight. It also helps in micro levelling and slight compaction necessary after sowing. Rollers are used mainly, to crush the hard clods and to compact the soil in seed rows. Planks are generally used behind cultivators on Indian farms during both dry seed bed preparation and puddling (Fig. 3.46).

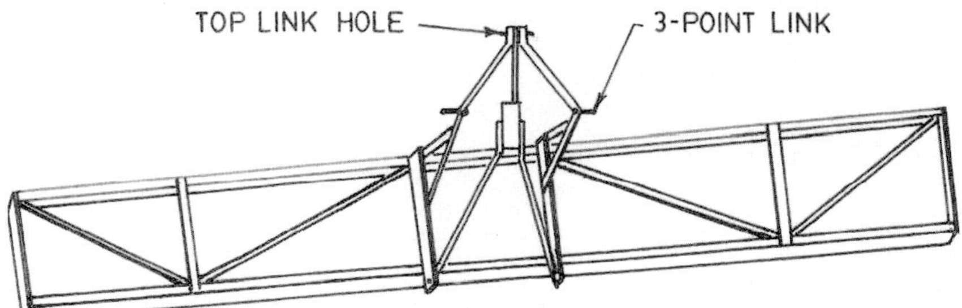

a) Isometric view of plank

b) Plank under puddling conditions

Fig. 3.45 : Tractor operated plank under puddling condition

Fig. 3.46 : Tractor operated cultivator with plank during puddling

Sweep cultivator: In stubble-mulch farming, it is difficult to prepare the land with ordinary implements due to clogging. Sweep cultivator is the implements useful under this condition. It consists of large inverted V shaped blades attached to a cultivator frame (Fig. 3.47). These blades run parallel to soil surface at a depth of 10 to 15 cm. They are armed in two rows and staggered. Sweep cultivator is used to cut up to 12 to 15 cm depth of soil during first operation after harvest and shallower during subsequent operations. It is used frequently to control weeds. It can also be used for harvesting groundnut.

Fig. 3.47 : Tractor operated sweep cultivator
Source: https://www.google.co.in/search?q=spring+tooth+harrow

Tractor operated cultivator with pulverizing roller attachment: A pulverizing roller attachment for tractor-drawn cultivator has been developed at Punjab Agricultural University, Ludhiana (Fig. 3.48) for breaking the clods and preparing a smooth seedbed (Garg and Singh, 2002; Singh and Pandey, 2008). It consists of an axle on which different numbers of star wheels are mounted depending upon the length of roller (Fig. 3.49). On the star wheels there are six spikes, through each of which pulverizing member passes. The pulverizing members run in a helix from one star wheel to next star wheel. The roller can be attached or detached from the cultivator with the help of a few nuts and bolts. The roller is made of mild steel flats. It has been found very useful for wetland and dry land cultivation. It pulverizes the soil thoroughly and makes it good for proper germination. It has also been found quite useful for puddling the field for rice transplanting (Fig. 3.50). It slightly increases the draft and power requirements. But it reduces the number of operations of cultivator and plank. The use of roller with cultivator for puddling also reduces the number of irrigations required for paddy by cutting down the rate of infiltration. Moreover, attaching roller to it does not reduce the field capacity of the cultivator. Pulverizing rollers are used for puddling as well as dry seedbed preparation in two runs, and creates good puddle. It saves 20-35% fuel consumption and 20-30% water requirement in comparison to traditional method. This equipment is used by a 35 hp tractors and above. Pulverizing roller is an attachment to commercially available cultivator. Working capacity is 0.6 ha/h.

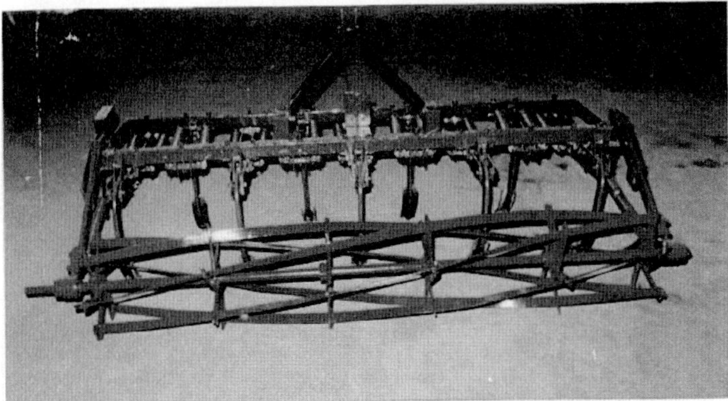

Fig. 3.48 : A view of pulverising roller attachment to cultivator

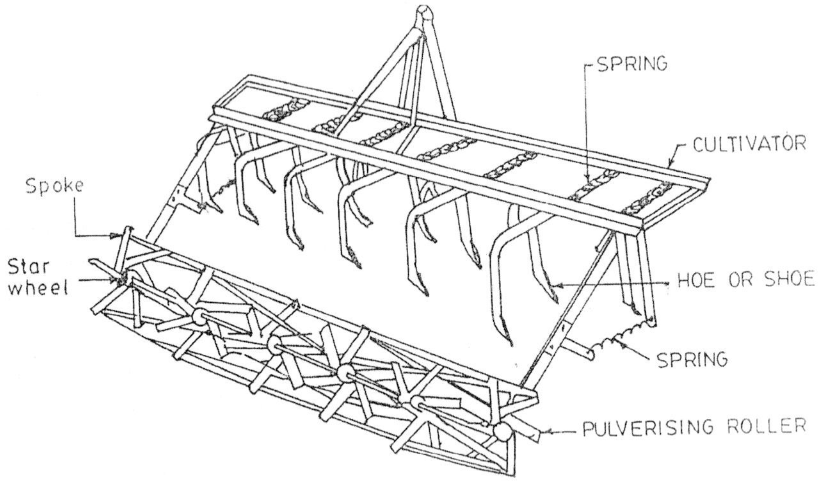

Fig. 3.49 : An isometric view of pulverising roller attachment to cultivator

Fig. 3.50 : Tractor operated pulverizing roller attachment to cultivator under puddling

Tractor operated roto–cultivator

M. B. Plough, cultivator and disc harrow are the conventional tillage implements. Prior to invention of rotavator, farmers were using mostly cultivator and disc harrow, for 3 – 4 passes or even more for good seed bed preparation. After introduction of rotavator, farmers use this implement for 2 passes to get their land prepared. The rotavator has also been found suitable for wet land tillage i.e. puddling operation and farmers frequently use rotavator for this purpose. The ability of rotavator to prepare seed bed in 2-passes and also to conduct satisfactory puddling are the main features behind liking of this machine by the farmer and that is the main cause of its large scale adoption. But, use of rotavator has its own limitations. The disadvantage associated with rotavator, mainly the compaction of soil below seed bed and formation of hard pan, more power requirement and therefore, stress upon tractor engine and its heavy purchase cost were the main factors, which forced the inventor to think for the development of new implement for dry land and wet land tillage in the name of "Roto-cultivator" for use by the farmers. Some characteristics of 'Roto–cultivator' are: Good and satisfactory seed bed preparation only in 2 – pass operation, no – soil compaction and suitable for high quality puddling operation. It can be operated by 45 hp tractor and above. Roto-cultivator has high quality discs, robust standard tynes with shovels, pulverizing roller, trailing board, frame and three-point hitch (Fig. 3.51). It is capable of deep ploughing (10-12 cm depth). Work quality is better than disc harrow, cultivator and at par with rotavator in two pass operation in light textured soil. It covers 0.4 ha/h at an average speed of 3.5 to 4.5 km/h. Another type of roto-cultivator has three rows of tynes with shovels and pulverizing roller attached behind (Fig. 3.52). This is also called tine cultivator. Pulverizing roller attached behind cultivator has also fingers for collection of loose straw. Tractor operated heavy duty puddler is also available commercially where wide spiral blades are used (Fig. 3.53). This gives better churning of soil being more in contact with soil. It has a very strong frame to support the spiral rollers.

Fig. 3.51 : Tractor operated roto cultivator
Courtesy : M/s Kisan Engineering Works, Muzaffarpur (Bihar)

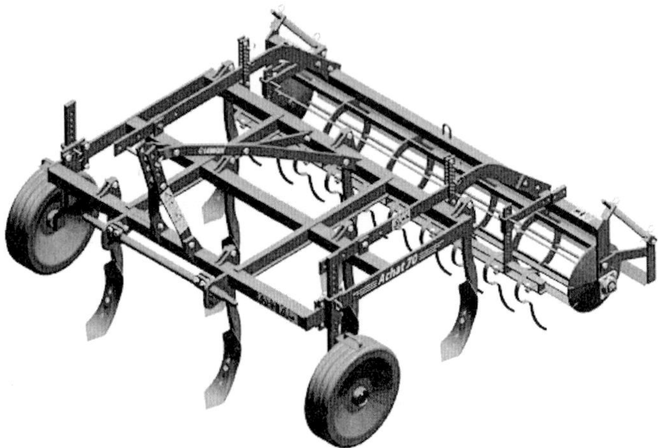

Fig. 3.52 : Tractor operated tine cultivator with pulverizing roller attachment
Courtesy: Lemken India Agro Equipment Pvt. Ltd., Nagpur (Maharashtra)

Fig. 3.53 : Tractor operated heavy duty puddler

Tractor operated duck foot cultivator

It is used for primary tillage operation, destruction of weeds and retention of soil moisture in vertisol soil. Five duck foot cultivator is used by a 25 to 35 hp and 7 duck foot cultivator by 35 to 50 hp tractors. The duck foot cultivator consisted of a box type steel rectangular frame, rigid tines and sweeps (Fig. 3.54). The sweeps are triangular in shape similar to foot of duck, hence called duckfoot cultivator (Pandey *et al.*, 1997). The sweeps are made from old leaf spring steel and joined to tines with fasteners, which makes them replaceable after being worn out or becoming dull. The tines are made of mild steel flat and forged to shape. It is a tractor-mounted implement and depth of operation is controlled by hydraulic system. The cultivator is popular in black cotton soils. Sweeps are attached to these tines. Three-point linkage of the tractor is attached to the implement. The implement is mostly used for shallow ploughing and in hard soils. Working capacity of cultivator is 0.4 ha/h.

Fig. 3.54 : A view of tractor operated duck foot cultivator.

Tractor operated rotavator (Rotary tiller)

It is an implement that cuts and pulverizes the soil by impact forces through a number of rotary tines or knives mounted on a horizontal shaft. It is also called "rotary tiller" (Fig. 3.55a). It is suitable for shallow cultivation and weed control. The rotary cultivator is widely considered to be the most important tool as it provides fine degree of pulverization enabling the necessary rapid and intimate mixing of soil besides reduction in traction demanded by the tractor driving wheels due to the ability of the soil working blades to provide some forward thrust to the cultivating outfit. Rotary tiller is directly mounted to the tractor with the help of three point linkage. It consists of a frame, a rotary shaft mounted with blades and power transmission system from the gearbox to the shaft (Fig. 3.55b). The power is transmitted from the tractor PTO (Power Take Off) shaft to a bevel gear box mounted on the top of the unit, through telescopic shaft and universal joint (Fig. 3.56). From the bevel gear box the drive is further transmitted to a power shaft, chain and sprocket transmission system to the rotor. The tynes are fixed to the rotor and the rotor with tynes revolves in the same direction as the tractor wheels. The number of tynes varies from 28 - 54. A leveling board is attached to the rear side of the unit for levelling the tilled soil. A depth control lever with depth wheel provided on either side of the unit ensures proper depth control.

a) A view of rotary tiller

Courtesy : Ganesh Agro Equipment, Vadpura (Gujarat)

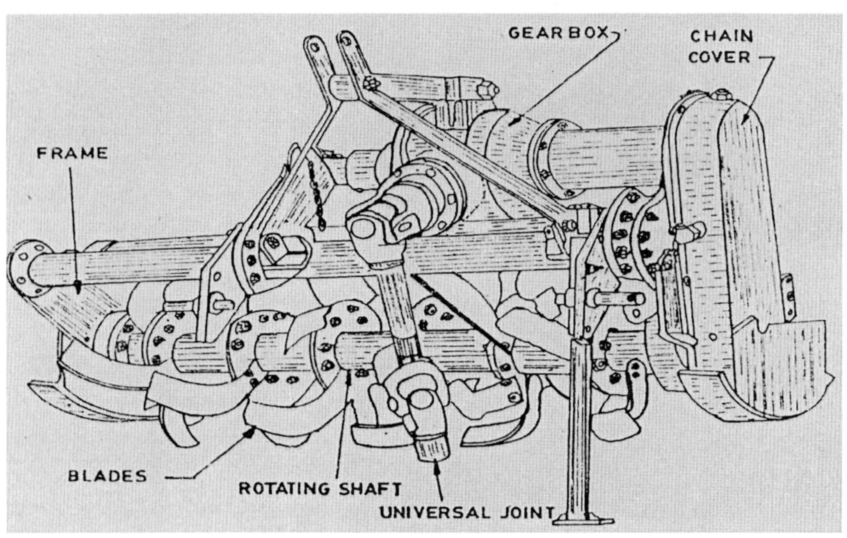

b) Details of rotary tiller

Fig. 3.55 : A view of tractor operated rotavator (Rotary tiller)

It consists of a power driven shaft on which knives or tines are mounted to cut the soil and trash. Rotor has got several types of tines fitted on the shaft having a speed of 200-300 rpm. The types of blades used with the rotor are spike blades, C-blades, J-blades and L-blades (Fig. 3.56). Generally, sharp edged L-shaped blades are used on the rotor which produces very fine tilth. Some rotors can be provided with C-type blades which require less power and provide coarse finish for better water penetration and retention. Twisted blade is suitable for deep tillage in relatively clean ground, but it has clogging and wrapping of trashes on the tynes and shafts. Straight blades are used only on mulchers designed mainly for secondary tillage. One or two operations of rotary tiller are sufficient for good pulverization of soil depending upon soil and crop conditions. It is not meant for sandy soil. It can plough up to 8-10 cm deep. However the depth of penetration can be adjusted up to 12.5 cm. The suitable protective cover is provided at the rear to prevent undue scattering of soil. It can cover about 0.2 to 0.4 ha/h depending upon the width of machine and condition of soil. Tractor operated rotavator is used by a 30 kW and above tractors. It has multispeed gear box in some of the rotavator. The use of rotavator can save 25-30% fuel for wheat cultivation and 30-35% for rice cultivation. Labour saving over conventional practices is about 40-45% for wheat and 25-30% for rice. Rotavator is available in different sizes starting from 1 m to 4 m and for average condition of the soil to heavy duty for sugarcane and cotton field. It has multi speed gear box. Working depth is adjustable. Different types of blades can be easily replaced on main haft to be used under various soil conditions. Rotary tiller is also suitable for incorporating straw (Fig. 3.57a) and green manure in the field (Fig. 3.57b). It is well adopted for preparing soil for rice transplanting under puddle condition (Fig. 3.58a) and for incorporating green manure under puddle condition (Fig. 3.58b).

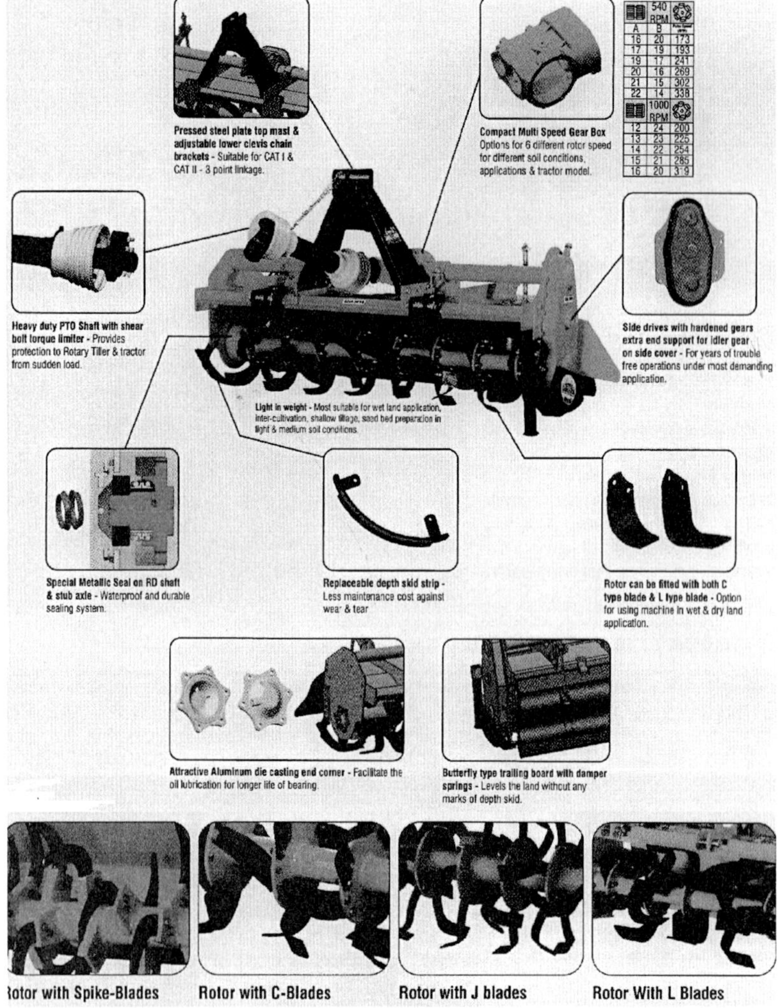

Fig. 3.56 : Various components of tractor operated rotavator with different types of blades
Courtesy: Tirth Agro Technology Pvt. Ltd. Rajkot (Gujarat)

a) Rotary tiller being used for dry seedbed preparation incorporating straw
Courtesy: Farm Implements (India) Pvt. Ltd., Chennai (Tamil Nadu)

b) Rotary tiller being used for dry seedbed preparation incorporating green manure

Fig. 3.57 : Tractor operated rotary tiller being used for dry seedbed preparation incorporating straw and green manure

a) Rotavator being used for puddling

b) Rotavator incorporating dhaicha under wet condition

Fig. 3.58 : Tractor-operated rotavator under wetland cultivation

Bullock-drawn engine operated rotary tiller is quite useful for timely preparation of seedbed particularly in rice-wheat rotation. Power tiller operated rotary tillers are also quite useful for hilly areas and small land holdings. It is suitable for preparing seedbed in a single pass both in dry (Fig. 3.59) and wetland conditions (Fig. 3.60). Under wet land cultivation pneumatic wheels are replaced with cage wheels for better traction. The lightweight power tiller is powered with an 8.3 hp diesel engine and above. The engine power is transmitted to ground wheels through V-belt-pulley. A tail wheel is provided at the rear to maintain the operating depth. The rotary weeding attachment does weeding. The rotary weeder consists of three rows of discs mounted with 6 numbers of curved blades in opposite directions alternatively on each disc. These blades help in cutting and mulching the soil. The width of coverage of the rotary tiller is 500 mm and the depth of operation can be adjusted to weed and mulch the soil in the cropped field. The field capacity is 0.12 – 0.15 ha/h. It is also used for mechanical control of weeds in crops such as tapioca, cotton, sugarcane, maize, tomato and pulses whose rows spacing is more than 450 mm. Attachments like sweep blades, ridger, trailer can be used with the machine. The lightweight power tiller can also be used for tillage under hill agriculture and terrace farming.

Fig. 3.59 : Power tiller operated rotary tiller for dry tilling
Courtesy: VST Tillers Tractors Ltd., Bangalore (Karnataka)

Fig. 3.60 : Power tiller operated rotary tiller for puddling
Courtesy: VST Tillers Tractors Ltd., Bangalore (Karnataka)

Tractor drawn Vishnu puddler in Godavari delta region

Degree of puddling is an important measure of loosening and softening of soil, which considerably reduces the strength of soil. Most of the transplanted rice fields are initially ploughed and puddled before seedlings is transplanted in the field. At present puddling operation is being done with cage wheels. Sometimes the puddling with cage wheels may not create semi impervious layer due to which deep percolation losses may occur. For effective puddling, tractor drawn Vishnu puddler has been developed at APWAM, Tirupati (Anonymous, 2013). It is pulled by 35 hp tractor and above. The frame is made of hollow rectangular mild steel and the length is 225 cm. It consists of 32 blades made of semi-circular mild steel (Fig. 3.61). The puddler rotates in the soil not from the power supply of the PTO shaft of the tractor but by the pull of the tractor. The implement is not an active one but is a passive type in terms of power supply.

A view of tractor operated Vishnu puddler   Field operation of tractor operated Vishnu puddler

Fig. 3.61 : Tractor operated Vishnu puddler

Tractor operated peg type puddler

The peg type puddler is a tractor-mounted implement using the three-point linkage of a tractor of 35 hp or more (Fig. 3.62). It is used to break the soil clods and puddle the soil for rice transplanting. It consists of a frame made of mild steel angle sections on which the three cross bars and the cleats for the three-point linkage are welded in place. It is operated when the soil moisture is near the saturation level to obtain a fine tilth and good puddle to facilitate mechanical transplanting. For achieving better performance, the tractor is fitted with cage wheels to improve traction and achieve higher field capacity. The working width is about 2 m and can work up to 11 cm depth. The field capacity is 0.4 ha/h.

Fig. 3.62 : Tractor operated peg type puddler with cage wheel

Cage Wheels

Cage wheels are attached to tractors (Fig. 3.63) and power tillers (Fig. 3.64) during puddling operation. They are made of a lattice of reinforced steel angles, flats or bars and are approximately three times the width of the normal steel or pneumatic tyre wheel in a power tiller or tractor. The wide cage wheels provide good floatation and traction for the power units pressing organic materials down into the puddle and the wheels themselves have a considerable puddling effect.

Fig. 3.63 : Puddling by tractor with cage wheel and cultivator

Fig. 3.64 : Power tiller operated rotary tiller with angle iron type cage wheel

Tractor operated power harrow

Tractor operated power harrow has a set of vertical rotating blades that works in the soil without creating any compaction on the soil (Fig. 3.65). The 260 mm long and 12 mm wide special alloy steel tines has better performance and long life. The arrangement of rotors on frame makes very smooth operation and uniform load on the tractor. Ach rotor has its own gear and bearings arrangement to make machine trouble free. Bolted tines can easily be replaced. The adjustable levelling bar (Fig. 3.66) behind the tines regulates the flow of soil. It helps in breaking the soil clods for better seedbed preparation. The side shields (Fig. 3.66) on each side equipped with special spring prevent the outer tines from creating ridges and also protects from flying stones. The machine has depth control tube bar roller to adjust the depth during ploughing. The roller without central axis provides better formation of soil even in the sticky soil. A shear bolt in PTO shaft coupler prevents against overload damages. The various components of machine are shown in Fig. 3.67. The machine is suitable for intensive mixing and crumbling action that results ideal clod distribution in the seedbed for better seed germination. For better clod breaking tractor operated power harrow is attached with rear spike roller (cone type clod breaker) that breaks the clods uniformly in the field (Fig. 3.68). It prevents formation of hard pan at the bottom layer. It is available in various sizes ranging from 1.0 m to 2.5 m. Number of tines depends on width of the machine e.g. 1.0 m width has 8 tines and 2.5 m width has 20 tines. It can be operated by 50 hp tractor and above.

Fig. 3.65 : Power harrow with leveling and pulverizing roller attachments
Courtesy: Lemken India Agro Equipment Pvt. Ltd., Nagpur (Maharashtra)

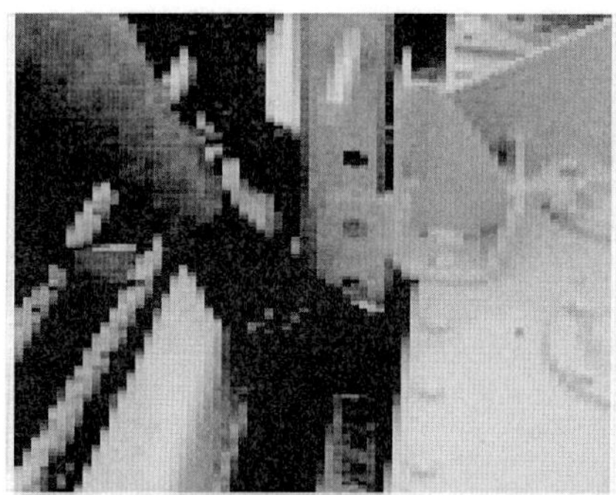

Fig. 3.66 : Power harrow with side shield and leveling bar for regulating the flow of soil
Courtesy: Lemken India Agro Equipment Pvt. Ltd., Nagpur (Maharashtra)

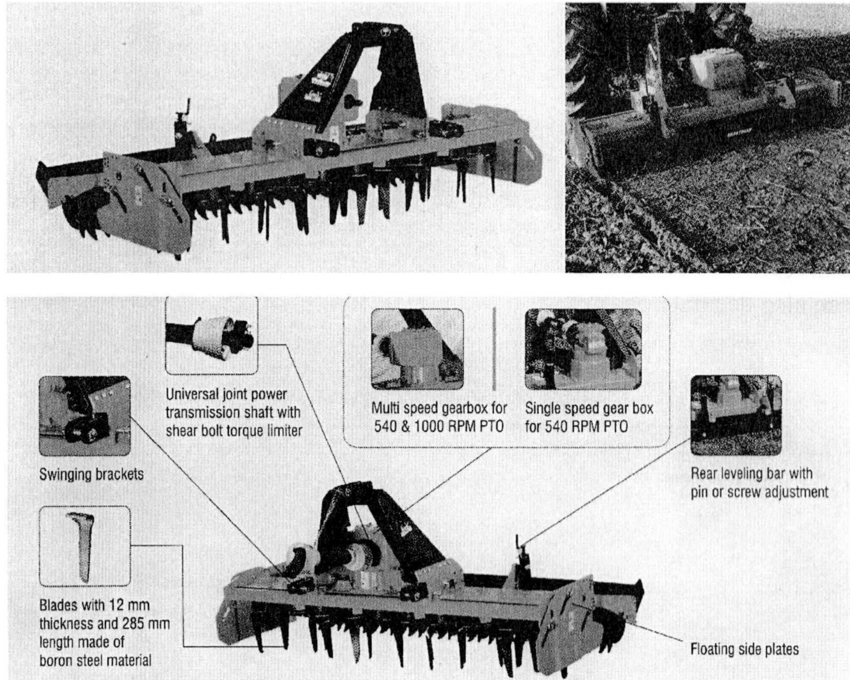

Power harrow with leveller (clock-wise a view of machine, machine in operation in field and details of machine)

Courtesy: Tirth Agro Technology Pvt. Ltd. Rajkot (Gujarat)

Fig. 3.67 : Tractor operated power harrow

Fig. 3.68 : Tractor operated power harrow with rear spike roller (cone type clod breaker)
Courtesy : Tirth Agro Technology Pvt. Ltd. Rajkot (Gujarat)

Tractor operated rotary plough

Tractor operated rotary plough is a primary tillage equipment and can be operated by 45 hp tractor and above. It is suitable for hard and clay soil. It can plough up to more depth with fine tilth in one pass. It can be used directly in barren land. It consists of swinging brackets, universal joint power transmission shaft with shear bolt torque limiter, multi speed gear box, moveable side plates, tines with 12 mm thickness and 285 mm length and rear leveling bar with pin or screw adjustment (Fig. 3.69). Rear spikes roller can also be attached for clod breaking and fine tilth.

Fig. 3.69 : Tractor operated rotary plough (stationery view and working in field)
Courtesy: Tirth Agro Technology Pvt. Ltd. Rajkot (Gujarat)

Tractor operated bund former

Tractor operated bund former is used for making bunds in prepared field (Fig. 3.70). This equipment is used by a 35 to 45 hp tractors. The bund former consisted of mild steel angle iron frame; hitch system, and two blades (wings). The blades are made by mounting mild steel sheet on an angle iron frame. The blades are adjusted in converging manner and has wider opening in the front in comparison at the rear end. The distance between blades can be adjusted according to size of bund required. The implement is mounted type and operated in tilled soil. Working capacity is 0.3 to 0.4 ha/h. Power tiller operated bund former is also available and generally used in orchard and small plots (Fig. 3.71).

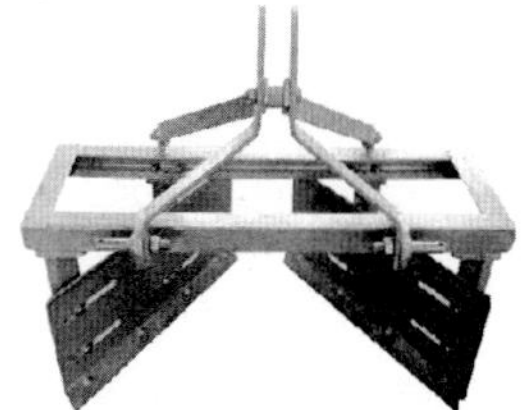

Fig. 3.70 : A view of tractor operated bund former

Fig. 3.71: Power tiller operated bund former
Courtesy: VST Tillers Tractors Ltd., Bangalore (Karnataka)

Tractor operated ridger

It is used for making furrows and ridges for sugarcane, cotton, potato and other row crops. The ridger is used in sugarcane growing area of the country. This equipment is operated by a 35 hp tractor and above. It consists of rectangular frame made of mild steel angle or channel section, 3-point hitch assembly, shanks and ridger body. The ridger body consists of two mould boards, share point and tie bars to vary the wingspan of ridgers (Fig. 3.72). The share point is made from medium carbon steel or low alloy steel, hardened and tempered to about 42 HRC. Upon wearing or becoming dull the share point can be replaced. The ridger is also of disc type (Fig. 3.73) and plate type (Fig. 3.74). The ridger is operated in tilled soil by a tractor, the share point penetrates in the soil, ridger body displaces the soil to both sides and a furrow is created. The soil mass between furrows forms a ridge. The depth of operation is controlled by hydraulic system of the tractor. Working capacity is 0.25 ha/h. Power tiller operated ridgers are also available and used on small forms and in orchard (Fig. 3.75).

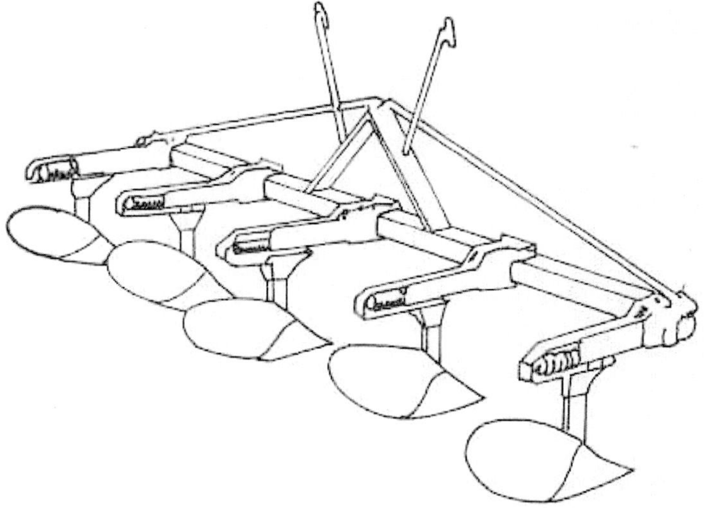

Fig. 3.72 : Tractor operated mould board type ridger

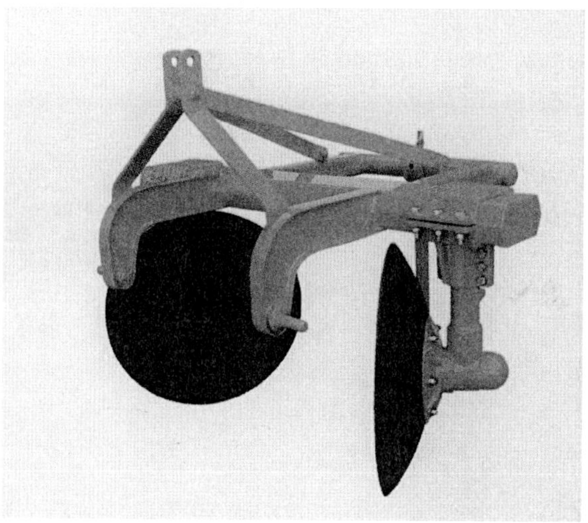

Fig. 3.73 : Tractor operated disc type ridger

Courtesy: Dasmesh Mechanical Works, Amargarh (Punjab)

Fig. 3.74 : Tractor operated plate type ridgers
Courtesy: Govind Industries Pvt. Ltd., Barabanki (Uttar Pradesh)

Fig. 3.75 : Power tiller operated ridger
Courtesy: VST Tillers Tractors Ltd., Bangalore (Karnataka)

Chisel plough

Chisel plough is an important implement used to cut through hard soils by narrow tines. It is used before using the regular plough. Chisel plough is used for breaking hard pans and for deep ploughing (60-70 cm) with less disturbance to the top layers. Its body is thin with replaceable cutting edge so as to have minimum disturbance to the top layers. It contains a replaceable share to shatter the lower layers. As stirring breaks the soil, it is not inverted and pulverize to the extent mould board and disc ploughs do. The chiseling and stirring operation does not throw enough soil to cover trash completely and hence, often used for stubble-mulch or sub-surface tillage practices. Special plough shovels, lister shovels, sweep and knife assemblies and seeding attachments make it possible to do many jobs with the chisel-type plough (Fig. 3.76). Chisel plough is a tool with rigid curved or straight shank with relatively narrow shovel points. It may be termed a heavy-duty deep cultivator. The standard or shank is made of nickel-alloy heat-treated spring steel that is given a long and gradual curve flat-wise to permit a slight spring action. The standards are arranged on heavy frames in two or three staggered rows to permit trash to pass between them without choking. Most chisel ploughs are provided with coil cushion springs in conjunction with clamps to allow the tool to swing back. Pneumatic-tired wheels control the depth as well as serve for transportation. Hydraulic lifts are provided for all sizes of chisel ploughs. Deep tillage shatters compacted sub soil layers and aids in better infiltration and storage of rainwater in the crop root zone. The improved soil structure also results in better development of root system and the yield of crops and their drought tolerance is also improved. The functional component of the unit include reversible share, tyne (chisel), beam, cross shaft and top link connection. Chisel plough can also be used with pulverizing roller attachment (Fig. 3.77) or peg type packer (Fig. 3.78). Heavy duty chisel plough are also available which can be used in very hard and stony soil (Fig. 3.79).

Fig. 3.76 : Tractor operated spring loaded chisel plough
Source: https://www.google.co.in/search?q=chisel+plough

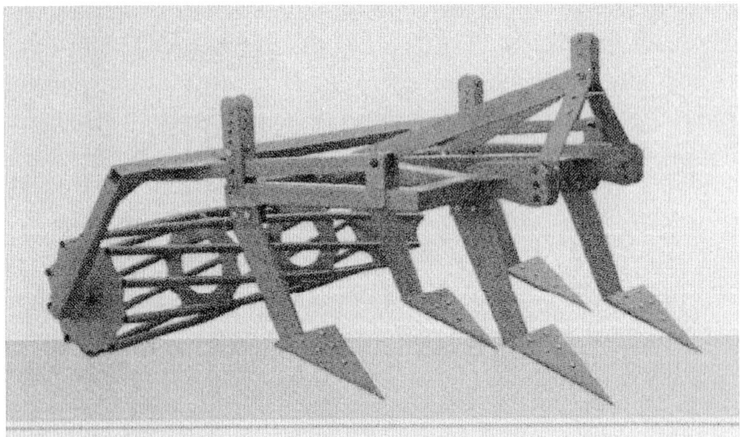

Fig. 3.77 : Tractor operated chisel plough with pulverizing roller attachment
Courtesy: Ganesh Agro Equipments, Vadpura (Gujarat)

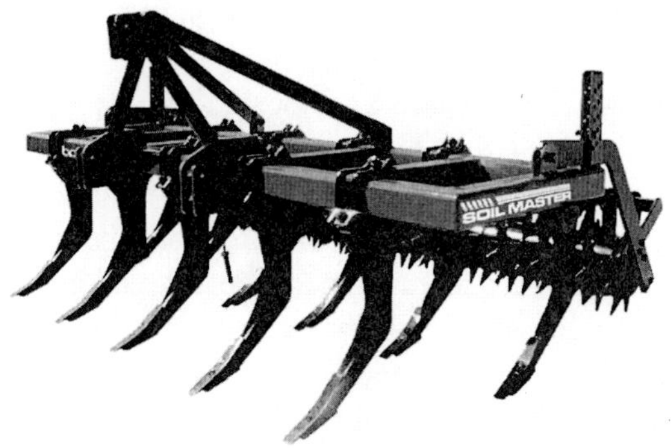

Fig. 3.78 : Tractor operated chisel plough with peg type packer
Source : https://www.google.co.in/search?q=chisel+plough

Fig. 3.79 : Tractor operated heavy duty chisel plough
Source: https://www.google.co.in/search?q=chisel+plough

Tractor operated sub-soiler

It is like a chisel plough with a single tine that may penetrate up to 100 cm depth or more and hence, made heavier than chisel plough. It is used to break hardpan of the soil, loosening of the soil and helps the water to seep into the soil for improving drainage (Fig. 3.80). Tractor of 60-85 hp is required for pulling single tine sub-soiler. One standard is used for deeper depth, but two or three can be used for shallower operations. It is used to plough the soil for orchards, vineyards and forest plantation. It can also be used to break the hard pan formed due to long cultivation or under-plough layer up to the depth of 60-80 cm. It consists of beam made of high carbon steel, beam supports which are flanged at upper and lower edges for rigidity, hollow steel adaptor welded to bottom end of the beam to accommodate share base, share base having square section, share plate made from high carbon steel and shank drilled and counter bored for set board which secures the base in the adaptor. Share plate is made from high carbon steel, hardened and tempered to suitable hardness. Two symmetrically located bolt holes allow reversibility of share. The working depth of the sub-soiler is controlled by hydraulic system and linkage of tractor. Ahead of plough bottom, a special chisel nose is installed that helps the plough to penetrate deeper into the soil and protects the share point against damage. Depth adjustment holes are provided on the sub-soiler. There is a screw type mechanism for adjusting plough depth and horizontal levelling of plough. Sub-soiler is available in both trailing and mounted type. It can cover about 1.2-1.5 ha/day. A mole ball can be attached to create a small tunnel in the soil, which serves as drainage channel for water. The sub-soiler consists of heavier tyne than the chisel plough to break through impervious layer shattering the sub-soil to a depth of 45 to 75 cm and sometimes called trencher also (Fig. 3.81).

Fig. 3.80 : A view of tractor operated sub-soiler (stationery & field operation)

Courtesy : M/s Massey Ferguson Ltd.

Fig. 3.81: Tractor operated Sub soiler/trencher in field operation

Para plough

Fig. 3.82 : Tractor operated para plough

Para plough is used for mulch tillage and moisture conservation under dry farming condition. It is operated by 45 hp tractor. It covers 0.20 ha/h. The para plough is a primary tillage implement used for deep ploughing without inversion (Anonymous, 2013). This enhances infiltration but reduces hazard of erosion. The implement consists of boxed frame structure with standard 3 point hitch and two bottoms one right hand and one left hand arranged facing outside (Fig. 3.82). The plough shares are fabricated out of 12 mm steel sheet. The plough bottom shanks are welded in such a way to make an included angle of 144°. The rake and bent angle of the plough are 22° and 32° respectively. Field trials with equipment showed that ploughing with para plough reduced the bulk density and increase the hydraulic conductivity. Increases the hydraulic conductivity of soil to an extend of 0.72 cm per hour. It looses the soil by 25% more at 32-45 cm depth than conventional treatment. It conserves more moisture at 45 cm depth and restricts moisture depletion up to 10 days after rain fall.

References

Anonymous. 2008. Research Highlight. AICRP on Farm Implements and Machinery, CIAE Bhopal. Technical Bulletin No.: CIAE/2008/141.

Anonymous. 2010. Research Highlight. AICRP on Farm Implements and Machinery, CIAE Bhopal. Technical Bulletin No.: CIAE/2010/151.

Anonymous. 2012. Directory of Successful Farm Machinery in SAARC Countries. SAARC Agriculture Centre. BARC Complex, Farmgate, Dhaka – 1215 (Bangladesh.

Anonymous. 2013. Research Highlight. AICRP on Farm Implements and Machinery, CIAE Bhopal. Technical Bulletin No.: CIAE/2013/158.

Bhardwaj K C; Pandey M M; Singh Gyanndra. 2004. Horticultural Tools and Equipment. CIAE Bhopal. Technical Bulletin No.: CIAE/2004/110.

Garg I K; Singh Surendra. 2002. Farm equipment for Punjab agriculture. Department of Farm Power & Machinery, Punjab Agricultural University, Ludhiana.

Pandey M M; Ganesan S. 2005. Farm Mechanization Package for Dryland Agriculture, Technical Bulletin No: CIAE/FIM/2005/117. Central Institute of Agricultural Engineering, Bhopal.

Pandey M M; Majumdar K L; Singh Gyanendra; Singh Gajendra. 1997. Farm Machinery Research Digest. Technical Bulletin No. CIAE/97/69, Central Institute of Agricultural Engineering, Bhopal.

Pandey M M; S Ganesan; R K Tiwari. 2006. Improved Farm Tools and Equipment for North Eastern Hills Region, Technical Bulletin No. CIAE/2006/121. Central Institute of Agricultural Engineering, Bhopal.

Rautaray S K. 2002. Improved puddling equipment and their performance. Proceedings of Training Programme on 'Advances in Tillage & Traction' under Center of Advanced Studies. Feb. 9 - March 1. Pages 45-55.

Saha K P; Singh D; et al. 2015. Status and utilization of draught animal power in Madhya Pradesh. AICRP on Utilization of Animal Energy with Enhanced System Efficiency. Technical Bulletin No. CIAE/UAE/2015.181. Central Institute of Agricultural Engineering, Bhopal.

Singh Surendra. 2007. Farm Machinery – Principles and Applications. Directorate of Information & Publication of Agriculture, Indian Council of Agricultural Research, Krishi Anusandhan Bhawan-I, Pusa Campus, New Delhi.

Singh Surendra; Manes G S; Dixit Anoop. 2015. Mechanization of cultivated crops. New India Publishing Agency, Pitam Pura, New Delhi.

Singh Surendra; Pandey M M. 2008. X Plan Achievements (2002-2007). AICRP on Farm Implements and Machinery, CIAE Bhopal. Technical Bulletin No.: CIAE/2008/137.

Singh Surendra; Verma S R. 2009. Farm Machinery Maintenance & Management. Directorate of Information & Publication of Agriculture, Indian Council of Agricultural Research, Krishi Anusandhan Bhawan-I, Pusa Campus, New Delhi.

4

Seeding, Planting and Transplanting Machinery

4.1 Seeding equipment

Sowing is an art of placing seeds in the soil to have good germination in the field and thereby gives better productivity. Seeding is one of the most important farm operations after soil preparation because the time of sowing and the way in which it is done decisively influence germination of the seed and growth of the seedling during its early stages. This in turn affects the subsequent time of weeding and inter-culture, harvesting and threshing and finally the yield. Seed needs an adequate supply of moisture, temperature, air and protected environment for good germination.

Crop sowing refers to broadcasting seeds on the surface of soil, placement of seeds in the soil or transplanting seedlings in the soil, under optimum conditions. The primary functions of any planting operation are to establish an optimum plant population and spacing, the ultimate goal being to obtain the maximum net return per unit area. Plant population and spacing requirements are affected by various factors such as the crop type and variety, soil type, soil fertility level, soil moisture and row spacing upon the cost and convenience of operations such as thinning, weed control etc. In order to get better returns, the right amount of seed should be placed at the right time and at a predetermined depth and spacing in the soil. Usually the depth of sowing depends upon the moisture availability and seed emergence capacity. The amount of seeds to be sown varies according to the size of seeds, percentage germination etc.

The basic objective of sowing operation is to put the seed and fertilizer in rows at desired depth and seed to seed spacing, cover the seeds with soil and provide proper compaction over the seed. The recommended row to row spacing, seed rate, seed to seed spacing and depth of seed placement vary from crop to crop and for different agro-climatic conditions to achieve optimum yields. Traditional methods include broadcasting manually, opening furrows by a country plough and dropping seeds by hand, known as *'Kera'*, and dropping seeds in the furrow through a bamboo/metal funnel attached to a country plough *(Pora)*. For sowing in small areas dibbling i.e., making holes or slits by a stick or tool and dropping seeds by hand, is practised. Multi-row traditional seeding devices with manual metering of seeds are quite popular with experienced farmers. Traditional sowing methods have following limitations;

a) In manual seeding, it is not possible to achieve uniformity in distribution of seeds. A farmer may sow at desired seed rate but inter-row and intra-row distribution of seeds is likely to be uneven resulting in bunching and gaps in field,

b) Poor control over depth of seed placement,

c) It is necessary to sow at high seeding rates and bring the plant population to desired level by thinning,

d) Labour requirement is high because two persons are required for dropping seed and fertilizer,

e) The effect of inaccuracies in seed placement on plant stand is greater in case of crops sown under dry farming conditions, and

f) During *kharif* sowing, placement of seeds at uneven depth may result in poor emergence because subsequent rains bring additional soil cover over the seed and affect plant emergence.

The quality and viability of a seed are important factors that affect germination. Seed quality is determined by conducting germination tests in the laboratory. Results of germination test will determine the adjustments in seed rate to account for non-viable seeds. Method of seed treatment is another seed factor that affects the germination. Seeds are treated to protect them while in storage and from soil borne diseases. They are also some times treated to facilitate proper seeding or planting, like in cotton, sulphuric acid treatment is recommended to remove lint from seed's surface. It also helps in removal of light and defective seeds, which stimulates germination.

Draft requirement of soil engaging tool is affected by soil and environmental factors such as bulk density, specific volume, porosity, cohesion, moisture, air, temperature, strength of soil etc (Singh, 2007). The bulk density of agricultural soil normally ranges between 1.3 to 1.6 g/cm^3. A bulk density of 1.0-1.2 g/cm^3 indicates freshly tilled soil. The specific volume of soil solids, water and air affect seedbed preparation. The average specific volumes of solid, water and air range between 50-55, 20-25 and 25-30%, respectively. Soil porosity is a very important property as it affects soil, air and water containing capacity. The high level of porosity is recommended to create favourable aerobic conditions for seed germination and seed growth. The heat supply to the seeds in the soil is the solar radiation or sunshine. Therefore, the temperature of the soil depends upon climate, season, solar elevation and changes in day light hours. Studies showed that plants facing south receive more incidental radiation than those facing south-west.

Factors affecting seed germination and emergence: Mechanical factors, which affect seed germination and emergence, are: i) seed damage during metering; ii) uniformity of depth of placement of seed; iii) uniformity of distribution of seed along rows; iv) transverse displacement of seed from the row; v) prevention of loose soil getting under the seed; vi) degree of soil compaction above the seed; vii) uniformity of soil cover over the seed; and viii) mixing of fertilizer with seed during placement in the furrow.

To achieve the best performance from a seed drill or planter, the above factors are to be optimised by proper design and selection of the components required on the machine to suit the needs of the crops. The seed drill or planter can play an important role in manipulating the physical environment. The metering system selected for the seed should not damage the seed while in operation. The speed of metering device is a very important factor with regards to damage. Seed damage can be avoided by selecting the proper spring loading rate of the cut-off device and knock-out device in case of plate type planters. Seeds should be handled in such a way that physical injury is avoided. Metering devices, seed conveying tubes and their location on the machine affect the seed distribution uniformity in the row. Their selection is affected by the furrow openers and the depth requirements for placement of the seed. The soil physical environment affected by the machine is the availability of soil moisture and mechanical resistance. The planter's performance is improved by manipulating the depth of sowing and thickness of soil cover over the seed as well as pressing the soil cover.

Under arid conditions the top soil becomes very dry; therefore, the seeds are placed 80-100 mm deep in the furrow. This requires the proper furrow opener. Further, the soil cover over the deep-placed seed should be lightly packed to achieve good emergence. In arid regions, dry seeding with deep placement of seed is recommended because the seeds will germinate only when there is sufficient rainfall or adequate moisture at the seed zone. The recommended fertilizer placement is 30-50 mm below and 30-50 mm from the side of the seed. Deep placement of fertilizer is possible if there is no constraint on the power supply to the machine. However, animal-drawn or manually operated planters have this limitation. The compromise is achieved by band placement of fertilizer 40-50 mm from the side of the seed at the same depth.

Functions of seed drills and planters: Seed drill is a machine for placing seeds in a continuous flow in furrows at a uniform rate and at right depth with or without the arrangement of covering them with soil. Seed drill with fertiliser dropping attachment, places seeds and fertilisers uniformly in the ground and is called seed-cum-fertiliser drill. A well designed seed drill or planter should be able to function as i) To open furrow in the soil; ii) To meter seeds of different sizes and shapes; iii) To place the seed in the acceptable pattern of distribution in the field; iv) To place the seed accurately and uniformly at the desired depth in the soil; and v) To cover the seed optionally and compact the soil around it to enhance germination and emergence. Depending upon climatic and soil conditions, seeds are sown on well-prepared and levelled fields, on ridges, in furrows or on beds. Flat seeding and planting refer to operation when the field being sown/planted is levelled and smooth. Seeds and tubers are planted on ridges either to improve soil drainage due to high rainfall or it may be a cropping requirement. Potatoes are usually sown on ridges. Seeding in furrows is done in arid regions to conserve soil moisture and improve plant growth. When two or more rows of seeds are planted in beds and separated by furrows, it is known as bed planting. Bed planting helps in conserving soil moisture, avoids soil compaction and promotes plant growth.

Seeding and planting methods: The following methods of sowing or planting crops are used:

a) **Broadcasting**: This method scatters the seed at random in the field and is usually done by hand but it can be done mechanically also. When broadcasting is done manually the uniformity of seeds depends upon the skill of man. Mechanical broadcasters scatter the seeds on the surface of seedbed at controlled rate. Soon after broadcasting seeds are covered by planking. Usually higher seed rate is obtained in this system.

b) **Dibbling**: It is the process of placing seeds in holes made in seedbed and covering them. In this method seeds are placed in holes made at definite depth and spacing by the equipment called dibbler. This is very time consuming operation and generally used for vegetable seeds.

c) **Drilling**: Drilling is the process of dropping seeds in furrow in a continuous flow and covering them with soil. Drilling may be manually as well as mechanically. The number of row drilled may be one or more depending upon the power source used. Drilling helps in achieving proper depth, spacing and correct amount of seeds to be sown in the field. It can be done by manual, animal-drawn and tractor-drawn seed-cum-fertilizer drills.

d) **Hill dropping or planting**: It consists of cells spaced at suitable intervals in the planter plate and large enough to hold sufficient seeds for one hill. The hill drops are located in the bottom of planter hopper and seeds fall through the seed tube to the soil. Seeds or tubers in clusters are dropped along the row at equal spacing between hills. In this system seeds are not placed continuously. Spacing between plant to plant in a row is constant.

e) **Check row or square planting**: Seeds or tubers are placed in rows in two directions perpendicular to each other, thus, facilitating intercultural operations in both directions. In this method of planting row to row and plant to plant distance is uniform. Seeds are placed precisely along straight parallel furrows. The machine used for check row planting is called '*Check Row Planter*'.

f) **Transplanting**: It is the process of growing seedlings in the nursery and then uprooting and planting in the field. It is commonly done for paddy, vegetables and flowers. It is done manually as well as mechanically. Equipment used for placing plants in the soil is called '*Transplanter*'.

Broadcaster: Broadcasting is scattering the seed on the surface of land either by hand or by a manual or tractor-operated broadcaster. Grass and other small seeds are generally broadcasted with hand seeders. Broadcasting of cereals is also carried out in some developing countries. It has however, a number of disadvantages. Seed may not be uniformly distributed in the soil that leads to uneven germination. There are many hand broadcasters, viz. hand fiddle, cyclone hand seeders and pushed broadcast seeders. The hand fiddle is a broadcasting implement consisting of a rotating horizontal disc that scatters the seed by centrifugal force, fed from bag or hopper (Fig. 4.1). The disc is mounted on a spindle that is driven by means of a bow and cord. The evenness of distribution is somewhat affected by the reciprocating action of the drive. A more even scatter of seed can be obtained by subsequent cross sowing. The sowing width is about 6-8 m.

Fig. 4.1 : Manual seed/fertilizer broadcaster

Courtesy: National Agro Industries, Ludhiana (Punjab)

Dibbler: It is used for seeding/dibbling of bold vegetable seeds in less area and for gap filling. The equipment can also be operated by farm women. The dibbler is manually operated hand tool (Fig. 4.2 a & c). It consisted of a seed hopper, cell type roller for metering of seeds, spring actuated jaws for penetration in the soil, pipe and handle (Fig. 4.2b). The diameter of roller is 45 mm. All the parts are made from mild steel except seed roller, which is fabricated from good quality wood. For its operation, the dibbler is held in both hands and jaws are pushed into the soil to the desired depth (up to 35 mm) at an angle of 20 degrees with vertical. The dibbler is given a jerk at the handle in the forward direction, which rotates the roller in seed hopper and releases one or two seeds depending upon the size of cell. At this moment the jaws also open and allow the seed to fall in the cavity created by the jaws. The dibbler is raised and moved to the next position of sowing. Upon raising of the dibbler the roller returns to original position and jaws also close. Working capacity of dibbler is 300 to 400 hills/h. Labour requirement is about 40 man-h/ha. Average missing with the use of this machine is about 6%.

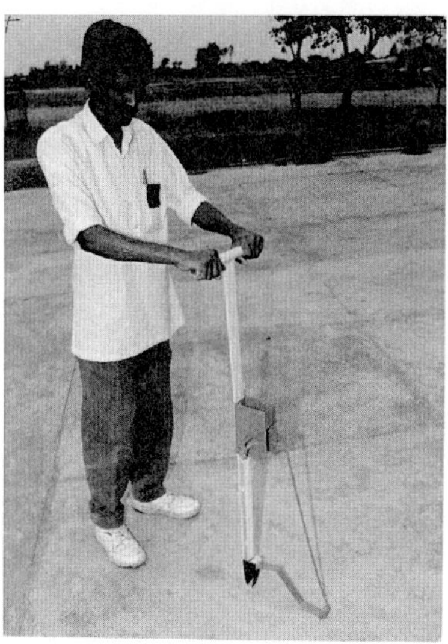

a) Hand operated Dibbler

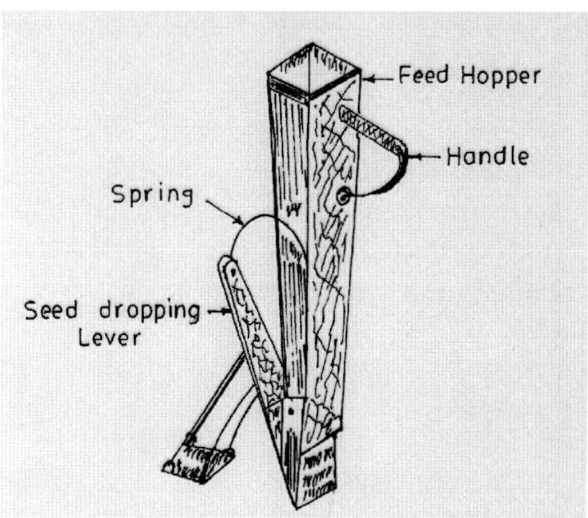

b) Details of hand operated dibbler

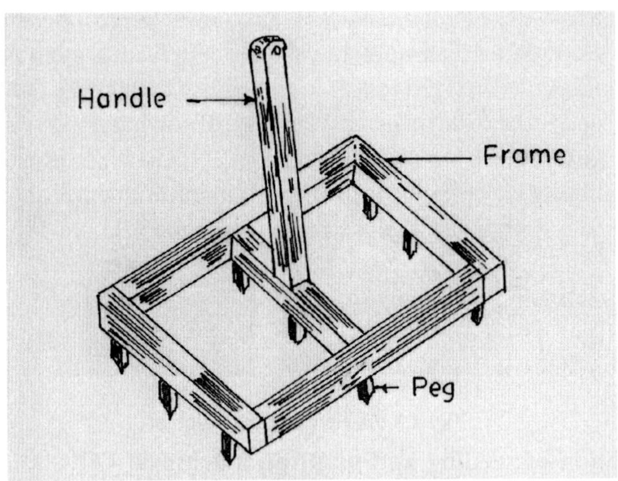

c) Peg type hand dibbler

Fig. 4.2 : Different types of hand dibbler

Rotary dibbler: The rotary dibbler is a manually operated push type device for dibbling of medium and bold size seeds of vegetable and other cereals (Fig. 4.3). The dibbler is manually operated hand tool. It consisted of a rotating dibbling head with penetrating jaws, covering-cum-transport wheel, seed hopper with cell type wooden roller and a handle. The diameter of roller is about 7 cm. Except seed roller, which is made of good quality wood, all the other parts are fabricated from mild steel. The number of jaws varies from five to eight among various designs, depending upon seed to seed distance. For its operation, the hopper is filled with seeds and transport-cum-covering wheel is drawn to rear side. The dibbler is then pushed forward in the direction of travel with covering-cum-transport wheel behind the dibbling head. The jaws penetrate into the soil up to 20 mm and automatically drop the seeds. The seed to seed distance depends upon size of the polygon plate to which jaws are attached. Working capacity is 0.042 ha/h. Labour requirement is about 27 man-h/ha. Average missing with the use of this machine is about 5%.

Fig. 4.3 : A view of rotary dibbler

Selection of seeding and planting machines: Different designs of improved seed drills/planters have been developed for sowing of crops. Basic difference in the design of these seed drills is mainly in the type of seed metering mechanism and furrow openers. Therefore, it is essential to select the machine with a metering unit and furrow opener suitable for the crop and soil conditions. For small seeds like rapeseed-mustard a seed drill or planter with vertical roller with cells, inclined seed plate with cells or small grooved fluted roller metering system is recommended. For medium seeds such as wheat, soybean, safflower and linseed, seed drills with standard fluted rollers are recommended. For bold seeds like groundnut and castor planters with inclined cell plate or cup feed type-metering system are recommended. Drills are used for sowing seeds in rows at 15-35 cm apart. This machine sows seeds and fertiliser in separate bands at specified rates in rows at proper depth and covers them with soil. This may be manually operated, animal-drawn or power operated. Manually operated drills are either single or multi-row seed drills operated by human beings by pushing or pulling it (Fig. 4.4). The seeds are dropped in the furrow formed by the furrow opener from seed hopper.

a) A view of seeding attachment to wheel hand hoe

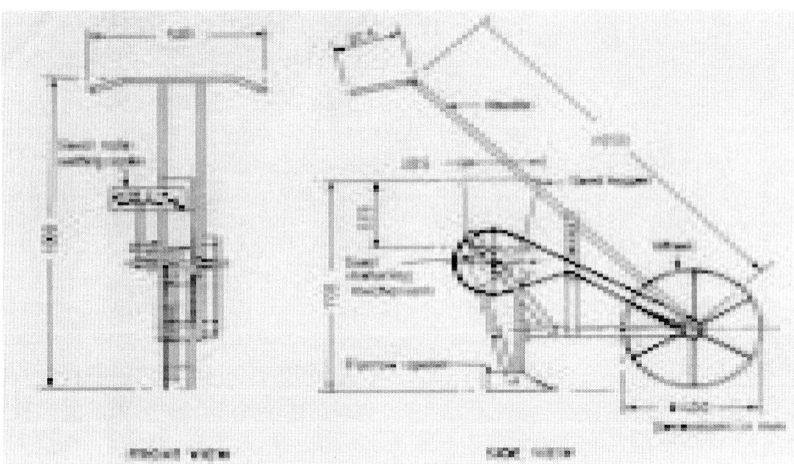

b) Details of seeding attachment to wheel hand hoe

Fig. 4.4 : Manually operated single row seeding attachment to wheel hand hoe

Animal-drawn seed drills are operated using draught animal power. The major components of seed drill include seed hopper, seed tubes, furrow openers, seed metering units, covering as well as depth controlling devices and transport-cum-power transmission wheels (Fig. 4.5). These components which are mounted on a frame, designed to be operated by a draft animal. Animal drawn seed drill consists of a wooden beam to which 3 to 6 tynes are fixed. These tynes open the furrows into which the seeds are dropped. Holes are made into these tynes and into these holes, the bottom ends of seed tubes are fitted. These seed tubes are connected at the top to a hopper filled with seeds. Fertilisers are placed at a depth of 5 cm and 5 cm away from seed rows for effective utilisation of fertilisers. Both operations *viz.* drilling seeds and fertilizers are done simultaneously by ferti-cum-seed drill. It is similar to seed drill, but with extra tynes and hopper for drilling fertilizers.

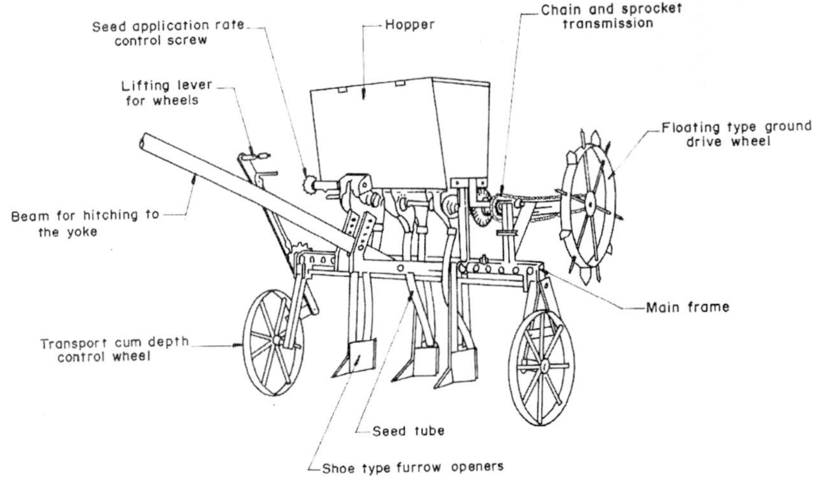

Fig. 4.5 : Animal drawn three row seed drill

Power operated drills are seven or more rows with fertiliser attachment, powered by tractor or power tillers. Tractor or power tiller operated seed drill consists of a seed drum with holes in the bottom plate corresponding to the number of seed tubes for passing the seed (Fig. 4.6). A rotating disc has holes in a circular path and it is kept over a bottom plate. When the holes of rotating disc and bottom plate coincide, seed falls into the tube on its way into the soil. The distance between two holes in rotating disc is proportional to the inter-row spacing of crop. For sowing seeds of different sizes, rotating discs with different sized holes are used. There is provision for altering the distance between the rows by changing distance between the tynes. Inter-row spacing can be changed by using rotating discs with more space between the holes. Seed drills with different mechanisms for automatic drilling of seed are also available.

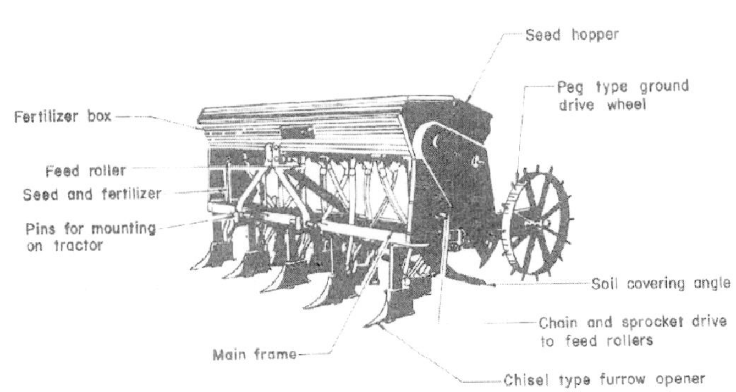

Fig. 4.6 : Tractor operated seed-cum-fertilizer drill

Seed metering devices: Seed metering device is the heart of seeding machine and its function is to distribute seeds uniformly at the desired application rates. In planters it also controls seed spacing in a row. A seed drill or planter may be required to drop the seeds at rates varying across wide range. To make the continuous flow of seeds into seed tube, metering devices are provided just below the seed box. The metering devices used in seed drills are either positive metering type or gravity type. Common type of metering devices used on seed drills and planters are i) adjustable orifice with agitator; ii) fluted roller (standard); iii) fluted roller (small flutes); iv) vertical rotor/roller with cells; v) plate with cells; vi) internal double run type and vii) cup feed (Fig. 4.7). All these metering devices are driven by ground wheel. Seed flow is regulated by changing the size of opening provided at hopper bottom in adjustable orifice with agitator type of metering device. An agitator fixed above seed opening helps in continuous flow of seeds. This does not give precise control over the seed rate and uniformity of distribution in rows. Many designs of conventional animal and tractor operated machines have adopted this mechanism on account of its simplicity and low cost. Axial or helical flutes are machined or cast on an aluminium, cast iron or plastic roller. Rotation of fluted roller in housing, filled with seeds, causes the seeds to flow out from roller housing in a continuous stream. Seed rate is controlled by changing exposed length of fluted roller in contact with seeds and fairly accurate seed rate can be achieved for a variety of medium size seeds like wheat, soybean, sunflower and safflower etc (Fig. 4.8). There is also an adjustable gate on the discharge side of fluted wheel. The gate opening can be changed to fit the size of the seed. Fluted feed roller type metering mechanism is favoured for sowing small seeds. The fluted feed mechanism is more positive in its metering action than the gravity metering mechanism. Normal size flutes are designed for seed rates of 20 to 120 kg/ha. For sowing of small seeds like rapeseed-mustard and sesamum fluted roller with small flutes has been developed at GBPUAT Pantnagar and PAU Ludhiana (Singh, 2007; Pandey *et al.*, 1997). These can be fitted by replacing the standard fluted rollers on the seed-cum-fertilizer drill. The roller is provided with 10 small flutes of 2 x 2 mm size. Low seed rates of 3 to 5 kg/ha can be achieved with this metering roller with an accuracy of ± 10 per cent.

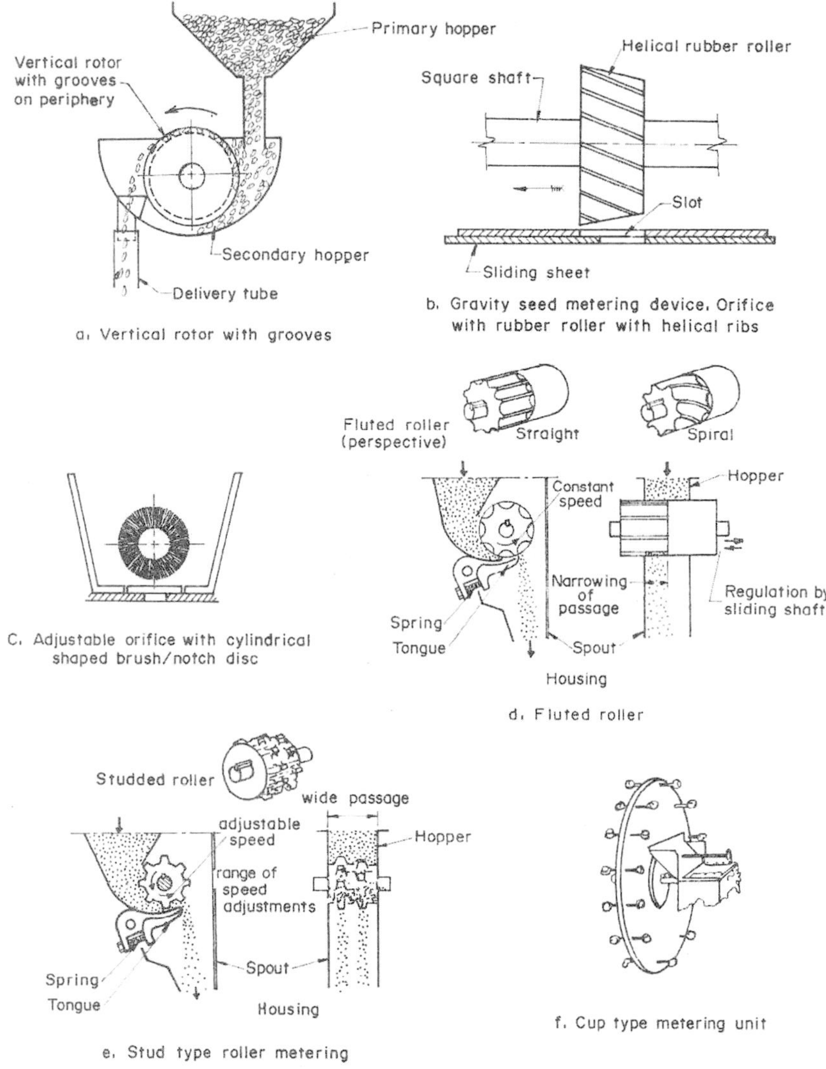

Fig. 4.7 : Different types of seed metering devices used in seed drills

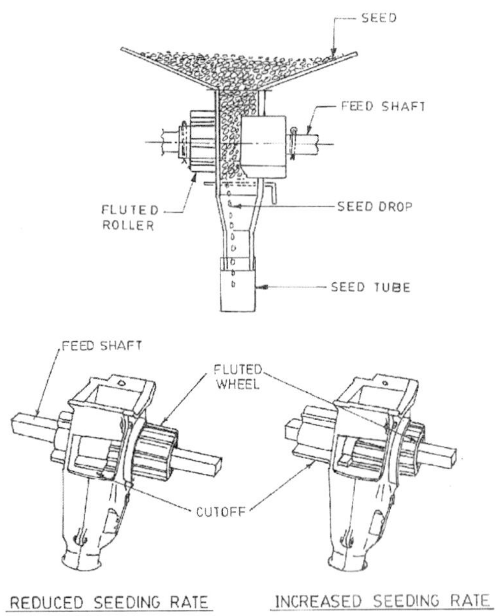

Fig. 4.8 : Fluted roller type seed metering device

Vertical rotors with cells are suitable for metering individual or hill of seeds (Fig. 4.9). The rotor with grooves or cells on its periphery is fixed in the hopper. The size and number of cells on the rotor are according to the size of seed and desired seed rate. A cut-off device is provided above the rotor for regulating the flow of seed to cells. In some designs seed rotor is fixed in a secondary hopper and rotor lifts the seeds in cells and drops these into seed funnel. For varying the seed rate and sowing different seeds, separate rotors are required. Horizontal, inclined or vertical plate with cell type metering mechanism picks and drops individual seed or a hill of seeds depending on design of cell on the plate (Fig. 4.10). Spacing between seeds/ hills is controlled by drive ratio and number of cells on plate. Separate plates are required for sowing different crops. It is desirable that seeds be graded and has high germination percentage for achieving recommended plant population and uniform seed spacing. Seed picking cups or spoons are provided on periphery of a vertical plate. When the plate rotates, cups pick seeds from seed hopper and drop them in seed funnel. Size of cups depends on size and number of seeds per hill. This type of metering is used for seeds, which are easily damaged by mechanical devices.

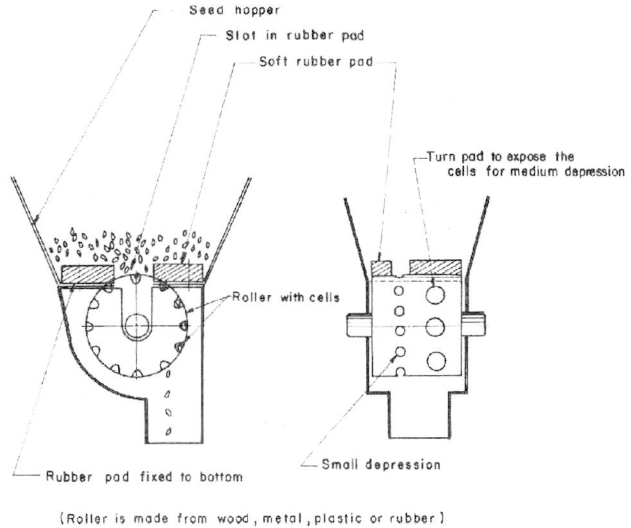

Fig. 4.9 : Vertical roller with cells seed metering device

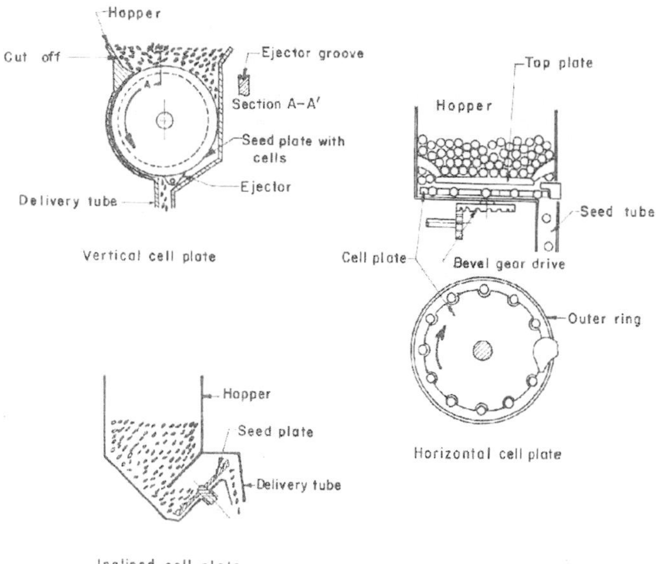

Fig. 4.10 : Horizontal, inclined or vertical plate with cell type metering mechanism

Furrow openers: Furrow openers are used to place the seed at the desired depth with minimum dispersion. The depth of placement at which seed is placed in the soil depends on the crop variety and the soil moisture level. Furrow openers form a neat groove in the moist zone of soil with minimum soil disturbance to avoid mixing the top dry soil with the under lying moist soil at seed level. Furrow openers used in the seeding equipment includes rotating or fixed type. The design of furrow openers of seed drills varies to suit the soil conditions of particular region. Most of the seed-cum-fertilizer drills are provided with pointed tool to form a narrow slit in the soil for seed deposition. There are two types of furrow openers viz. rotating type and fixed type. The rotating type furrow opener includes single and double disc furrow openers (Fig. 4.11). These type of furrow openers are not been used in seed-cum-fertilizer drills being manufactured in India. Single disc type furrow openers are widely used on seed drills for cereals. It consists of a concave disc made of hardened steel. It is used in the sticky soil where stone, debris or trash is available. The disc is set an angle, which during operation shifts the soil to one side making small ridge. The disc is kept clean by toe shaped scraper at convex side and T-shaped at the concave side. In this system, disc penetrates well in the soil and cuts trashes and clods in the field. Double disc type furrow openers consist of two flat discs set at an angle. The discs open furrow and leave a small ridge in the centre. The seeds are dropped in between discs and placed more accurately. It is suitable for trashy and stony land.

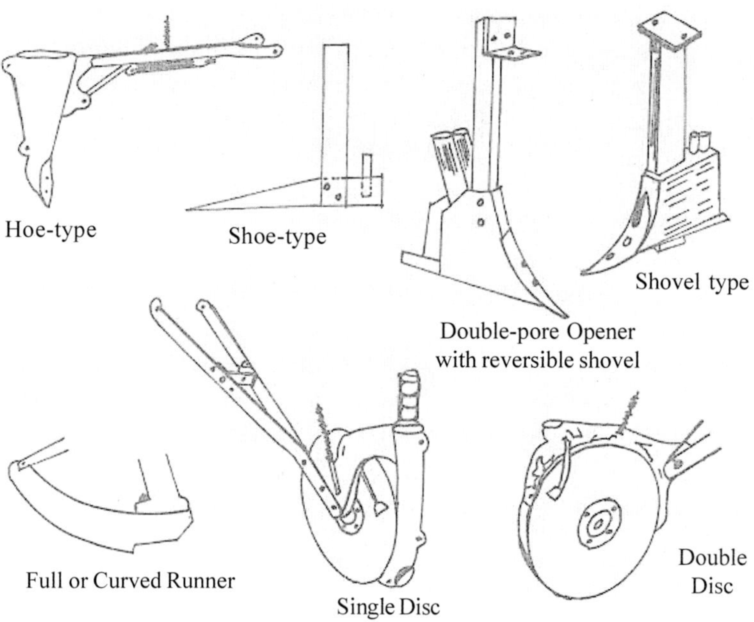

Fig. 4.11 : Different types of furrow openers used on seed-cum-fertiliser drills and planters.

There are many fixed type furrow openers available commercially (Fig. 4.11). **Hoe type furrow opener** is probably the most primitive type of furrow opener (Singh, 2007; Singh *et al.*, 2015). It consists of a narrow pointed shovel followed by a delivery boot. A pointed hoe digs according to the set furrow depth. It lifts and pushed the top layer soil toward the sides and forms V-shaped groove. It works well in many soil conditions but not in the field with large stubble. **Runner type furrow opener** also called sword opener and works well in the clean seedbeds. It is used mostly for shallow seeding. Due to its length it compacts the bottom of the furrow. **Shovel type furrow opener** is a narrow pointed shovel. The leading edge of opener is a sharp pointed triangle. The shovel is bolted to the boot for easy replacement. At the rear, there is a boot with one or two small tubes for seeds and fertilisers. The length of shovel varies from 100-200 mm. These are easier to fabricate. **Shoe type furrow opener** drops the seeds and fertilisers simultaneously in separate bands at the same depth. Its boot is protected at the top and bottom by metal plates to prevent clogging. The seeds and fertilisers are placed 50 mm apart through the boots using two compartments.

Furrow openers should be selected according to the type of soil and depth of seed placement (Table 4.1). For trashy, stony and light to medium soils, shovel type openers are used. The depth of seed placement from 50 to 100 mm is achieved with these openers. Small shoe or shovel type openers are also used for shallow (20 to 50 mm deep) placement of seeds in dry farming areas. Shoe type openers with single or twin boots are used for sowing in heavy and medium soils for seed placement at 20 to 70 mm depth. Runner type opener is widely used for placement of seeds at shallow depth where soil disturbance required is minimum. Soil cover over seed is also minimal. Covering chains and wooden planks are widely used to cover and compact the soil over seeds in the furrows and level the fields after sowing operation. Pointed bar type (diamond shaped) furrow openers are used for forming narrow slit under heavy soils for placement of seeds at medium depths.

Table 4.1: Type of furrow openers used in seed-cum-fertilizer drills and planters

Type of Furrow Opener	Description	Suitability
Hoe type	Single or double pointed shovel with one or two pores/boots.	Suitable for light, medium soils. Soil free of excessive trash has good penetration.
Stub or full covered runner	Resembles a curved sword with a thin sharp cutting edge with single or double boots at the rear.	Widely used on row crop planters. Suitable for shallow sowing. Sharp blade cuts through the clods and sod. Low draft and minimum soil disturbance.
Single disk opener	One disk slightly curved, fastened to the boot and set to run at an angle.	Good penetration cuts the trash and does not clog.
Double disk opener	Two disks facing each other placed at a slight angle.	Suitable for deep sowing at relatively higher speeds. Ideal for sowing small seeds in trashy seedbeds.
Chisel-type furrow opener	A body with bar shape has the share point projecting over the shoe.	Especially suited to very hard and cloddy soils typical in the black soil belt in the rainfed areas.

Seed tubes: All seed drills require a delivery passage leading from the hopper to the furrow opener. When the furrow opener is lifted independently of the seed box, the passage must be telescopic or ribbon construction. Transparent delivery tubes or windows in the side of the passage are sometimes used to enable the operator to see whether the seed is dropping properly. The delivery tube should have smooth walls to reduce bounce and may be of straight tube of minimum length and diameter. The inclination of tubes from the vertical is normally kept below 20 degrees. Seed tubes may be made of different materials and can have different shapes. It may be transparent plastic tube, spiral steel tube, tapered plastic tube, funnel shaped tube or corrugated rubber tube (Fig. 4.12). A tube size of 25 mm in diameter can accommodate a majority of seed types. The velocity of seed at the end of the tube may be low to minimise bouncing and rolling of seeds in furrow.

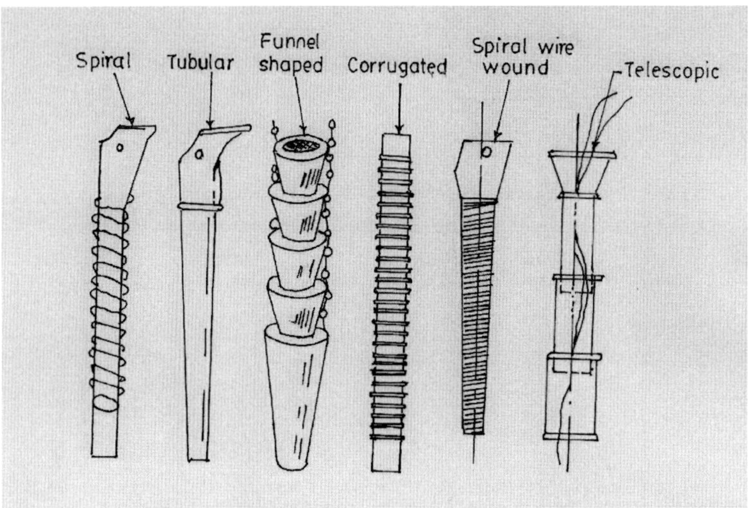

Fig. 4.12 : Different types of seed tubes used on the drills.

Covering and compacting devices: For proper germination seed must be in contact with moist soil and covered with a layer of soil through which sprout can penetrate. Covering devices employed on seeders are drag chain; drag bars, scraper blades, steel press wheels, rubber covered on zero pressure pneumatic press wheels, disc hillers and various combinations of these units (Fig. 4.13 and Fig. 4.14). A covering device should place moist soil in contact with the seeds, press the soil firmly around the seeds, cover them to proper depth and leave the soil directly above the row loose enough to minimise crusting and helps easy emergence. In loose, sandy soil or for furrow drilling of grain in heavy residue, narrow press wheels with steel or rubber rims may be used behind the openers. Compacting devices are mostly used with planters (Fig. 4.14).

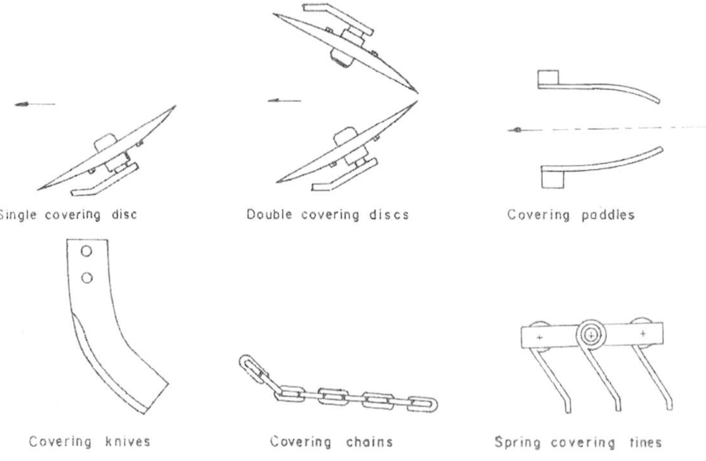

Fig. 4.13 : Soil covering devices

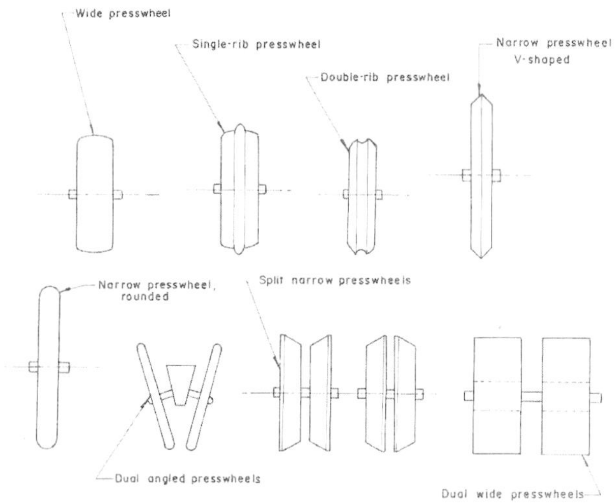

Fig. 4.14 : Soil compacting devices

Power transmission unit: The metering device is usually actuated by the ground driven wheel which transmits power by means of a suitable drive viz. chain-sprocket, belt-pulley or gear drive to achieve uniform delivery rates. There are different types of ground wheels to be used on drills depending on the ground conditions (Singh, 2007). When the ground wheel carries the weight of machine as well as the seeds and fertiliser stored in boxes, the load on the wheel is sufficient to enable the power transmission from the wheel to the metering device. Two types of wheel i.e. pneumatic and rigid steel wheels are mostly used in seed drills and planters. Rigid steel wheels are most commonly used because of their low cost, low maintenance and long life. The steel wheels can be categorised into three types plain wheel, lugged wheel and pegged wheel (Fig. 4.15). **Plain wheels** are used on seed drills and planters with width varying from 75 to 100 mm, diameter from 400-700 mm and number of spokes ranging from eight to twelve. It runs smoothly and provides good contact with soil surface and develops better traction for the drive mechanism. It is used mostly in light soils. In the **Lugged wheel,** small lugs are provided on the periphery of the wheel. The lugs are 25 mm in height and are welded at an angle 20-25 degree with the axis of rotation to reduce slip, wear, vibration and rolling resistance. The diameter of the lugged wheel ranges from 350 to 450 mm and these are suitable to develop better traction or grip in the soil. **Pegged wheels** are suitable for use under wet or sticky soils, where plain, lugged or even pneumatic wheel fail to work. The diameter of wheel ranges from 500 to 800 mm and number of spokes varies from 12 to 30 depends on the size of wheel.

Fig. 4.15 : Different types of ground wheels used on seed-cum-fertiliser drills and planters

Calibration of seed drills: For precise metering of seeds and fertilisers seed drill should be calibrated in the laboratory as well as in the field under actual operating conditions. A preliminary check-up of seeding mechanism setting for a preset rate is of great practical value since it eliminates the possibility of over or under sowing. Ground wheel should be jacked up first and the diameter of ground wheel measured. The effective width of seed drill can be determined by measuring distance between two furrow openers and multiplying it by number of furrow openers. If distance between two furrow openers is 'd' and number of furrow openers 'n', the effective width of the seed drill (w) in meter shall be n d. Area covered in one revolution of drive wheel shall be p D w, where D is diameter of drive wheel in m. The number of revolutions of drive wheel to cover one hectare be 10000/(p D w). Assuming 10 per cent positive slip during operation, actual number of revolutions of drive wheel shall be 9000/(p D w). Place the seeds in hopper and make the feed-cup setting on drill for desired seed rate. Turn drive wheel for one-hundredth revolution of actual revolutions and collect the seeds under each seed tube. Weigh the seeds collected separately for inter-row variation and then add and multiply by 100 to get the seed rate. If the seed rate obtained does not equal the desired seed rate, repeat the test after adjusting the feed cup/metering device till the desired seed rate is obtained. Seed drill needs to be checked in the field also before doing the actual sowing operation. Fill the seeds in hopper and mark. Run the drill in the strip marked and find the seeds used. If quantity of seeds is more than calculated above, then drill is sowing more seeds and feed roll length should be reduced. If quantity of seeds is less than calculated one, then drill is sowing fewer seeds and feed roll length should be increased.

Manually operated seed drill

This is a small manually operated single/double row seed-cum-fertiliser drill in which fluted roller metering mechanism is provided (Singh, 2007; Anonymous, 2013). The machine consists of a seed box attached to the mainframe of a hand wheel hoe and shovel type furrow opener (Fig. 4.16). A fluted roller assembly is provided at the bottom of the seed box. Fluted roller is rotated with the help of chain and sprockets from the ground wheel. The seed rate can be adjusted with the help of a lever provided in the seed box. The fluted roller used for sowing rapeseed and mustard has 8 flutes. Each flute is 3 mm wide and 2 mm deep. The diameter of the fluted roller is 50 mm and its length, 32 mm. A ground wheel is provided to drive the metering rollers. Seed and fertiliser are stored in a small hopper and a long beam is provided by which the implement could be pulled by one operator. For operation, the machine is pulled by rope attached to the hook of machine by one man and other person steers the machine by holding it by the handle.

Due to the provision of fluted rollers, it is suited for drilling soybean, maize, pigeon pea, sorghum, green gram, Bengal gram, wheat etc. It is also used for sowing small seed like rapeseed and mustard and grass seeds by changing the suitable fluted roller. Shoe type furrow openers are provided for easy operation. Working capacity of drill is 0.043 ha/h. It is also suitable for inter-row sowing.

a) Fluted roller type manual seed drill

b) Commercial manual seed drill with fluted roller
Courtesy: Khedut Agro Engineering Pvt. Ltd. Rajkot (Gujarat)

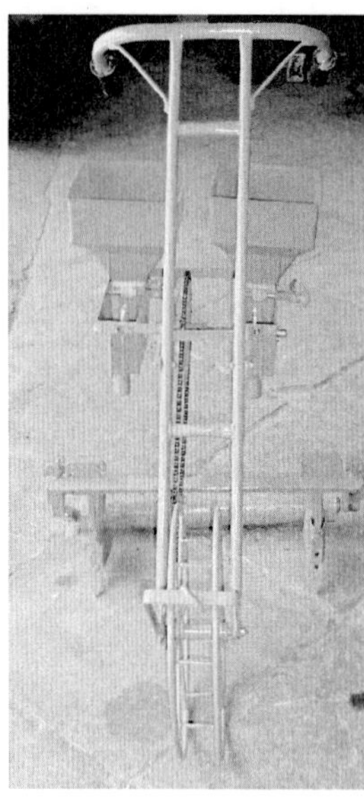

c) Two row manual seed drill with wheel hand hoe

Fig. 4.16 : Different types of manual seed drill with fluted roller

Animal drawn seed drill

The simplest seed drill, which small farmers use widely, consists of a seed funnel and a vertical seed tube fitted to indigenous plough (Fig. 4.17). The vertical tube is either fitted to the shoe of plough or it is tied with the body to drop seeds just behind the plough in the furrow (Singh, 2007; Singh *et al.*, 2015). Either the ploughman or another man walks along with the plough and drops the seeds in the funnel. There are animal-drawn seed drills in use with one, two, three, four and six seed tubes (Fig. 4.18). In manually metered seed drills, the uniformity of seed distribution depends upon the skill of man dropping the seeds in funnel. There is no seed-metering device to control the desired amount of seeds to be sown. In another type of country seed drills, the fertiliser and seed boxes are attached to the soil stirring plough. Three-row animal-drawn seed-cum-fertiliser drill has become very popular among the farmers because it has a fluted roller type metering devices for metering seeds (Fig. 4.18d)

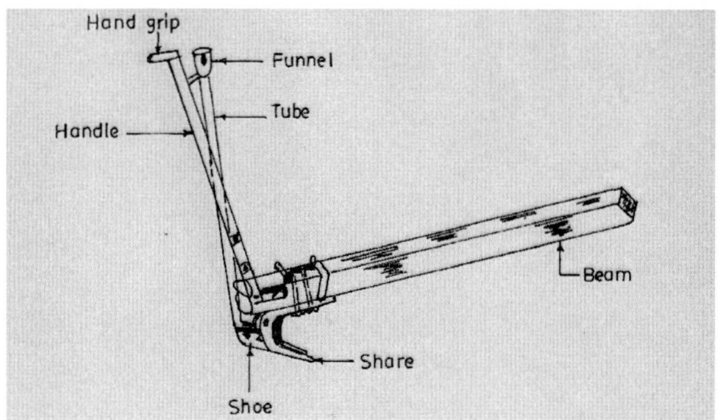

Fig. 4.17 : Single row animal-drawn indigenous seed drill

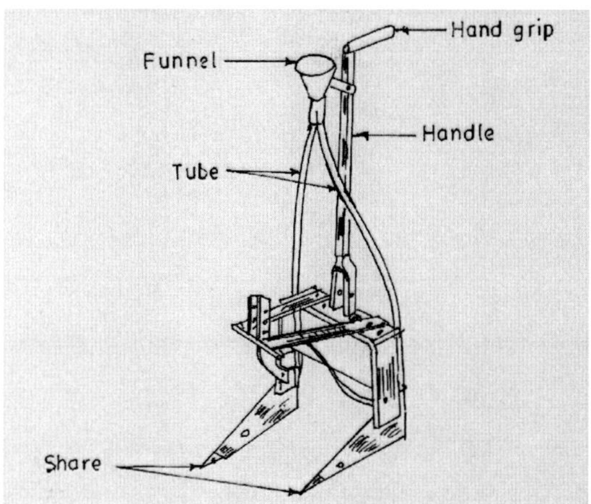

a) Two row animal-drawn seed drill

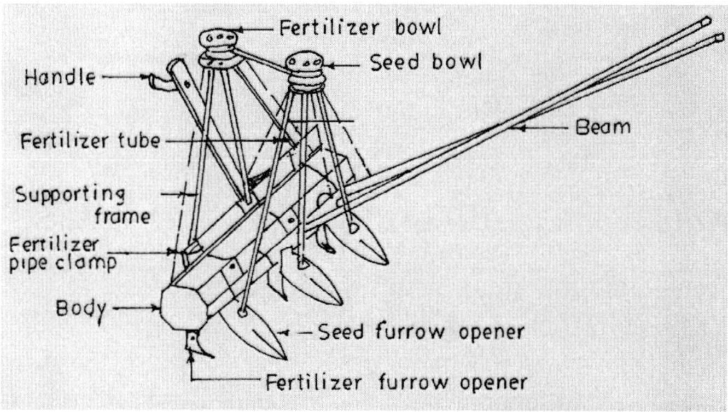

b) Three row animal-drawn seed-cum-fertiliser drill for dry land

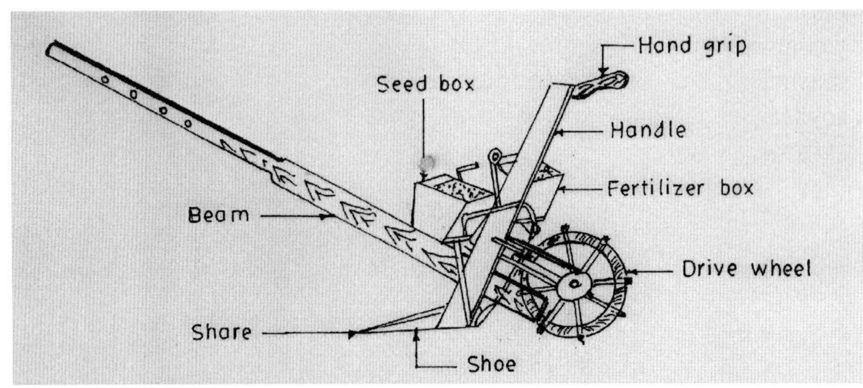

c) Animal-drawn country plough with seeding and fertiliser attachment

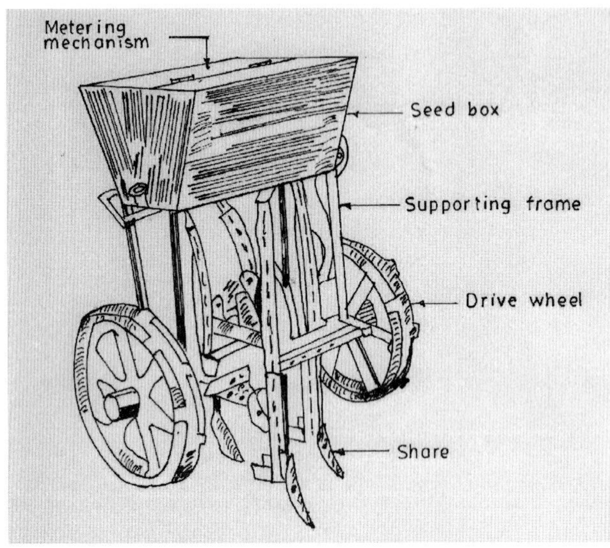

d) Animal-drawn 3-row seed-cum-fertiliser drill

Fig. 4.18 : Animal drawn one, two and three row seed drills

Animal drawn low cast seed drill

It is suitable for simultaneous drilling of seeds and fertilizer in two rows (Singh, 2007). The equipment is operated by a pair of bullock of low body weight (Fig. 4.19). It consists of a hopper with partitions for seeds and fertilisers. Fluted rollers are used for metering seeds. Use of mild steel strips for the frame and the ground wheel have helped reduce the cost of the unit. Shoe type furrow openers are provided for drilling two rows. Common bicycle chain and sprockets are used for transmitting motion to the metering unit form the ground wheel. It is suitable for drilling seeds of soybean, maize, pigeon pea, sorghum, Bengal gram, wheat, sunflower, safflower, mustard etc. Working capacity of drill is 0.06 ha/h.

Fig. 4.19 : A view of animal drawn low cast seed drill

Animal drawn automatic seed drill

It is used to drill seeds in three to five rows. The equipment is operated by a pair of bullock (Fig. 4.20). It is a standardised animal drawn seed cum fertiliser drill which is suitable for crops like wheat, gram, sorghum, soybean, lentil, pea, sunflower, safflower etc. It is simple, light in weight, and compact in construction for drilling fertiliser or seeds. It has one drive (ground) wheels. Box section frame having many holes helps in adjusting the row to row spacing at the desired value. The fluted roller metering mechanism, fitted in the unit, gets the drive from ground drive wheel through chain and sprocket. The profile cutting type furrow openers with non-clogging boot place the seed at desired depth. Working capacity of drill is 0.10 to 0.20 ha/h. Row spacing can be adjusted.

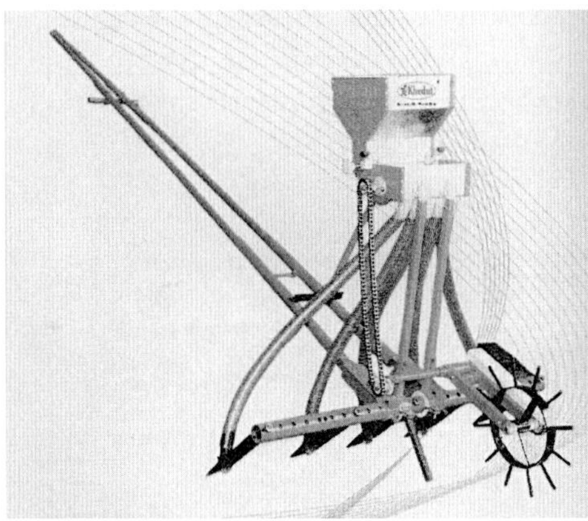

Fig. 4.20 : Animal drawn automatic seed drill

Courtesy : Khedut Agro Engineering Pvt. Ltd. Rajkot (Gujarat)

Animal drawn seed drill attached on tool frame

The seeding attachment is suitable for sowing wheat, gram, pea, soybean, sorghum and pigeon pea (Fig. 4.21). It can apply granular fertilizers like urea and DAP. The equipment is operated by a pair of bullock. It is an attachment made for the bullock drawn CIAE multipurpose tool frame (Pandey *et al.*, 1997). It has fluted roller type seed metering device. Diameter of fluted roller is 40 mm. The hopper has compartments for fertiliser and seed and the ground wheel is a floating type thus enabling uniform seed placement even when the soil surface is not properly levelled. Depth of placement of seeds can be adjusted. Separate side wheels allow accurate adjustment of the seed drill attachment and are also useful for transportation. It saves 73 per cent labour and operating time and 55 per cent on cost of operation compared to conventional method of sowing behind country plough or seeding by broadcasting. It also results in 10 to 18 per cent increase in yield compared to sowing by conventional method. Working capacity of machine is 0.1 ha/h for wheat, 0.12 ha/h for soybean, 0.25 ha/h for pea & 0.28 ha/h for sorghum. Row to row spacing is 22 cm and above.

Fig. 4.21 : A view of animal drawn seed drill attached on tool frame

Animal drawn two row mustard seed drill

The animal drawn two row seed drill can be used for sowing mustard and other small seeds (Pandey *et al.*, 1997). The equipment is operated by a pair of bullock. It consists of tubular steel section frame on which other components are mounted (Fig. 4.22). The seed box is of mild steel and the metering mechanism uses aluminium fluted rollers for the fertiliser and a rotor with cells on the periphery for the seeds. The furrow openers are of shoe type and are made of medium carbon steel hardened and tempered for opening the furrow. The ground wheel provides the power needed for operating the seed metering mechanism and a pair of idler wheels on either side help in proper adjustment of depth of seed placement. It also serves to transport the drill on the farm roads. A lever mechanism is also provided for raising and lowering the ground wheel on turns. Some of the other components are the seed pipes, steel beam and the power transmission system. The row spacing of the seed drill can be adjusted as per the requirement of the crop to be sown. Working capacity of seed drill is 0.12 to 0.16 ha/h.

Fig. 4.22 : A view of animal drawn two row mustard seed drill

Animal drawn multi crop seed drill

Animal drawn multi crop seed drill consists of rectangular frame made of MS angle iron square section on which a seed-cum-fertilizer drill is mounted (Fig. 4.23), Anonymous, 2015. The seed and fertilizer box is of two sections i.e. storage box and feed box. Feed box consists of plastic discs having grooves/cups on the periphery. Two pneumatic wheels are provided for transportation on either side of main frame through axle. One of these wheels supply power to metering device through chain and counter shaft. A clutch is provided to engage or disengage power to metering shaft. The field capacity of machine is 0.24 ha/h and field efficiency 83%.

Animal drawn seed drill for intercropping

Animal drawn seed drill for intercropping consists of rectangular frame made of MS angle iron square section on which four seed boxes are mounted (Fig. 4.24), Anonymous (2015). Feed box of seed box consists of plastic discs having grooves/cups on its periphery. Two pneumatic wheels are provided for transportation on either side of main frame through axle. One of these wheels supply power to metering device through chain and counter shaft. A clutch is provided to engage or disengage power to metering shaft. Plastic discs with different size grooves are fitted in individual boxes as per required intercropping pattern. The seed rate is governed by adjusting the opening between storage and feed box through rack and pinion arrangement. The field capacity of machine is 0.4 ha/h and field efficiency 82%. The power requirement of machine is 230 N.

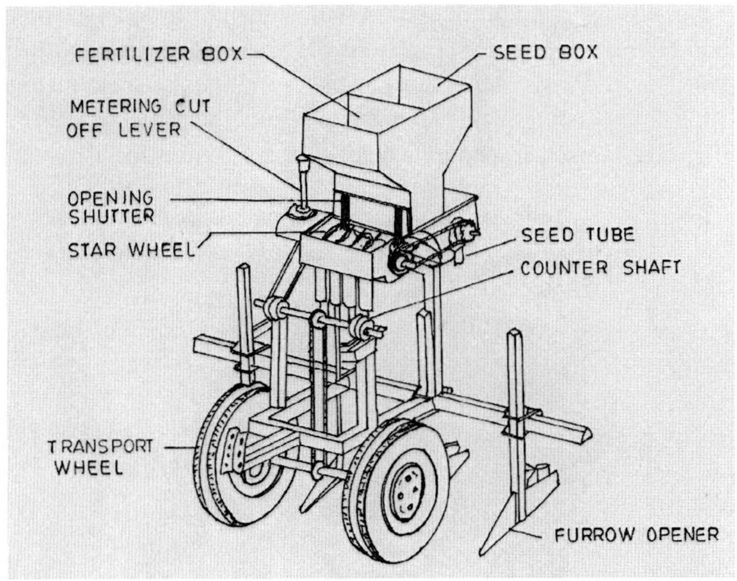

Fig. 4.23 : Animal drawn multi crop seed drill

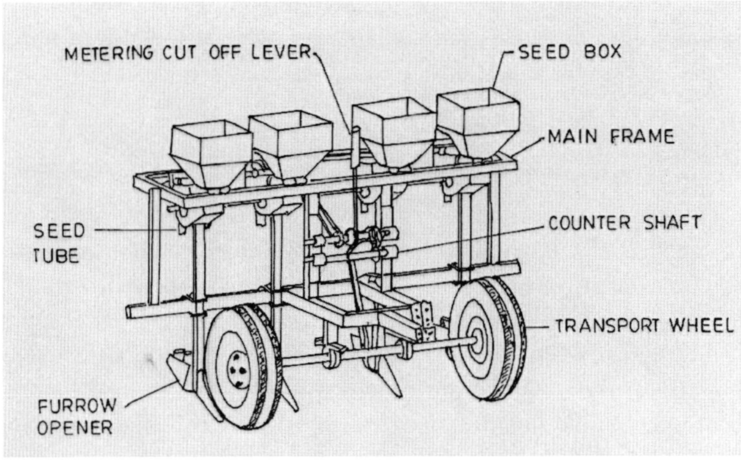

Fig. 4.24 : Animal drawn seed drill for intercropping

Power tiller operated seed-cum-fertilizer drill

It is suitable for drilling seeds (wheat, soybean, Bengal gram, sorghum etc) and fertilizer in six rows in medium and heavy soils. This machine had been specially designed for operation with a power tiller of 8-10 hp to drill seed and fertilizer in 6 rows (Fig. 4.25). It is provided with depth control gauge wheels and a ground wheel for operating the metering mechanism. Some of its major components are the main frame, seed and fertilizer boxes, metering mechanism, transport wheel, furrow openers, hitch system etc. Working capacity of machine is 0.20 to 0.25 ha/h.

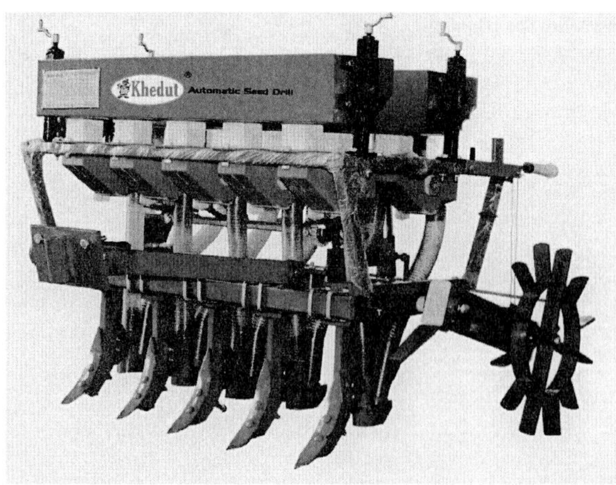

Fig. 4.25 : A view of Power tiller operated seed cum fertilizer drill

Courtesy: Khedut Agro Engineering, Rajkot (Gujarat)

Tractor operated seed-cum-fertilizer drills

It is used for sowing of wheat and other cereal crops in already prepared field (Singh, 2007). The equipment is operated by 35 hp tractor. It consists of seed box, fertilizer box, seed metering mechanism, fertilizer metering mechanism, seed tubes, furrow openers, seed rate adjusting lever and transport cum power transmitting wheel (Fig. 4.26). The fluted rollers are driven by a shaft. Fluted rollers, which are mounted at the bottom of the seed box, receive the seeds into longitudinal grooves of fluted roller and expel them in the seed tube attached to the furrow openers. By shifting the rollers sideways, the length of the grooves exposed to the seed, can be increased or decreased and hence the amount of seed sown is changed. The seed-cum-fertilizer drill is popular in northern region of the country. Working capacity is 0.4 ha/h. It has shovel and hoe type furrow openers. Row to row spacing is 200 mm. It has gravity type metering device for fertilizers.

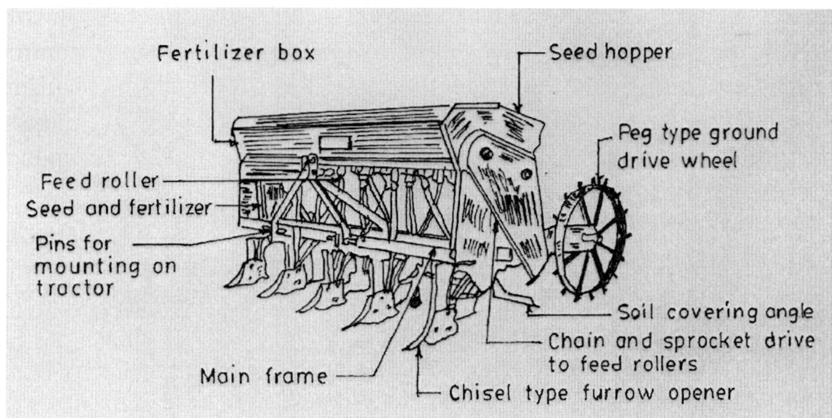

Fig. 4.26 : Tractor mounted 9-row seed-cum-fertiliser drill with fluted feed seed metering mechanism for seeds

A tractor mounted 9-row seed-cum-fertiliser drill with grooved periphery disc metering mechanism for seeds and fertilisers is shown in Fig. 4.27. Drive wheel is connected through chain and sprocket to seed and fertiliser metering devices (Singh, 2007; Singh and Verma, 2009; Chauhan *et al*., 1984). The rate is adjusted by means of a lever provided with seed and fertiliser metering devices. The seed and fertiliser falling through the box are conducted through the transparent tubes and finally dropped into the furrow made by the plough. The field capacity of these seed drills ranges between 0.2-0.6 ha/day. Tractor operated seed-cum-fertilizer drill with side markers are also available which avoids overlapping and gives clear row spacing (Fig. 4.28). Fluted type seed metering devices are used for small seeds/oil seed crops such as rape seeds, mustard, raya etc (Fig. 4.29). Table 4.2 gives the details of different types of seed-cum-fertiliser drills used in the country.

Table 4.2: Details of different types of seed-cum-fertiliser drills used in the country.

Type of Machine	Power Source	Metering device		Type of drive	Type of furrow opener	Output ha/day
		Seed	Fertiliser			
1-row seed drill	Human	Variable orifice with curved disc agitator	-	Chain & sprocket	Single shoe with single pore	0.3-0.4
1-row seed-cum-fertiliser drill	Human	-do-	Variable orifice with curved disc agitator	-do-	Single shoe with double pore	-do-
3-row seed drill	Pair of bullocks	Fluted roller with distribution funnel	-	-do-	Single shoe with single pore	1.0-1.2
3-row seed-cum-fertiliser drill	-do-	-do-	Variable orifice with curved disc agitator	-do-	Single shoe with double pore	-do-
4-row seed-cum-fertiliser drill	-do-	-do-	Auger system	V-belt & pulley	-do-	1.2-1.8
6-row seed-cum-fertiliser drill	-do-	50 mm perforated MS pipe having a concentric perforated wooden roller both for seed and fertiliser		V-belt & pulley	-do-	1.8-2.0
7-row seed-cum-fertiliser drill	-do-	Seed agitator & circular disc with metering holes	Variable orifice with curved disc with curved disc	Ground wheel	-do-	1.8-2.0

8-row seed-cum-fertiliser drill	-do-	Fluted rollers	MS plate with holes at hopper bottom & agitator	Crank arrange-ment	-do-	2.0-2.5
8-row seed-cum-fertiliser drill	-do-	-do-	Variable orifice with curved disc agitator	Chain & sprocket	-do-	2.1-3.2
9-row seed-cum-fertiliser drill	Tractor	Fluted roller	Variable orifice with agitator	Chain & sprocket	-do-	4.0-5.0
9-row seed-cum-fertiliser drill	Tractor	Rubber roller with variable orific	Variable orifice with agitator	Chain & sprocket	Chain &	
9-row seed-cum-fertiliser drill	Tractor	Laterally notched circular plate in seed box	Variable orifice with agitator	Chain & sprocket	-do-	4.0-5.0
11 & 13-row seed-cum-fertiliser drill	Tractor	Fluted roller	Fluted roller with agitator	-do-	-do-	5.0-6.0

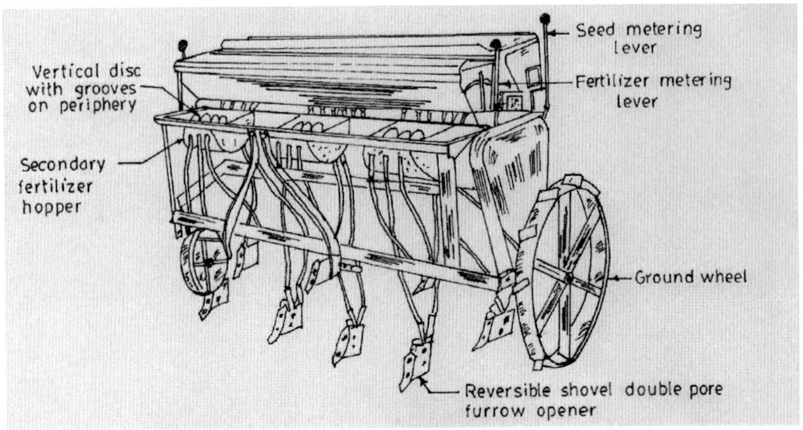

Fig. 4.27 : Tractor mounted 9-row seed-cum-fertiliser drill with grooved periphery disc metering mechanism.

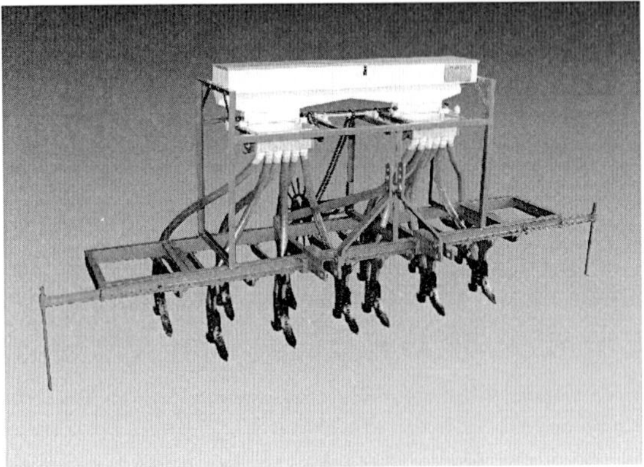

Fig. 4.28 : Tractor operated seed fertilizer drill with markers

Courtesy: Khedut Agro Engineering, Rajkot (Gujarat)

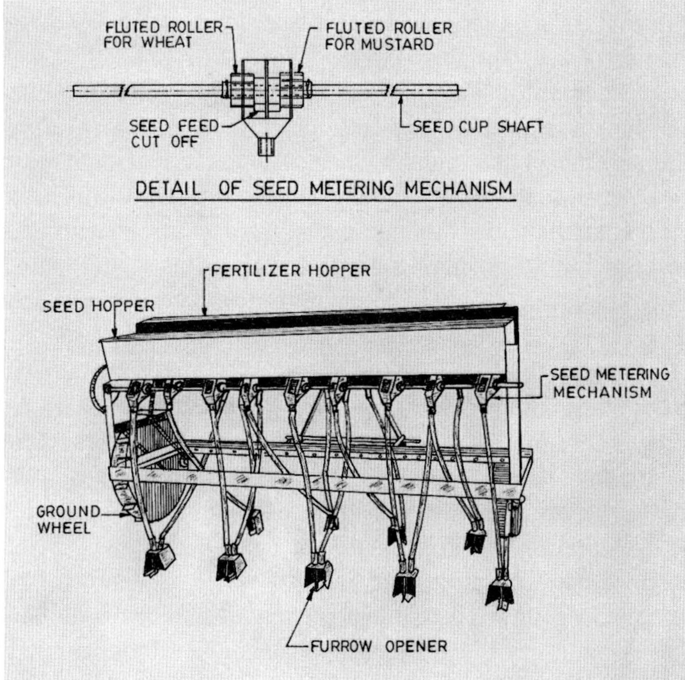

Fig. 4.29 : Tractor operated seed-cum-fertilizer drill for oil seed crops

Zero-till seed-cum-fertilizer drill

Wheat sowing by conventional methods requires large number of tillage operations to prepare a fine seed bed after paddy harvesting. Generally, 6-8 or sometimes even more tillage operations are required which cost both time and money for the farmers. Moreover shortage of time from paddy harvest to wheat sowing creates uncertainty and delay in sowing operation, which results in poor crop yields (Garg and Singh, 2002; Singh and Pandey, 2008; Anonymous, 2008, 2010, 2010c). Keeping in view the limitations of time and high expense of energy required in the conventional tillage system for wheat sowing, the tractor operated no-till drill for sowing of wheat crop was developed at GB Pant University of Agriculture and Technology, Pantnagar. The zero-till seed-fertilizer drill has been developed to sow wheat directly in rice-harvested fields without preparing the seedbed (Fig. 4.30a). It is a 9/11/13-row unit consisting of fluted rollers for seed metering and agitators over adjustable openings for fertilizer metering (Fig. 4.30b). The ground drive wheel supplies power through sprockets and chain for metering of seed and fertilizer. The furrow openers are of inverted 'T' type, spaced at 200 mm (adjustable). The machine is found very successful under the condition when the rice harvesting gets delayed and soil moisture is too high to facilitate seedbed preparation for wheat. The field capacity of the machine is 0.3-0.4 ha/h with about 75% efficiency. Use of zero-till drill for direct sowing of wheat after rice is found to be advantageous in terms of 50-60% saving in time and 40-50% saving in cost of sowing as compared to the conventional practice of seed bed preparation and sowing with seed-cum-fertilizer drill. The machine is operated by tractor of 35 hp (26 kW) or above. The crop yields obtained are at par with farmer's practice or even sometimes higher. Timeliness in sowing by no-till drill saves 7-8 days as compared to traditional method. Saves 60-75 litres HSD per hectare which results in saving of natural resources and environment. It gives 30-40% less infestation of weeds and saves irrigation water up to 10-15% during first irrigation, early and uniform germination and better plant stand than traditional practice, no crust formation after rains hence, no effect of rains on germination, improvement in soil structure and fertility, no lodging of crops at the time of maturity in case of heavy rains and winds and machine is simple and easy to operate.

a) Stationery view and machine in field operation

Courtesy: National Agro Industries, Ludhiana (Punjab)

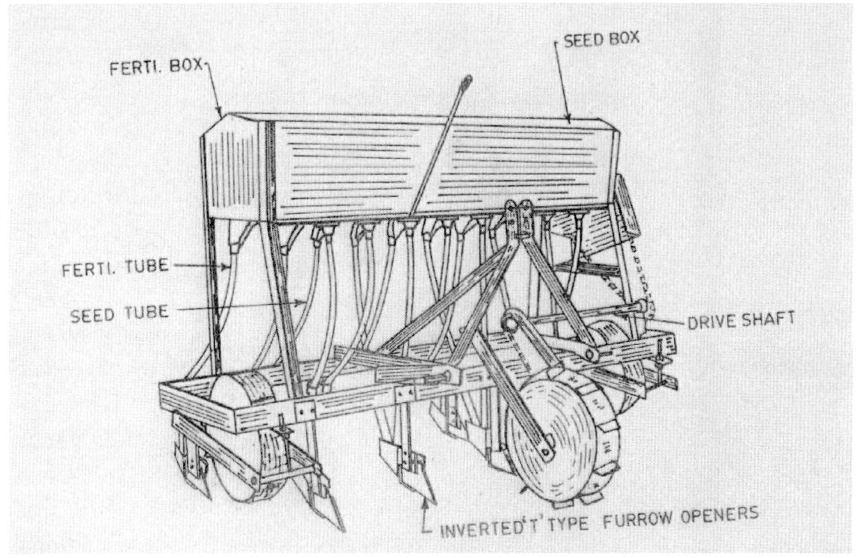

b) Details of zero till drill

Fig. 4.30 : Tractor operated zero-till drill

Tractor operated zero till multi crop planter is used for sowing bold grains like maize, groundnut, peas, cotton, sunflower etc (Fig. 4.31). The planting metering mechanism can be changed easily without dismantling the seed hopper's main shaft. Number of tines can vary from 6 to 11. Seed metering device is of nylon type inclined plate and fertilizer metering device is of cell type. Wheel is provided for depth control. The tine used is of inverted T-type furrow opener. Power to the metering shaft is given through ground wheel. Row-to-row and plant-to-plant spacing are adjustable. Row spacing can be changed from 15 to 60 cm. Missing of grains and grain damage is negligible. In the tractor operated zero till multi crop raised bed planter planting of bold grains like maize, groundnut, peas, cotton, sunflower etc is done on the two raised beds formed by three ridgers (Fig. 4.32). Seed metering device is of inclined rotating disc type and fertilizer metering device is of fluted roller type. Row-to-row and plant-to-plant spacing are adjustable. Missing of grains and grain damage is negligible. Roller is provided for shaping the raised bed properly and covering the seeds.

Fig. 4.31 : Tractor operated zero till multi crop planter
Courtesy: Dasmesh Mechanical Works, Amargarh (Punjab)

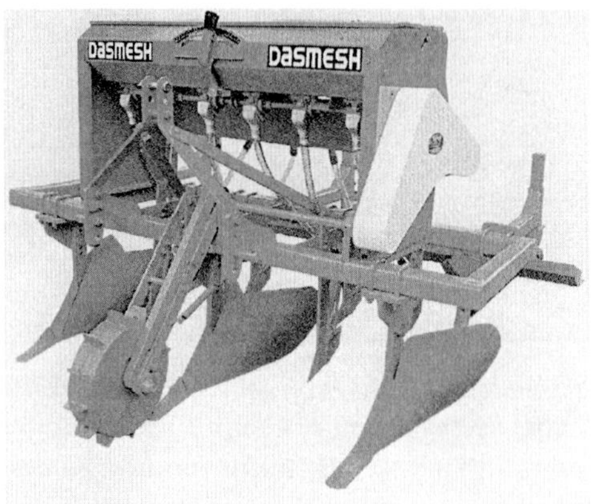

Fig. 4.32 : Tractor operated zero till multi crop raised bed planter
Courtesy: Dasmesh Mechanical Works, Amargarh (Punjab)

Power tiller operated zero-till drill for hilly region

In hilly areas, the farming is practiced in terraces and sloppy land. The soil resource has suffered from degradation over the past many years. The major contributor to this trend in soil organic matter loss is the tilling of soil with disc plough and cultivator in preparing the land for seeding. If we are to counter the effects of soil degradation caused by excessive tillage of the soil, we must find and adopt new methods of annual cropping. In zero tillage crops are planted with minimum disturbance of soil by placing the seeds in a narrow slit 3-4 cm wide and 4-7 cm deep without land preparation. A prototype of power tiller operated zero-till drill was designed and developed at Himachal Pradesh Agricultural University, Palampur under ICAR AICRP on Farm Implements and Machinery for sowing wheat crop (Fig. 4.33), Anonymous (2010). The effective field capacity of machine is 0.09-0.10 ha/h at a forward speed of 2.1-2.2 km/h with field efficiency 56-62%. The cost of operation with zero-till drill is 60-63% lower as compared to traditional method.

Fig. 4.33 : Power tiller operated zero-till drill

Power tiller operated zero till multi crop planter is used for sowing various crops such as maize, wheat, rice grain, pulses etc (Fig. 4.34). This is available in four rows. Row-to-row spacing is 20 cm and can be adjusted. Seed metering mechanism is of rotating disc type with cells on its periphery (inclined plate type) and fertilizer metering mechanism is sliding orifice type with agitator.

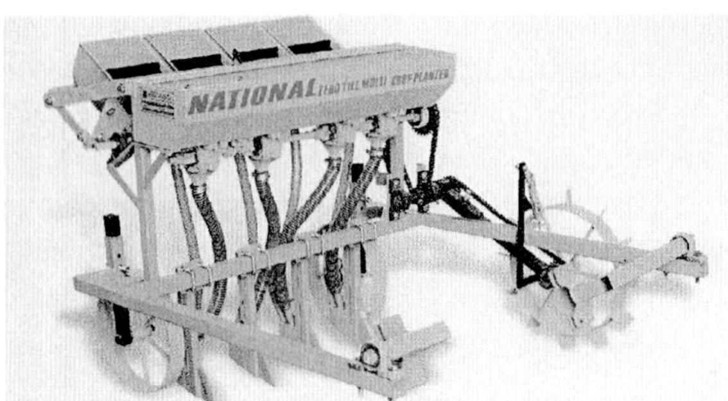

Fig. 4.34 : Power tiller operated zero till multi crop planter
Courtesy : National Agro Industries, Ludhiana (Punjab)

Animal-drawn zero-till drill

Animal-drawn zero-till seed-cum-fertilizer drill developed at GBPUA&T, Pantnagar is used for direct seeding of wheat and placement of fertilizer without preparatory tillage under high soil moisture condition (Chaudhuri and Singh, 2013). The traditional practice of farmer is generally 3 tillage operations when the soil is friable and seeding behind country plough. In high soil moisture condition after harvest of rice the farmers normally wait for the soil to become suitable to do the field operations due to which the seeding gets delayed resulting into loss of yield. By use of zero-till drill the seeding is done timely at reduced cost of cultivation. It consists of inverted T type furrow opener, fluted roller (8 flutes) for seed metering and stationary opening with agitator for fertilizer metering, power transmission through ground wheel with chain and sprocket (Fig. 4.35). The machine can cover 0.02 to 0.06 ha/h and can be used for 75 - 100 h/year. It could be two rows or three rows.

Fig. 4.35 : Animal drawn zero till drill

Tractor operated strip-till Drill

Tillage is one of the major farm operations and is an important contributor to the total cost of production. Excessive tillage is energy and time consuming and costly operation. It is considered harmful to the soil structure. It also contributes to wind and water erosion of the soil. The rising cost of hydrocarbon fuels which are bound to be exhausted sooner or later, availability of herbicides coupled with the motive of timely sowing and reducing the cost of production has provided enough incentive to the researchers all over the world to investigate tillage operations more closely. Development of a suitable tractor operated strip-till drill for sowing wheat without any prior field preparation was taken up at Punjab Agricultural University, Ludhiana (Shukla, 2001; Shukla *et al.*, 1990 & 1996). The drill is essentially a 9 row seed-cum-fertilizer drill with a rotary blade attachment for minimum soil manipulation running ahead of the normal furrow openers (Fig. 4.36). An 11-row machine is also available. The strip till drill is operated by a 35 kW tractor (Garg and Singh, 2002; Anonymous, 2008, 2008d & 2010). The rotary attachment consists of a frame with rotor having 9 flanges. Each flange has 6 L or C-type blades which prepare a 75 mm wide strip in the front of every furrow opener (Fig. 4.37). Thus with every row, 125 mm of the strip is left untilled and only 40 percent of area is tilled. The spacing between flanges is the same as the row spacing for crop to be sown. Power to rotor shaft is provided from tractor PTO through speed reduction gearbox and chain and sprocket drive. The rotor revolves at a speed of 300 rpm. The machine is provided with a MS sheet cover to protect the power transmission system and also helps to reduce the soil cover over the seed. It cuts and tills 7.5-cm wide strip in front of each furrow opener. It consists of a standard seed drill with shovel type furrow opener and fluted roller type seeding device. It has gravity feed type metering device for fertilizer placement. It can give 3-5 cm depth. Tilling and sowing is done simultaneously. Working capacity of strip drill is 0.25 to 0.4 ha/h. The use of machine can save about 40 litre of diesel per hectare and can save 65-70% time as compared to conventional method. It also helps to reduce the soil cover over the seed.

Fig. 4.36 : Tractor operated strip till drill

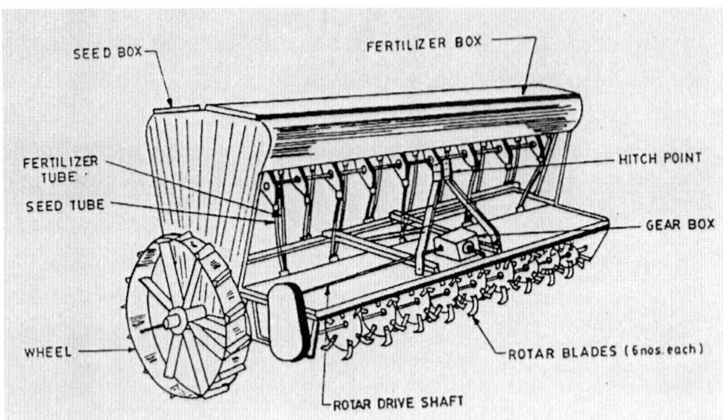

Fig. 4.37 : Details of tractor operated strip-till drill with rotary blade attachment

Tractor operated roto till drill

Timely sowing of wheat after paddy harvest by combines is a big problem faced by farmers. Consequently the sowing of wheat is either delayed or done in poorly prepared seedbed which lowers the yield. Tractor operated roto till drill which is a combination of a rotavator and a seed drill was considered good alternative and thus developed (Singh and Pandey, 2008; Garg and Singh, 2002; Anonymous, 2008 & 2010). The machine in a single operation can prepare seed bed and sow the seeds. Thus machine is very useful for farmers for timely sowing of wheat after harvesting paddy. The machine has 9-row seed cum fertilizer drill mounted on a 1.6 m wide rotavator (Fig. 4.38). Roto till drills are available in different sizes i.e. 1.6 m,

1.8 m, 2.0 m with 9-row, 11-row, 13-row drills depending upon the tractor power availability. The seed drill has a fluted roller mechanism for metering seed. The metering of fertilizer is done by gravity with adjustable holes and agitator in the hopper. The drive to metering mechanism is given by ground wheel through chain and sprocket. The furrow openers for the placement of seed and fertilizer are mounted at the back of rotavator in two rows in a staggered way. The row to row spacing has been kept as 17.5 cm. The rotavator has L-type 36 blades mounted on 6 flanges and each flange has 6 blades. The drive to the gear box on the rotavator from tractor is supplied through PTO shaft (540 rpm). The gear box having set of bevel and pinion change the direction of power from PTO to horizontal shaft and power from this shaft to rotary is supplied either through gears or chain and sprockets. The field capacity of machine is 0.4 ha/h. The major advantages of this machine is sowing can be done in one go so there is substantial saving of time (Fig. 4.39). The cost of operation is less as compared to traditional practice. The sowing can be done in relatively wet field thus avoiding delay.

Fig. 4.38 : Tractor operated roto till drill
Courtesy: Dasmesh Mechanical Works, Amargarh (Punjab)

Fig. 4.39 : Tractor operated roto till drill in operation
Courtesy: National Agro Industries, Ludhiana (Punjab)

Tractor operated happy seed drill

Until recently, direct drilling of any crop into combine harvested rice stubbles from a reasonable rice yield has not been possible without prior burning or removal of straw. To solve the problem, tractor operated double disc direct seeder was developed (Fig. 4.40). The machine has double disc type furrow opener which cuts the straw in the field and drilling is done by opening the furrow behind that. A new machine called as 'Happy Seeder' capable of direct drilling wheat into heavy rice residue loads, without burning in a single operation by managing only that part of straw which is coming just in front of furrow openers has been developed by Punjab Agricultural University, Ludhiana in collaboration with CSIRO, Land and Water, Griffith, Australia. Happy Seeder combines the stubble mulching and seed drilling functions into the one machine (Anonymous, 2013, Sidhu, 2008). Happy seeder consists of a rotor for managing the paddy residues and a zero till drill for sowing of wheat (Fig. 4.41). Turbo Happy Seeder is the most successful implement for sowing wheat in rice residue without burning rice residue. This gives the possibility of sowing wheat crop just after rice harvesting i.e. option for long duration wheat and rice varieties; possibility of sowing wheat in the residual moisture i.e. saving of one irrigation; timely sowing wheat even after long duration basmati rice varieties; crop residue as much helps in moisture and temperature conservation; improved soil health and environment friendly technology to check air pollution.

Flail type (Gamma) straight blades are mounted on the straw management rotor which cuts (hits/shear) the standing stubbles/ loose straw coming in front of the sowing tine and clean each tine twice in one rotation of rotor for proper placement of seed in soil. The rotor blades/flails guide/ push the residues as surface mulch between the seeded rows (Fig. 4.42). This PTO driven machine can be operated with 45 hp tractor and can cover 0.3-0.4 ha/h. It has been observed that wheat yield in case of happy seeder is nearly 10% more than conventional sowing practice. Weed matter is nearly 50% lesser on happy seeder plots compared to conventionally sown plots. Happy seeder sows wheat directly in paddy residue in combine harvested field hence, prevents residue burning thus reduces air pollution. Mulched crop residue improves the soil health and adds organic matter to the soil. Soil temperature regime is conducive for adoption of long duration wheat varieties due to increased window period. Substantial water saving has also been recorded due to avoidance of first irrigation and mulching.

Crop residue helps in moisture and temperature conservation. This gives fewer weeds and improves soil health. Seed metering mechanism is aluminium type fluted rollers or nylon type inclined plates and fertilizer metering mechanism is nylon type fluted rollers.

Fig. 4.40 : Tractor operated double disc direct seeder

Courtesy: Dasmesh Mechanical Works, Amargarh (Punjab)

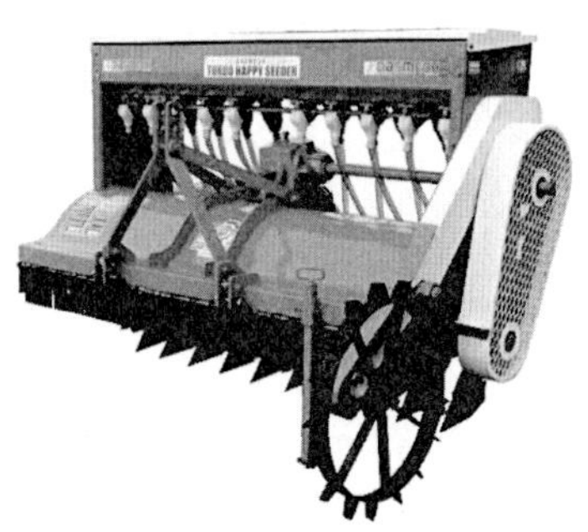

Fig. 4.41 : Tractor operated Turbo happy seeder for wheat sowing

Courtesy: Dasmesh Mechanical Works, Amargarh (Punjab)

Fig. 4.42 : Tractor operated happy seeder for wheat sowing in rice field

Courtesy: Dasmesh Mechanical Works, Amargarh (Punjab)

4.2 Planting equipment

Planter is a device that places the seed at desired depth, at equal distance in rows at specified rates and covers them with soil. Row planting facilitates inter-row cultivation. The planter can also work as a drill if required. Planters are used for row planting or check row planting of bold seeds, when plant to plant and row to row distances are to be maintained as per recommended practice (Singh, 2007; Singh & Verma, 2009; Singh *et al.*, 1985 & 1988). They give more accurate results with larger seeds like maize, cotton, groundnut, sunflower etc. Most of the row-crop planters are adaptable to a variety of crops by changing the seed plates, hopper bottom or some other element of metering device. Planters use many combinations of seed boxes, metering devices, covering and compacting units, seed-tubes, transmission unit etc.

Seed box: Seed and fertiliser boxes are provided in the planter, as in the seed drills. Seed hoppers are provided for each row. According to the seed-metering device, the shape of the hopper may be triangular, conical or cylindrical. Inclined plate planter uses triangular seed base to facilitate easy changing of metering plates. The height of seed box in a planter is lower than that of seed drill. These boxes are placed close to the ground level to reduce the time of travel from metering unit to furrow opener to deposit seeds at a low terminal velocity. Height of seed box above the ground varies from 30 to 50 cm. The capacity of seed box may be designed as per requirement of seed.

Seed tubes: Seed falls freely from feed cup through the tube into the furrow. In planters uniform seed to seed spacing is achieved when all seeds are released by metering device from same height with the same velocity. The type of tubes to be used for seed dropping can be selected from Fig. 4.12.

Metering of single seed or precision planting: Planters are designed to drop seeds in the furrow one by one at uniform distance. The height of fall must be short to keep the seed in place in the furrows. Device for metering single seeds usually have cells on a moving member or an arrangement to pickup single seeds and lift them out of a seed mass. Metering devices for planters are classified as horizontal plate, inclined plate, and vertical plate and rotor type (Fig. 4.9 and Fig. 4.10). These devices have slots, notches, cavities or cells on discs to catch seed from hopper and release it into a seed tube. The plates are driven by ground wheel. Seed rate can be changed by changing the speed of rotation of the disc or by changing the number of cells on it.

Horizontal plate metering device has plates with cells at the edge of a disc. It is the most popular mechanism where accuracy requirement of single seed selection is achieved. A stationery outer ring that forms the base for cylindrical hopper surrounds the metering plate. The outer ring also carries a cut off device and a knockout device for the seed located near the mouth of the seed tube. Wide range plates with varying sizes and number of cells are provided to meet the requirements of different seeds and crops (Fig. 4.43). The edge-drop carries the seed on edge in the cell of plate. The flat-drop carries the seed flat in the cell of plate. The full-hill plate has cells around the outer edge large enough to admit several seeds at the same time. Sufficient seeds for one complete hill are dropped together. The cells normally fill with one seed each and pass over a cut off mechanism which stops additional seeds from dropping in the furrow. The cell passes over the discharge spout, and a positive knock out is used to ensure the seed dropping out of the cell. This type of cell plate is widely used for maize planting. The risk of crushing unevenly sized seed is greater in this plate than other devices. Different plates are required for planting of different crop.

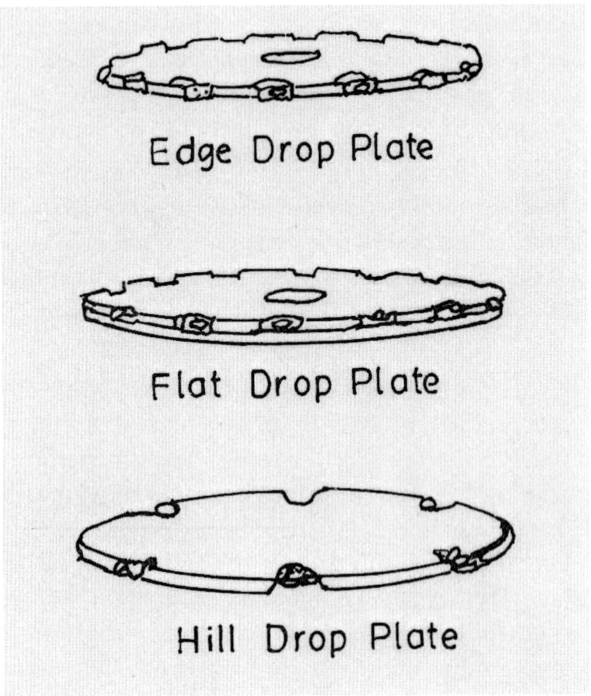

Fig. 4.43 : Seed metering plates.

Vertical plate metering device is positioned vertically at the base of the seed hopper. The cells in the rotor plate are made according to the size of seed and desired spacing. The seeds are picked up into the cells and are dropped one by one, into the seed tube. A positive knockout device ejects seed from cells at the bottom of the wheel. **Inclined plate metering device is a**n inclined seed plate with cells in the edge that dips into a hopper lifts the seeds and drops them into a delivery tube. This type of mechanism is simpler than above discussed systems. It does not use cut off device which otherwise cause seed injury and is affected by hopper tilt and rough ground. The inclined plate normally rotates at an angle of 60 degrees from the horizontal. This type of metering mechanism can be used for various crops such as cotton, groundnut, sunflower, gram, pea etc. In **belt with cell-type metering device** the metering hole is equal to the size of the seed and provided at equal spacing on an endless belt. The seeds from the main hopper enter a small chamber above the metering belt through an opening and remain at controlled level. The counter rotating seed repelling roller pushes back surplus seeds so that only one seed occupies each cell. This system is used to sow all types of seeds.

Cup type or spoon type metering device is used for those crops where seeds can be easily damaged due to rough mechanical handling. A series of cups on the rim of a vertical rotating wheel dip into a shallow pool of seed, lifts a few at a time and carrying them over top, where they are dropped into a delivery channel. It is suitable for planting of crops like potato, garlic etc. There is no restricted gate area through which the seed must pass, thus eliminating a possible source of seed damage. The accuracy of this type of mechanism is adversely affected by tilt angle of the hopper. In **picker wheel mechanism** (Fig. 4.44) a vertical plate is provided with radially projected arms that drop large seed like potato in furrow. This type of drop is also called reverse-feed drop because the agitator in bottom of hopper revolves in one direction and small picker located under outer edge of hopper bottom revolves in the opposite direction. Exposing more or less of picker wheel to the seeds by means of sliding gate shutter regulates the quantity of seed to be picked up by the picker wheel. Twin cup feed type conveyer for sowing potato seeds are also used on automatic potato planter (Fig. 4.45).

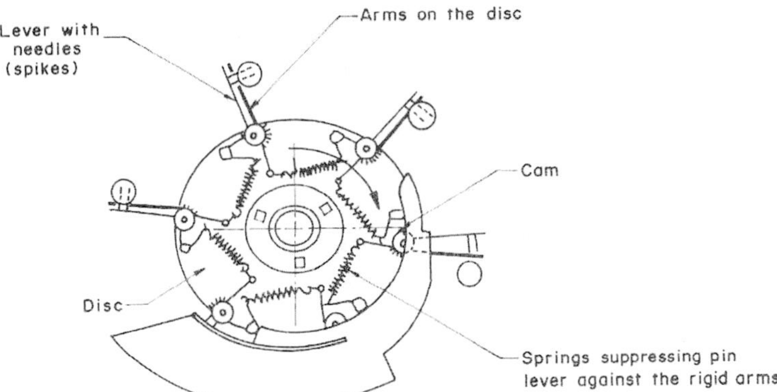

Fig. 4.44 : Picker wheel type seeding device

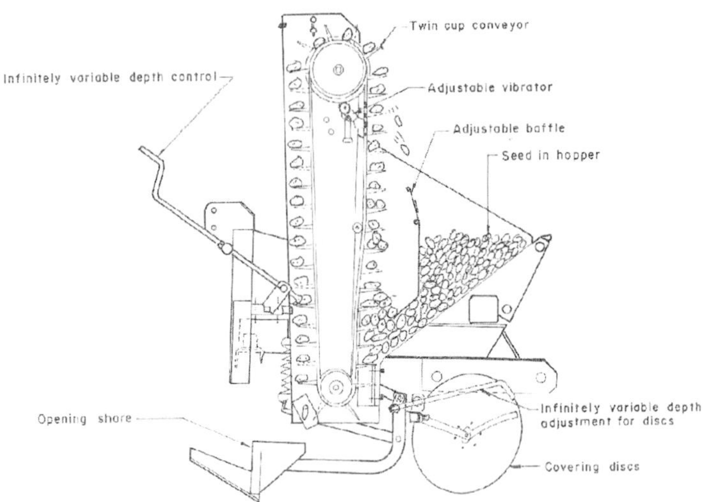

Fig. 4.45 : Twin cup feed type conveyer for sowing potato seeds

Manual garlic and multi-crop planter: It is used for sowing of garlic, maize, moong, peas, groundnut etc. Garlic planter is labour saving equipment and requires 90 man-hours to plant one hectare (Singh *et al.*, 2015; Pandey *et al.*, 1997; Anonymous, 2013a). The equipment is manually operated hand tool. This is a hand wheel hoe, used for inter-culture operation on which planting mechanism is mounted. The planting mechanism consists of a vertical plate with spoons, and receives drive motion from the ground wheel through chain and sprockets (Fig, 4.46). For operation of the planter, a person pulls it from rope attached to the hook and other person steers the machine by holding it by the handle. Upon pulling the planter forward, ground wheel starts rotating transmitting motion to the vertical plate fitted with spoons in the hopper. The hopper is filled with seeds or garlic cloves. As the vertical plate rotates, the spoons pick up the bulb/seed, which is discharged, in the small hopper connected to the furrow opener through a tube. The seed is then dropped in the furrow created by the shovel type furrow opener. It is provided with markers for maintaining the row-to-row spacing. Varying the number of spoons on vertical plate can vary plant spacing. Planting spoons for other crops like peas, sunflower, cotton, okra (*bhindi*), maize, and soybean are also available. It can plant up to 30 mm depth. Working capacity of planter is 0.045 ha/h. It can save 60-70% labour as compared to traditional method. Manually operated rolling injector planters are also available for planting maize and other bold grains. This machine consists seed box, seed cut off flap, handle, press wheel, soil covering flap and hexagonal shaped wheel to operate the feed roller (Fig. 4.47). This also covers 0.04 to 0.06 ha/h.

Fig. 4.46 : A view of Garlic and multi-crop planter.

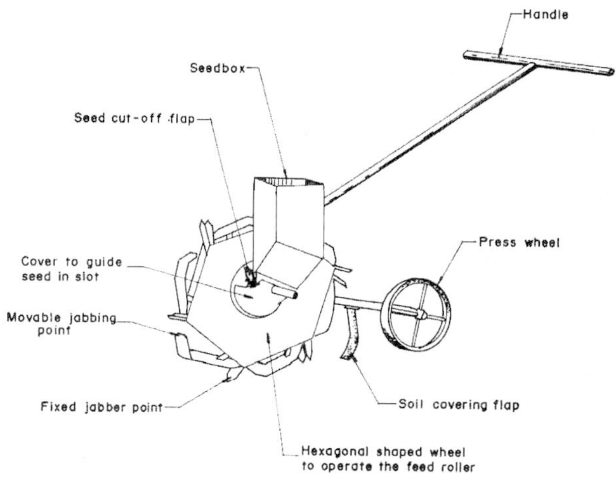

Fig. 4.47 : Manually operated rolling injector planter for maize

Bullock drawn 4 row groundnut planters: Groundnut is one of the important oilseed crops in India. The farmer has to perform sowing of groundnut at right time and as early as possible to take advantage of favourable situation in dry land conditions. The groundnut farmers are adopting the traditional practice of farm operations with the available bullock drawn implements and labour intensive operations. It is estimated that about 30 to 40 percent production cost can be reduced by introducing mechanization in dry land conditions. The 4 row automatic bullock drawn groundnut seed drill developed by ANGRAU Hyderabad (Anonymous, 2008, 2010, 2010a) consists of seed hoper (8 kg capacity), trough feed type seed metering mechanism, ground wheel of 30 cm diameter and sprocket and chains for power transmission (Fig. 4.48). The width of the seed drill is 120 cm. The row to row and plant to plant spacing are 30 cm and 10 cm respectively. The field capacity varies from 2.0 to 2.4 ha/day.

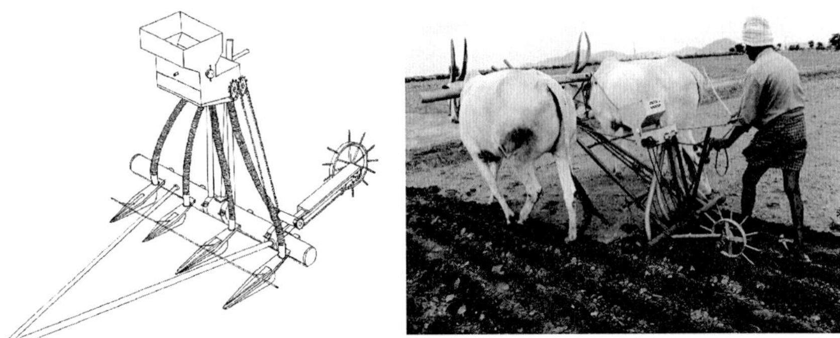

Fig. 4.48 : Bullock drawn 4-rows groundnut planter during field operation.

Animal drawn Jyoti multi-crop planter

It is suitable for planting groundnut, sunflower, safflower, soybean, pigeon pea, bengal gram, sorghum, wheat and maize. The equipment is operated by a pair of bullocks. The equipment can also be used for intercropping. It consists of separate hoppers for seeds and fertiliser (Fig. 4.49). Seeds are metered by a cell type rollers and fertiliser by agitator and orifice in the fertiliser box. Drive is obtained directly from the ground wheel and gears for seed metering and through chain and sprocket for the fertiliser metering. It has three rows and row to row distance can be easily adjusted and the unit has been accepted well for bold seeds. The capacity of seed hopper is about 12 kg. It has chisel type furrow opener. It can sow seeds up to 30 mm depth. It saves 75 per cent labour, 30 per cent operating time and 72 per cent on cost of operation compared to conventional method of sowing behind the country plough. Working capacity of multi crop planter is 0.1 ha/h.

(a) Stationery view

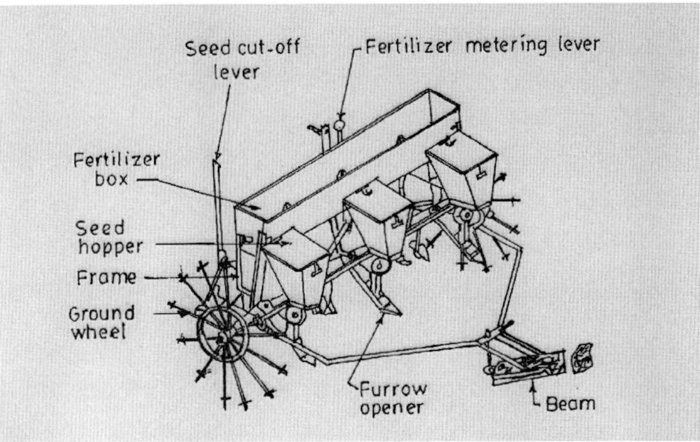

(b) Components of planers

Fig. 4.49 : Animal drawn Jyoti multi-crop planter

Animal drawn planter

The animal drawn 3-row planter is suitable for sowing of bold seeds like groundnut, maize, Kabuli gram, etc and very small seeds such as mustard, sorghum, etc which otherwise cannot be satisfactorily sown by conventional seed drills (Pandey *et al.*, 2006; Chaudhuri *et al.*, 1999). The planter is also suitable for sowing of intercrops (Fig. 4.50a & 4.50b). The equipment is operated by a pair of bullock. It consists of a frame with tool bar, modular seed boxes; furrow openers and ground wheel (Fig. 4.50c). It has three independent seed boxes with inclined plate seed metering mechanism. Seed plates for sowing different seeds can be selected and easily changed. The plate thickness, number and size of cells on seed plate vary according to seed size and desired plant-to-plant spacing. Shoe type furrow openers ensure deeper seed placement in moist zone for sowing under dry land conditions. Modular seed box-furrow opener units are adjustable for sowing seeds at different row-to-row spacing. Drive to seed metering mechanism is transmitted from ground wheel through chain and sprockets. Ground wheel and power transmission system are fixed on the main frame. An optional fertilizer box with fluted roller type metering system can also be mounted on the main frame for application of granular fertilizers. The planter is also suitable for sowing of intercrops as different seeds can be filled in different boxes. Power from ground wheel is transmitted to the counter drive shaft through a set of chain and sprockets. Another set of sprockets on the counter shaft transmits the power to main drive shaft. Main drive shaft drives the individual drive shafts of modular seed boxes through sets of chain and sprockets and these shafts in turn rotate the inclined seed metering plates through a set of bevel gears. Drive ratio between ground drive wheel and seed plate can be changed, by selecting appropriate size of sprocket on wheel axle or on counter and main drive shafts. Working capacity of planter is 0.12 to 0.15 ha/h. It can place seeds up to 100 mm depth.

(a) Three row small seed planter

(b) Vegetable planter developed at ANGRAU Hyderabad

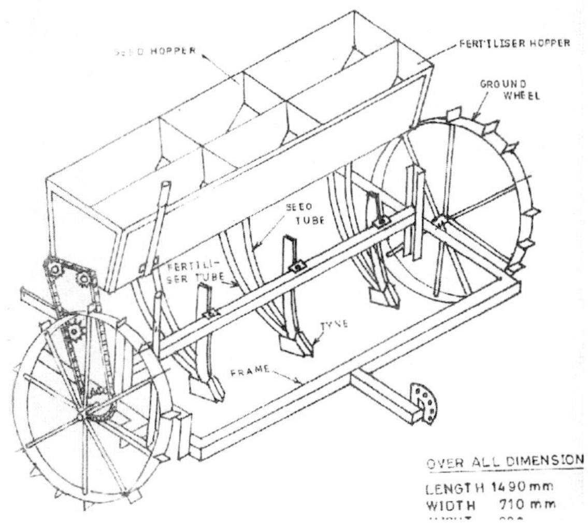

(c) Details of vegetable planter developed at ANGRAU Hyderabad

Fig. 4.50 : A view of animal drawn planter

Power tiller operated multi crop planter: It is used for sowing of wheat, maize, groundnut, and other bold seeds (Pandey *et al.*, 1997). A power tiller operated multi-crop planter (3-rows for wheat & 2 for maize) is suitable for sowing of different crops and applying fertilizer with the help of metering rollers (Fig. 4.51). Depth of operation is adjustable up to 100 mm by fixing furrow opener standard up & down through nuts and bolts. The power to metering mechanism is available from ground wheel through sprocket and chain arrangement. The machine requires 10-15 hp power tillers for its operation. The machine is simple in construction, light in weight and easy to operate on terraces. Thus, it is suitable for farmers of the hills as well as plains having power tiller. Working capacity of multi crop planter is 0.10 ha/h for wheat and 0.2 ha/h for maize.

Fig. 4.51 : A view of power tiller operated multi crop planter

Tractor operated garlic planter

Fig. 4.52 : Tractor operated garlic planter

Cultivation of garlic is increasing to a large extent in many parts of the country because of large local demands and export of garlic products. Manual planting of garlic is highly laborious (Fig. 4.46). MPUAT, Udaipur has developed a tractor operated 13 and 15 row garlic planter (Fig. 4.52). The seed rate of garlic cloves can be varied between 560-900 kg/ha depending upon clove size (Singh and Pandey, 2008; Anonymous, 2010). The row to row and plant to plant spacing can be adjusted. The field capacity of garlic planter is 0.4-0.5 ha/h for 13 & 15 row machine. The time and cost saving has been found as 95 and 85 per cent respectively. The details of tractor operated garlic planter are given in Fig. 4.53. Tractor operated garlic planter is a multipurpose machine also used for groundnut, cumin, wheat, sesamum, mustard, gram, pigeon pea etc (Fig. 4.54). Fluted roller is used for seed metering and for seed dropping vertical rotating disc with cells on its periphery are used. Number of tines could be five to eleven.

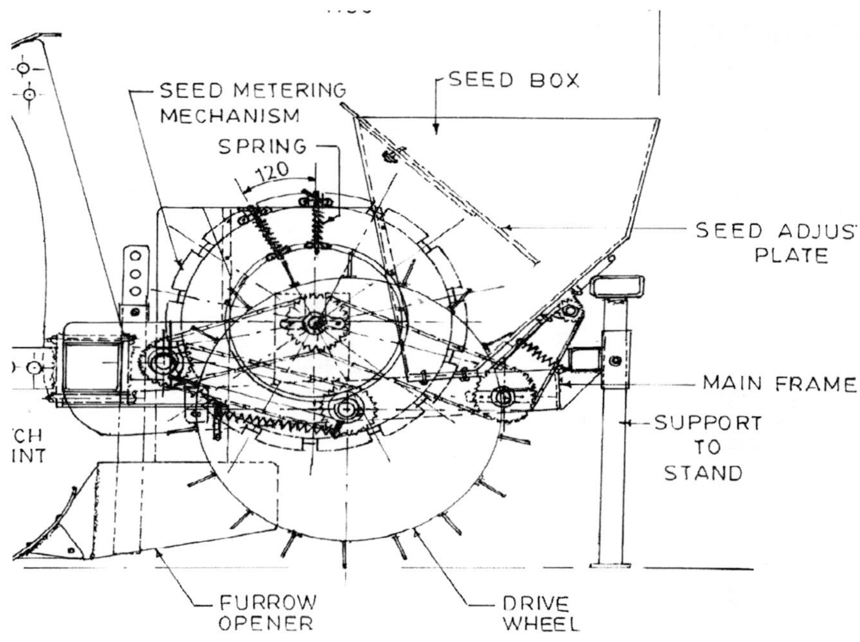

Fig. 4.53 : Details of tractor operated garlic planter

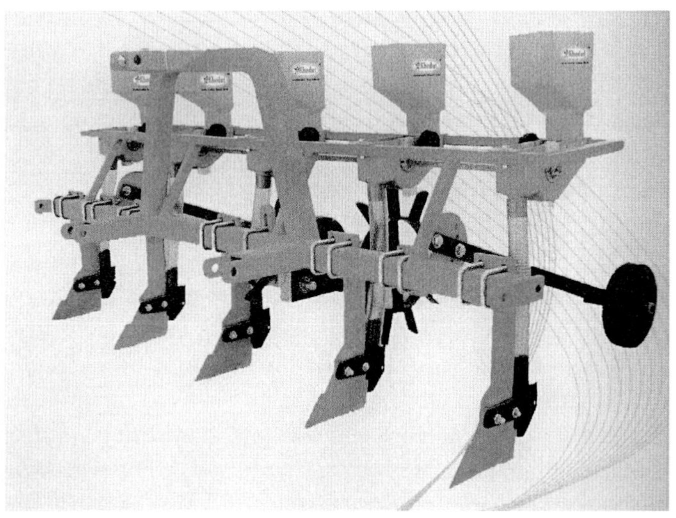

Fig. 4.54 : Tractor operated garlic planter

Courtesy: Khedut Agro Engineering Pvt. Ltd. Rajkot (Gujarat)

Tractor operated inclined plate planter

Intercropping is an effective system of crop production in rain-fed areas to cushion the impact of failure of one crop due to vagaries of nature. Growing crops on raised/broad beds is advantageous over flat sowing due to moisture conservation, improved drainage and easier inter-culture operations. CIAE 6-row tractor mounted inclined plate planter has been adopted for sowing intercrop on broad beds and a bed shaping/forming attachment has been added as an integral part in the refined design of the planter (Anonymous 2008 & 2010; Singh and Pandey, 2008). It is suitable for sowing of seeds like groundnut, maize, Kabuli gram, etc and very small seeds like mustard, sorghum, etc. The equipment is operated by a 29 kW (35 hp) tractor and above. The planter consists of a frame with tool bar, modular seed boxes; furrow openers and ground drive wheel system (Fig. 4.55). It has six modular design seed boxes with independent inclined plate type seed metering mechanism. Each seed box (capacity 15 kg) is provided with inclined plate (120 mm diameter) type seed metering mechanism (Fig. 4.55). The seed metering system is driven by a spiked ground drive wheel, fitted on front side of the frame through sets of chain and sprockets and bevel gears. The drive ratio (1:1) can be changed at different stages. A common fertilizer box with 6 units of fluted roller assemblies for metering granular fertilizer is fixed on the main frame. The drive to fertilizer metering shaft is given through the main drive shaft of the planter. The provision has been made for different crops by selecting seed plates and by changing the transmission ratio. The size of cells and inclined plate thickness depends on the seeds to be sown. The row spacing (225-450 mm) between furrow openers can be changed by sliding the furrow openers on rear tool bar of mainframe. The depth control of planter is performed by tractor hydraulic system. Shoe type furrow openers ensure deeper seed placement in moist zone for sowing under dry land conditions. The planter is also suitable for sowing of intercrops as different boxes can be simultaneously used for planting different seeds. Seeds are filled in the compartments of seed box. Flow of seeds to seed metering compartment is controlled trough the adjustable opening so as to keep the seed level in metering compartment up to centre of seed plate for effective picking of seeds. For inter crop sowing, different seed boxes can be loaded with different seeds with appropriate seed plates. Working capacity of inclined plate planter is 0.45 to 0.65 ha/h (for row spacing of 25-45 cm). Field efficiency is 64.0 percent. The planter gives seed rate of about 20 kg/ha for maize, 105 kg/ha for kabuli gram and 14 kg/ha for cotton. The field capacity is 0.5-0.6 ha/h for maize and 0.4-0.5 ha/h for sowing kabuli gram.

Fig. 4.55 : A view of tractor operated inclined plate planter with inclined plate seeding device

Tractor operated inclined plate planter *(PAU design)* is also used for sowing bold grains like maize, soybean, groundnut and cotton etc (Singh and Pandey, 2008; Garg and Singh, 2002). The equipment is operated by a 26 kW (35 hp) tractor. In the inclined plate planter, planting attachment has been attached to the commercially available seed-cum-fertilizer drills (Fig. 4.56a). The planting attachment consists of hopper, inclined plate metering mechanism fitted in the hopper, furrow openers, ground wheel, power transmission mechanism and seed tubes (Fig. 4.56b). For operation, the seed is filled in the hopper, seeds are picked up by the cells of inclined plate and delivered in the opening connected to furrow opener through seed tubes. It can plant 6-rows of groundnut at a spacing of 30 cm in addition to number of other crops like maize, cotton, soybean, sunflower etc. Seed-metering mechanism in planting attachment is of inclined plate type with notched cells for each row. For metering fertilizers adjustable opening with agitators are provided. Row to row spacing and plant-to-plant spacing is adjustable. Plant to plant spacing can be varied by changing the transmission ratio. The drive to the metering mechanism is given through the ground wheel by means of chains, sprockets and bevel gears (Fig. 4.56c and Fig. 4.56d). Working capacity of planter is 0.4 ha/h. It gives 25 kg/ha seed rate for maize, 85 kg/ha for groundnut, 20 kg/ha for cotton and 3.5 kg/ha for sunflower. The use of machine saves about 60% labour and 10-20% in cost of operation. It gives 5-10% increase in yield by way of precision planting. Power tiller operated inclined plate planters are also used in Odisha and some other parts of country (Fig. 4.57), Anonymous (2008a).

a) A view of inclined plate planter attachment with conventional seed-cum-fertilizer drill

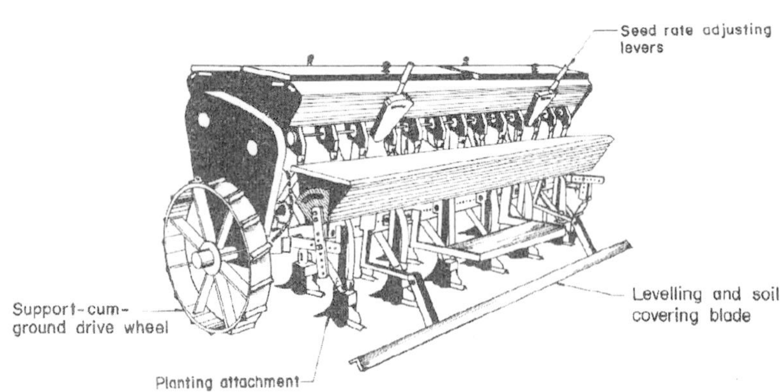

b) Details of inclined plate planter attachment with conventional seed-cum-fertilizer drill

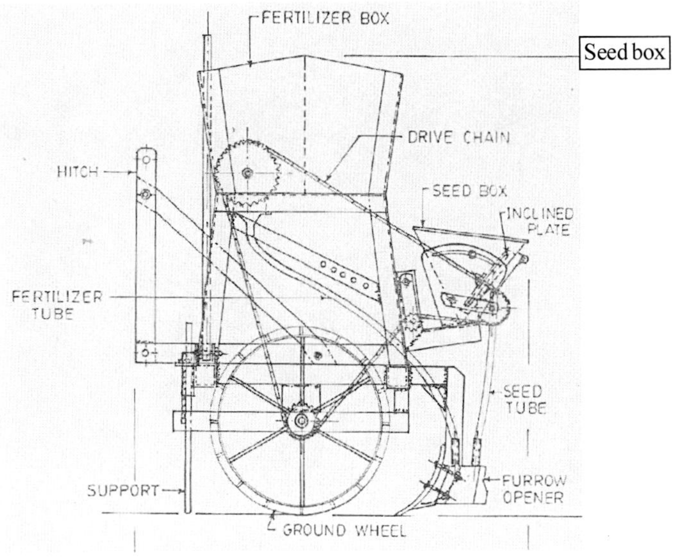

c) Side view of inclined plate planter

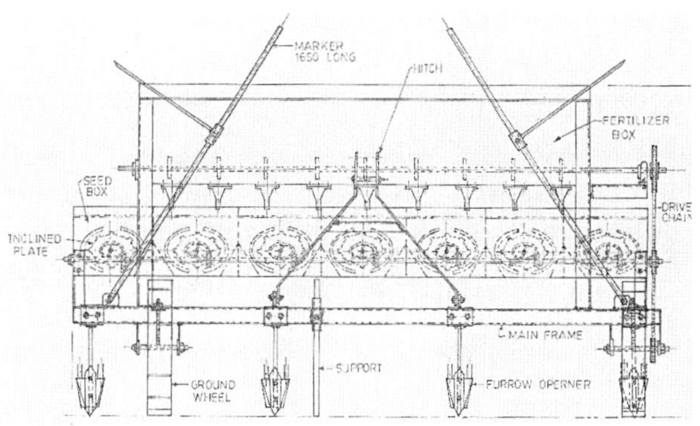

d) Front view of inclined plate planter

Fig. 4.56 : A view of tractor operated inclined plate planter (PAU design)

Fig. 4.57 : Power tiller operated inclined plate planter

Cotton planter

Cotton is grown under climatic conditions that range from dry land conditions requiring irrigation to humid conditions. The soil type ranges from light, sandy-loam to heavy clay. There are many types of farming practices and methods for planting of cotton and they vary greatly with the climatic conditions and soil types. Cotton is usually planted on beds in humid regions and in furrows in the sub-humid regions. Irrigated lands are prepared flat for planting of cotton. These different planting practices require different types of planting equipment. The main differences in different planting equipment are in the manner of mounting the planter on tractor, equipment for making furrow for planting seeds and device used for covering seeds.

Cotton planters range from one row bullock-drawn to six-row tractor-drawn or mounted models (Fig. 4.58). Cotton planters are also used for planting other crops like pea, gram, beans etc. Cotton is traditionally drilled thicker than the required and thinned by hand. Mechanical thinners for cotton crop are also available which chops the extra plants with chopping blade. Horizontal plate or inclined plate planters are quite popular for cotton sowing. For successful single cell selection of seeds, the seeds must be delinted mechanically or by using acid or sometimes ash is also used for delinting. If lint from seed is not removed, single or double selection of seed with planter will not be possible. Acid or mechanically delinted seeds can be metered properly. It increases the accuracy of planting. This type of seed is generally used for planting purposes.Inclined plate planter also gives more accuracy of planting, when the seeds are properly delinted with acid or mechanically. Ash delinted seed give lesser accuracy but is economical. The row spacing and seed spacing can be adjusted as required.

One of the constraints in adopting improved and efficient implement for cotton cultivation is the difficulty being faced in mechanical sowing, which otherwise makes easy the operations like inter-culture, weeding and plant protection. Cottonseeds bear considerable quantity of fuzz around them, which chokes the seed metering and dropping mechanism of sowing machines. Viable and healthy seeds cannot be separated from the lot of fuzzy seeds with the result that there is a poor crop stand. The fuzz around the seeds can be removed by acid delinting, which involves treating the cottonseeds with concentrated sulfuric acid in specific quantities and time and subsequently washing the mix in abundant clean water. In this process, the light and impotent seeds start floating and easily separated. Heavy and healthy seeds, which sink, are later dried and sown.

Small quantity of cottonseeds can be acid-delinted in a plastic bucket. The seeds to be delinted are put into bucket and required amount of sulfuric acid is added and mixed till all the fuzz is burnt. The seeds then are transferred to a sieve tray washed thoroughly in water. The capacity of this system is very low. The hand operated small size delinter with an enameled drum (Fig. 4.59) can be used by small farmers (Garg and Singh, 2002; Singh, 2007; Anonymous, 2010c). Agitators provided on a vertical shaft and rotated manually mix the seeds and acid in the drum properly. The hand operated commercial grade cotton delinter is available for large-scale acid delinting of cottonseeds. It consists of a mixing drum, which is eccentrically mounted on a horizontal shaft. Agitators are fixed to the shaft inside the drum for thorough mixing of acid and seeds. A wheel mounted on the shaft rotates the mixing drum. A suitable gate is provided in the drum for filling and taking out seeds. Fuzzy cottonseeds are put into the drum and specific quantity of concentrated sulfuric acid is added to it. The machine rotated manually till the fuzz of cottonseeds is burnt. The seed is taken out from drum, washed with clean water, rinsed with limewater and dried under the sun.

Fig. 4.58 : Tractor operated inclined plate type cotton planter

Acid delinting of cottonseeds is a precise process and requires precautions. If the quantity of acid is more or mixing time is more, the whole of seed may get damaged because of overreaction of acid. The outer coat of seed may get ruptured and result in reduced germination. On the contrary, if less acid is added or mixing time is short, delinting may be incomplete and some fuzz may still remain on the seed. Every care should be taken to ensure required quantity of sulfuric acid to be added to seed and treated till delinting is complete. The person working with this operation should be careful in handling sulfuric acid. The residual water after cleaning the treated seed should be disposed off to a suitable place with necessary care.

Fig. 4.59 : Hand operated small size delinter
Courtesy : Khedut Agro Engineering Pvt. Ltd. Rajkot (Gujarat)

The picker wheel mechanism is the most common for handling fuzzy seeds that are planted thickly for subsequent thinning. It consists of a horizontal spider plate with raised radial fingers at its periphery which carry cotton seed over small vertical toothed picked wheel that turns in opposite direction of the spider plate and picks seed while moving downward into the seed tube. The seed rate is regulated by adjusting the height of gate bar above the spider plate at picker wheel location that controls the thickness of layer of cottonseeds exposed to picker wheel.

Tractor operated inclined plate planters are also available specifically for cotton crop in cotton belt (Fig. 4.60a). Groundnut and maize crops can also be sown with this machine. Average seed missing is between 1 to 2 per cent. Speed of operation is within 1.5 to 2.5 km/h to avoid scattering of seeds. The average capacity of the machine is 0.3 ha/h. Tractor operated vertical disc planter is operated by a tractor of 30 kW or above (Garg and Singh, 2002). It consists of a planting attachment mounted on a 9-row seed-cum-fertilizer drill (Fig. 4.60b). The planting attachment has seven hoppers. The planting mechanism is of vertical disc type with spoons on its face. The planting mechanism in all the hoppers gets the drive through a common shaft driven by ground wheel of the seed drill through chain and sprocket. Planting plates for different crops like cotton, groundnut, maize, soybean, sunflower etc can be mounted for a particular crop by nut and bolts. It can plant 7-rows of any crop at a row spacing of 30 cm. The row spacing and number of rows for different crop can be varied. Plant spacing can be varied by varying the number of spoons on the disc or by changing the sprockets. Furrow openers are of reversible shovel type. Working capacity of planter is 0.4 - 0.6 ha/h. Planter can be operated at 2.0-2.5 km/h speed. It can give 12-15 kg/ha seed rate for cotton, 87-90 kg/ha for groundnut, 20-25 kg/ha for maize, 4.5-5.0 kg/ha for sunflower and about 7.5 kg/ha for ladyfinger (bhindi). It gives about 1-5% missing.

a) Tractor operated inclined plate type cotton planter used in cotton belt areas of Punjab and Haryana
Courtesy: Gyani Agricultural Works, Malout (Punjab)

b) Tractor operated vertical disc planter with spoon type metering device

Fig. 4.60 : Tractor operated planters

Raised bed planter

Bed planting system is referred to the planting and cultivation of crops on raised beds. Generally wheat and some other crops are planted on raised beds. Researchers from several organizations (DWR, Karnal; PAU, Ludhiana; CIAE, Bhopal; PDCSR, Modipuram; CCS HAU, Hisar, RWC-IGP and CIMMYT etc) have reported that planting wheat on raised beds improves yield, increases fertilizer use efficiency, reduces herbicides dependence, facilitates better weed management and mobility in the crop field for other cultural operations, less lodging of crops and saves seed, fertilizer and irrigation water. The total production cost of wheat compared to flat sown although found reduced marginally in the first planting on fresh beds but reduces by 25-35 per cent subsequently when the beds are reused. Bed planting technique is gaining acceptance by farmers because of more benefit-cost ratio compared to the flat sown crop and is being assessed for its suitability in different parts of the country for different cropping systems. The making of beds on well-tilled soil, planting of seeds, basal application of fertilizer and covering and dressing of planted beds are done in one operation.

Animal-drawn raised bed planter has also been developed at CIAE, Bhopal (Pandey *et al*., 1997) which is used for planting of vegetable seeds (Okra). The conventional practice is manual dibbling in tilled soil which is time consuming and requires more water for irrigation. Due to bed and furrow configuration by raised bed planter the water requirement is low as it could be retained for longer period in the furrow, besides increase in yield of okra being on beds. The field capacity is 0.05 ha/h.

Tractor operated raised bed planter consists of 3 mould boards, 2 beds and a seeding mechanism to sow 2, 3 or 4 rows of wheat on each bed simultaneously (Fig. 4.61), Singh and Pandey (2008). The machine can make three beds in single run and the width of each bed is adjustable (35 to 45 cm). A Planting attachment has also been made with the machine for sowing maize, groundnut, cotton etc on the beds (Anonymous, 2008, 2008d, 2010 & 2010b). The frame is made of mild steel sections. The furrow openers are ridger type and have mould board and share point. The wingspan of the mould board can be adjusted. The share is made of medium carbon steel or alloy steel, hardened and tempered to suitable hardness. Machine can sow two or three rows of wheat on each bed. Machine has seed metering unit of inclined plate type. The machine is also provided with a shaper after seeding tynes to shape and compact the beds. Working capacity of machine is 0.2 to 0.25 ha/h. The machine is operated by 34 kW (45 hp) tractors and above. The machine can be operated at a forward speed of 2.5 to 3.5 km/h. The time, cost and operational energy use per hectare of bed planting with preparatory tillage have been found 28.5, 26.2 and 25.0 percent higher respectively compared to the flat sowings. The early irrigation is required for proper germination of wheat. By reuse of beds, the bed planting operation is found to be energy efficient and cost effective compared to the conventional flat sowing of wheat with preparatory tillage. The technique saves about 20% seeds, 25% fertilizers and 30-35% irrigation water. Vegetables seeds can also be sown by the planter.

Courtesy : National Agro Industries, Ludhiana (Punjab)

Fig. 4.61 : Tractor operated raised bed planters with different types of shapers

Tractor operated multi crop raised bed planter with inclined plate is used for planting different seeds on two raised beds formed by the machine (Fig. 4.62). Bed making and planting of seeds on the bed is done in single operation. This machine is available in 5 rows two on each bed (Fig. 4.62). It is also available in four beds (Fig. 4.63). Seed metering device is of fluted roller type and fertilizer metering device of sliding orifice type with agitator. Tractor operated multi crop ridge planter is used for planting maize on ridge (Fig. 4.64). It is available in 2, 3, 4, 5 rows. It makes ridges and plant seeds simultaneously (Fig. 4.65). Row to row spacing is 60 cm but can be adjusted. Seed metering device is of rotating disc type with cells on its periphery i.e. inclined plate type. Fertilizer metering device has sliding orifice type with agitator. Some of the manufacturers have started manufacturing rotavator with wide bed former (Fig. 4.66) which can be used for sowing of small seeds such as raya, rapeseed, mustard etc and vegetable seeds on beds.

Fig. 4.62 : Tractor operated multi crop ridge planter
Courtesy: National Agro Industries, Ludhiana (Punjab)

Fig. 4.63 : Tractor operated raised and furrow seed-cum-fertilizer drill and planter
Courtesy: Khedut Agro Engineering Pvt. Ltd., Rajkot (Gujarat)

Fig. 4.64 : Tractor operated raised bed maize planter
Courtesy: National Agro Industries, Ludhiana (Punjab)

Fig. 4.65 : Tractor operated raised bed planter in operation

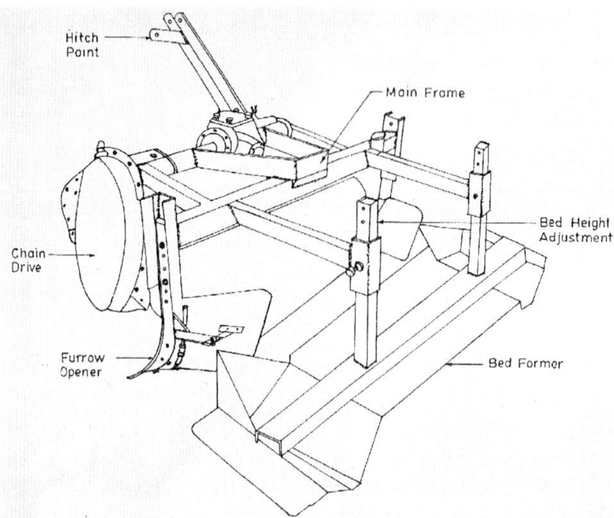

Fig. 4.66 : Tractor operated rotavator-cum-wide bed former

Tractor operated multi crop planter for seed spices

Seed spices cumin, fenugreek, fennel coriander and ajwain are the seed spices largely cultivated in Rajasthan, Gujarat and other part of the country. The sowing of seed spices is mainly done by broadcasting or drilled in small plots at a spacing of 25-30 cm and depth of 1-1.5 cm. A 7 row planter with individual hopper boxes has been developed by MPUAT Udaipur as a multi crop planter with seed metering mechanism mounted on common frame (Anonymous, 2008b, 2010, 2013a). The power transmission is through chain and sprocket and drive wheel (Fig. 4.67). The seed metering mechanism of star wheel (plastic) is made of circular rotor of 90 mm diameter with 10 cells of 20 mm length. The height of seed dropping from hopper is kept at 40 cm to get the accurate placement of seeds at depth between 10-15 mm. The inverted T type furrow opener of smaller size is fitted in comparison with other commercial drills. The machine has also fertilizer drilling attachment and variable row to row spacing arrangement. The seed rate of 6.5 -7.5 kg/ha, 12-15 kg/ha and 9-10 kg /ha is achieved for cumin, fenugreek and coriander respectively. The field capacity of machine is 0.28-0.30 ha/h with depth of seed placement as 12-15 mm. A two row hand operated multi crop seed spices planter has also been developed by MPUAT Udaipur (Fig. 4.68).

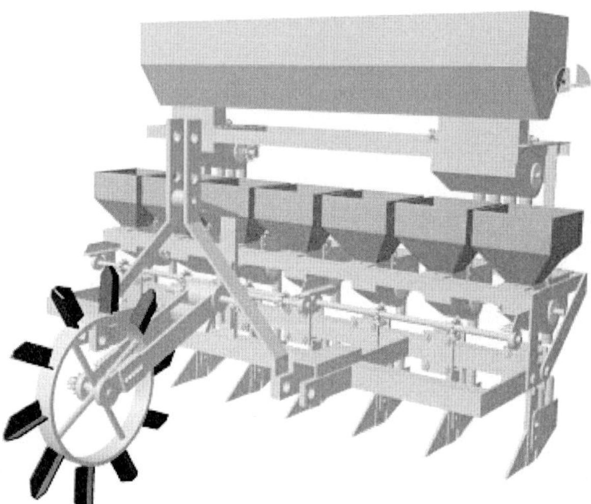

Fig. 4.67 : CAD diagram and operation of seed spices planter for sowing of cumin

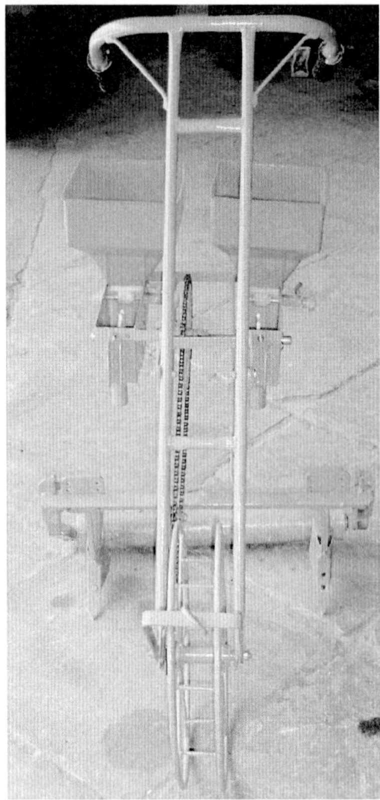

Fig. 4.68 : Two row hand operated multi crop seed spices planter

Tractor operated hill drop planter

Conventional seeders cannot maintain uniform spacing between the plants in the row. Non-uniform spacing on the row demands excessive thinning operations. Choosing the plants to be thinned on the row so as to bring about uniform plant spacing is also difficult. Precision planting requires 100 per cent viable seeds which is not feasible. To avoid the complexity and the associated cost of precision planting, the best alternative is a hill drop planter. A hill drop planter has been developed by TNAU Coimbatore which has a main frame, seed metering mechanism, seed hopper, drive to seed metering mechanism, furrow opener, check valve system, toggle mechanism and furrow closer (Fig. 4.69). The cup feed seed metering has been adopted in this machine, because of its capacity to meter seeds of different size and shape as well as its constructional simplicity (Anonymous, 2008c and 2010).

Fig. 4.69 : Tractor operated hill drop planter

The drive to the cup feed rotors is from the seeder's right hand end wheel through a chain and sprocket. They are mounted on hinged drag arms, with adjustable skids provided on either side of the runner type opener. Depth of seed placement can be adjusted between 25 to 60 mm as required for different crops. Since each drag arm with opener is independently hinged to the frame, the field undulation is precisely followed and placement intricately done. The spring pressure on the drag arm can be adjusted according to the hardness of the tilled soil to obtain the desired depth of operation. A check valve system is provided to accumulate the seeds coming through the delivery tube of each opener and drop them in a hill at pre-determined spacing. A toggle mechanism with an activating pegged disc on the other end wheel opens the check valves in all the furrow openers at pre-determined interval of time, with respect to the travel of the seeder's end wheel. Cable drives are provided to actuate these valves. The number of pegs on the end wheel is changed to vary the hill to hill spacing on the row. The crop stand hence allows easy thinning. The unit can save 87.0% and 95.0% time and labour respectively compared to manual sowing in hills.

Tractor operated check-row planter

Planters are usually adjustable for row spacing as well as plant spacing. Precise planting rate is possible in maize because of size and nature of seed. Accurate single seed selection of graded seeds with properly fitted seed plates is possible. Maize is sown on both flat land and ridges. It can be check-rowed also for cross cultivation. This practice facilitates weed control, especially in very wet years. Mostly it is a horizontal or inclined plate type of planters and single seed selection can also be made. Planting rates depend on the number of cells per plate and the drive rates. Check row planters traditionally drop two, three or four seeds per hill. At higher speeds, scattering of seed is pre-dominant due to bouncing. The accuracy of a planter depends upon the uniformity of kernels, shape of hopper bottom, speed of plate, shape and size of cell and fullness of hopper. A cone-shaped hopper bottom causes the seed to gravitate into the cells. The yield cut off pawl acting under spring pressure pushes the extra seed back as the cell passes under the plate cover. The height of the planter box from the ground level should not be more than 45-60 cm. The MPKV, Rahuri has developed a tractor operated check-row planter with row spacing of 100 cm (Anonymous, 2008c). It consists of main frame, three seed boxes with metering device, three furrow openers, power transmission unit and marker (Fig. 4.70). The seed plates with two cells are fitted in aluminium housing in each seed box to accommodate 2-3 seeds per cell. The furrow opener of 520 mm length is supported by tie rod of 10 mm at the front soil-working end. The machine is provided with two spiked wheels fitted with an eccentric weight to maintain the orientation of seed cell, which help in maintaining check row spacing after every turn at headlands. For maintaining the line spacing after every turn, markers are also incorporated in the unit. The working width is 2.7 m. The depth of planting, row spacing and seed rate are 52 mm, 902 mm and 1.21 kg/ha, respectively. The field capacity, and field efficiency are 0.613 ha/h and 79.09% respectively. The number of hills and plant population are 7572 and 11995/ha, respectively.

Fig. 4.70 : Tractor operated check row planter

Tractor operated pneumatic precision planter

The precision planters are used for accurate metering of single seed. Two types of designs are used on commercial units (Anonymous, 2008, 2010). In one type, a blower supplies air into a seed drum where an air pressure of 4.1 kPa is maintained (Fig. 4.71). Air pressure holds the seeds in pockets on a drum until the seeds are conveyed up to a stationary brush near the top where they are knocked off and dropped into the seed tubes. The flow through tube conveys seeds to the furrow openers and deposits them in the furrow. This system is appropriate for crops like corn, beans, etc. A different type of drum is used for each kind of seed. Seed spacing is controlled by ground wheel drive. Number of rows can be varied in the planter. The machine can plant 0.4-0.6 ha/h.

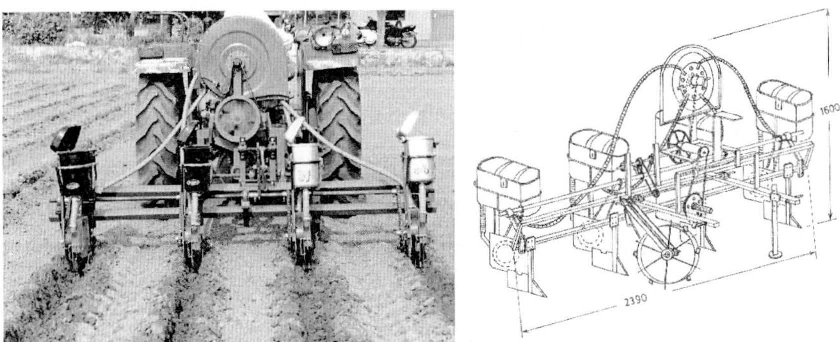

Fig. 4.71 : Stationary view of tractor-operated pneumatic planter

In another type of pneumatic planter, air suction is used for picking the seed (Fig. 4.72a). Single row metering devices are used for this type of planter. Tractor-operated pneumatic planter consists of main frame, aspirator blower disc with cell type metering plate, individual hopper for each row, furrow openers, PTO driven shaft and ground drive wheel (Fig. 4.72b). The machine works on the air suction principle and a suction pressure of 2 kg/cm^2 is developed by the machine. Air is sucked through a rotating plate having various holes radially. Seed plates with hole angle of 90°, hole size 3 mm and with 4 holes are used for sowing of cotton seeds. A seed coming in contact with the plate gets stuck to the holes on the plate and falls when suction is cut-off at the lowest position near the ground. The fall of the seed is synchronized with the predetermined seed spacing. Since the seed is lifted under suction no mechanical damage takes place. Row to row spacing is adjustable. The drive to the metering mechanism is given through the ground wheel by means of chains and sprockets. The machine is suitable for sowing cotton, soybean, groundnut, sorghum, pigeon pea, maize etc by changing the seed plates. An aspirator blower is used to create a suction pressure of 3-5 kPa in the blower inlet chamber connected to the vacuum disc of the planter unit. The seed-metering disc has holes on its periphery. Seeds are conveyed from the main hopper near the metering disc. The vacuum disc is connected to the inlet chamber of the aspirator through a tube. Air is sucked through holes of the metering disc, thus, forcing the seeds to fall on the seed plate. As the machine moves forward, the seed disc is rotated counter-clockwise. Seeds move up and released when the pockets pass a baffle that cuts the vacuum from the disc near the opener. The absence of suction allows the seed to drop into the furrow. Seed velocity at the delivery end is low; therefore, uniform seed distribution is achieved in the furrow. Vacuum type metering devices have 99-100% cell filling efficiency and 99-100% single seed picking efficiency. They are operated at higher speeds compared with plate type planters. Variation in seed size affects the picking efficiency. The field capacity of the pneumatic planter is 0.49 ha/h and field efficiency 77%.

a) A view of pneumatic planter in field operation

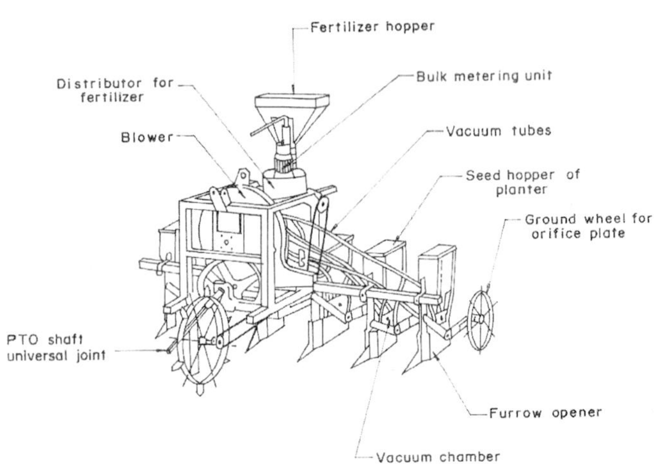

b) Details of pneumatic planter

Fig. 4.72 : Tractor operated pneumatic planter

Tractor operated vegetable seeder/pneumatic planter

The planting of small seeds like onion, mustard etc are planted in Maharashtra using Pune seed drill (Anonymous, 2010). Pune drill consists of local manual 8-row seed drill which hand meteres onion seed in funnel with tubes on raised bed (Fig. 4.73). The 8-row vegetable pneumatic planter (Fig. 4.74) imported by CIAE Bhopal was tested and evaluated at DOGR, Rajguru Nagar, Pune for direct onion seeding on 1.20 m wide and 15 cm high raised beds. The machine can plant seeds at 200 mm row spacing with average seed to seed distance of 94 mm and at an average 27 mm depth, in 6 rows along the beds. The average field capacity of the machine is 1.26 ha/h. The manufacturing of this machine has started locally reducing the cost drastically (Fig. 4.75).

Fig. 4.73 : Pune manual 8-row seed drill that hand meters onion seed in funnel with tubes on raised bed

Fig. 4.74 : Tractor operated pneumatic vegetable seeder in operation for direct seeding of onion seeds at DOGR, Pune

Fig. 4.75 : Tractor operated pneumatic multi crop planter
Courtesy: Khedut Agro Engineering Pvt. Ltd., Rajkot (Gujarat)

Self-propelled single row vacuum planter

Manually operated single row vacuum precision planter has been developed in the country. This is very good machine for small plots, hilly areas and horticultural crops. It consists of 4 hp power unit, transmission box, row marker, seed disc, funnel, seed hopper capacity 5 litres, adjustable coulter, handle, all controls, two drive wheels and one support wheel (Fig.

4.76). The machine is mainly used for planting maize, soybean, sunflowers, rapeseeds, beans, fennel, tomatoes etc. Vacuum pump activated by the engine generates vacuum inside the distributor which then after distribute seeds of various sizes to seed plate. Belt transmission allows adjusting the seed-to-seed distance.

Fig. 4.76 : Self-propelled single row vacuum planter

Courtesy: Falcon Garden Tools (P) Ltd. Ludhiana (Punjab)

Sugarcane planting system

Sugarcane is an important cash crop in India. It provides raw material for sugar, khandsari and jaggery industries (Srivastava, 2001). Its by-products and other items are used for production of alcohol, paper and many other products. Of the total production of sugarcane about 48% is utilised for sugar, 40% for jaggery and khandsari and about 12% for seed, feed and chewing purposes. As per the studies of the AICRP on Energy Requirement in Agriculture Sector, the energy requirement for cultivation of sugarcane crop varies from 70,000 to 100,000 MJ/ha in different areas of which 20,000-40,000 MJ/ha is operational energy and remaining for seed, fertilizer and plant protection chemicals (Singh and Singh, 2014). The distribution of the operational energy for different field operations are: tillage 18.3%, planting 20%, interculture 2.7%, irrigation 39.5%, harvesting 14.8%, others about 4.5%. Use of improved implement and machinery can save quite an appreciable percentage of these energies and also the energy requirement in terms of seed, fertilizer and plant protection chemicals by their judicious and proper applications.

The cost of production of sugarcane is going higher every year due to increase in labour wages and increase in prices of agricultural inputs. In order to increase the profits of the farmers it is necessary to economise on the cost of production by timely farm operations and use of efficient and high capacity sett cutting machines and planters. Sugarcane is propagated through its cuttings called setts. Its planting consists of two operationally different but sequentially connected operations of sett preparation and later, their placement in furrows. Both these operations are labour intensive and get delayed when there are shortages of labour at peak planting times. For both these operations machines have been developed. Method of sugarcane planting differs from place to place in two ways; on the method of planting setts in the soil, and on the spatial arrangement of one sett from the other. In the first case 5 methods are employed:

i) ***Flat planting*** - In this case shallow furrows of 10-12 cm are made, setts are planted and covered fully with the soil and the field becomes flat. This type of planting is common in lighter soils.

ii) ***Furrow Planting*** - In this method of planting ridges and furrows are made of 15-20 cm depth. After planting the setts in the furrows, they are covered with about 3-5 cm soil cover leaving the furrows partially open, which helps in economising in irrigation water. This type of planting is common all over the country.

iii) ***Trench Planting*** - This is similar to (ii) above except that the trenches are made 30 to 45 cm deep. It is not common in India.

iv) ***Pit Planting*** - In this case circular or square shaped pits of 45-60 cm diameter or length and 15-25 cm depth are dug at 90 or 100 cm apart from both side and in each pit 3-4 or even more setts are planted. Ten to twelve thousand pits are required to be dug per ha. This method is also not common in India.

v) ***Wet planting in furrows*** - This is similar to (ii) above. Furrows and ridges are made using tractor/bullock operated ridgers. Irrigation water is allowed in the furrows and setts are dropped in the wet soil and pressed in the mud.

In the second case the setts are planted end to end, 25% overlap, 50% overlap and in some cases 100% overlap. Normally 30,000 setts (3 budded) are used in one ha. The seed rate varies from 5-7 tonnes/ha. Sugarcane has physiological phenomena of top dominance i.e. the top buds which are comparatively young, succulent and immature germinate first and try to suck food from the whole cane, if it is not cut into pieces. This adversely effects the germination of lower buds. This phenomenon is called "Top dominance". To overcome this problem the cane is cut into setts. Three budded setts are recommended for planting but at many places 2 budded and single budded setts are also used.

Sett cutting machine

About 30,000 three budded setts/ha are required for planting which require about 15-20 man days to prepare setts. A sett cutting machine has been developed at IISR, Lucknow and PAU Ludhiana for cutting setts which can be operated either by a 5-7 hp engine or even by a tractor on low throttle (Fig. 4.77). With a team of 4 persons -2 for cutting the canes and 2 for help in lifting the canes, an output of approximately 12-13 thousand setts/h are obtained (Srivastava and Sharma, 1977; Shukla *et al.,* 1978). The machine consists of two circular saws mounted on a platform. Suitable guards are provided over the blades for safety. The canes are fed in bundles (5-6 at a time). Cut setts are dropped into a fungicidal tank from where they are removed and planted. It saves 55 per cent labour and operating time and 25 per cent on cost of operation compared to conventional method of manual cutting by hand tools.

Fig. 4.77 : Tractor or engine operated set cutting machine

Sugarcane planters

Sugarcane planting requires 6 operations to be done simultaneously namely opening of furrows, dropping of setts in the furrows, applying fertilizer, applying fungicidal/insecticidal solution on the setts and soil surface, covering of setts and slightly pressing of the soil cover (Shukla *et al.*, 1978 & 1984; Singh and Sharma, 2008; Ravinndra *et al.*, 2005). Machines have been developed to do all these six operations simultaneously in one pass. The sugarcane planters can be classified into two broad categories viz. Drop planter and Cutter planter. In the first category cut setts are planted in the furrows while in the second category the setts are also cut in the same machine. Drop planters are more common as it provides an opportunity of selection of individual setts and requires comparatively less power. Cutter planters are heavy but reduce one separate operation of sett preparation. In both types of machines semi-automatic and automatic versions are available.

Different designs of sugarcane cutter planters are available in the country. Overall effect of mechanical planting have been found encouraging because sugarcane sett moisture is not lost with the use of planters rather germination is better in case of mechanical planting. Human drudgery is reduced. Desired seed and fertilizer and other chemicals are placed at the desired depth and quantity. Different types of sugarcane planters are manufactured and used in the country and some of the discussed here.

Semi-automatic animal drawn sugarcane planter: Bullock-drawn sugarcane planter consists of a frame mounted on three wheels (Fig. 4.78). The front wheel pivots for turning of machine. A wooden box for carrying the seed cane setts is placed on each side of the operator. This arrangement enables the operator to use both hands for picking the setts and dropping them in furrow. The fertiliser box, tunnel shaped chute and chemical tank are mounted on the rear frame that is hinged on the front of the machine. The furrow opener is of hoe type with sweeping type wings, which makes 15-cm deep furrow. Side cover plates are provided to prevent the collapse of soil over the cane set and allows the application of fertiliser in bands or both side and the spraying of a chemical solution over the setts. The covering device covers the cane with soil and levels the surface. It is suitable for planting straight cane setts and simultaneous application of granular fertilizer. The equipment is operated by a pair of bullock. It had separate provision for placing the cut setts and fertiliser. Metering mechanism gets its drive from the ground wheel and a set of chain and sprockets. Furrow opening, placement of setts and covering of the setts is accomplished in a single pass. It saves 60 per cent labour, 65 per cent time of operation and 55 per cent on cost of operation compared to conventional method of opening of furrow by tractor drawn ridger or two passes of bullock drawn country plough, manual sett cutting, dropping of setts, manual application of fertilizer and insecticides and sett covering by country plough and planking. It also results in 10 per cent increase in yield. It is a single row machine. It has belt conveyor type sett metering and fluted roller type for fertilizer metering. Working capacity of sugarcane planter is 0.1 ha/h at row to row spacing of 90 cm. Three persons are required to operate the machine - one for guiding the bullocks, another for dropping setts and third for guiding the implement.

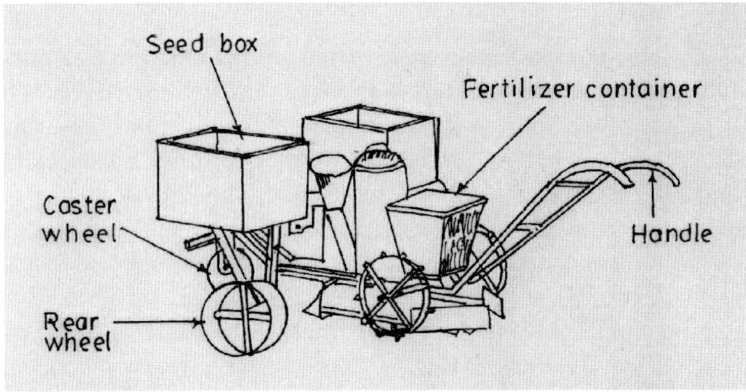

Fig. 4.78 : Details of animal drawn sugarcane planter

Drop type tractor operated sugarcane planter

In drop planters, pre-cut sugarcane setts of desired length are fed into the machine. The Indian Institute of Sugarcane Research, Lucknow has done pioneering work in developing drop type sugarcane planters. Drop type planters are more suited to Indian conditions. These planters are being manufactured commercially and are being used by the farmers. One of the problems with drop type planter is the difficulty in achieving a uniform sett placement. To overcome this problem, a two-row rota-drum type semiautomatic sugarcane planter has been developed at Punjab Agricultural University, Ludhiana (Shukla *et al.* 1984). It consists of a frame, two furrow openers, two seed chutes, two vertical rotating drums with 12 circular vertical compartments in each drum, seed hopper, fertiliser box, operator's seat and chain and sprocket drive from ground wheel (Fig. 4.79). The main frame is made of channel iron. The diameter of the ground wheel is 75 cm. Length and diameter of the vertical rotating drum is 30 and 29 cm, respectively. The ground wheel through chain and sprocket powers the rotating drum and fertiliser metering units. For end to end placement, speed ratio between ground wheel and central shaft is 1:0.71 and for about 20 per cent overlapping 1:0.88. The bevel gears are used for driving the drums. The speed ratio between central shaft and rotating drums is 1:0.61. The fertiliser is covered by a small covering attachment. The sugarcane setts are covered using tines followed by compacting rollers. The furrow openers are standard ridgers same as used in potato planters. The depth of furrow as well as covering is adjustable. Markers are provided on the machine to guide the tractor operator in maintaining uniform row to row spacing. Fertiliser cut off mechanism has also been provided.

The tractor-mounted planter is hitched with 3-point linkage of tractor (Fig. 4.80). It has mouldboard type bottom for making deep furrows in the soil. On the top of the frame, a wide hopper is mounted to keep the sugarcane setts. The tool frame extends to the rear to carry fertiliser boxes, chemical tanks and soil covering and compacting device. The fertiliser cover serves as seat for two workers to meter sugarcane sets manually. The sets are dropped through chute into furrows. The fertiliser is applied in bands and chemical is sprinkled over the cane. The setts are then covered with soil that is compacted lightly with a press wheel. Sugarcane setts are placed end to end with an overlap of 10-20%. In order to achieve proper placement of setts in the furrow, skilled workers are seated for metering purposes. A 90 cm row spacing planter has field capacity of 2-3 ha/day and requires 25 man-h at a forward speed of 2-3 km/h.

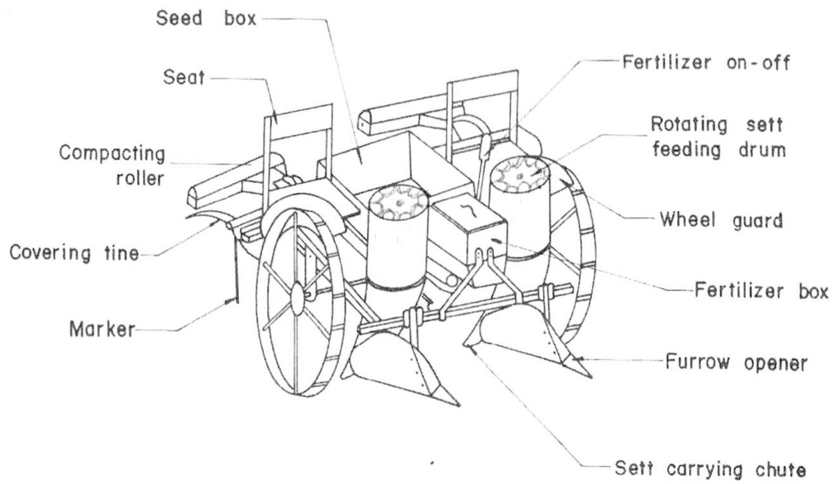

a) An isometric view of semi-automatic sugarcane planter

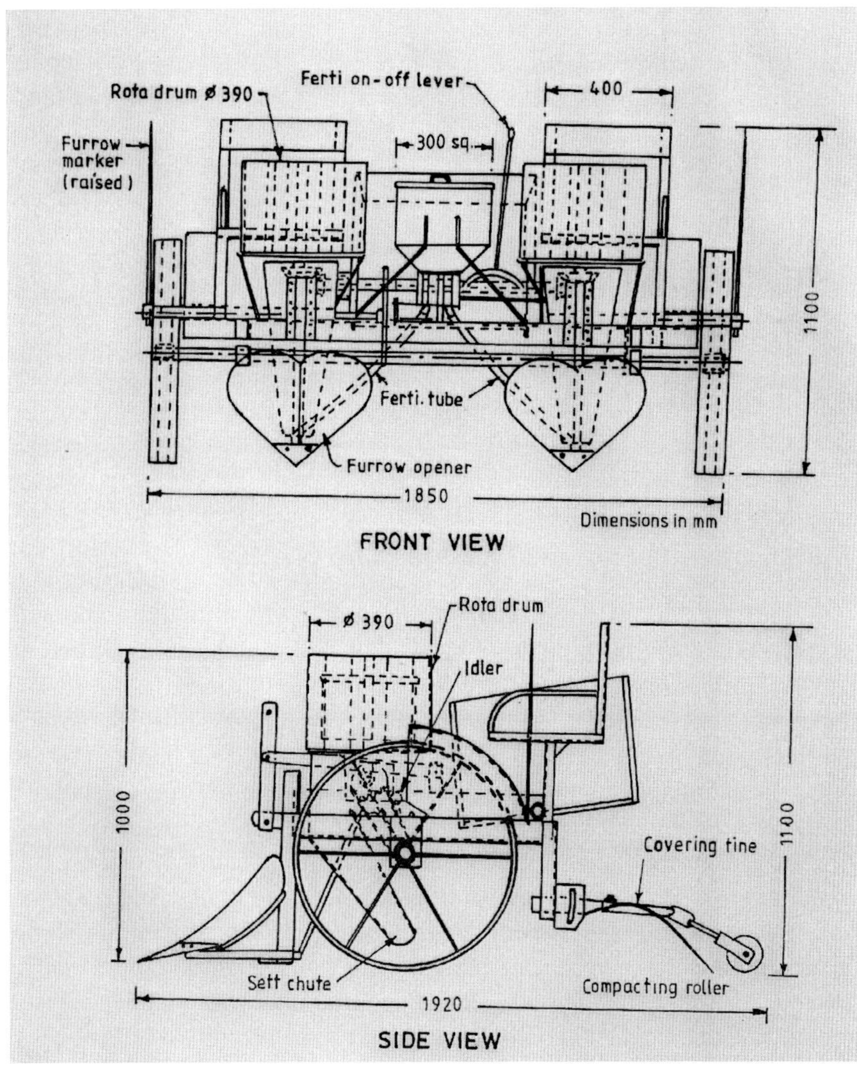

Front and side views of rota-drum type sugarcane planter

Fig. 4.79 : Tractor operated semi-automatic rota-drum type sugarcane planter

Fig. 4.80 : A view of semi-automatic sugarcane planter during field operation

Tractor operated automatic sugarcane planter

This machine is similar to the drop planter with only one additional feature that in this machine the whole cane is fed, which is cut into pieces and get dropped into the furrows. Other operations like opening of furrows, applying of chemical and fertilizer, covering and tamping of soil cover are similar to that of drop planter (Singh, 2014; Yadav, 2003 & 2004). Tractor operated automatic sugarcane planter cuts the canes to a predetermined length and carries it to the furrow. It is also of two types viz. ridger type and disc type. Generally two row machines are commercially available. The spacing between the rows can be adjusted from 60-90 cm. The machine has two sets of cutting units, one for each row, consisting of pair of cutting blades mounted on a vertical shaft. PTO drives the shaft and these blades rotate horizontally and cut the whole cane into predetermined length and drop into the furrows (Fig. 4.81). The distance between two setts of cane depends on the forward speed of machine. It can cover about 0.8 ha/day when operated at a speed of 1.5 km/h. About three persons are required beside one to operate the machine; two for feeding the cane and one for filling the box (Fig. 4.82).

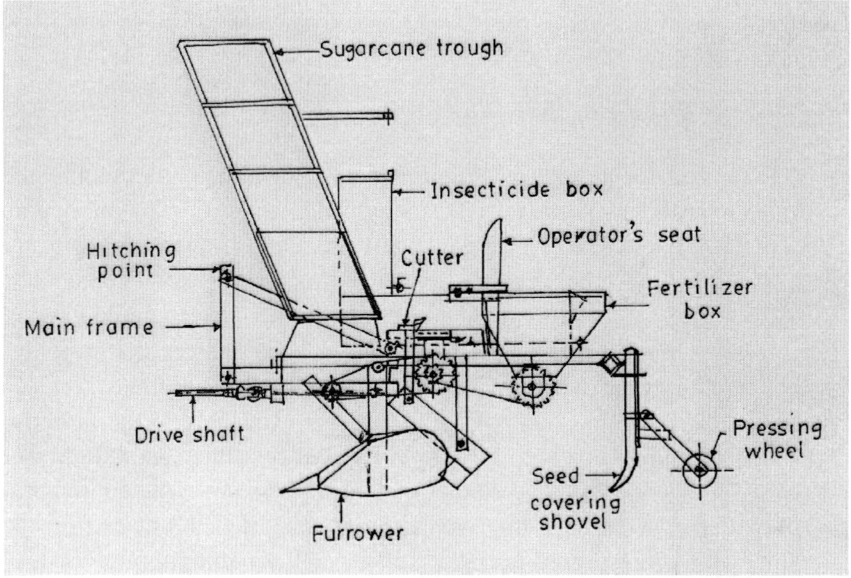

Fig. 4.81 : An isometric view of sugarcane cutter planter

Fig. 4.82 : Tractor operated whole cane planter working in field

Wet Planting

In southern Indian conditions wet planting of sugarcane setts is in vogue where setts are planted in furrows in wet soil. Under such situations these machines cannot be used. However several trials have shown that if the planting is done in dry or moist soil and then the field is irrigated, same yields are obtained. In this process the above planters can be used in those situations also.

Tractor operated twin-auger-digger

It is used for sowing of sugarcane setts in pits. The equipment is operated by a 30-35 kW tractor PTO (Singh and Pandey, 2008; Anonymous 2008 & 2010). A twin auger digger sugarcane planter can dig two pits at a time. The machine consists of a heavy duty frame made from a channel of size 125 x 62.5 x 4.5 mm on which two rotary units having double helical auger blade (EN-32) has been fixed in opposite direction around a high pressure circular pipe of 98 mm diameter (Fig. 4.83). A set of three triangular shaped blades made of EN-45 of size 105 x 75 x 11mm is fitted at the end of each helical auger for digging of soil. The helical auger is made from a sheet of 5 mm thickness. The diameter of auger is 685 mm and depth 28-35 cm. The distance between the two augers is 1.22 m. The power from the tractor PTO is transmitted to the auger through a reduction gear of ratio 22:12 and from gear box to augers through V-belt (C-87) and pulley consisting of three grooves. Working capacity of tractor operated twin-auger-digger is 0.020 - 0.025 ha/h.

Fig. 4.83 : Tractor operated twin-auger-digger for sugarcane planting

Tractor operated multipurpose implement for sugarcane

Sugarcane planting is very labour intensive job and involves considerable human energy. Planting creates the foundation for a crop and plays an important role in its growth and yield. Being costly machine the need was felt to develop multipurpose implement, which could facilitate interculture and earthing up operations in addition to planting of sugarcane setts. The equipment with all the three attachments has been developed at IISR Lucknow (Singh, 2014). It reduces human drudgery and provides adequate time for subsequent field operations. The machine consists of a frame, feeding chute, coulter, ridger, seed box for sugarcane, seat for the operator, covering and packing device (Fig. 4.84). It also has a provision for attaching cultivator, seed drill and puddler. In three-row machine the operator lifts the cane from seed tray and puts the same in the slanting chute. Cane slides down and a cutter cuts the cane into setts automatically. When cane comes to end another cane is placed in the chute. In this way, there is some free time to operator. The three persons feeding cane in to the cane cutting units feel comfortable and appreciates the new mechanism. The equipment in interculture mode has six standard tynes with reversible shovels that provides soil cover in three rows of cane placed in opened furrows. Three extra tynes with shovels facilitates interculture operation in three inter rows of cane in single pass. The earthing up attachment gives field capacity of 0.81 ha/h with 3.5% cane damage.

This machine can cut whole sugarcane into pieces of uniform size, place in the furrow prepared by ridger, setts are covered by soil and then soil is pressed in one operation (Fig. 4.84). Fertilizers and chemical can also be applied. The field capacity of machine is 0.20 ha/h. For land preparation the equipment can be used as tractor operated nine-tine cultivator. For interculture the equipment can be used for intercultural operation in sugarcane field. The field capacity in interculture mode is 0.72 ha/h. Three inter row spaces are intercultured in a single pass. For earthing up, three-rows of cane can be covered by this equipment with a field capacity of 0.66 ha/h. For puddling, beater type sub-unit is mounted with the main frame of the equipment having the field capacity of 0.35 ha/h.

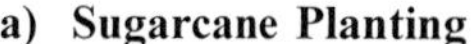

a) Sugarcane Planting

b) Interculture

c) Seed drill mode

d) Puddling

Fig. 4.84 : Tractor operated multipurpose implement for sugarcane in operation

Tractor operated trench planter

Tractor operated trench planter has been developed at IISR Lucknow during 2012-13 for paired row planting of sugarcane at a spacing of 30 cm (Singh, 2014). With the help of this equipment all the unit operations involved in cane planting viz., sett cutting, furrow opening, placement of setts into opened furrows, application of fertilizer and insecticide solution and covering of soil over the seed-setts are performed simultaneously in a single pass of the tractor (Fig. 4.85). An attachment has also been designed, developed with this planter for facilitating laying of sub-surface drip laterals underneath the centre of planted paired furrows. The equipment is able to lay the drip laterals at 7.5-10 cm underneath the centre of the planted furrows along with the paired row planting of sugarcane.

Fig. 4.85 : Tractor operated trench planter with sub-surface drip laterals laying attachment

Single row sugarcane cutter planter

This unit has provision to plant sugarcane in a single row (Yadav, 2003 & 2004). The planter can be operated with the help of 25 hp tractor. The cutting mechanism of the machine is operated by ground wheel (Fig. 4.86). This planter unit has been designed and developed by Vasantdada Sugar Institute (VSI) Pune under NATP on Sugarcane Mechanization.

Fig. 4.86 : VSI-NATP single row sugarcane cutter (ground wheel driven) planter

Two row adjustable tractor PTO driven sugarcane cutter planter

The VSI Pune under the NATP project developed a two row tractor mounted type machine with an adjustment of 150, 120 and 90 cm row-to-row spacing (Singh, 2014; Yadav, 2003 & 2004). The planter is operated by 45 hp tractor and above. In order to achieve uniform length and better cutting of the setts the rotary cutting along with the cylindrical feeding (positive feeding) mechanism is provided (Fig. 4.87). The placement of sugarcane setts is adjustable as per the requirement i.e. end to end, overlapping and maintaining suitable spacing between the setts.

Fig. 4.87 : Two row adjustable tractor PTO driven sugarcane cutter planter

Tractor operated ridger type sugarcane cutter planter

This machine performs operations such as sett cutting, furrow opening, placement of seed setts, fertilizer and liquid chemical dispensing, soil covering over setts and tamping of soil in a single pass (Singh, 2014; Yadav, 2003 & 2004). Ridgers are used to open the furrows (Fig. 4.88). The machine plants two rows of furrows at a row spacing of 75/90 cm. It is operated by a 35 hp tractor or above. The machine can cover 0.20 ha/h.

Fig. 4.88: Tractor operated ridger type sugarcane cutter planter

Tractor operated 3-row sugarcane cutter planter

This machine performs all operations involved in sugarcane planting in a single pass (Singh, 2014). It has silt type furrow openers, suitable for flat planting of cane (Fig. 4.89). The field capacity of this machine is 0.25-0.30 ha/h. With the use of this machine there is saving of 65% in cost of operation as compared to conventional method.

Fig. 4.89 : Tractor operated 3-row sugarcane cutter planter

Tractor operated raised bed seeder-cum-sugarcane planter

This machine developed by IISR Lucknow performs planting of two rows of sugarcane in furrows and sowing of two rows of seeds of companion crop like wheat, pulses etc on the raised bed and fertilizer dispensing operations simultaneously in a single pass (Fig. 4.90), Singh (2014). The capacity of this machine is 0.20-0.25 ha/h.

Fig. 4.90 : Tractor operated raised bed seeder-cum-sugarcane planter

4.3 Vegetable planting and transplanting machinery

Vegetable planting and transplanting is very labour intensive activity. It requires 15-20 man-days per ha for transplanting (Singh, 2001). The mechanization of vegetable cultivation has been considered as one solution to these problems. Up to now, several kinds of machine and equipment have been developed. Consequently, most of the work is now partly mechanized, and the number of required work-hours has decreased. However, there is still a need for more mechanization. Most growers desire more labour-saving machines and facilities. Vegetable planters and transplanters are thus very important if labour requirement is to be reduced.

India has the distinction of being one of the prominent growers of fruits and vegetables in the world. It ranks second in the world in the field of vegetable production. There is wide variety of vegetable crops grown in India. Some common types of vegetables are:

i) Root, bulb and tuber vegetables, such as potato, onion, radish, turnip, carrot, sweet potato, rabid (asparagus) etc.

ii) Fruit and pod vegetables, such as tomato, brianna (egg plant), green peppers (capsicum), beans (flash beans, cluster beans), peas, jack fruit, orate etc.

iii) Vine vegetables, such as, gourds (bitter gourd, bottle gourd, snake gourd, etc.), cucumber, pumpkin, portal, squash melon, summer squash etc.

iv) Leaf and stem vegetables, such as, spinach, cabbage, lettuce, celery, sweet corn, kolkhoz etc.

There is a wide disparity among the seed size, seed rate and agronomic practices, apart from the soil and climatic conditions best suited for optimum growth and yield. Table 4.3 shows the propagating methods for different vegetables. The crop is raised either by planting the seeds/tubers or by transplanting the seedlings. In advanced countries, for most of the vegetables, appropriate equipment and practices have been developed. However, in India, with the exception of few vegetables such as potato, the level of mechanization for most other vegetables is low or almost negligible. This is true for both planted and transplanted crops. With the exception of potato, for almost all other crops, the planting of the seeds, seedlings and the stems and roots is carried out manually. Planting/ transplanting is considered as a fairly labour intensive, time consuming and expensive operation. According to a survey conducted by Chaudhuri *et al.* (1999), the labour requirements for vegetable seedling transplanting range between 15 and 100 man – days/ ha. In order to reduce the cost of production and to enhance the quality of produce, it is necessary that the production of vegetable crops is fast mechanized. In India planting/transplanting and harvesting are the two most labour intensive operations in the production of vegetable crops. Interculture and plant production are also labour and time consuming.

Table 4.3 : Common planting practices and recommended equipment for selected vegetables

Vegetable	Equipment recommended for Planting
A. Vegetables produced from seeds	
Radish	Mechanical/pneumatic planter/precision drill
Turnip	-do-
Carrot	-do-
Tomato	Seed planter & transplanter
Brenda (Egg plant)	-do-
Green pepper	-do-
Peas	Seed planter
Beans	-do-
Okra	-do-
Gourds	-do-
Cucumber	Seedling Transplanter
Pumpkin	-do-
Summer squash	-do-
Squash melon	Seed planter
Spinach	Broadcaster
Celery	Broadcaster
Lettuce	Broadcaster
B. Vegetables produced from seedlings	
Onion	Seedling Transplanter
Tomato	-do-
Brinjal	-do-
Green Peppers	-do-
Cabbage	-do-
Cauliflower	-do-
Knoll-khol	-do-
C. Vegetables produced from tubers, stems etc.	
Potato	Planter
Sweet Potato	Transplanter

Types of planting and transplanting equipment used for various vegetables

The common types of sowing and planting/transplanting equipment can be divided in two categories viz. hand planters/dibblers and power operated vegetable planters and transplanters. There are large number of hand planters and dibblers. Dibblers and hand planters punch a hole in the ground to the required depth and the seeds are dropped by hand or by a mechanical seed-metering device, which gets activated with the tilting of the handle of the dibbler/hand planter. Since these devices are operated in a standing posture, they reduce the drudgery and increase the work output. Simplicity of construction and low cost is their main advantage. Vegetable planters work on same principles as the grain planters. They have seed and fertilizer hopper(s), cell-type metering device (horizontal, inclined or vertical plates/rotors), seed delivery tubes, furrow openers (runner, shoe, hoe or rolling coulter), seed covering and compacting device (as discussed in Chapters 4.1 and 4.2). The seed metering plates have well designed cells conforming to the shape and size of the seeds. The planters have requisite arrangement for adjusting the row-to-row and seed to seed spacing as well the depth of planting. Furthermore the seed planter design should be such that it facilitates planting on flat surface, furrows, ridges or beds. Seed to seed spacing is altered by selecting the seed plate with requisite number of cells as well as with the help of distance changing sprockets.

Pneumatic planters are now widely used for planting the seeds of vegetable crops. By selecting the right type of seed plate, these can plant with utmost precision, from fairly small sized seeds as that of reddish and turnip, to fairly large sized seeds as that of peas and beans. The basic principle of a vacuum type vegetable planter is that seeds are placed in a hopper and a revolving vertical seed plate with holes of desired size (depending upon size of seed) is held against a vacuum chamber. Due to vacuum in the chamber the holes/apertures on the seed plate suck a single seed from the seed hopper and hold it till the seed holding portion of the revolving vertical seed plate enters into the zone with atmospheric pressure. This leads to the release of the seeds into the seed tube and to furrow. Fig. 4.91 shows a cut-section of the vacuum type pneumatic vegetable planter.

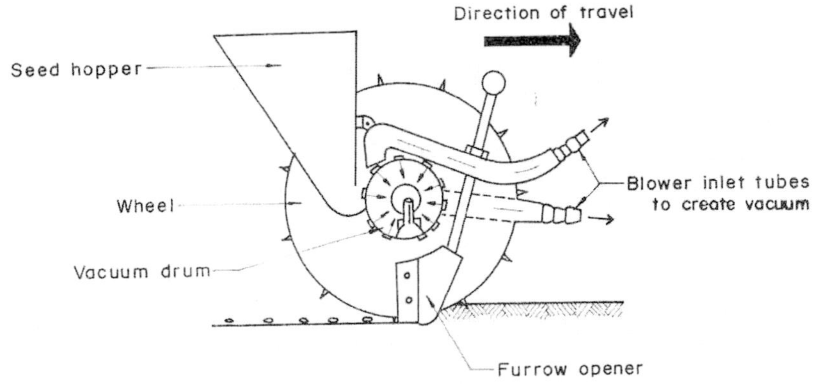

Fig. 4.91 : Vacuum type single seed planter

Design considerations for vegetable planters

The main considerations in the design of a vegetable planter are:

i) The planter should be able to plant the seeds in the desired manner by maintaining the desired seed rate, specified depth, seed and row spacing. It is to be noticed that vegetables have a wide variation in the seed size, inter and intra-row variations, agronomic practices of planting on flat land, beds, ridges or in the furrows (lister planting) at recommended depths for optimum seed germination, plant growth and yield.

ii) Keeping in mind the aforementioned factors and variation in the soil conditions, the size of the machine ought to be decided keeping in view the power source available, viz. manual, animal, tractor, power tiller or self-propelled. For better manoeuvrability, if one is to design a fully mounted machine, its weight must be considered against the given hydraulic lift capacity of the tractor. For fully mounted machines, the criterion used in the design is to assume 3 to 8 hp per row. The draft requirement varies between 40 and 60 kg_f/row.

iii) A vegetable planter might use a mechanical cell type or a pneumatic (pressurized or vacuum type) seed metering system. For cell type-metering system, the cell size must be carefully designed to conform to the seed size and shape. As a matter of fact, the cell size is kept slightly larger (about 10 per cent more) than the seed size. Similarly, for vacuum type pneumatic planter, the seed plate is designed with holes of requisite size keeping in view the seed size.

iv) To attain the desired seed rate, there should be provision in the planter to select a seed plate with desired number of seed cells as well as distance changing sprockets, which could be fixed easily and quickly.

v) The seed metering system should be able to perform within a reasonable range of change in the forward speeds without inter and intra row variation in seed rate not exceeding ± 5%.

vi) A vegetable planter is designed to work as a modular or unit planter. The units are mounted on a tool bar with the provision to alter the row-to-row spacing and plant population as recommended for a crop.

vii) A vegetable planter is designed to plant a given number of rows. The number might vary from one to twelve rows with a complete planter.

viii) A vegetable planter must be able to open the soil, meter the seed, place the seed at the right depth in a right manner, cover the seed and firms up the soil around the seed.

ix) An optimum type of furrow opener should be selected for a vegetable planter. The most common type of opener used on the vegetable planters in overseas countries is the runner type opener. The planting depth is one of the most critical factors due to inability of vegetable seeds to germinate when planted too deep. The runner type opener can be adjusted up or down with the help of depth bands to control the planting depth.

x) Due to relatively smaller size of the vegetable seeds, a relatively smaller seed hopper is provided in the vegetable planters.

xi) The seed plate is one of the critical components of the planting mechanism. Seed plates for vegetable planters are designed for different seed sizes and shapes of seed cells to accommodate the wide variety of sizes and shapes of vegetable seeds.

xii) A vegetable seed planter is designed with the provision to change the seed spacing by choosing the driving and driven sprockets with right number of teeth and the seed plates with given number of cells. The driving sprocket is provided on the ground wheel and the driven sprocket on the seed hopper shaft.

xiii) For 90 to 110 per cent seed cell fill, the speed of seed plates of the vegetable seed planter should be about 15 m/min.

xiv) The planter should be a multipurpose machine and should enable the user to plant different types of vegetable seeds in various types of field and soil conditions.

xv) The percent errors in planting by way of skips (missing) and doubles should not exceed 5%.

xvi) The planter should be a multi-row machine to make full use of the available power source.

xvii) The planter must have provision for easy adjustment and calibration.

xviii) The design should be simple and it should be possible to manufacture it with available materials and manufacturing processes.

Requirements for precision planting with cell type metering

The following parameters are critical for proper functioning and design of cell-type metering devices for vegetable planting:

i) seeds of uniform size and shape,

ii) provision of proper size of cells in the seed plates,

iii) high degree of precision and stringent tolerances in the construction of seed plates and other parts of the seed metering system,

iv) required plate speed and adequate exposure time for the cells to get filled with the seeds to avoid misses,

v) efficient cut-off devices to avoid multiple cell-fills and seed damage,

vi) positive unloading of seeds from the cells by efficient "Knock out" mechanism,

vii) proper mechanism to eliminate seed damage during metering to ensure high seed germination,

viii) smooth conveying of the seed from the metering system to the bottom of the furrow to obtain desired seed spacing, and

ix) proper placement of seeds at desired depth in the furrow.

There are many factors influencing cell filling such as seed size conforming to cell size, seed size range, shape of seeds, shape of cells, exposure time of cells for filling and linear speed of cells. These factors have a bearing on the cell filling efficiency and these must be kept in view while designing a planter. It has been observed that at a plate speed of about 15 m/min, nearly 100% cell fill could be achieved. It has been found that the cell diameter should be 10% larger than the maximum seed diameter and the depth of the cell should be equal the average seed diameter. In order to get 100% average cell fill, optimum combination of seed size, cell size and metering plate speed are necessary. Percent cell fill can be calculated as

$$\text{Percent cell fill} = \frac{\text{Total no. of seeds discharged}}{\text{Total no. of cells passing the discharge point}} \times 100$$

To minimize drop time variations, there should be direct discharge of the seed from the metering device to the furrow bottom through short, smooth, small diameter delivery tubes with discharge end close to the furrow bottom.

Potato planters

Potato is an important cash crop, which can be cultivated in wide variety of soils and weather conditions (Verma *et al.*, 1992; Anonymous, 2010c, 2012). One of the main bottlenecks in increasing the area under potato cultivation has been its high labour requirement for planting, earthing and inter-culture (Anonymous, 2012). Potato crop yield also responds to improved soil aeration and deep tillage. Potato planting is done in rows of 50-60 cm with 15-20 cm plant-to-plant spacing. Potato planter performs the function of furrow opening, seed metering, seed placement at proper depth and forming of the ridges to cover the seed tubers. It is observed that manual planting and earthing of potato require 300-350 man-h/ha. Non-availability of adequate labour during the planting season in winter when labour work is in one shift only necessitates the mechanization of potato planting. As a result, large numbers of semi-automatic, automatic two, three or four row potato planters and diggers have been developed and are being used by the farmers.

There are different types of metering devices used to regulate the potato seeds such as belt cup type, magazine type and picker wheel type. Magazine type metering device is rotated in the horizontal plane in an open housing. Operator sitting in front of rotating magazine keeps on filling the seed from trough. Rotating magazine carries the seeds to seed tube, which in turn places it in the furrow. Seed spacing within the row is maintained by maintaining proper speed of magazine. Belt cup type metering mechanism is used in semi-automatic potato planters. The cups are fastened to an endless belt. Belt moves horizontally and passes through hopper, picks up seed in cups as it moves and carries it to the seed tube. During this process some of the cups remain unfilled, which are filled by the operator.

Design considerations for potato planters

The design considerations for a potato planter can be enumerated as follows:

i) The first and foremost consideration is the type of potato planter to be designed. One must be clear about the power source viz. animal power, tractor, power-tiller or self-propelled. Animal drawn and power-tiller driven machines are usually single-row machines. However, tractor operated/self-propelled potato planter are usually of 2, 3 or 4-row-type suited to power ranging from 35 to 75 hp or even more.

ii) A tractor operated planter can be fully-mounted, semi-mounted or trailed type. Mounted units have better manoeuvrability resulting in higher field efficiency and are easy to transport.

iii) The next design consideration for a potato planter is the seed type and size. In India, all potato planters use whole tubers as seed.

iv) Keeping in view the horse power, draft, hydraulic lift capacity and longitudinal stability of the tractor, one should design a 2-row machine for a 35-40 hp tractor, 3-row machine for a 45-50 hp tractor and 4-row machine for tractor exceeding 55 hp.

v) The planter should be suitable, for the recommended agronomic practices in terms of row and tuber spacing, depth of planting as well as planting on ridges or beds. For a potato planter suited to

Indian conditions a minimum row spacing of 60 cm is required to permit the use of the tractor. However, recommended row spacing varies between 60 and 90 cm. Tuber spacing of 20 to 40 cm is common. The planting depth of tuber varies from 5 – 10 cm and should be uniform.

vi) The planter could be semi-automatic or automatic type. In a semi-automatic machine, the tubers are manually fed or assisted to be fed into the metering device. On the other hand, in automatic planters graded seed is used and the feeding and metering are automatic and fully mechanized. Depending upon the initial cost, labour availability, complexity of the device, its maintenance and repair cost, one can decide about the feeding and metering system to be used in the planter to be designed.

vii) The metering device should permit the planting of the tubers at the desired seed spacing (25-50 cm) and have the provision to alter it easily either by changing the speed of the ground wheel or that of the metering elements. Variable diameter ground wheels are usually provided.

viii) The planter should maintain the seed metering accuracy at different forward speeds. The intra-row and inter-row seed variation should not exceed ± 5%.

ix) No internal or external injury be caused to the tubers during metering, conveying and laying of the tubers in the furrows. The damage in no case should exceed 3%.

x) The planter must have a well-designed furrow and ridge/bed forming system to cover the tubers and compact the ridges/beds.

xi) In a semi-automatic machine, there should be provision for ergonomically well-designed operator seats to minimize the physical and mental fatigue to the operators.

xii) In semi-automatic machines, to reduce the mental fatigue due to visual acuity, the speed of the rotary magazine should not exceed 10 rpm.

xiii) The planter should be versatile enough to work in different types of field and soil conditions and be able to plant different types of seeds as far as possible.

xiv) The weight of the planter and its rearward overhang in mounted machines should be well within the permissible hydraulic lift capacity of the tractor.

xv) The planter must be simple in design, construction and fabrication and utilize the standard graded material for its various components.

xvi) Once the number of rows is decided, the effective field capacity of the planter would depend upon the forward speed of the planter. A machine which can accurately feed and meter the tubers at higher forward speeds with due accuracy and uniformity should be preferred.

xvii) An optimum design should be the best compromise between metering accuracy, versatility, simplicity, low repair and maintenance and initial cost.

xviii) The material of construction for various components must be in conformity with the recommendations laid down in national/international codes.

xix) The height of drop of the tubers into the furrow must be minimum to prevent spreading and rolling of the tubers and to minimize irregularity in seed spacing.

Semi automatic potato planter

Two types of tractor operated semi-automatic potato planters have been developed at Punjab Agricultural University Ludhiana, viz. Belt and cup type (Fig. 4.92) and Revolving magazine type (Fig. 4.93); Anonymous (2012) and Garg and Singh (2002). In the first type the seed tubers are placed in a hopper with its two sides slanting. It consists of a frame, furrow openers, seed box, two belt-conveyers with cups, seats, seed tubes and a ground wheel for transmitting the power to the shaft (Fig. 4.92a). The planter utilizes two endless canvas belts with cups riveted onto them. Each cup picks up a tuber. Two persons, one for each row, sit on the planter and observe the seed metering belts to ensure that each cup contains a tuber (Fig. 4.92b). if not; then correction is made by removing/placing one of the tuber(s). Field capacity of the machine is 0.12-0.15 ha/h. The second type of potato planter uses a revolving magazine having 10-12 compartments. It consists of a frame, furrow openers, seed box, two revolving magazines, seats, seed tubes and a ground wheel for transmitting the power to the shaft (Fig. 4.93a). Depending upon the number of rows (2/4) of the planter, one person is employed for each row to place (feed) the tubers into the seed compartments of each revolving magazine. A stationary plate is provided under the magazine with a slot directly over the delivery chute to drop the tuber. The output capacity of a two-row planter with revolving magazine varies from 0.12-0.15 ha/h. Machine is operated by a 35 hp tractor (Fig. 4.93b). Two persons sitting on the machine fill the seeding cups by picking tubers from the hopper. In rotary magazine type planter, cups are filled manually with any size of tubers. However, in case of belt-conveyer type machine graded tubers are used and 70-80% of cups are filled automatically and remaining manually. Damage to the potato tubers is negligible. Tuber missing is about one per cent. Row to row and plant to plant spacing can be adjusted. It can save 40-50% labour requirement and 15-20% cost of operation in comparison to traditional method. It can plant up to 20 cm depth.

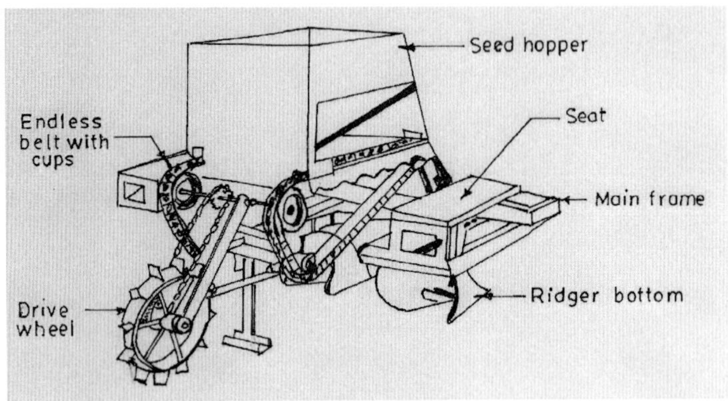

a) Details of belt and cup type potato planter

b) A view of endless belt and cups type semi-automatic potato planter in field operation

Fig. 4.92 : Tractor operated endless belt and cups type semi-automatic potato planter

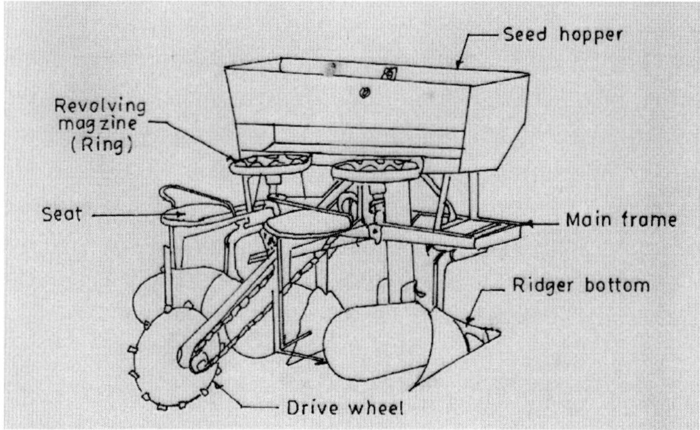

a) Details of revolving magazine type semi-automatic potato planter

b) A view of revolving magazine type semi-automatic potato planter in field operation

Fig. 4.93 : Tractor operated revolving magazine type semi-automatic potato planter

Tractor operated automatic potato planter

Tractor operated potato planters are used for sowing of potato crop. The equipment is operated by a 30 to 40 hp tractor and above (Anonymous, 2012). It consists of a hopper, two picker wheels for picking the tubers, seed tubes, furrow openers, three bottom ridger to form two ridges, a fertilizer metering system and a frame (Fig. 4.94a). Hopper is rectangular in shape at the top with sides slopping towards bottom. At the bottom of the hopper, agitators are provided to improve the delivery of potato tubers to feeder. The agitators and the screw conveyors ensure proper feed of the tubers to the picker unit. The picker units of the planter are set in motion from the ground wheel. The plant spacing can be varied by suitably modifying the gear ratio. The average speed of operation for automatic potato planter varies from 2.5 to 3.5 km/h and the field capacity between 0.35 to 0.37 ha/h (Fig. 4.94b). In automatic potato planter the labour requirement is around 8 man-h/ha. The planter has fertilizer distributors for the application of fertilizers during planting. Automatic potato planter saves about 70% of labour in comparison to semi-automatic potato planter. The machine is found to be highly satisfactory when uniform size of the potato tubers is used.

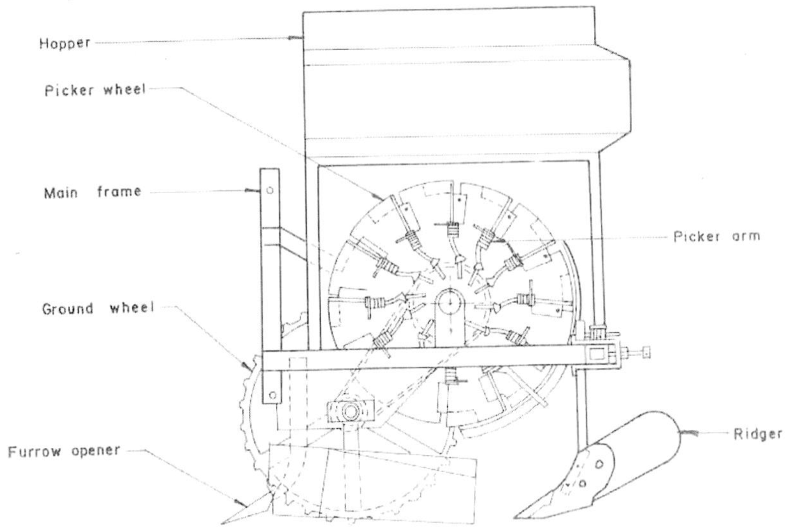

a) Details of automatic potato planter

b) Automatic potato planter in operation

Fig. 4.94 : Tractor operated automatic potato planter

Power tiller operated potato planter

It is used for sowing of potato. It is operated by power tiller (Fig. 4.95). A power tiller semi-automatic potato planter is suitable equipment for sowing potato, applying fertilizer and making ridges in single operation (Anonymous, 2012). A rotating device is used for placing the potato and metering rollers for applying fertilizer. Depth of operation is adjustable up to 80 mm by fixing furrow opener standard up & down through nuts and bolts. The power to metering mechanism is drawn from ground wheel through sprocket and chain arrangement. The machine is simple in construction, light in weight and easy to operate on terraces. Working capacity is 0.03 ha/h.

Fig. 4.95 : A view of power tiller operated potato planter

Tractor operated vertical belt paired row potato planter

The paired row planting of potato also helps in increasing the field capacity of the potato digger/harvester by uprooting two rows per bed in comparison to single row, single bed system (Anonymous, 2012, 2013). Tractor operated vertical belt paired row potato planter consists of two vertical rubber belts fitted with metal cups in paired rows for picking potato tubers, provision for fertilizer application and shovel type furrow openers (Fig. 4.96a). Arrangement of cups on the belt is in Zig-Zag manner. Number of cups per belt is 40. Planter is equipped with multispeed gears to adjust the seed spacing. Provision to adjust the row spacing from 609-711 mm and also to spread the manure on both the sides is also provided. The machine is suitable to be operated by 45 hp tractor and above (Fig. 4.96b). This has ability to make the bed with disc and mould board plough with quick and easy adjustments. Gear adjustment makes it possible to choose seeding distance up to 12 positions. Elevator belts have two lines of interposed/ parallel scoop cups and adjustable mechanical wheel vibrators behind them to guarantee a perfect picking. Reduction cups are provided with different sizes to suit potatoes below 30 mm. The machine has mould type/disc type bed covering system with adjustable skids. Centre to centre row widths is adjustable from 50 to 75 cm and are adjustable by repositioning the planting units and ridging bodies. This is available in 2, 3 and 4 rows. Seeding depth varies 4-5 cm. Planting capacity is 0.25 ha/h and fuel consumption 4.5-5.0 l/h (Fig. 4.96c).

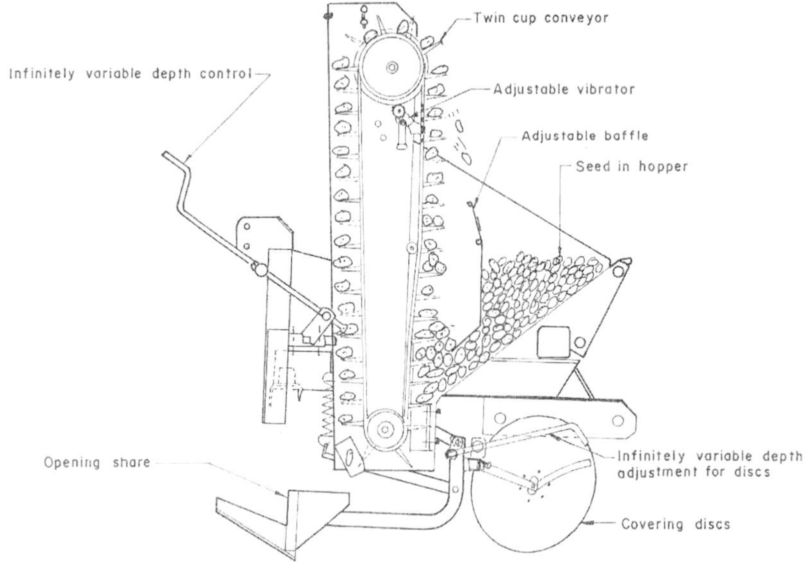

a) Details of vertical belt paired row potato planter

b) Stationary view of vertical belt paired row potato planter
Courtesy : Ganesh Agro Equipments, Vadpura (Gujarat)

c) A view of vertical belt paired row potato planter in operation

Fig. 4.96 : Tractor operated vertical belt paired row potato planter

Manual operated onion seeder

Generally, the onion seeds are sown in nursery and transplanted with row to spacing of 15 cm and plant to plant spacing of 7.5 cm to get optimum yield. During onion cultivation, transplanting of seedlings, weeding and harvesting are the most labour intensive operations that are presently done manually. The labour requirement in manual transplanting of onion seedlings is as high as 100 – 120 man- days/ha as 0.89 million seedlings per hectare are to be transplanted. Because of high requirement and shortage of labour, the area under onion cultivation is low and can be increased by mechanization of this crop. Onion can also be grown by direct seeding method which is most economical. The direct seeded onion crop sown in *kharif* season gives higher yield as compared to transplanted crop. Among various seed sowing methods, manual sowing and Pune seed drill (Fig. 4.73) produces higher yield. The seed of onion is of very small size, having low density and irregular shape which poses problem in precision planting. To overcome these difficulties 'Seed Pelleting Technique' is being used. Efforts have been made in developing drills/planters for small seeds such as onion, radish, turnip, carrot, etc which could have belt feed mechanism or precision seeder cell wheel or plate feed mechanism. A two row, manually operated small seed planter has been developed at PAU Ludhiana (Anonymous, 2010). A new seeding plate (inclined) having grooves according to the size and shape of onion seed has been developed (Fig. 4.97). Six rows having row to row spacing of 15 cm can be planted on the bed. The onion seeder can be operated at a forward speed of 1.2 to 1.5 km/h. The effective field capacity is observed as 0.04 ha/h and labour requirement 50 man-h/ha. The yield for direct sown onion is slightly less as compared to transplanted onion. The loss in yield is attributed to the weed competition being faced by the direct sown crop at the early growing stage.

Fig. 4.97 : Manual operated two row onion seeder

Vegetable transplanter

A vegetable transplanting machine must handle a seedling with or without soil, place it in a vertical position and press the soil around its roots and simultaneously release the seedling. In addition to above some transplanting machines also apply water to freshly transplanted seedling. Water is only sprinkled around the seedling by opening a valve when nozzle is above the just transplanted seedling. Vegetable transplanting machines may be broadly grouped into 'hand-fed', mechanical and automatic. In the first type, operator sits on seat provided on the machine and drops the plants by hand directly into a furrow of suitable depth and width. A hollow coulter generally makes the furrow. A pair of press wheels compresses the soil around the plants. Hand-fed transplanter is pulled very slowly (1-1.5 km/h) to allow operator to drop the seedlings at proper spacing. With mechanical transplanter, two or more people can work on each row. Operator places the seedlings in trays or between fingers on endless conveyor from where seedlings are picked up and transplanted mechanically. The conveyor takes the seedlings down to an open furrow, release them in furrow at correct position and place the soil around. With some mechanical transplanters a device for watering each plant just before soil is pressed around may be fitted for use in dry soil. Fully automatic transplanting generally adopts the methods of either 'bandolier' or 'chain-pot' systems specific to machine and crop. These systems require expansive resources for growing the seedlings in a nursery and handling them to transplanting machine. These systems are well suited for greenhouse enterprises where maximum output is desired from fixed cost overheads without excessive demand for costs such as human labour.

A two row vegetable transplanter has been developed by the PAU Ludhiana and TNAU Coimbatore for transplanting seedlings of brinjal, cauliflower, tomato, chillies etc (Anonymous, 2008 & 2010). To achieve vertical position of the seedling relative velocity of the seedling with ground should be zero. Rotating or moving the planting tool in direction opposite to advance of the machine obtains this. Peripheral speed of the planting disc should be equal to forward speed of the transplanter. Furrow openers in the transplanters generally open a near rectangular furrow that gives best results.

Transplanting permits a higher level of stabilization of vegetable harvests as compared to direct seeding. Using transplants shortens the time that crops are in the field. For second crops, transplanting may allow growers more time for land preparation. This may be very advantage where growers have a limited time to produce for a given market. Raising and transplanting of seedlings, however, requires much labour and cost. Several kinds of machines and equipment have thus been developed where labour availability is a problem. A simple transplanting aid, pulled through the field and used to carry the plant and workers, reduces labour requirements by about 20 per cent.

There are also various kinds of seedling, which can be used with a mechanical transplanter, including bare seedlings, and seedlings with soil, which are cell plug, paper pot and soil block seedlings. Recently, walking or riding type, semi or fully automatic mechanical transplanter, some equipped with devices for feeding the seedlings, are available (Singh, 2001). These machines enable high speed and efficiency in planting work.

Design considerations for vegetable transplanters: The following points are kept in mind before designing the vegetable transplanter:

i) Whether the machine is required to transplant bare-root seedlings or seedlings grown in soil-blocks or paper/pulp pots,

ii) Whether the transplanter is to be a "hand fed" or mechanically fed device. In the hand-fed machine, the operators sit on the machine and drop the plants by hand direct into a furrow of suitable depth and width made by a furrow opener. A pair of press wheels usually follows and compresses the soil around the plants. Hand - fed planters are normally worked at fairly low forward speeds of 1-2 km/h. Mechanical planters employ two or more persons to work for each planted row. They place the plants in holding pockets or fingers attached to an endless conveyor and from this point onward the plants are dealt with entirely mechanically. The plants are carried down and placed in an open furrow in an erect position and soil around them is compacted,

iii) It must be decided before hand whether the machine to be designed, is to be an automatic, semi-automatic or a simple transplanting aid,

iv) It must be decided whether the machine is to be a self-propelled, tractor-mounted or trailed,

v) Whether the plants have to be placed in the furrows or in the holes made by the openers of the machine,

vi) The seedlings should be transplanted vertically and the permissible deviation from the vertical should not exceed 15°,

vii) In order to ensure that the seedlings remain near vertical after being released by the gripping jaws/picking fingers/holding pockets, the forward speed of the machine and the peripheral speed of the picking fingers must be equal. In other words, speed index (λ) is

$$\lambda = \frac{Wr}{V} = 1$$

where,

w = angular velocity of holding/gripping finger or grab, (rad/s)

r = radius of the transplanting arm, m

V = forward speed of the machine, m/s

viii) The machine must have provision to release a small quantity (0.4 - 0.6 litre) of water to irrigate each seedling to prevent mortality of the plants,

ix) The transplanter must plant the seedlings in straight rows with inter row variation not exceeding ± 2-4 cm,

x) Variation of individual seedlings from the row axis must not exceed ± 3 cm,

xi) Deviation in plant spacing must not exceed ± 1 cm, while planting with a step up to 20 cm and ± 2 cm while planting with a step greater than 20 cm,

xii) Operating width of the boots of furrow openers should be between 5 and 15 cm with a deviation of ± 2 cm,

xiii) Viability of the transplanted seedlings must not be below 95%,

xiv) Design of the transplanter must be such that the seedling and water stocks suffice for continuous working of at least 200 m length,

xv) Furrow openers must be able to open the furrow up to a depth of 8 cm and width 15 cm,

xvi) A road clearance of at least 25 cm must be provided in trailing type machines,

xvii) While designing the transplanter, it must be kept in mind that the person's employed for feeding the seedlings into the gripping (holding) jaws or fingers can feed at an average rate of 30-40 seedlings per minute. If the plant spacing is known, the distance travelled per minute or the forward speed of the machine can be found out. To be able to transplant at a faster rate, two persons are usually employed to feed the seedlings for each planting unit,

xviii) The press wheels are invariably used in compacting the soil around the seedlings. These wheels are used in pairs, pitched and gathered for efficient operation. The press wheels have tilted rims to gather more soil around the plants. The value of compactness index (K) is usually 2-3. It can be calculated as

$$K = \frac{G}{B\sqrt{D}}$$

Where, G = vertical load on the press wheels, kg

B = rim width, cm

D = diameter of the press wheel, cm

xix) The transplanting unit must have the provision to adjust the planting depth by adjusting the shoe type opener by moving it up or down as well as forward or backward,

xx) The transplanter must have the provision to adjust the plant spacing within the row by regulating the number of teeth on the spacing sprocket and number of plant holding pockets around the conveyer chain or disk,

xxi) The design of the transplanter should be such that it should not damage the seedlings, be suitable for transplanting different types of crops and maintain proper depth of planting, row spacing and the spacing between plants should be uniform, and

xxii) The transplanter must be able to transplant the seedlings by the square as well as the drill - seeding method.

Semi-automatic transplanter

Semi automatic transplanters consist of the transplanting mechanism, manual-feeding mechanism, drives system, etc. Rate of planting is limited by the speed of one by one-manual seedling feeding. Generally, it takes about two second to feed a seedling by hand, and this work is very tedious in a large field. This type of transplanter, however, has some advantages, such as it is simple mechanism, low price, and wide adaptability to various seedling types. Semi automatic transplanters are thus widely used. To keep the seedling vertical, the discs must release it when it is vertically below the axis of the discs. A premature release results in backward bending of the seedling because it is just held planting when released. Delayed release results in forward bending because horizontal component of peripheral speed of planting disc starts reducing and seedling gets some relative motion w.r.t. soil.

Walking type semi automatic transplanter with engine of 2.3 kW is also available. This planting device consists of a bill type opener driven by the engine, press wheels, a revolving type manual feeding apparatus with seedling cups. Some transplanters have a synchronized device for burning or cutting holes through plastic mulch. These types of transplanters can transplant seedling with soil, which are cell plug, paper pot or soil block seedlings. Rate of planting is about 2000-2500 seedlings per hour, and rate of work is 2 to 5 h/ha depending on plant spacing.

Tractor operated two row semi-automatic vegetable transplanter (PAU Model)

A two row vegetable transplanter with picker wheel type metering mechanism has been developed (Fig. 4.98a) at Punjab Agricultural University, Ludhiana for transplanting seedlings of brinjal, cauliflower, tomato etc (Anonymous, 2010; Singh *et al.*, 2015). The machine consists of a frame, two lugged ground wheels, seedling tray, seat for the operators, furrow opener, compaction wheels, finger guide tunnel, picker wheel type metering mechanism and water tank (Fig. 4.98b). Picking fork has spring mounted rubber flapper, which closes while passing through the tunnel. Again these flappers open at the bottom end of the tunnel to release the seedlings in a furrow. The inclined wheel compact the soil around the seedlings. Power from the wheel is supplied to the planting mechanism through shaft, chain and sprockets. Two persons, one for each row is required to place the seedlings in the flappers when these open at the top position. The root side of the seedlings is kept towards the operator. Water is sprayed on each row by the shower provided behind the picker wheel for temporary relief to the plant. The plant-to-plant spacing and row spacing are adjustable to suit for transplanting requirements of different types of vegetables/crops. The plant-to-plant spacing can be varied as 30, 45 and 60 cm by changing sprockets or number of fingers. The machine can be used for transplanting vegetable seedlings of chilli, brinjal, cabbage and cauliflower (Fig. 4.98c). The plant missing is about 2 to 3.5 percent at a speed of about 0.8-1.0 km/h depending upon the plant to plant spacing and skill of operator. Machine saves about 70-80 percent labour in comparison to manual sowing depending upon crop spacing. Due to low capacity of machine the saving in cost is up to 10 percent. The machine can cover about 0.7-0.8 ha/day. The machine saves 50-60% labour as compared to convention methods. The equipment is operated by 30 kW tractors and above.

a) Stationary view of vegetable transplanter

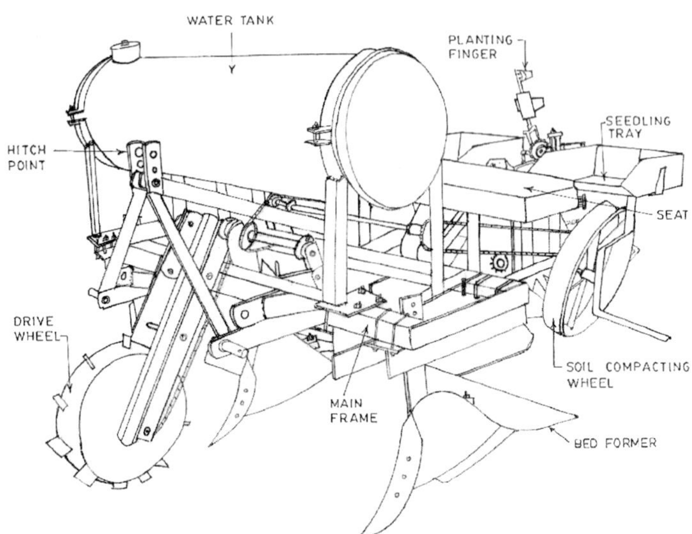

b) Components of vegetable transplanter

c) Vegetable transplanter being used in field

Fig. 4.98 : Tractor operated picker wheel type vegetable transplanter (PAU Model) with bed former

Tractor operated three row plug type vegetable transplanter (TNAU Model)

Tractor operated plug type vegetable transplanter consists of main frame with hitching system, ground wheel, shoe type furrow openers, compaction wheel, operator's seats, two depth control wheels and plug type metering mechanism (Fig. 4.99). It employes press wheels inclined at an angle of 15° with the vertical as soil covering device (Anonymous, 2008, 2008d & 2010). The plant spacing can be adjusted by changing sprockets/ pulleys or by changing in alternate holes. The power is transmitted from ground wheel (750 mm diameter) to gearbox through chain and sprockets. For transmitting power from gearbox to discs, belt and pulley are employed. In transplanting three discs of 460 mm diameter are used. The power is reduced at a ratio of 2:1 from gearbox to discs. The auxiliary frames are provided to mount furrow openers, furrow closures, operator's seat and two depth control wheels. Presently nursery preparation including soil preparation, seeding, and maintenance and transplanting operations is performed manually. The mechanization of bare root vegetable planting is complex task because the seedlings of many species are extremely tender which requires special care in handling. The advantages of mechanized nursery and transplanting are control of the plant growth during propagation and the survival and growth of the plants after transplanting. For vegetable crops like chilli, tomato, brinjal etc media mix is optimized. The best medium observed is cocopiat and vermin-compost in 1:1 ratio. The nursery raising system is the optimized portable seedling tray with 98 cells. The components of the system are media bin, bucket elevator, media sieve, screw conveyor, media mixer and tray conveyor. The capacity of nursery raising system is 750 trays/h. The impact of the seedling with soil block helps in its placement. The average field capacity of transplanter is about 0.14 ha/h. Average field efficiency of machine is 75% for chilli, tomato and brinjal with 450 mm row spacing. The working width of machine is 1.35 m.

Fig. 4.99 : Plug type tractor operated vegetable transplanter (TNAU Model)

Revolving magazine type vegetable transplanter

The revolving magazine type transplanting mechanism has been developed (Anonymous, 2010 & 2010b). For growing plug type seedlings for different vegetable crops, soil less media is used in plastic trays. Mixture of soil less media is filled in the different types of plastic trays. Seeds of different vegetable crops are sown at a shallow depth after pressing the media with finger in a gentle way into the plugs. The machine consists of single furrow opener shoe, packing wheel, main axle, frame, three point hitch system, ground wheels, seedling trays and power transmission (Fig. 4.100). Based on the performance of the developed revolving magazine type metering mechanism, a two row vegetable transplanter with revolving magazine type metering mechanism has been also developed (Fig. 4.101). The average labour requirement for transplanting is 32.76 man-h/ha for brinjal with machine transplanting and 230 man-h/ha in manual transplanting. For tomato, the average labour requirement for transplanting is 35.08 man-h/ha with machine transplanting and 254 man-h/ha in manual transplanting. The average field capacity of the machine is 0.12 ha/h for brinjal and 0.115 ha/h for tomato crop with field efficiency of 80.4 per cent and 81.6 per cent respectively.

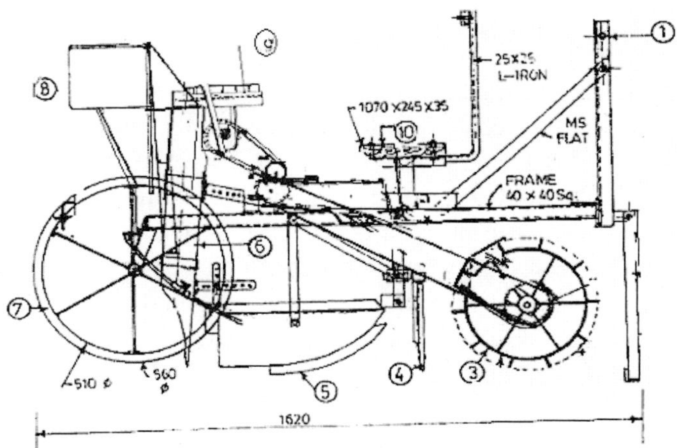

1. Hitch system 2. Stand 3. Ground wheel 4. Peg 5. Furrow opener shoe 6. Drop pipe 7. Packing wheel 8. Seedling tray 9. Revolving magazine type metering mechanism 10. Seat.

(a) Side view of a single row vegetable transplanter fitted with revolving magazine

a) A view of machine

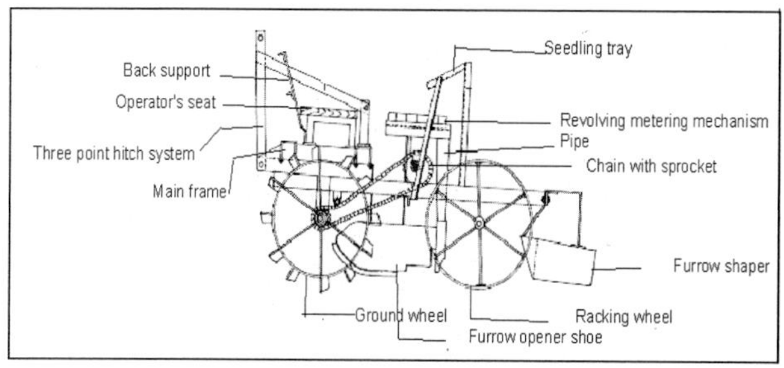

b) Details of machine

Fig. 4.100: Revolving magazine type metering mechanism

Fully automatic transplanters

It consists of the transplanting mechanism, automatic seedling feeding mechanism, drive system, etc. Transplanting is fully automatic, and rate of planting is thus high (Bao, 1984; Shaw, 1997). The operator needs only to steer the machine and replace the seedling trays. However, this type of transplanter can utilize only an appropriate seedling type and seedling arrangement on tray. Naturally, some transplanters synchronize burning or cutting holes through the mulch plastic for full-bed plastic mulch culture, which is now popular for lettuce. Planting depth is controlled automatically. Fig. 4.101 shows an automatic transplanter for cell plug seedlings. This transplanter is a row type with four-cycle 1.67 kW engine. A flexible plastic cell tray (the under-tray is the same as that used for rice) is employed. The cells are 30 mm square with 128 cells (8x16) per tray, and 25 mm square with 200 cells (10x20) per tray. Cell plugs are shaped like an inverted pyramid. The fingers, which operate almost like chopsticks, pick up a cell plug from the tray and throw it in the bill type opener, which plants the plug after pre-pressing the soil surface with wheels. Finally conical wheels press the soil around both sides of the plug. Hill spacing is between 28-56 cm, and suitable ridge height is 0-25 cm. Rate of planting is about one seedling per second, and rate of work for a single operator is 2 to 2.5 hour per ha depending on hill spacing.

Fig. 4.101 : Automatic vegetable transplanter for cell plug seedlings

Bare root transplanter is the standard all purpose transplanter for setting bare root plants or pots and cells no large than 38 mm wide (Fig. 4.102). Plant spacing in the row are adjustable from 14 cm and up. There is a potato planting attachment also available for planting potatoes, garlic and onion sets. Multi-row planters can plant 45 cm rows and up on single tool bars, and 22 cm rows and up on tandem tool bars frames. Plants are hand set into the furrow opened by the shoe. Packing wheels then pack the soil in around the plant. It is preferred by strawberry growers for accurately setting the crown depth. It can also set tall plants of any height. In these transplanters the seat of the operator setting the plant seedlings is so close to the ground that plants are set with utmost ease by the operators.

Fig. 4.102 : Automatic vegetable transplanter for pulp mould cell pot seedling

Fig. 4.103 shows an automatic transplanter for chained pulp pot seedlings. This transplanter is a row type with four-cycle gasoline 1.7 kW engine. The seedlings are grown in soil pods or paper pulp pods and transferring of the pot containing a seedling and fixing it into a hole or furrow is done mechanically. In this type of devices the operator places the plant in the flexible rubber discs held together by spring fingers and operated by small rollers at the top to allow insertion of the plants and another pair at the bottom for release into the furrow in front of the press wheels. These machines though reduce the labour requirements drastically they are expensive and involve sophisticated mechanism. The chained paper pots are in a row of 264 pots. The paper pot is 30 mm in diameter and 5 cm high. Shape is cylindrical. A rotating type planting finger picks a paper pot, then guides it through the slider and into the soil. The planting finger splits the paper pot one by one before transplanting. Conical press wheels press soil against both sides of the seedling. Hill spacing must be between 24-64 cm, and suitable ridge height is 0-25 cm. Rate of planting is about two seedlings per second, and rate of work for a single operator is 1.5 to 2.4 hour per ha depending on hill spacing.

Fig. 4.103 : Automatic vegetable transplanter for chained pulp pot seedlings

Plastic mulch laying machine

Plastic mulch-laying machine has been developed at CIAE Bhopal for cultivation of Kabuli gram, and vegetables; like peas and capsicum (Anonymous, 2008, 2010). Mulching with the plastic mulch-laying machine could be carried out using white colour plastic films of 15 micron for pea and capsicum and 25 micron thickness for Kabuli gram crops. Prior to laying of plastic mulch, plastic sheets are re-rolled on 100 mm diameter PVC pipes (Fig. 4.104). Machine is adjusted and set for different sizes of plastic mulch bed required for different crops. Perforations on the polythene sheet for placement of seeds are made manually after plastic mulching. The size of perforations is 30 mm. The performance of crops is excellent on plastic mulch beds, in terms of plant growth, no disease, high yield etc. It gives 33.2% and 35.8 % higher yield for pea and Kabuli gram respectively as compared to the control.

The field capacity and labour requirement are 0.17 ha/h, and 11.3 - 12.5 man-h/ha, respectively for mulch laying with the machine, where as in manual mulching the respective values are 0.003 ha/h and 228-246 man-h/ha. There is 96% reduction in labour requirement over manual mulching. Existing design of plastic mulch laying machine can be adopted for laying of plastic mulch on two beds of 40-50 cm top width simultaneously (Fig. 4.105). Provision is made for mounting two rolls of plastic sheet for laying on beds. Additional pneumatic wheels are provided for holding plastic sheets laid on beds and scrappers for putting the soil on plastic sheets (Fig. 4.106). Developed tractor operated punch planter could plant on plastic mulched raised beds at 30 cm row spacing with plant-to-plant distance of 14 ± 2 cm satisfactorily. Use of machines results in saving of labour from 81 to 92% in groundnut and 72 to 75% in French bean over manual methods. Also there is 28 to 32% saving in irrigation water, 62 to 64% labour saving in weeding operation and 1.89 to 2.51 times higher yields in French bean and 25 to 30% saving in irrigation water, 72 to 77% saving of labour in weeding and 22 to 29% higher yield in groundnut over the cultivation of unmulched raised beds (Fig. 4.107). Field capacity of machine is 0.21 ha/h.

Fig. 4.104 : Single row plastic mulch-laying machine

Fig. 4.105 : Two row plastic mulch-laying machine

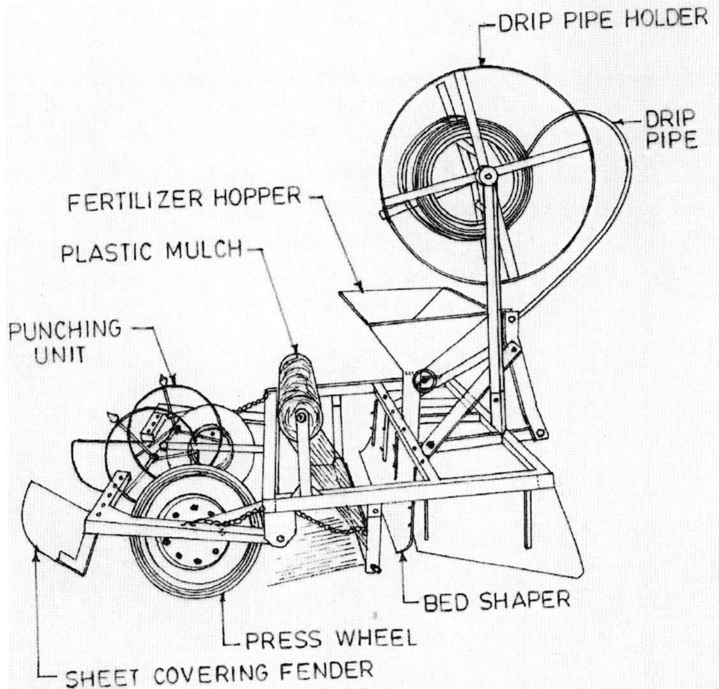

Fig. 4.106 : Isometric view of plastic mulch laying machine

Fig. 4.107 : Planting of French bean on raised beds with plastic mulched

4.4 Rice seeding and transplanting machinery

In India, rice is grown under widely varying conditions of altitude and climate. The climate of India is difficult to lay due to the country's large geographic size and varied topography. Therefore, the rice growing seasons vary in different parts of the country, depending upon temperature, rainfall, soil types, water availability and other climatic conditions. In eastern and southern regions of the country, the mean temperature is found favourable for rice cultivation throughout the year. Hence, two or three crops of rice are grown in a year in eastern and southern states. In northern and western parts of the country, where rainfall is high and winter temperature is fairly low, only one crop of rice is grown during the month from May to November. Major constraints to rice production that India faces are land, water, labour and other inputs such as fertilisers, pesticides and insecticides, and even high quality germplasm, without affecting the already degraded and stressed agricultural environment. The problems/constraints in rice production vary from state to state and area to area and some of them are:

i) About 78% of the farmers are small and marginal in the country and they are poor in resource,

ii) The problems of flash floods, water logging/ submergence due to poor drainage are very common in East India,

iii) Continuous use of traditional varieties due to the non-availability of seeds and farmers lack of awareness about high yielding varieties particularly in remote rural areas,

iv) Low soil fertility due to soil erosion resulting in loss of plant nutrients and moisture,

v) Low and imbalanced use of fertilizers, low use efficiency of applied fertilizers particularly in the North-Eastern and Eastern States,

vi) The Eastern region experiences high rainfall and severe flood almost every year which lead to heavy loss,

vii) Heavy infestation of weeds and insects/pests,

viii) Delay in monsoon onset often results in delayed and prolong transplanting and sub-optimum plant population mostly in rainfed lowlands, and

ix) In the years of scanty or adverse distribution of rainfall, the crop fails owing to drought etc.

Rice farming is practiced in several agro ecological zones in India. No other country in the world has such diversity in rice ecosystems than India. Because cultivation is so widespread, development of four distinct types of ecosystems has occurred in India, such as Irrigated Rice Eco System, Rainfed Upland Rice Eco System, Rainfed Lowland Rice Eco System, and Flood Prone Rice Eco System (Singh and Singh, 2014). In India, rice is grown in different types of soils. Experts point out that in India, rice is grown in such varied soil conditions that it is difficult to point out the soil on which it cannot be grown. However, soils having good water retention capacity and good amount of clay and organic matter are considered ideal for rice cultivation. It grows well in soils having a pH range between 5.5 and 6.5.

The method of rice cultivation depends upon the rainfall pattern, availability of irrigation water, timely land preparation and availability of labour. Two methods are used for raising rice crop in India, namely Upland cultivation (direct seeding) and Wetland cultivation also called low land cultivation (direct seeding and seedling transplanting). Rice transplanting by hand is very arduous, expensive and labour consuming operation. Many attempts have been made to develop manual as well as self-propelled rice transplanters in rice growing countries such as Japan, China, Korea and India (Garg, 1992). It is used for transplanting of rice seedlings. Paddy is grown generally by transplanting or direct sowing. In spite of considerable progress made in the area of direct seeding of paddy with chemical weed control, the transplanting remains the most common method of paddy sowing. Transplanting essentially refers to the planting of 20-35 days old and 20-30 cm long seedlings under wet land conditions. Transplanting has a number of advantages over direct sowing like i) the time that crop occupies the land is reduced by 2-4 weeks resulting in less water requirements, ii) helps the plant a better start over the weeds hence lesser growth of weeds, iii) selection of more vigorous seedlings, and iv) less seed requirement.

The transplanting of rice crop completely depends on manual labour in India (Garg and Sharma, 1984). Transplanting which is done in bending posture is very tiresome and labour consuming as the operation is done under adverse conditions. Apart from this, the scarcity of labour during peak season of transplanting poses a big problem to complete the transplanting in time. The transplanting after sowing period causes progressive decrease in the yield. In northern regions, very late transplanting may result in complete failure of crop due to cold weather at the time of maturity. The contract labour, which is generally employed for this operation, transplants the seedling at spaces for wider than the recommended ones. In spite of 200-250 man-hours required for random transplanting of one hectare (Fig. 4.108a), the correct plant population is not achieved. Transplanting in rows requires much higher labour requirements and is therefore not generally practiced in India (Fig. 4.108b). However, it has been reported that the yields for line transplanting are higher. The weeding and intureculture for line transplanting is also much easier. A survey conducted by P.A.U. Ludhiana revealed that on an average there is only 22 single plants/sq. m in place of recommended 33-hills/sq. m each of 2-3 seedling (Garg and Sharma, 1984). The labourers generally keep plant density low in the centre of the field. The solution to the above problems is the mechanical transplanter. Apart from saving in time, cost of transplanting and removing the human drudgery, it can give uniform and desired plant density. Also, one can transplant the crop in line at no extra cost.

a) Random manual transplanting

b) Manual transplanting in rows

Fig. 4.108 : Manual rice transplanting in India

Transplanters in general can be classified under two groups depending upon the type of seedlings used viz. transplanting using washed root seedlings and transplanting using mat type or soil bearing type seedlings. These transplanters can be further classified under four groups depending upon the source of power used i.e. manually operated, animal drawn, tractor-mounted and self-propelled (Garg, 1992; Garg and Sharma, 1985). Transplanters using washed root seedlings, the seedlings having 4-6 leaves and 18-35 cm height are removed from the traditional nurseries just before transplanting and roots are washed with water. In some cases, the roots and tops are pruned and then neatly arranged on the seedling platforms or boxes of the machine for transplanting. In general, it has been found that the labour requirement for uprooting, washing and arranging of seedlings in the transplanter boxes is quite high.

Manual rice transplanter

It consists of frame, floats, seedling tray, operating handle, fingers (pickers), tray drive unit, tray drive and depth control mechanism (Fig. 4.109). To operate the equipment, a mat type nursery is raised (Garg *et al.*, 1982). The size of mat is 22 cm in width, 45 cm in length and thickness of soil of 15 cm. The worker movement in the operation of the rice transplanter is backward and worker pulls rice transplanter after every stroke. After lifting the operating handle, it is pushed down gently for transplanting. To maintain the plant to plant spacing of 15 cm, the machine is pulled by the operator. If required, gap filling can be done manually. Mat type seedlings are refilled when exhausted. After completion of each day of work, the transplanter is washed with water. The scaled down version of 2 and 4 row transplanter are suitable for operation by female workers. Mat type seedlings are placed in the inclined trays. Fingers pick up the seedlings when they are pushed downward and place them in the prepared soil. Plant to plant spacing can be controlled by the operator. Its working capacity varies from 0.3-0.4 ha/day and requires two persons, one for operating the transplanter and other for filling the tray with mat seedlings (Garg and Sharma, 1984; Garg and Singh, 2002). Two, four (Fig. 4.110), five (Fig. 4.109) and six row (Fig. 4.111) rice transplanters are commercially available. The equipment is manually operated hand tool. When the operator pulls the machine and operates the handle, the picker, gathers two or three seedlings and place them in the puddled soil. The fingers actuate with the help of a hand lever attached with the handle. The uprooted seedlings of 20 to 25 days old and 10 to 25 cm long of any variety grown in a traditionally prepared seedbed are packed in the detachable trays after washing and separation. A feed guide plate which can be moved up and down by a screw is provided to keep the seedlings packed. The fingers pick the seedlings from the tray and plant them into the puddled field. The tray sliding mechanism enables the fingers to pick the seedlings from a fresh place every time. The tray moves 6 mm in each planting stroke. Shallow, soft and settled puddle with 1 to 3 cm of standing water in the field is ideal for the operation of this machine. The machine is operated by one person. By pressing the finger-opening lever against the handle, the clamp is uniformly opened and is pressed against the seedlings. The seedlings are then picked which are planted into the soil in the forth stroke. The row to row distance can be maintained at 200 mm. Stainless steel floats are also available with transplanter.

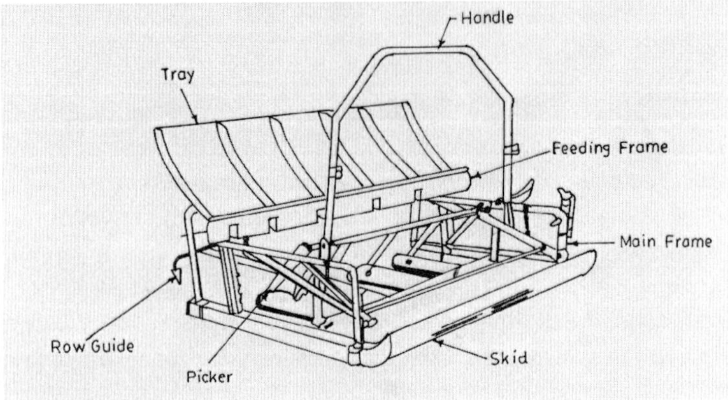

Fig. 4.109 : Details of manually operated 5-row paddy transplanter

Fig. 4.110 : Four row rice transplanter being operated by woman

Fig. 4.111: A view of manually operated 6-row rice transplanter.

Mat type seedings

Mat type seedlings are not currently grown in India because of manual transplanting. However, efforts have been made to evolve a system to grow mat type seedlings economically (Garg *et al.*, 1982; Garg and Singh, 2002; Garg, 1987). Detailed procedure for raising mat type seedlings is described below:

i) Select the nursery location in such a way that the transport of the seedlings to the field where transplanting is to be done is minimum. Nursery field should have a good soil. About 125 m^2 area is sufficient for sowing nursery for one hectare.

ii) Prepare the field by harrowing and levelling at about 10% moisture content. It is better to add few baskets of farm yard manure in the field before harrowing.

iii) Spread 50-60 gauge polythene sheet of 90-100 cm wide over a prepared level ground. It is better if the sheet has perforations.

iv) Place 3-4 steel frames having 14 compartments of 40 x 20 x 1.5 cm in each frame over the polythene sheet. About 350 g of polythene sheet spread to a length of about 20 meters is sufficient for preparing seedlings for 0.4 ha. Number and size of compartment can vary according to machine specifications.

v) Prepare soil and farmyard manure mixture as done in Fig. 4.112 and fill in the frames uniformly up to the top surface. The soil is lifted from both the sides of the frames by making shallow furrows/channels. Soil for filling in the frames can also be prepared separately by mixing one fourth quantity of farm yard manure. Care must be taken at the time of sowing that there are no stones in the soil which is filled in the frames.

vi) Spread about 700 g of pre-germinated seed evenly in each frame to achieve seedling density of 2 to 3 plant/cm^2 in the mats. About 10 kg seed is sufficient to sow 200 mats required for transplanting 0.4 ha.

vii) Cover the seeds by a thin layer of soil and sprinkle water by hand sprayer for proper setting of the soil.

viii) Lift the frames and put these at the next place.

ix) Repeat the above procedure for sowing required number of seedling mats. Two persons can sow seedlings for 1.2 to 1.5 ha in a day. For proper irrigation, it is better if 7-8 frames are grown on a single bed.

x) After sowing, irrigate the field by flooding so that the seeds are not submerged or washed away. Care must be taken that the seedling mat area is always wet.

xi) Spray or broadcast the seedlings after about 10 days at the rate of 300 g urea/200 mats after irrigating the nursery.

xii) The seedling mats become ready after 25-30 days of sowing. Drain the water from the nursery field a few hours before nursery uprooting.

xiii) The seedling mats are uprooted by giving a cut with a sharp blade/

sickle along the boundaries of the mat. Lift the nursery mat and dress it to the proper size of the tray of the transplanter, if needed. One person can uproot the seedlings for 2 ha in a day. The uprooted mats are transported to the filed either by tractor trailer or by any other means.

Fig. 4.112 : Soil and FYM mix sieving for nursery bed

Nursery sowing seeder

For uniform distribution of seed a nursery-sowing seeder has been development by PAU Ludhiana (Garg, 1987 and 1992). The seeder has a diameter of 19 cm and width 100 cm which is equal to the width of the nursery sowing frame and is made of GI sheet (Fig. 4.113). A long shaft passes through the center of the seeder and the ends of the shaft are connected with the wheels. The seeder is also connected with a handle for its movement. The seeder has 2 partitions and in each compartment, two agitators have been provided on the shaft for agitating the paddy seed. This has been done to reduce the chocking of seed at the opening holes. The seeder has the openings of 1 cm diameter on full length of the roller for uniform distribution of the seed. The holes are staggered for better distribution. Two feeding gates have been also provided on the seeder for filling the seed in each box. The nursery sowing seeder is calibrated in the laboratory, using pre-germinated paddy seed having different sprouted length (1-2 mm, 2-3 mm and 4-5 mm). The seeder is filled to 2/3 rd of its capacity with pre-germinated seeds, which weighs about 5.5 kg. The seeder is then operated over 4 nursery-sowing frames, which are placed on polythene sheet. The seed dropped in each frame is collected and weighs. The same procedure is repeated thrice for 3 different lengths. In general 25-30 kg seed is used for raising mat type seedlings/ha (Fig. 4.114). Nursery mat is lifted and dress it to the proper size of the tray of the transplanter. One person can uproot seedling mats for 2 to 2.5 ha in a day. Before transplanting field is puddled and allow the soil to settle for 1-2 days depending upon the type of soil. However, sandy fields are transplanted after 2-3 hours of puddling. The water level in the field at the time of machine operation should be preferably 3-4 cm.

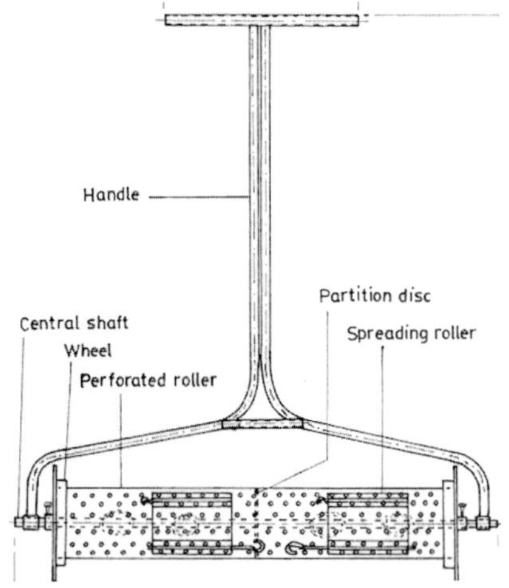

Fig. 4.113 : Nursery sowing seeder

Fig. 4.114 : Nursery sowing seeder being used in field

Engine operated rice transplanter

Japan, China and Korea have already developed engine operated rice transplanters and are being used extensively (Anonymous, 2009, 2010, 2013). It consists of air-cooled gasoline engine, main clutch, running clutch, planting clutch, seeding table, float, star wheel, accelerator lever, ground wheel, handle and linkage mechanism (Fig. 4.115). Seedlings are grown in special seedling trays in controlled environment called mat seedlings. When seedlings are 25-30 days old, they are uprooted and placed in slanting seedling trays. Power from the engine is transferred to main clutch from where it is transferred to planting and a running clutch. The fingers on four bar linkage mechanism catch 3-4 seedlings at a time separate them from the mat and place it in the puddled soil. A float supports the machine on the water while working in the field. There are two end wheels that facilitate the movement of the transplanter. A marker is provided to demarcate the transplanting width during operation. The machine maintains row to row and plant to plant spacing. Engine operated rice transplanter is used for transplanting of mat type rice seedlings in puddle field. The equipment is operated by 4.0 kW diesel engines. The machine is riding type and it transplants seedlings from mat type nursery in four/six/eight rows in a single pass. The drive wheel receives power from the engine through V-belt, cone clutch and gearbox. A propeller shaft from the gear box provides power to the transplanting mechanism mounted over the float. The float facilitates the transplanter to slide over the puddle surface. The tray containing mat type nursery is moved sideways by a scroll shaft mechanism, which converts rotary motion received from the engine through belt-pulley, gear and universal joint shaft into linear motion of a rod connected to the seedling tray having provision to reverse the direction of movement of tray after it reaches the extreme position at one end. Fixed fork with knock out lever type planting fingers (cranking type) are moved by a four bar linkage to give the designed locus to the tip of the planting finger. Working capacity of transplanter is 0.13 to 0.2 ha/h.

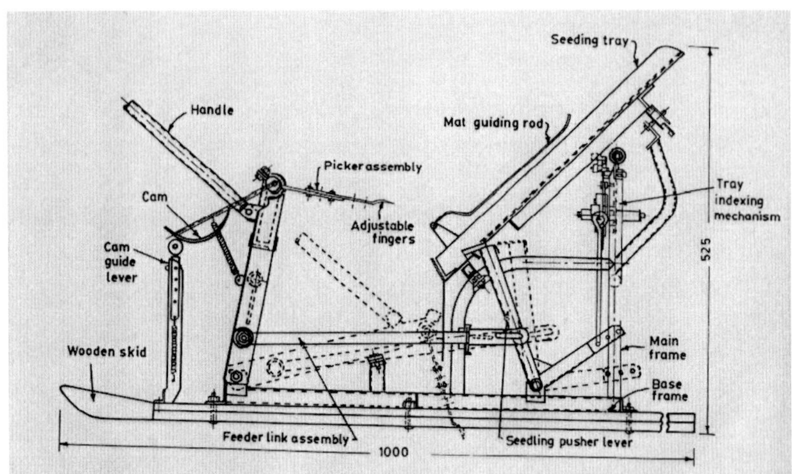

a) Details of engine operated six row rice transplanter

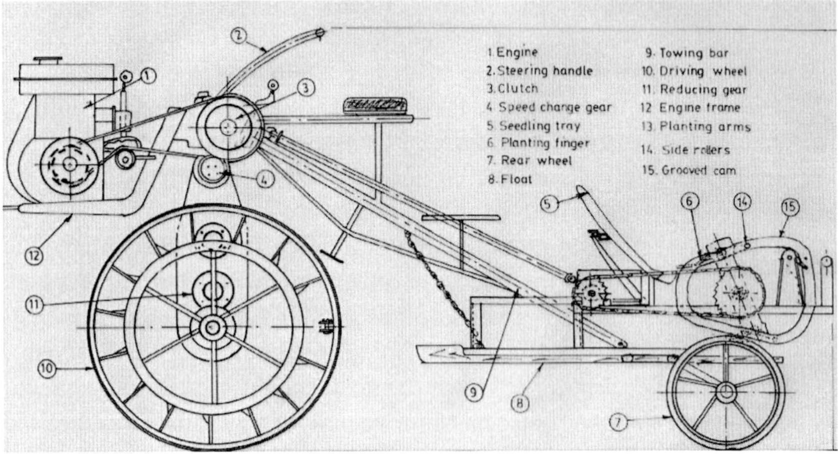

b) Details of engine operated eight row rice transplanter

Fig. 4.115 : Engine operated rice transplanter

8-row riding type engine operated paddy transplanter

An 8-row riding type engine operated paddy transplanter has been developed at PAU Ludhiana with double planting arm with 8 fingers on each arm working in rotational cycle instead of single arm to increase the field capacity and to avoid end row coverage with loose soil (Fig. 4.116). This is basically the modifications of Yanji paddy transplanter (Fig. 4.117a). An overload clutch has been provided in the transmission shaft to avoid fingers damage on account of stones etc in the nursery (Anonymous, 2009, 2010, 2013). The machine consists of engine, power transmission system, main frame, float and rice planting system (Fig. 4.115b). The engine is mounted in front of the machine that drives the driving wheel and the rice planting system. Power transmission system from the engine to the wheel and the rice transplanting system is the same as in case of 6-row machine and no change has been made. The rice transplanting system is mounted on a frame which is fitted on the float with the help of two pivots and a depth adjusting lever. The whole frame can be raised or lowered with the help of a depth adjusting lever. The transplanting system consists of a transmission box, curved tray with 8 partitions, holding platform over which tray moves and two double planting arms with 8 fixed types of fingers on each arm. The two planting arms are attached to the central shaft the ends of which are fixed on the frame with the help of brackets and bearings to avoid jerkey motion. Power from the speed change gear is transmitted to the transmission box of the rice transplanting system with the help of a universal shaft which in turn supplied power to the scroll shaft with the help of gears. The extended shaft from the transmission box supplies power to the central shaft of the planting arms through a chain. A jaw type spring loaded safety clutch is provided in the extended shaft of the transmission box to take care of hard materials in the nursery. The tray sets over the holding platform which is supported at the back through two grooved pulleys for smooth linear motion. The scroll shaft rotates a shaft with the help of a key which is connected to the tray through side brackets for its to and fro motion. Tray movement is 21 cm on each side which is equal to the width of the tray compartment minus the nursery cutting potion. A mat pusher is provided to hold the sliding mats. It is operated with the help of a cam provided on the extended scroll shaft. The motion of the planting arm is controlled with the help of side rollers provided on the two planting arms which work in a controlled groove. Float system in the modified machine is the same as or in case of 6-row machine except that in the new prototype one row on each side of the float have been extended out of float to take care of the covering of already transplanted row with loose soil.

Principles of working: Before field operation of the machine, the tire ring from the wheel and two side wheels from the float are removed. Machine is loaded with seedling mats. Eight mats are placed on the tray and eight on the two side platforms. The engine is started and the person sitting on the seat takes the machine in the field by engaging the clutch. The machine starts planting when the power to the rice planting unit is supplied. The machine is operated in high/low gear at about half to three fourth throttle depending upon the field conditions. The planting arms move in rotational cycle and in each circle the planting arms work twice. Each planting finger on each arm cut a small cluster of soil along with few seedlings and transplants them in the puddle soils. The machine is operated in the adjacent tuns in such a manner that the markers move on the already transplanted row which maintain the required row to row distance. Provision of double planting

Fig. 4.116 : 8-row riding type engine operated paddy transplanter

a) A view of 8 row engine operated Yanji rice transplanter

b) 8 row engine operated Yanji rice transplanter in field operation

Fig. 4.117 : A view of 8 row engine operated Yanji rice transplanter

arms in one cycle helps to increase the planting capacity The machine uses mat type seedlings which are grown according to procedure developed at PAU, Ludhiana and discussed above. Seedlings up to maximum of 45 days can be used for better machine performance. Also, for proper machine functioning, transplanting in heavy soils i.e. clay type of soil is done 2-3 days after puddling and in medium soils 1-2 days after puddling. In sandy soils, transplanting is done 8-10 hours after puddling. Machine capacity varies from 0.14-0.17 ha/h while working at a speed of 1.15 to 1.40 km/h. The difference in the machine capacity is mainly due to wide variations in field conditions. Hill population varies from 21-28/m^2 (Avg. 24/m^2) in high gear and 28-35 hills/m^2 (Avg. 30/m^2) in low gear. Plant mortality is observed to be about 3-4% . Plant to plant spacing varies from 12-15 cm. Number of plants/hill varies from 1 to 6 (Avg. 3). Number of workers requires working on the machine varies from 5-6 including nursery up-rooting, transport and feeding in the machine. Machine can save both labour and cost of transplanting in addition to increase in yield.

6-row riding type engine operated paddy transplanter

The machine consists of a 5 hp light weight diesel engine, power transmission system, main frame, wooden float and rice planting system (Fig. 4.118). The engine is mounted in front of the machine that drives the driving wheel and the rice planting system. Power from the engine is transmitted through V-belts to the clutch belt pulley of the power transmission system. The engine pulley and clutch belt pulley are double grooved with different diameter for use on road and in the field operation. Power transmission system consists of engine frame, clutch, speed change gear, reducing gear, driving wheel, steering handle, seat for the operator and towing bars (Anonymous, 2010a, 2010b). Power transmitted from the engine to the speed change gear can be engaged or disengaged with the help of a hand clutch. The speed change gear transmits or breaks off power to the rice planting system and reduces the forward speed of the machine. It is also fitted with a speed change lever (low/high) for changing forward speed of the machine. The reducing gear receives power from the speed change gear through a universal joint and transmits it to the driving wheel through reduction in three stages. The driving wheel is worked as lugged wheel while operating in the field and smooth tire wheel on the road. This is done by mounting the tire ring on the lugged wheel with the help of a belt.

Rice transplanting system comprises of a curved tray with six partitions, bolding platform over which tray moves, planting arm with six fixed type of fingers and transmission box. The holding platform is mounted on the transmission box with the help of brackets. The tray sets over the holding platform which is supported at the back with the help of a guide. The guide also helps the tray for its linear motion. Tray movements are 22.5 cm which is equal to the width of compartments of the tray. The scroll shaft also operates the planting arm with the help of cams and levers. At the rear of the mechanism, a spring is provided on the frame and planting arm to reduce the additional forces of the mechanism that arise during operation of the machine.

The towing bars of the power transmission system are connected to the wooden float with the help of two pivots. The angle of the float can be adjusted by changing the length of chains provided on the float w.r.t. chain hooks on the towing bars. The whole rice planting system along with frame is also mounted on the float with the help of two pivots and one depth adjusting screw. The planting system can be lowered or raised w.r.t. float with the help of this screw. The float makes it possible for the machine not to get begged in the paddy fields. It also helps to smoothes the rough surface. Better surface of the float is corrugated which helps to move the machine straight forward. Two small wheels provided on both sides of the float are removed while transplanting. These wheels help in machine transport. A provision has also been made on the towing bars for a person to sit for feeding the seedling mats. Two platforms on both sides of the float have been provided to carry additional nursery mats. Two row markers have been provided on the engine frame. These row markers can be folded/ removed, if required. These help in steering the machine in adjacent runs to maintain row to row distance.

Machine transplants 0.150 ha/h in high gear and 0.116 ha/h in low gear, the variations in field capacity are mainly due to variations in forward speed of travel (1.26 km/h in high gear) and (1.01 km/h in low gear). Total labour requirements of machine transplanting system are 67 man-h/ha as compared to 218.0 man-h/ha for the traditional manual method. This represents a reduction in labour requirements of about 68.0% which are substantial savings.

Fig. 4.118 : Six row engine operated paddy transplanter

A 6-row riding type paddy transplanter using conventional washed root seedlings has been imported from Korea (Anonymous, 2010, 2010b, 2013). It comprises of a transplanting unit, engine, power transmission system, and power driven lugged wheel, machine frame, a wooden float, and a seat for the operator and a steering handle (Fig. 4.119). Power from a 4.0 hp petrol engine is given to the transplanting unit as well as the powered wheel through a gearbox. The forward speed can be changed through suitable selection of gears, which would also change the plant to plant spacing. A neutral position exists so that the power to the wheel alone during transportation and the transplanting system can be given independently. Power from the gear box is transmitted to another gear box on the transporting unit through a universal coupling. Gearbox of the transplanting unit has a clutch and a scroll shaft. Scroll shaft moves the nursery boxes in a to and fro motion. It also operates the fingers arm with the help of cams mounted on the shaft and other linkages. The linkages also operate the seedlings pusher for proper nursery picking by the fingers. A steel ring fitted with rubber surface can be fitted on the lugged wheel for machine transportation on hard ground/roads. This can be done very easily in the field itself. Also, two rubber wheels are mounted on the float at its rear end for machine transportation, which are removed during transplanting. Machine is operated and controlled by the operator sitting on the machine with the help of a steering. Provisions also exist for changing the depth of transplanting, plant to plant spacing and number of plants per hill. The field capacity of the machine in low gear is 0.12 ha/h and 0.19 ha/h in high gear. Plant to plant spacing in low gear and high gear are kept 15 cm and 21 cm respectively. Plant missing are mainly because of non-picking of the seedlings by the fingers or due to non-release of the seedlings by the fingers. Missing is observed higher in case of old nursery (about 30 days and above) because it has comparatively thicker stems and allowed only one or two plants to be picked up by fingers. On the other hand, younger seedlings are comparatively thin and allow more seedlings to be picked up by the fingers, thus reducing the chances of plant missing. Average number of seedlings per hill varies from 2 to 3. Variations in number of seedlings per hill within the rows are not significant and could not be attributed to any particular machine factor. Labour requirements for machine transplanting, nursery washing and arranging on the machine are 35 man-h/ha in low gear and 27 man-h/ha in high gear.

Fig. 4.119 : A 6-row self-propelled riding type rice transplanter using wash root type seedling

4-row engine operated walking type rice transplanter: A 4-row walk behind type transplanter has been imported from Japan to study its performance and for its adaptability in the local conditions (Anonymous, 2009, 2010, 2013). The paddy transplanter is operated by a 1.7 hp petrol engine and controlled by an operator walking behind it. It consists of an engine, float, power unit and wheels and the planting unit through scroll shaft after speed reduction (Fig. 4.120a). The forward speed of the machine varies from 1.4 to 2.1 km/h. Planting unit consists of crankshaft, planting arms (left and right), shaking arms, fingers seedling trays and mat pusher mechanism etc. A pair of fingers are fitted on each planting arm and actuated by the rotary motion of crankshaft through cam, push rod and a pair of spring loaded pinions. The finger tips follow an elliptical path during one revolution of crank shaft and fingers come closer just before hitting mat with a jerk and get apart at the lower most point i.e. a hill of soil is sheared off by each pair of fingers from nursery mat during downward movement of fingers and pushed into the puddled soil; but while coming back, fingers open with a jerk leaving plant hill into the soil. The process of planting is continuous as tray moves to and from over rails by the movement of scroll shaft (Fig. 4.120b). The power from the scroll shaft is provided to the mat pusher mechanism, which pushes the mat downward at the beginning of reverse motion of tray. An automatic arrangement to stop the planting unit is also provided, if any overloading occurs due to blocking of fingers caused by stones in the nursery mat. Arrangement has also been provided for adjusting the planting depth, number of plants per hill and plant to plant distance. By adjusting the stroke length of planting arm, the number of plants can be increased or decreased. For this purpose, a slot is provided in the shaking arm. Three variations for plant to plant distance viz. 140, 160 and 180 mm are provided. This is achieved with the help of adjustable pulleys. Planting clutch for starting and stopping the mechanism and side clutches for operating the machine have been provided. Side clutches help in easy turning of the machine. There is about 2-3 percent missing of hills but this is mostly due to the uneven seedling emergence. This could easily be avoided, if proper care is taken at the time of sowing of nursery, so that the seed is distributed uniformly at all the places.

a) A view of four row engine operated walk behind rice transplanter
Courtesy: Khedut Agro Engineering, Rajkot

b) Engine operated 4-row walk behind rice transplanter

Source: https://www.google.co.in/search?q=kubota+rice+transplanter&biw

Fig. 4.120 : Four row engine operated walk behind rice transplanter

Rice drum seeder

It is used for sowing pre-germinated paddy seeds directly on puddled fields (Garg and Singh, 2002; Anonymous, 2010 & 2010a). The equipment is manually operated hand tool (Fig. 4.121). It consists of a seed drum, main shaft, ground wheel, floats, and handle (Fig. 4.122). Joining smaller ends of frustum of cones makes the seed drum. Nine numbers of seed metering holes of 10 mm diameter are provided along the circumference of the drum at both the ends for a row-to-row spacing of 200 mm. Flat spikes 12 mm wide and 25 mm long are joined to ground wheel parallel to its axis of rotation. The slopes of the cone facilitate the free flow of seeds towards the metering holes. Two floats are provided on either side to prevent and restrict the sinkage and to facilitate easy pulling of the seeder. Working capacity of seeder is 0.09 to 0.16 ha/h depending upon the number of rows. Rice seeder for sowing of the paddy can be adopted as it saves time, labour and cost. The paddy seeder saves about 90% labour as compared to the traditional method. It is available in 2-row; 4-row, 6-row and 8-row. Small size of 2-row and 4-row mostly used on small plots and in hilly areas.

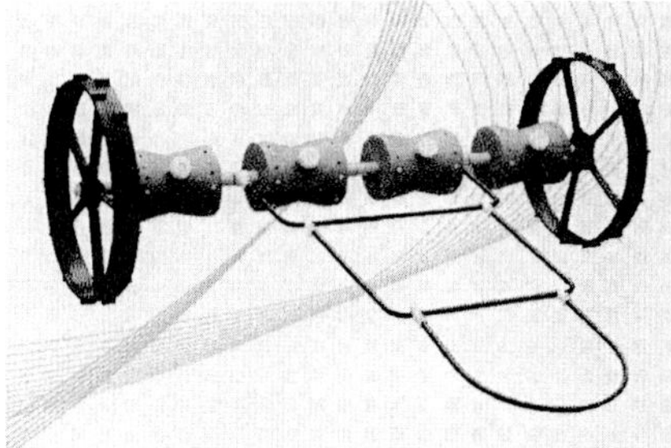

a) 8-row manual paddy seeder

b) 2-row manual paddy seeder

Fig. 4.121 : Manual direct paddy seeder

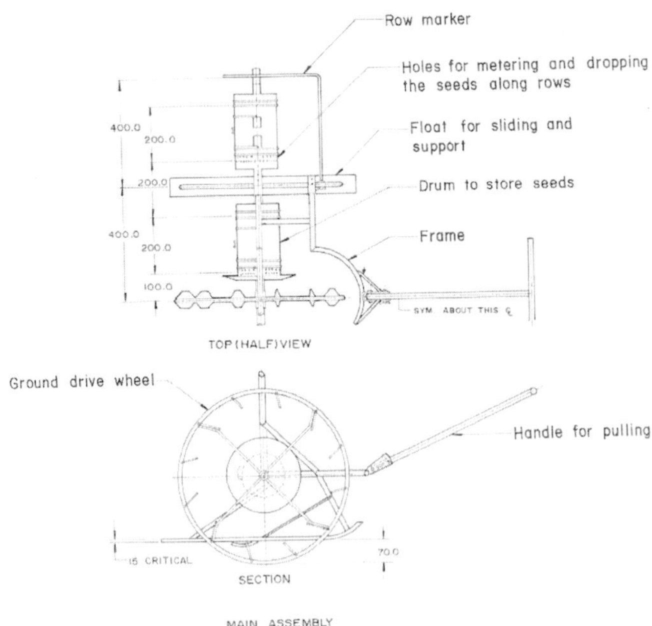

Fig. 4.122 : Details of 4-row manual paddy seeder

Improved direct paddy seeder

Transplanting of rice seedlings is a highly labour intensive and a costly method. To find out substitution for this method, an 8-row direct paddy seeder has been developed at TNAU Coimbatore. The seeder is an eight-row unit and is useful for sowing of pre-germinated paddy seeds in puddled field in wetland cultivation (Fig. 4.123). The main feature of the seeder is its lightweight i.e. 8 kg only as against 20 to 30 kg of other similar seeders available in the country. The row-to-row spacing is 200 mm. The shape of drum is hyperboloid and the diameter is 200 mm.

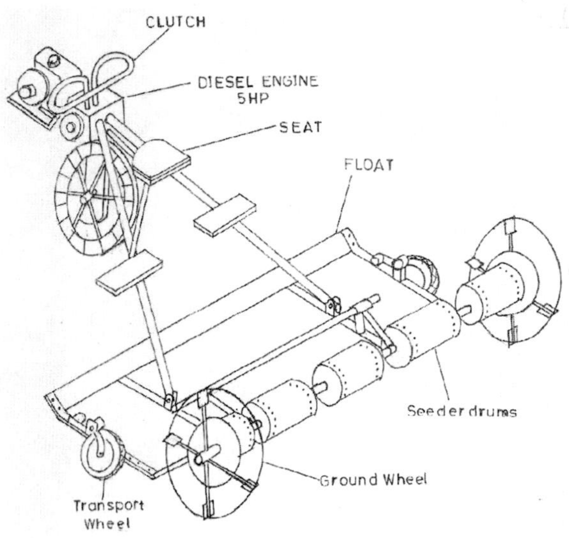

Fig. 4.123 : A 8 row riding type engine operated rice seeder

Tractor operated and animal drawn sprouted rice seeders have been developed at many places such as at PAU Ludhiana (Fig. 4.124), CIAE Bhopal (Fig. 4.125) and IGKVV, Raipur (Fig. 4.126) and found useful for faster row seeding of sprouted rice compared to manual broadcasting (Anonymous, 2013b & 2014). The row seeding is advantageous in terms of uniform plant population and subsequent weeding in rows by use of improved weeders which otherwise is not possible in broadcasted fields being of random distribution.

Fig. 4.124 : Tractor operated paddy seeder/multi crop planter
Courtesy: National Agro Industries, Ludhiana (Punjab)

Fig. 4.125 : Animal drawn Indira paddy seed-cum-fertilizer drill
Source: Chaudhuri and Singh (2013)

Fig. 4.126 : Animal drawn sprouted rice seeder

Animal drawn inter row crop seeder for rice and sesbania has been developed at CIAE, Bhopal (Chaudhuri and Singh, 2013) and used for row seeding of rice simultaneously intercropped with sesbania (Fig. 4.127). For biasi operation the seeding unit is removed and the same equipment with two-row biasi tool is operated in the row spaces where the sesbania get chopped and mixed to the soil. In conventional row seeded rice the sesbania seeds are either broadcasted or drilled separately in row spaces being an additional operation. The biasi plough is used thereafter for biasi operation (Fig. 4.128). The inter row crop seeder serves tripple purpose of seeding of rice and sesbania, placement of fertilizer and biasi operation with choping of sesbania.

Fig. 4.127 : Animal drawn inter row crop seeder

Fig. 4.128 : Power operated Biasi machine

Manually operated check row planter for dry seeding of rice has been developed at Indira Gandhi Krishi Vishwavidyalaya, Raipur (Anonymous, 2013b & 2014), Fig. 4.129. Row-crop planter (25 cm x 25 cm) enables operator to perform hill planting at definite spacing (in checks or squares). Seed hopper bottoms for row crops have rotating agitator feeding mechanism that rotates in small circular pipe. A small cell is attached below every hopper where seeds drop at desired intervals. The full hill drop receives 2 or 3 seeds in the cell at a time. After receiving the seeds, the cell opens and closes under a cut off mechanism. Seeds usually drop on a 'valve' and wait until the valve is tripped like a trap door to drop the seed into the furrow or hill. A wide range of spacing can be obtained by using metering roller and star wheel which is fixed on side of the hopper and rotate with the help of metering shaft. This star wheel performs the cut-off of seed by adjusting the numbers of blade on star wheel to maintain the cut off operation by which plant to plant distance is maintained. The placement of paddy seeds are at uniform depth under a range 2.4 cm to 3.25 cm. For sowing one ha of land the planter requires 5.44 man days which is much less as compared to manually transplanting in SRI and Dry seeding manual method. The planter saves 66.7-72.4% money over transplanting dry seeding manually.

Fig. 4.129 : Manually operated check row planter

System of Rice Intensification (SRI)

"SRI is now being seen more than a set of practices (young seedlings, wider spacing, less water etc), as an opportunity for maximizing production potentialities and even as a philosophy. It encourages farmer participation and innovation to make it more appropriate to local conditions and more owned by the users" (Satyanarayana and Babu, 2004; Kumar *et al.*, 2008; Anonymous, 2007). System of Rice Intensification (SRI) is a combination of several practices that include changes in nursery management, time of transplanting, nutrient, water and weed management. SRI uses less external inputs, less seed, less plants per unit area, less chemical fertilizer, less or no pesticides and give a higher yield for low cost of production per kg. SRI is referred to as a methodology, rather than a technology. Key cultural practices followed in SRI cultivation are soil nutrient management through application of farmyard manure or other decomposed biomass, transplanting young seedlings (preferably 8 to 12 days old), transplanting seedlings carefully, with a soil clump (along with seed sac) attached to their roots, transplanting at wider spacing and in a square pattern (25 x 25 cm). Weed management in rice field is a laborious and cumbersome work. Hand weeding is still the most common practice in low land rice fields. Though the hand weeding is effective it requires 150-175 man-h/ha which is about 20 per cent of the labour requirement for the entire crop production. Benefits of SRI method are:

i) Less utilization of inputs, more output thereby more returns in SRI than traditional method.

ii) Less seed rate, (5 kg/ha), thereby it is very convenient and easy to get foundation seed for best quality.

iii) Transplanting 8-12 days old seedlings resulted into increase the tillering capacity and reduced duration to 10-15 days.

iv) Less irrigation water is utilized upto panicle initiation stage.

v) Less incidence of pests & disease due to more spacing of 25 cm x 25 cm.

vi) Less cost of cultivation due to low input use like fertilizer & pesticides.

vii) Profuse & robust root systems lead to healthy plant growth and uptake of nitrogen is more.

viii) Due to operation of cono weeder, good root system developed thereby more tillers / hill.

ix) More number of grains / panicle 300 – 400 grains / panicle.

x) Grain weight is more & quality of seed is good.

xi) Increased yield, thereby increase in returns.

SRI method is being followed manually in general and row to row distance and plant to plant distances are marked manually using threads/ rope (Anonymous, 2010a). This is very cumbersome process especially in puddled fields. Special markers have been developed for line transplanting in cross rows at 25 x 25 cm row-to-row and plant-to-plant spacing (Fig. 4.130a). Lines are marked with this device in the field (Fig. 4.130b) and rice nursery is transplanted at cross-sections by male/female workers under bending posture.

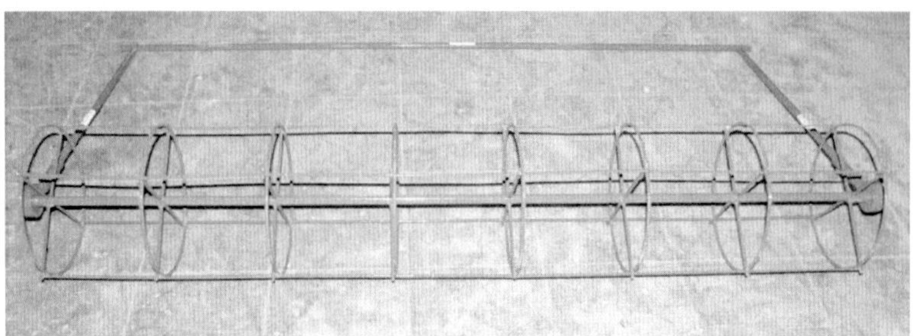

a) Eight row marker

(b) Line marking by row marker in the field

Fig. 4.130 : Row marker for manual transplanting of rice seedlings under SRI.

Efforts have been made to develop 8 row rice transplanter for SRI which can transplant paddy as a SRI requirement and the single row manually operated power weeder to reduce drudgery and to increase the labour productivity (Anonymous, 2010; Anonymous, 2010a). The existing engine operated paddy transplanter (VST make) has been modified to have plant spacing suited to SRI cultivation of rice. It is a self propelled, riding type, 8 row machine operated by a 2.94 kW diesel engine (Anonymous 2010 & 2010b). The machine consists of an engine, power transmission system, and seat for operator, main frame and rice transplanting tray, float and four pairs of transplanting units (Fig. 4.131). It has only one lugged traction (front) wheel and the weight of the machine rests on the lugged wheel and float at the time of transplanting. For transportation the front lugged wheel is replaced by pneumatic wheel and two small wheels are fixed at the rear of the float to support the weight of the machine. Power from the engine is transmitted to front traction wheel through gear train and to the transmission housing of transplanting unit through universal shaft. Rest of the functions is same as of 8-row riding type engine operated paddy transplanter.

Fig. 4.131: Stationary view paddy transplanter (VST make)
Courtesy: VST Tillers Tractors Pvt. Ltd. Bangalore

REFERENCES

Anonymous. 2007. Farmers Experiences in Rice Intensification (SRI). ANGRAU Hyderabad.

Anonymous. 2008. Research Highlight. AICRP on Farm Implements and Machinery, CIAE Bhopal. Technical Bulletin No.: CIAE/2008/141.

Anonymous. 2008a. Annual Report (2006-08). AICRP on Farm Implements and Machinery, OUAT Bhubneshwar.

Anonymous. 2008b. Annual Report (2006-08). AICRP on Farm Implements and Machinery, MPUAT Udaipur.

Anonymous. 2008c. NAAS Report on the evaluation of Plan Scheme Central Institute of Agricultural Engineering, Bhopal for the X Five Year Plan (2002-2007). NAAS New Delhi

Anonymous. 2008d. Success Stories. AICRP on Farm Implements and Machinery, CIAE Bhopal. Extension Bulletin No.: CIAE/FIM/2008/80.

Anonymous. 2010. Research Highlight. AICRP on Farm Implements and Machinery, CIAE Bhopal. Technical Bulletin No.: CIAE/2010/151.

Anonymous. 2010a. Annual Reports (2008-10). AICRP on Farm Implements and Machinery, ANGRAU Hyderabad.

Anonymous. 2010b. Annual Reports (2008-10). AICRP on Farm Implements and Machinery, PAU Ludhiana.

Anonymous. 2010c. Success Stories. AICRP on Farm Implements and Machinery, CIAE Bhopal. Extension Bulletin No.: CIAE/FIM/2010/83.

Anonymous. 2012. Research work on Potato cultivation. Department of Farm Machinery & Power Engg. PAU, Ludhiana

Anonymous. 2013. Research Highlight. AICRP on Farm Implements and Machinery, CIAE Bhopal. Technical Bulletin No.: CIAE/2013/158.

Anonymous. 2013a. Success Stories. AICRP on Farm Implements and Machinery, CIAE Bhopal. Extension Bulletin No.: CIAE/FIM/2013/40.

Anonymous. 2013b. Annual Report (2012-13). Niche Area of excellence (NAE), Farm Mechanization in Rainfed agriculture. Department of Farm Machinery & Power, IGKV, Raipur.

Anonymous. 2014. Annual Report (2013-14). Niche Area of excellence (NAE), Farm Mechanization in Rainfed agriculture. Department of Farm Machinery & Power, IGKV, Raipur.

Boa W. 1984. The design and performance of an automatic transplanter for field vegetables. *J. Agric. Engng. Res.* 30:123-130.

Chaudhuri D; Singh R C. 2013. Improved Technology for Utilization of Draught animals. AICRP on Utilization of Animal Energy with Enhanced System Efficiency, CIAE Bhopal. Technical Bulletin No.: CIAE/UAE/2013/160.

Chaudhuri, D., V.V. Singh and A K Dubey. 1999. Final Report, Survey of Existing Status on Seeding/Planting Methods and Determination of Mechanization Needs for Important Vegetable Crops in India

Chauhan A M; Singh Surendra; Rawal G S. 1984. Role of seeding and planting equipment in farm mechanisation. Paper presented at the all India seminar on,"Role of Agricultural Engineering in Rural Development" organised by the Institution of Engineers, Agril. Engg. Division, New Delhi.

Garg I K. 1987. Annual Report of ICAR Co-ordinated scheme for Research and Development of Farm Implements and Machinery. Department of Farm Power and Machinery, PAU, Ludhiana. Pp 47-69.

Garg I K. 1992. Investigations on the design and operational aspects of manually pulled engine operated paddy transplanter using mat type seedlings, unpublished Ph.D. Thesis. PAU Ludhiana.

Garg I K; Sharma V K. 1984. Development and evaluation of a manually operated paddy transplanter. *J. Agric. Engg. ISAE* XXI (1-2) 17-24

Garg I K; Sharma V K. 1985. Design, Development and Evaluation of PAU Riding type Engine Operated Paddy Transplanter Using MAT type Seedlings. *Proc. ISAE. SJC.* 1(2):57-63

Garg I K; Singh C P; Sharma V K. 1982. Influence of selected seedling mat parameters and planting speed on performance of Rice Transplanter. *AMA*, 13(2):27-33.

Garg I K; Singh Surendra. 2002. Farm equipment for Punjab agriculture. Department of Farm Power & Machinery, Punjab Agricultural University, Ludhiana.

Kumar R M et al. 2008. Package of Practices for SRI. Directorate of Rice Research (ICAR), Hyderabad. DRR Technical Bulletin No. 31/2008.

Pandey M M; Ganesan S. 2005. Farm Mechanization Package for Dryland Agriculture, Technical Bulletin No: CIAE/FIM/2005/117. Central Institute of Agricultural Engineering, Bhopal.

Pandey M M; Majumdar K L; Singh Gyanendra; Singh Gajendra. 1997. Farm Machinery Research Digest. Technical Bulletin No. CIAE/97/69, Central Institute of Agricultural Engineering, Bhopal, 328 p.

Pandey M M; S Ganesan; R K Tiwari. 2006. Improved Farm Tools and Equipment for North Eastern Hills Region, Technical Bulletin No. CIAE/2006/121. Central Institute of Agricultural Engineering, Bhopal.

Ravindra Dnyansagar, *et al.* 2005. Innovative Management Information System for Integrated Cane Development and Management. The Sugar Technologists Association of India. Proceedings of the 8th Joint Annual Convention of STAI, DSTA. SISSTA, 13-15 August, 2005, Hyderabad.

Satyanarayana A; Babu K S. 2004. Manual on System of Rice Intensification (SRI) – A revolutionary Method of Rice Cultivation. ANGRAU Hyderabad.

Shaw L N. 1997. Automatic transplanter for vegetables. *Proc. Fla. Stat Hort. Soc.* 110:262-263.

Shukla L N. 2001. Development and Field Performance of Strip-Till-Drill. Proceedings of Summer School on 'Advances in Seeding and Harvesting Machinery'. Dept. Farm Power & Machinery, PAU Ludhiana. May 21 to June 19. Pages 52-59.

Shukla L N; Chauhan A M; Dhaliwal I S. 1990. Minimum tillage machine for development countries for early 21st century. Proceddings of the International Agricultural Engineering Conference and Exhibition, Bangkok, Thailand, Vol. 1, pp: 381-388, Dec. 3-6, 1990.

Shukla L N; Chauhan A M; Dhaliwal I S; Verma S R. 1996. Development of Minimum Till Planting Machinery. Published in *AMA*, Vol. 27(4):15-18

Shukla L N; Tandon S K; Verma S R. 1984a. Development and field evaluation of coulter attachment for direct drilling. Published in *AMA*, Vol. XV(3), 19-21, 24, Summer 1984, Tokyo, Japan.

Shukla L N; Verma S R; Singh Amar. 1978. Development of a semi-automatic tractor drawn roto drum sugarcane planter. *J.Agric. Engg.* XV(4) : 211-214.

Shukla L N; Verma S R; Singh Surendra. 1984. Tractor-Drawn Rota-Drum Sugarcane Planter. *Punjab Agricultural University, Ludhiana.*

Sidhu H S. 2008. Happy seeder- An effort for rice residue management. *Indian Journal of Air pollution Control.* Vol. III(i): 68-75.

Singh A K; Sharma M P. 2008. IISR sugarcane cutter - planter. Operation Manual No. AE/08/01, IISR Lucknow: p 6.

Singh Jaskaran. 2001. Equipment for Vegetable Planting and Transplanting. Proceedings of Summer School on 'Advances in Seeding and Harvesting Machinery'. Dept. Farm Power & Machinery, PAU Ludhiana. May 21 to June 19. Pages 260-266.

Singh P R. 2014. Souvenir, Tractor and Agricultural Machinery Manufacturers' Meet (TAMM 2014). Held at IISR Lucknow during February 8-9.

Singh Surendra. 2007. Farm Machinery – Principles and Applications. Directorate of Information & Publication of Agriculture, Indian Council of Agricultural Research, Krishi Anusandhan Bhawan-I, Pusa Campus, New Delhi.

Singh Surendra; Manes G S; Dixit Anoop. 2015. Mechanization of cultivated crops. New India Publishing Agency, Pitam Pura, New Delhi.

Singh Surendra; Pandey M M. 2008. X Plan Achievements (2002-2007). AICRP on Farm Implements and Machinery, CIAE Bhopal. Technical Bulletin No.: CIAE/2008/137.

Singh Surendra; Sharma V K; Gupta P K; Chauhan A M. 1985. A ridge planter for winter maize. *J. of Agril. Engg.* 22(2): 115-18.

Singh Surendra; Sharma V K; Gupta P K; Chauhan A M. 1988. Design, development and evaluation of two-row tractor mounted ridge planter for winter maize. *Indian J. of Agricultural Sciences.* 58(1): 45-47.

Singh Surendra; Singh R S. 2014. Energy for Production Agriculture. Directorate of Information & Publication of Agriculture, Indian Council of Agricultural Research, Krishi Anusandhan Bhawan-I, Pusa Campus, New Delhi.

Singh Surendra; Verma S R. 2009. Farm Machinery Maintenance & Management. Directorate of Information & Publication of Agriculture, Indian Council of Agricultural Research, Krishi Anusandhan Bhawan-I, Pusa Campus, New Delhi.

Srivastava N S L. 2001. Advances in sugarcane planting equipment. Proceedings of Summer School on 'Advances in Seeding and Harvesting Machinery'. Dept. Farm Power & Machinery, PAU Ludhiana. May 21 to June 19. Pages 97-100.

Srivastava N S L; Sharma M P. 1977. A power operated sett cutting machine. *Journal of Agril. Engg.* Vol. 14(3):119-120.

Verma S R; Sharma V K; Singh Surendra; Ahuja S S; Singh Santokh (eds.). 1992. Proceedings of National Colloquium on Potato Mechanization in India. *Punjab Agricultural University, Ludhiana.*

Yadav R N S. 2003. Sugarcane Production Machinery, CIAE, Bhopal. Bulletin No. CIAE/NATP/-SM/2002/95.

Yadav R N S. 2004. Present status and future need of mechanization of sugarcane planting in India. Power Machinery Systems and Ergonomics, Safety and Health, etc 2004. Proceedings IIT Kharagpur. December 14-17, 2004.

❑❑❑

5

Weeding and Fertilizer Application Equipment

5.1 Weeding and Interculture Equipment

The aim of weeding and intercultivation operation is to provide best opportunity to crop to establish and grow vigorously up to time of maturity. The main objectives of weed control are to improve the soil conditions by reducing evaporation from the soil surface, improve infiltration of rain or surface water, and reduce runoff to maintain ridges or beds on which the crop is grown and to reduce competition of weeds for light, nutrients and water. Mechanical methods of weed control are simple and easily understood by farmers. The tools and implements for mechanical weed control are mostly manual and animal operated. Manual method is most effective but is slow. It is popular in regions where labour wages are low and labour is easily available during the season. The additional cost of weeding using implements is comparatively less than the gains due to extra yields obtained. First weeding operation is mostly done between and along the rows. Remaining operations are done mostly between the rows. Hand hoes are generally used for removing weeds between plants in a row.

Weed control during kharif (rainy) season is a big problem for the farmers. To get a good kharif crop the weeds should be suppressed for at least up to first 40 days for 120-130 days crops and for 70 days for the crops like cotton; otherwise the production declines. Hand weeding if done in wet fields, results into increase in soil density due to trampling. It is costly and many times human labour is not available at critical stages. Hand weeding operation becomes inefficient if weeds are not removed completely, only shoots are removed and root portion remains in the ground. Weeds again come up after 3-4 days and earlier efforts made become futile. The work output in moist and weedy fields is very low requiring more labour per unit area. The intermediate technology generally adopted is to scratch the land with hoe or harrow and then complete the uprooting and collection of weeds in loose soil. This improves the work efficiency of labours and output increases.

The first tool used for control of weeds was, perhaps, the hoe. During ancient time crops were sown by broadcasting and hoe was the only tool used to kill weeds among plants (Singh, 2007). When crops are sown in rows, the ploughs or cultivators can be used to destroy the weeds between the rows and throw soil over young weed seedlings in the crop row and smothered them to death. There has been much advancement in weed-control equipment/technologies. Many types of equipment are in use now a days, ranging from small hand-pushed garden tools suitable for family garden to large tractor-mounted cultivators capable of cultivating tens of hectares per day. The type and size needed depends upon the area to be covered, crop for which it is to be used, soil type and conditions, rainfall, type of farming practiced and kind of power source available. Typical work rates of hand hoe (Khurpi) is varying from 300-500 man-h/ha. For hand hoeing between rows, by chopping hoe, labour requirement varies from 200-300 man-h/ha (Sharma *et al.*, 1987). Operation of the push-pull type weeder along the row in typical conditions requires 100-125 man-h/ha. For animal drawn weeding tools (blade hoe and blade harrow) labour requirement varies from 6-20 man-h/ha. The benefit of using improved weeding tools are, reduction in time requirement, reduction in human effort and effectiveness of operation. The time saved by use of these implements may be utilized in better care and management of crop. Thus higher yields can be obtained. The choice and selection of implements depends on operational and economic factors. The aim should be to introduce new tools and implements to reduce drudgery and achieve timeliness of operation.

Types of weeding tools

The weeding tools and equipment are categorized based on their power source, i.e. manual, animal drawn and power or tractor operated. Small weeding tools or aids are traditional hand held type hoes like *"Khurpi"* (Fig. 5.1) used by the farmers (Singh, 2007; Singh *et al.*, 2015; Bhardwaj *et al.*, 2004a). These tools are operated in squatting posture and have very low work output. Different designs of these tools are being used by the farmers of different regions. These tools are suitable for removing the weeds between plants in both row-sown and broadcast fields and are quite efficient. Spades (Fig. 5.2) or chopping hoes work on the principle of impact and have straight, curved or pronged blades. Weeds are removed by digging, cutting and uprooting. These are operated in the bending posture. The operation is normally slow and tiring. Long handle tools/weeders have a soil working tool mounted at the end of a 1.5 to 2 metre long wooden/bamboo handle (Fig. 5.3 and Fig. 5.4). These tools are operated in push or push-pull or pull mode and in standing posture. The soil working tool consists of one or more blades of different shape and size mounted on tines which in turn are fixed on a socket for fitting to the handle. The common shapes of blades on these weeders are straight, convex, V-shape, sweep, serrated, etc. These weeders weigh 1.5 to 2.5 kg. These are designed to work under friable soil moisture conditions and give high work output at the early stages of crop growth when weeds are small. The shape and size of hoe vary from place to place. The most commonly used are blade hoes and tined hoes. The wheel hoe with bicycle wheel rim (Fig. 5.5) gives the highest work rate of about 0.6 ha/day followed by wheel hoe (0.4 ha/day), kasola (0.2 ha/day) and khurpa (0.08 ha/day). The high work rate in case of wheel hoe with bicycle wheel rim is because of its comparatively high operating speed and lower effort requirements in operating the wheel hoes. The work rate of 'kasola' is higher than 'khurpa' because 'kasola' is operated in a standing posture, which renders the operator more mobile. Low work rate with 'khurpa' is mainly because of its low working speed due to sitting working posture. Table 5.1 gives the details of various hand tools used in weeding and their comparative performance (Sharma *et al.,* 1987).

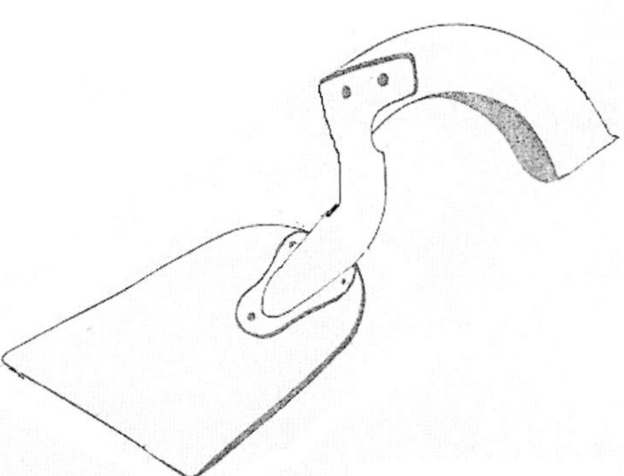

Fig. 5.1 : An isometric view of 'Khurpi'

Fig. 5.2 : Spade

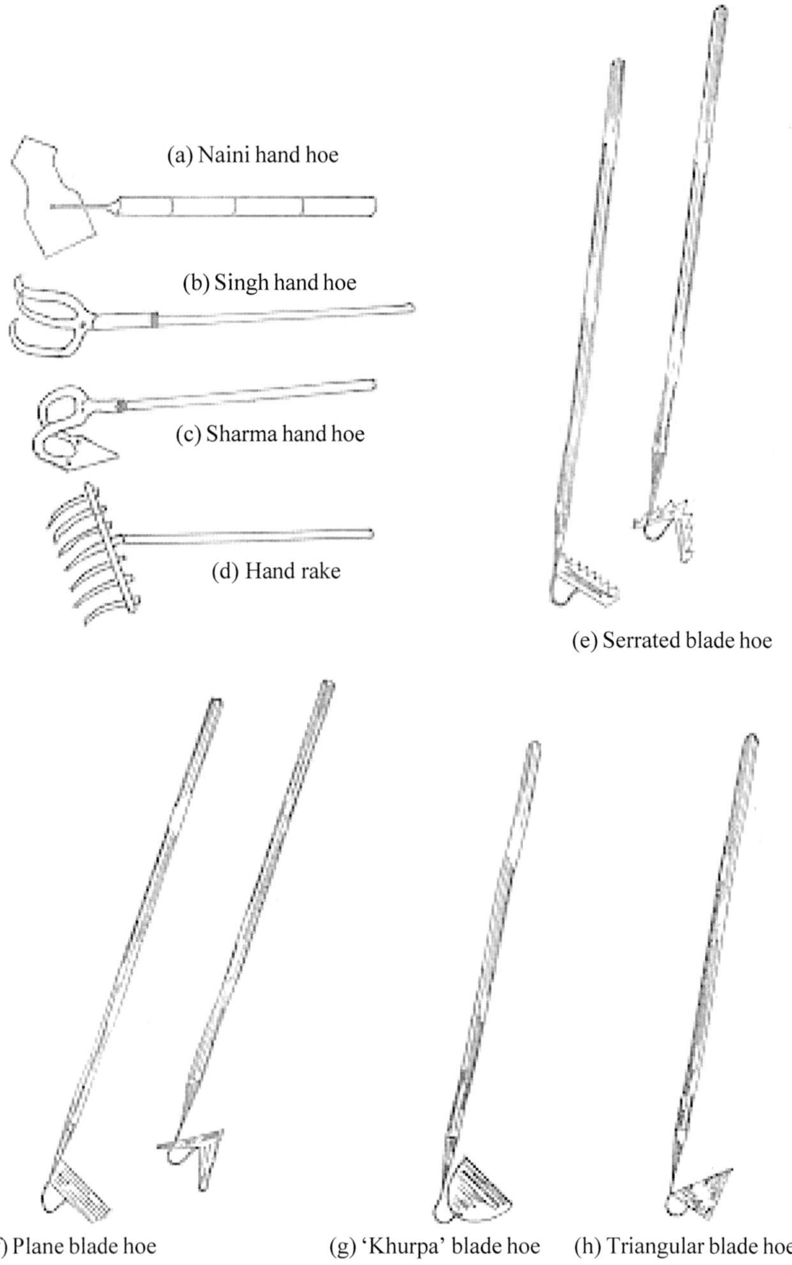

Fig. 5.3 : Different types of hand hoes

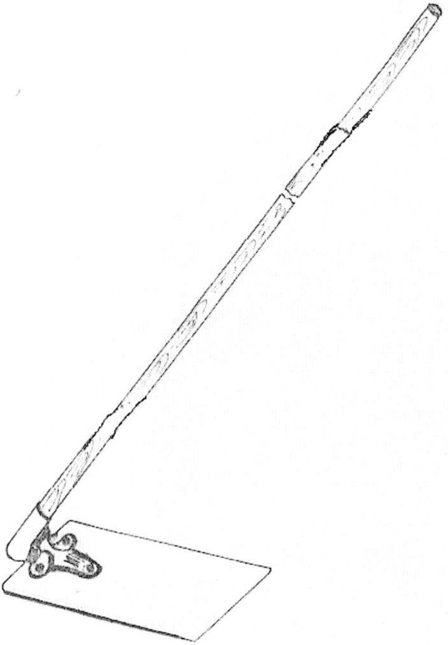

Fig. 5.4 : A view of Kasola

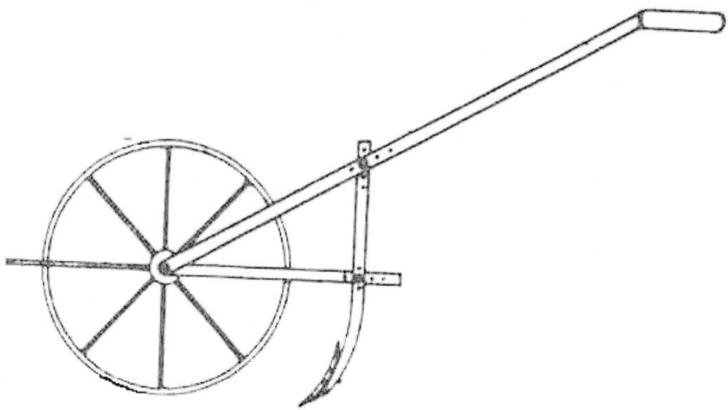

Fig. 5.5 : Wheel hoe with bicycle wheel rim

Table 5.1: Details of various equipment used in weeding and their comparative performance.

Weeding Technology	Blade width Cm	Weight kg	Wheel diameter cm	Overall length cm	Labour require-ment man-h/ha	Speed of operation km/h	Width covered/ run, cm	Yieldkg/ ha
Knapsack sprayer	-	7.0 (empty)	-	-	13.6	0.65	135.0	3900
Khurpa	11.2	0.35	-	28.2	101.7	0.53	22.5	3984
Kasola	10.0	1.0	-	150.0	48.3	0.98	22.5	4034
Wheelhoe with bicyclerim	11.0	7.0	65.0	115.0	12.5	3.93	22.5	4150
Wheelhand hoe	14.0	8.0	15.0	175.0	16.7	3.09	22.5	4050

Three tined hoes (Grubber)

It is used for weeding, interculture and breaking of the soil crust in vegetable gardens, in flower crops and nurseries (Pandey and Ganesan, 2005; Bhardwaj *et al.*, 2004). It is manually operated. The three tined hand hoes (grubber) is one of the widely used hoes for weeding and interculture in horticultural crops. It is a long handled weeding tool and can be easily manoeuvred between the rows and around the plants. The tool essentially consists of three curved tines, which are bent to appropriate shapes (Fig. 5.6). The working blade tips are forged, flattened and sharpened. The section of these tines is round or square depending upon the design. The tines are made from medium carbon steel and the tips are hardened to 40-45 HRC. The tines are either welded to the ferrule or secured in the ferrule by I-bolt, which helps in adding the number of tines. The ferrule is made from mild steel in which the wooden handle is inserted. The handle is made from good quality wood or bamboo. The hoe is of sweep type. Tine spacing can be changed but generally kept at 65 mm. Width of the blade is 100 mm. It can till up to 34 mm depth. For operating the tool it is held in standing position and pulled towards the operator, which causes suction at the tips/ edges of the tines and penetrate into the soil. The pulling action uproots the weeds and also aerates the soil. In some of the designs, small V-shape blades are welded to the tips. Working capacity of machine is 0.005 ha/h and weight of the machine 2.0 kg. Labour requirement is about 188 man-h/ha. Weeding efficiency is about 55-65% and plant damage about 10%.

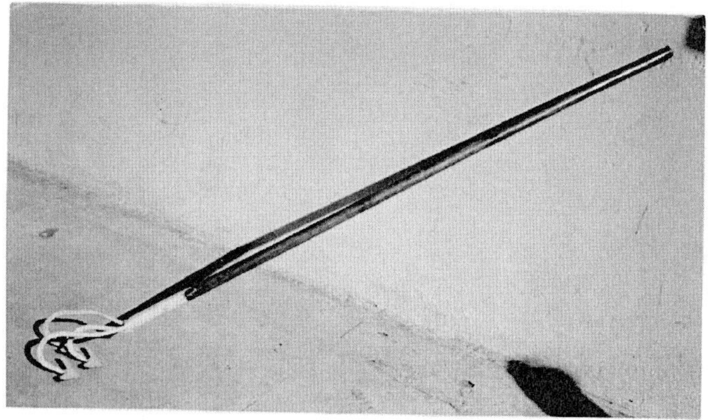

Fig. 5.6 : Manually operated three tined hoe (Grubber).

V-Blade hand hoe

The hoe is used for weeding of the vegetable crop planted in the rows and earthing operation (Bhardwaj *et al.*, 2004; Singh, 2007). The V-blade hand hoe is a long handled weeding tool for operation in between the crop rows. The hoe consists of a V-blade, arms, ferrule and a wooden handle (Fig. 5.7). The arms are welded to the ferrule. Wooden handle is inserted in the ferrule. The blade of the hoe is the important component, which enters into the soil and performs the cutting and uprooting of weeds. The blade is fabricated from medium carbon steel and the edges of the blade are hardened to 40-45 HRC. The width of blade varies from 150 to 225 mm. The arms and the ferrule are made from mild steel. The handle is made from high quality wood or bamboo. Length of handle varies from 1400 to 1600 mm depending upon local requirement. Being a long handled tool, the hoe is operated in the standing posture by pulling action towards the operator. The pulling actions cause penetration of the blade into the soil and cut or uproot the weeds. Because of the V-shape, the blade creates small furrows between the crop rows and also earthing of the plants. This operation facilitates the flow of irrigation water to the root zone of the plant.

Fig. 5.7 : A view of V-Blade hand hoe

Wheel hand hoe

It is used for weeding and interculture of vegetables and other crops sown in rows (Bhardwaj *et al.,* 2004; Singh, 2007; Singh and Verma, 2009). The wheel hoe is a widely accepted weeding tool for weeding and interculture in row crops (Fig. 5.8). It is manually operated. It is a long handled tools operated by push and pull action. As the name implies, the general construction of wheel hoe comprises of wheel assembly, miniature tool frame, a set of replaceable tools and handle assembly (Fig. 5.8c). The number of wheel varies from one to two and the diameter depends upon the design. The frame has got a provision to accommodate different types of soil working tools such as straight blade, reversible blades, sweeps, V-blade, tine cultivator, pronged hoe, miniature furrower, spike harrow (rake) etc which can be operated by a single person. The handle assembly has a provision to adjust the height of the handle to suit the operator. All the soil working components of the tool are made from medium carbon steel and hardened to 40-45 HRC. The other assemblies of the wheel hoe are made from structural mild steel and thin walled mild steel pipes. The working depth of the tool can be adjusted with the help of clamp or through the plate with multiple holes provided in the frame and welded to the tool assembly. The handle height is also adjustable. Some of the designs are provided with a pulling ring in the frame, which enables it to be operated by two persons. For operation, the working depth of the tool and handle height is adjusted and the wheel hoe is operated by repeated push- pull action which allows the soil working components to penetrate into the soil and cut/uproot the weeds in between the crop rows. With this action, the weeds also get buried in the soil. Working capacity of wheel hand hoe is 0.4 ha/day. The weight of the machine varies between 4 to 12 kg. Number of tines depends on crop and row spacing. For example; one for wheat and 3 for wider row crops like maize, cotton etc. Tine width is approximately 13.5 cm. Depth of penetration can be changed using holes provided in the shank.

a) Wheel hand hoe in operation

b) A view of machine in stationary as well as in field operation

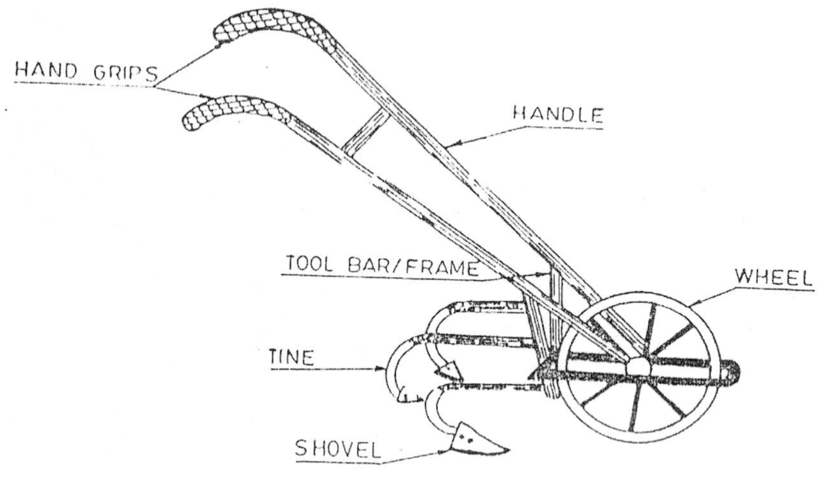

c) Components of machine

Fig. 5.8 : Wheel hand hoe

Twin wheel hoe

It is used for weeding and interculture in upland row crops in black soil region (Pandey *et al.*, 1997). It is operated by human being. It consists of two wheels, frame, V-blade fixed on a tyne, U-clamp and a handle (Fig. 5.9). The cutting and uprooting of weeds in field is done through push and pull type action of the equipment. The equipment is operated at optimum soil moisture condition and preferably after 20-25 days of sowing i.e. when the weeds are small i.e. 1 to 3 cm height for better weeding performance. It works on principle of pull and push action. Width of coverage is about 210 mm and depth of tilling 20-30 mm. Weeding efficiency is about 80% and labour requirement about 100 man-h/ha. Working capacity of twine wheel hoe is 0.015 ha/h.

Fig. 5.9 : A view of twin wheel hoe

Peg type dryland weeder

The dry land weeder (peg type) is one of the useful manually operated weeders for operation between the crop rows (Pandey and Ganesan, 2005). The weeder is used for removing weeds in vegetable gardens, basins of orchard trees and vineyard plantations. It is also used for breaking the soil crust and creation of soil mulch. It consists of a roller, which has two mild steel discs joined by mild steel rods (Fig. 5.10). The axle passes through the centre of discs and is mounted on the two arms, which also constitutes the frame. The small diamonds shaped pegs are welded on the rods in a staggered fashion. Diameter of pegged wheel is about 22 cm. The complete roller assembly is made of mild steel. The V-shaped blade follows the roller assembly and is mounted on the arms. The blade is fabricated from medium carbon steel and forged to shape. The cutting edges are hardened to 40- 45 HRC. The height of the blade can be adjusted according to the working depth. Working width of weeder is 15 cm. The arms are joined to the handle assembly, which is made from thin walled pipes. The height of the handle can also be adjusted according to the operator. For operation the weeder is repeatedly pushed and pulled in between the crop rows in the standing position. The diamonds shaped pegs penetrate into the soil and the rolling action pulverize the soil. The blade in the push mode penetrates into the soil and cuts or uproots the weeds. Shovel type dry land weeders are also popular and used by farmers (Fig. 5.11).

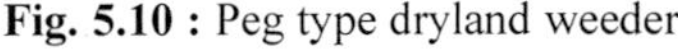

Fig. 5.10 : Peg type dryland weeder

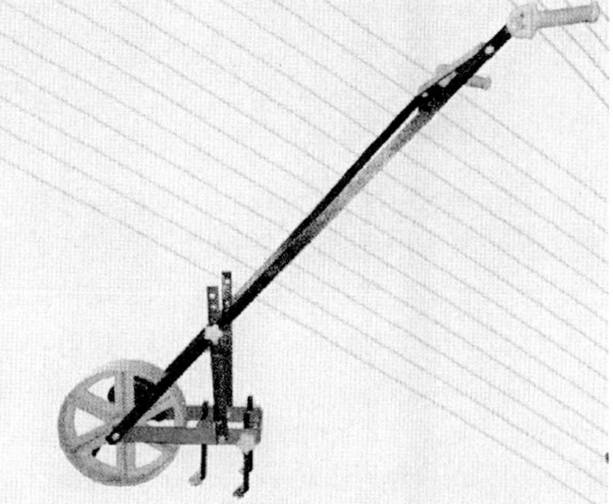

Fig. 5.11 : Shovel type dry land weeder
Courtesy: Khedut Agro Engineering Pvt. Ltd. Rajkot (Gujarat)

Paddy weeder

It is important equipment for inter-culture used in paddy cultivation. It is used for uprooting weeds and burying them in puddled soil between rows of standing paddy crop (Pandey *et al.,* 1997). It improves aeration of soil. It consists of frame, weeding roll, the tines, float and handle (Fig. 5.12). The frame is made of mild steel to which float, front and rear weeding rolls and angle regulator are attached. Weeding roll with finger like projections does the weeding operation in water. There are two weeding rolls one in front and other at rear and both are made of mild steel. Float, made of mild steel, is placed in front of weeding roll that helps in maintaining easy sliding motion during operation.

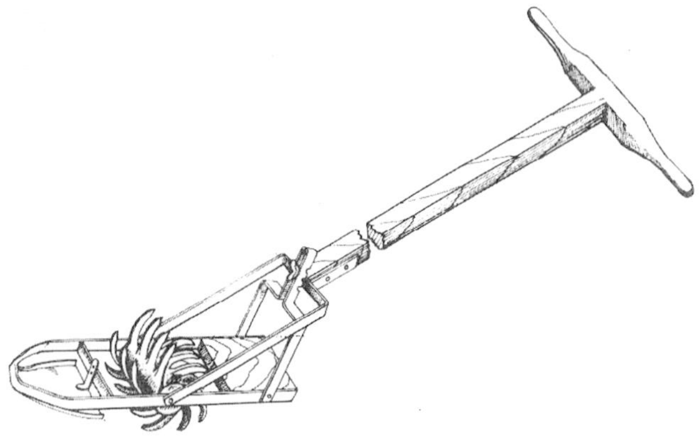

Fig. 5.12 : Paddy weeder

Cono weeder

It is used for crushing the weeds between rows of paddy crop efficiently (Pandey *et al*., 1997; Singh *et al*., 2015, Anonymous, 2008 & 2010). It is easy to operate, and does not sink in the puddled field. It is manually operated. It consisted of two rotors, float, frame and handle (Fig. 5.13). The rotors are cone frustum in shape, smooth and serrated strips are welded on the surface along its length. The rotors are mounted in tandem with opposite orientation. The float, rotors and handle are joined to the frame. The float controls working depth and does not allow rotor assembly to sink in the puddle. The cono weeder is operated by pushing action. The orientation of rotors create a back and forth movement in the top 3 cm of soil. Working capacity of cono weeder is 0.02 ha/h.

Fig. 5.13 : A view of cono weeder (Stationary and field operation)

Animal drawn weeding tools

Animal-drawn hoes are also commonly used in various parts of country. In row crops, interculture and weeding operations could be done quickly and efficiently by using improved animal drawn implements. It is essential to provide wider row spacing (above 30 cm), for movement of animals and implement, if animal drawn weeders are to be used (Pandey *et al.*, 1997). Animal drawn single row hoes are most widely used by the farmers of different states. Straight or slightly curved blades are commonly used in single row hoes. The size of blade can be changed as per crop row spacing. Multi-row units are also used in some states for wide coverage and timely weeding operation. The three tine cultivator or '*Triphali'*, 'Akola' hoe, 'Bardoli' hoe and animal drawn sweeps of different designs are some of the new animal drawn weeders.

Animal drawn tool frame for weeding

It is suitable for interculture and weeding operations in row crops. The equipment is operated by a pair of bullock (Chaudhuri and Singh, 2013). It is an attachment to CIAE multipurpose animal drawn tool frame (Fig. 5.14). Sweep type tynes are mounted on a steerable tool bar behind the tool frame.

The number of rows is 4. It saves 90 per cent labour and operating time and 88 per cent on cost of operation compared to conventional method of using hand hoe (*Khurpa*). Working capacity is 0.15 to 0.22 ha/h.

Fig. 5.14 : A view of animal drawn tool frame for weeding

Animal-drawn weeder

The faster intercultivation with low soil compaction during the limited period of available time results into higher crop growth and yield due to aeration at root zone and nodulation. Animal-drawn weeder (Blade hoe type) developed at CIAE Bhopal is useful for faster weeding and interculture in soybean and other wide spaced crops compared to the manual weeding by hand hoe (Chaudhuri and Singh, 2013). The animal-drwan weeder can cover 0.08 ha per hour (Fig. 5.15). IGKV Raipur has also developed two row animal drawn weeder (Fig. 5.16). It can cover 0.06 ha/h (Chaudhuri and Singh, 2013; Anonymous, 2013b & 2014).

Fig. 5.15 : Animal-drawn weeder (Blade hoe type)

Fig. 5.16 : Animal drawn IGKV weeder

Animal drawn biasi plough

Animal drawn improved biasi cultivator developed at CIAE, Bhopal and IGKVV, Raipur is used for biasi operation in row-seeded rice. In traditional practice, country plough is used in broadcasted rice, which is time consuming and causes high plant mortality. The animal drawn 5-tine biasi plough (Fig. 5.17) is being used in rice crop (Anonymous, 2013b & 2014; Anonymous, 2015). It works well in crop height ranging from 20 to 30 cm. The field capacity of plough is 0.17 ha/h and weeding efficiency 66%. The two row improved biasi cultivator (Improved wedge plough) gives higher work rate and reduces plant mortality compared to the traditional practice (Fig. 5.18). The effective field capacity of two row biasi cultivator is 0.06 ha/h.

Fig. 5.17 : Biasi operation with 5 row plough

Fig. 5.18 : Biasi operation with 2 row cultivator

Power operated weeding tools

With increasing row crop cultivation at various row widths, it often becomes necessary to use different sizes and shapes of blades to intercultivate growing crops. The implements for intercultivation can therefore be designed with quick attaching working parts like exchangeable blades to make them versatile. The work rate for various weeding implements vary due to variation in crop growth, row and plant spacings, weed intensity, soil conditions and other factors. Some designs of small engine operated tools have been developed for inter-row cultivation; however, the cost of operation on small farms with a power operated weeder is higher than that with push-pull type weeder. Therefore, their usefulness is limited. Tractor operated implements can be used for intercultivation but these require wider row spacing and leaving of space at the headlands for allowing the tractor to operate and turn before entering into the rows.

Rotary weeders

The rotary hoe is a cultivating implement used to cultivate and destroy weeds and grass around young plants. When rain causes a hard crust over soil and hinders the emergence of young seedlings, rotary hoe is excellent tool for breaking the crust. Where there is a large amount of crop residue on the surface in a fluffy condition, the rotary hoe is useful in packing the residue down into the surface soil. This makes the use of field cultivators and rod weeder easier. Rotary hoes work fairly well at higher forward speed.

Engine operated dry land weeder

Engine operated dry land weeder is suitable for row crops like maize, cotton, vegetables, pulses, medicinal plants etc (Pandey & Ganesan, 2005;

Pandey *et al.*, 2006). It consists of engine run by diesel/petrol, rotary blades, and two tine holders, oil lubricated gear box, handle and two supporting wheels (Fig. 5.19). Single row and two row models are available. It has working width of 30 cm for single row and 38 cm for two row suitable for weeding in 30 and 45 cm inter crop distance. Row spacing is adjustable. Rotor speed is 170 rpm. It has 8 tines for single row machine and 18 tines for two row machines. It is light weight, easy to operate and has weeding efficiency more than 90% depending upon crop and soil conditions. Single row weeder is operated by 1.7 hp petrol engine and two row weeder by 1.9 hp petrol engine.

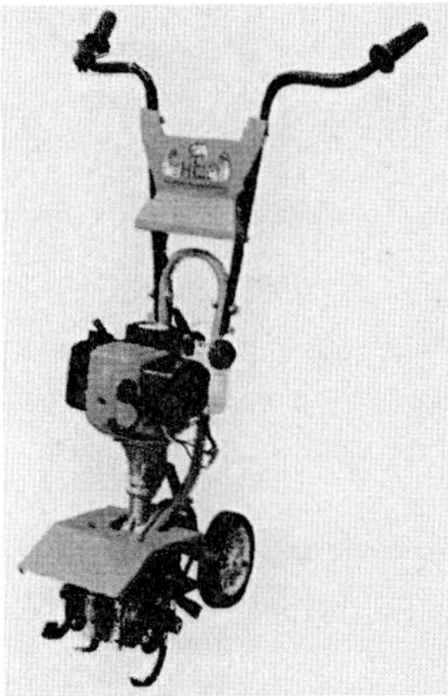

a) Engine operated single row weeder

b) Engine operated two row weeder

Fig. 5.19 : Engine operated rotary dryland weeder

Courtesy: Renaaissance Power Products Pvt. Ltd. Coimbatore (Tamil Nadu)

Engine operated manually controlled low land paddy weeder

The machine consists of an engine run by diesel/petrol, rotary blades, casing, handle, and float (Fig. 5.20). It can be single row, two row and three row machine (Anonymous, 2008 & 2010). Two row weeder has 1.75 hp petrol engines operating at 6500 rpm (Fig. 5.20a). Rotary attachment is operated at 170 rpm. Weeding width can be adjusted between 120 to 200 mm and suitable for row spacing 18 to 30 cm. This is light weight machine and has 6 tines on each rotor giving better weeding performance. A view of three row machine is given in Fig. 5.20c.

a) Engine operated two row paddy weeder
Courtesy: Renaissance Power Products Pvt. Ltd. Coimbatore (Tamil Nadu)

b) Engine operated single row paddy weeder

c) Engine operated three row paddy weeder

Fig. 5.20 : Engine operated manually controlled low land paddy weeder

Self-propelled power weeder

Power tillers are commercially successful only in areas for wetland cultivation. The use of power tillers in upland cultivation is negligible. In paddy cultivation especially on small and medium farms and for the farm operations in hilly areas, orchards and forestry, power tiller is the only solution. The existing power tillers are too heavy to be used for hilly area. The equipment has to be lifted for shifting from one field to another. So there is a need for developing a lightweight power tiller, which can be used for doing various farm operations (Singh, 2007; Singh and Pandey, 2008; Garg and Singh, 2002). Also, cost of this power tiller should not be very high. It should be able to help small farmers to mechanize the farm operations at lower initial cost. The operational cost of this power tiller for doing various jobs should be less in comparison to traditional power system being used presently. It would also reduce the labour requirement and would enhance efficiency and effectiveness of operation. The development of machine would help to mechanize hilly area, small land holding, and plantation crops and also wide row crops like cotton, sunflower, sugarcane etc. Also, development of this machine helps to mechanize weeding operation under the horticultural trees, which at present is a big problem. In hilly area the cultivation of land is generally done by using bullock-operated implements and intercultural operation is carried out manually. Also the weeding in row crop is carried out using manual hand tools like wheel hand hoe, *khurpa*, *khaola* etc., or by using weedicides. The manual weeding is very labour intensive. Some farmers uses tractor with cultivator or harrow for interculture operation in orchards. This method is not only costly but also in effective due to non-weeding near the plant row due to obstruction of branches of the plants.

It is engine operated machine used for weeding purposes (Anonymous, 2012; Anonymous, 2013; Garg and Singh, 2002). A light weight self-propelled power weeder consists of a 4.1 kW diesel engine mounted on the power tiller chassis, power transmission system, two MS wheels, a frame and a rotary (Fig. 5.21a). The width of the rotary varies from 40-45 cm. The power from the engine is provided to traction wheels and rotary with the help of a gear trains and belts and pulleys. The machine is operated at rotary speed of 160-180 rpm to achieve the constant forward speed. The rotary has been provided with 16 blades fitted on high-pressure pipe of 37.5 mm diameter with the help of nuts and bolts to the flanges. For depth adjustment two skids made of flat are provided on both sides of power tiller. A power cut off device is provided to engage or disengage the power to the rotary system. The rotary blades are made of high carbon steel (EN-31), which acts as working tool for weeding or seedbed preparation. The working width of the machine is about 45 cm (adjustable). The wheels with lugs are provided for traction. The unit is employed with simple mechanism of power transmission as the gear train is replaced with the help of pulley and sprockets, resulting in less cost of maintenance. It is used for weeding in wider row crops and for tillage in orchard crops. The clutch is also provided on both sides for turning the machine to right or left. The rotary blades mostly of L-type are the soil working tool for weeding or seedbed preparation. The wheels are made from MS flat and have lugs. The depth of cut varies from 4-7 cm. The weed control is observed to be 88-94%. The field capacity of the machine is 0.13 ha/h at operational speed of 1.8 - 2.0 km/h for weeding in sugarcane and cotton crops (Fig. 5.21b). The weeding index is 78.0 - 85.0% at 6 cm depth of operation and 83.0 - 88.0% at 10 cm depth of operation.

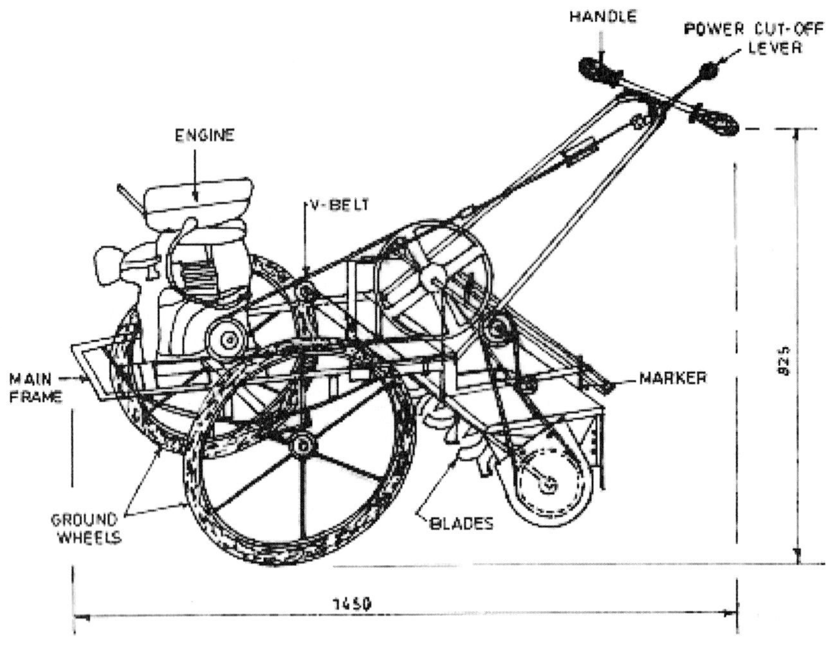

a) Components of rotary power weeder

b) Light weight power in field operation

Fig. 5.21 : Power weeder in operation

Power tiller operated sweep cultivator

It is used for performing interculture operations (Bhardwaj *et al.*, 2004 & 2004a). This machine has been specially designed for operation with a power tiller of 6-8 hp to perform interculture operations in standing crop in soybean, sorghum, Bengal gram, pigeon pea etc where the row spacing is wide enough for the power tiller to pass without damage to the plants (Fig. 5.22). One depth control wheel is provided at its rear for adjusting the depth of operation. It is suitable for medium and light soils. It is provided with depth control gauge wheels. Some of its major components are the main frame with hitch system, handle, and drive wheel, tynes. Working width is about 135 cm and depth of operation 5 cm. With the use of this machine weeding efficiency is 50-60% and plant damage 0.5-1.5%. It can be operated at the speed of 1.5-2.5 km/h. It can cover 0.18-0.25 ha/h. It requires 4-6 man-h/ha.

Fig. 5.22 : Power tiller operated sweep cultivator

Engine operated power weeder

The CIAE engine operated power weeder is suitable for weeding and intercultural operations in upland row crops like groundnut, maize, soybean, pigeon pea, etc. sown at row-to-row spacing of more than 30 cm (Bhardwaj *et al.*, 2004a). The engine operated weeder consists of a box section chassis, 3-hp petrol – start- kerosene- run engine, transmission system, drive wheels, and tool mounting bar of 70 × 70 mm size (Fig. 5.23). The engine, gearbox and tool mounting bar are fixed to main chassis. The engine is mounted on front side of the chassis on proper foundation. The gearbox with transmission system is mounted in the middle of chassis. Two steel lugged drive wheels of 500 mm diameter and 110 mm width are mounted on both ends of hexagonal shaft connected to transmission box. The wheel tread (spacing between two wheels) can be adjusted from 400 to 650 mm to suit the row-to-row spacing of different crops. The V-shaped weeding sweeps of 150 and 200 mm size are fitted on tynes. These tynes are mounted on the tool bar with clamps. The tynes are raised or lowered for adjusting the depth of operation and locked in position by the clamp. The spacing between sweeps is easily adjusted by sliding the tynes on tool mounting bar to suit the crop row spacing. Power from engine is transmitted to gearbox through V-belt and pulleys. Inside the transmission box power is transmitted through chain and sprockets to two hexagonal drive shafts/axles. A lever operated clutch/ idler pulley is used for tightening/ loosening of V-belt, transmitting the power from engine to transmission box. For operation, the tynes are adjusted to the raised position so they do not dig into the ground and the weeder is transported to the field. The tynes are adjusted according to the row spacing of the crop and depth of operation. It gives weeding efficiency of 60-70% at field capacity of 0.15 ha/h.

Fig. 5.23 : Engine operated power weeder

Power weeder for horticultural crops

The power weeder is suitable for interculture operations in orchard crops (Singh and Pandey, 2008). It consists of 4 blades, mounted on an adjustable shaft suitable for weeding in various horticulture crops (Fig. 5.24). The blades are rotated by 1.4 hp Mitsubishi engine operated with petrol. The effective with of the weeder is 25 cm. Depending on the requirement; the weeder can be operated on 4, 3 or 2 wheels with 25 cm, 20 cm or 16 cm operating width. The weeder is of 1m height and has 10.4 kg weight. This is suitable for weeding in chillies, turmeric, sugarcane, banana, vegetable crops, sweet orange, lemon, papaya, medicinal plants and other row crops like jowar, groundnut and mulberry.

Fig. 5.24 : Power weeder in operation

Tractor operated rotary weeder

The purpose of weeding and inter-culture operation is to provide best conditions for crop to establish and grow vigorously. Both mechanical and chemical methods are effective for weed control but mechanical method is preferred to chemical method because weedicides are not only expensive but some of the weedicides are injurious to crops and human beings. On the other hand, mechanical weeding keeps the soil surface loose which results in better aeration and moisture conservation thus enhances quality of produce. Mechanical weeding at present is done manually by khurpa, wheel hoe and tri-phalis. On the other hand rotary power weeder can accomplish better job. It can mix the weeds thoroughly. The self propelled rotary power has been developed for weeding in orchards, cotton and other row crops. But the capacity of the machine is very low. So there is scope for tractor operated rotary weeder for wider row crops be introduced due to higher field capacity and less human fatigue (Anonymous 2008, 2010; Singh and Pandey, 2008). It consists of a main frame, gearbox, three rotary weeding blade assemblies, a square shaft for transmission of power from gearbox to rotary assemblies and sets of sprockets and chains (Fig. 5.25). Main frame consists of a 2540 x 80 x 80 mm square box made from C-channel. A standard 3-point hitch arrangement had been provided to mount the frame to tractor. Power from tractor PTO is transmitted to main square shaft through gearbox mounted on main frame, set of sprockets and chain. The speed reduction at gear box is 9:5. The weeder encompasses three sets of rotary blade assemblies (Fig. 5.26). Power to these assemblies is provided from main shaft with the help of chain and sprockets. Power from gearbox to blade assemblies is provided by 40 mm square shaft. This facilitated the adjustment of row to row spacing from approximately 675 to 1165 mm. Working capacity of weeder is 0.3-0.8 ha/h. The machine is operated with tractors of 35 hp and above (Fig. 5.27). It could be operated in 1st low and 2nd low gears. The machine tills soil strips of approximately 50 cm between two rows of wider row crops. Three soil strips are tilled simultaneously. Approximately 4.5 cm strip remained untilled in some cases when soil is very firm otherwise the soil in this untilled region gets fractured. The machine uproots and mixes all the weeds in strip which is tilled by rotors. However the weeds in untilled soil have to be removed manually. Approximately 60 man-h/ha are required for removing weeds in crop sown at row to row spacing of 75 cm. Generally a group of 20-25 persons complete weeding of one hectare sugarcane in 10-12 hours.

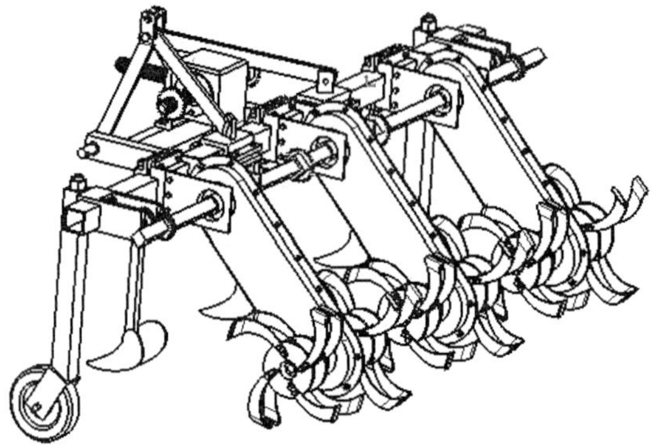

Fig. 5.25 : Isometric view of tractor operated rotary weeder

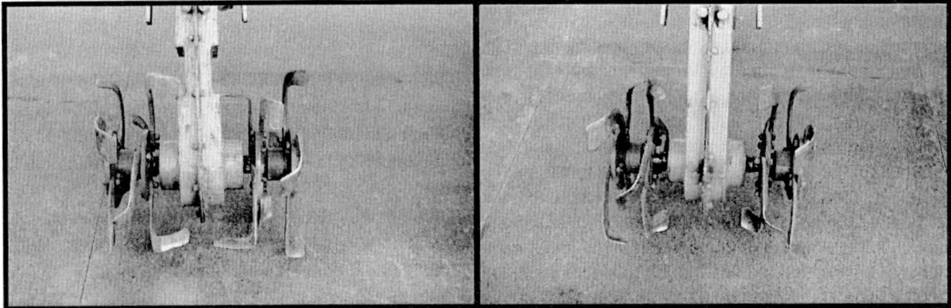

Fig. 5.26 : A view of rotary attachments

Fig. 5.27 : Tractor operated 3-row rotary weeder operation in sugarcane and cotton fields

Row crop cultivators

Row crop cultivators also called field cultivators (Fig. 5.28) have working width of 3-6 m. It has working capacity of 2-5 ha/h. The machine is intended for complete tillage of soil i.e. loosening to a depth of up to 12 cm and killing of weeds in low moisture and arid regions (Singh, 2007; Singh *et al*, 2015). It is designed for light, stony and heavy soils. A vertical knife, mounted ahead of every sweep, cuts the soil slice, stalks and roots of weeds. This prevents the clogging of sweep shanks and improves the cultivation of soil. The working members of cultivator are grouped in sections possessing a high transversal stability. The land wheels are arranged ahead of working members under the main frame that ensures proper cultivation of wheel tracks in operation and increases possibility of implement in transit. The working members are lowered and raised by tractor operator with help of hydraulic cylinder actuated from tractor hydraulic system. A tractor-drawn rotating type of interculturing tool does the job of scratching, loosening and uprooting weeds in one operation. The machine has a rectangular frame of cultivator with five tines. The gearbox provided on the machine reduces the rpm and changes the direction of rotation of PTO shaft. The power from the gearbox is provided through chain and sprocket. The machine works at 6-8 cm depth and about 200 rpm. It covers one hectare cross hoeing in two hours. The roots are uprooted and soil is well pulverized because of rotating action.

Fig. 5.28 : Tractor operated six tine cultivator weeding cotton crop

CIAE has developed weeding attachment with four/seven sweeps for weeding of soybean crop sown at 35 cm row spacing (Fig. 5.29). The weeding operation is carried out at the crop age of 32 - 38 days after sowing and the average plant population and plant height are 12 per metre row length and 32.6 cm, respectively (Anonymous, 2010). The mean values of forward speed, effective field capacity, hourly fuel consumption, weeding efficiency and plant damage are 2.63 km/h, 0.25 ha/h, 1.33 l/h, 64% and 5.5%, respectively.

Fig. 5.29 : Tractor operated four/seven row crop cultivator

Weeders for SRI (System of Rice Intesification)

Weed growth is generally more in SRI cultivation than conventional method due to wider spacing and alternate wetting. Effective and timely weed control is crucial for the success of SRI. At present farmers are using weed collector-cum-surface looser (Fig. 5.30), manual cono weeders for weeding in SRI (Fig. 5.31) which is consuming more time and causing muscle pains to the labour. The power weeder for SRI cultivation of paddy has been developed by ANGRAU Huderabad (Anonymous, 2008 & 2010; Anonymous, 2013a). It consists of 1.5 hp loona engine, 2 liters plastic petrol tank, handles, power transmission system and rotary wheels with cutting blades (Fig. 5.32). The rotating wheel has 140 mm diameter and 8 cutting blades of 4 teeth of 30 mm height and 50 mm width. The power transmission system consists of worm and worm gear arrangement. The rotary wheels are rotated by the power transmission system of the engine. The working width of the weeder in the field is 200 mm. The weeder is provided with two floats of 190 mm length, 75 mm front width and 120 mm rear width arranged in front and back with the help of 550 mm length and 25 mm x 25 mm 3 mm M.S. angle to avoid sinkage in the field. The depth of operation can be adjusted by float height. The field capacity of machine is 0.1 ha/h with field efficiency 70-80% and weeding efficiency 80-90%.

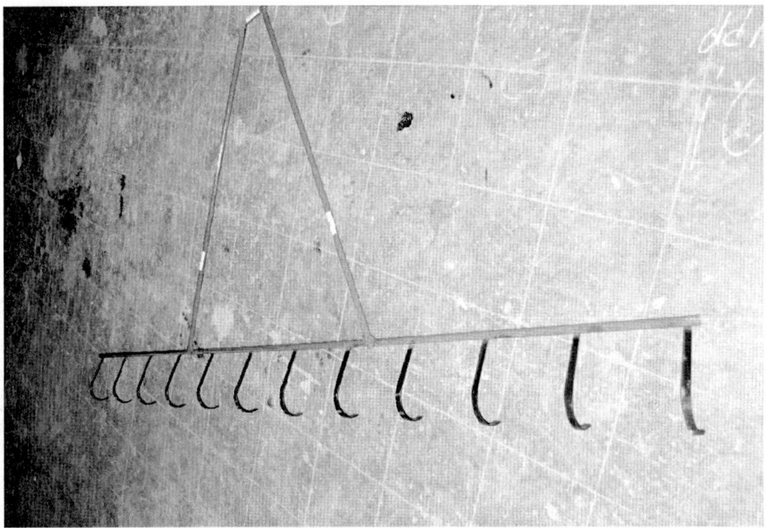

Fig. 5.30 : Weed collector-cum-surface looser

Fig. 5.31 : Single row cono weeder

a) Stationery view b) Field operation

Fig. 5.32 : Power weeder for SRI

TNAU multi-row paddy weeders

For weeding in rice fields, 1/2/3 row finger type rotary weeder is used (Fig. 5.33). The field capacity of the single row weeder is very low (0.01 ha/h) and hence time consuming (Anonymous, 2006). Two rows and three row weeders have also been developed (Anonymous, 2008, 2010, 2013a). Star weeder is a small implement pushed by manual labour. It consists of a long wooden or iron vertical rod with a small horizontal rod for holding the implement (Fig. 5.34). To the other end, two star like wheels and a small blade of 10 cm are attached. The pointed teeth of rotating wheels loosen the soil and help in easy mobility of the implement while the blade helps in cutting the weeds. It is useful to control small weeds in close growing crops like groundnut, foxtail millet etc. TNAU, Coimbatore centre has developed 3-row finger type rotary weeder (Fig. 5.35), which consists of three rotary members fitted with six numbers of L shaped blades mounted on a frame with hub. It has provision for adjusting the row spacing from 200 to 250 mm. It can be operated in a standing position with the help of the long handle (1140 mm). Each rotary member has an adjustable float. The ground clearance is 450 mm to ensure operation of the weeder upto a crop height of 300 mm. Considering the field coverage and ergonomic considerations, the two-row rotary paddy weeder gives the best results with saving of 82% in labour and 84% in cost of operation. Field capacity of two row weeder is 0.02 ha/h and that of three row weeder 0.03 ha/h with field efficiency of 85% and 80% respectively.

Fig. 5.33 : Manually operated single/two row/three row weeder

Fig. 5.34 : Two row star wheel type weeder in field operation

Fig. 5.35 : Manually operated three row weeder in field operation

Tractor mounted earthing-cum-interculture equipment

It is a mould board type small tractor operated implement suitable for earthing and interculture operation in sugarcane where planting is done at row spacing of 700 mm or more (Fig. 5.36). The equipment is useful for areas where lodging of crop is pronounced. This implement has been developed at GBPUAT, Pantnagar (Pandey *et al.*, 1997). It saves 25 per cent on cost of operation and it also results in 8 per cent increase in yield compared to conventional method of using manual spade for earthing and cultivator for interculture.

Fig. 5.36 : Small tractor operated earthing-cum-intercultural equipment

5.2 Fertilizer Application Equipment

Fertilizers are required where soils are deficient in plant food elements. When land is planted to crops over a long period of time, the plant food elements slowly get reduced and productivity of crop goes down. Sandy soil looses plant food elements rapidly because of heavy rainfall or application of irrigation water leaches these elements out from the soil. Some clay soils in low rainfall area loose nutrients slower than the sandy soil. Adding chemical fertilizer increases fertility of soil. Plant nutrients are needed for better growth and yield of a crop. Principally nitrogen, phosphorous and potassium are added to the soil to promote greater yields. Uniform distribution and proper placement of fertilizer in the soil have become increasingly important factor in providing maximum crop response at minimum cost.

Since the fertilizer application rate is quite high in most of the crops and movement of most fertilizer in the soil is very limited, proper placement in relation to seeds or plants roots is important for better response and efficient utilization of nutrients. Placing the fertilizer near the seeds at the time of planting helps stimulation of seedlings and results in more effective utilization of nutrients. At the same time quantity of fertilizer should not be more than the required, otherwise it will have adverse effect on the growth of plants. In the row crops side dressing of fertilizer is done which has immediate benefit when placed in the moist soil within the root zone.

The fertilizer (N.P.K.) are essential inputs directly affecting crop yield. The effective use of these nutrients by crop is of utmost importance. Fertilizer use efficiency is seldom high and depends on method of application. The fertilizer use efficiency ranges from 20 to 50 percent for nitrogen, 10 to 20 percent for phosphorus and 80 to 90% for potassium (Goswami & Kamath, 1982). Fertilizer is used in solid, liquid and gaseous forms. However, in India fertilizers in solid form are mostly used. These fertilizers are either broadcasted over the surface or drilled into the soil. The use of fertilizers through irrigation water and foliar application has also been adopted recently. Broadcasting of granular fertilizer by hand leads to a non-uniform distribution and low fertilizer use efficiency (Khan, 1982-84). Drilling or band placement of fertilizer ensure uniformity of distribution and accuracy of placement which leads to better plant response and growth and better fertilizer use efficiency. Fertilizers are applied at different stages: partly at the time of sowing/planting or sometime even before, and partly in one-two dozes in the standing crop in the form of side or top dressing. Therefore, the choice of the method of application depends on time of application during crop growth, soil, crop, fertilizer type, quantity to be applied, availability of water and application machinery.

Fertilizer can be applied to the soil either by broadcasting or by drilling with the help of mechanical device. The uniformity of distribution is difficult to maintain by broadcasting. Urea is highly hygroscopic and has a tendency to leach down and other fertilizers remain on the surface without being available to the crop. Thus, metering and placing of proper amount of fertilizers is important for modern agriculture. The uniformity and ease of drill ability of fertilizers through openings depends on flow ability, clod formation under the influence of agitators, moisture, size and shape of the particles and bulk density. Any slight change in air humidity changes shape, makes them clody, opts to bridging and thus, changes the flow behaviour through openings. To prevent bridging for uniform and free flow of fertilizers, agitator of different shapes (Fig. 5.37) is located inside the fertilizer box. Different types of agitators viz. star, rod type, circular rubber or canvas flaps, and vane and notch discs are commonly used. Circular shape openings give higher discharge rate than square openings. In general star type agitator give higher discharge rate than other types of agitators. Proper placement and uniform distribution of fertilizer can only be achieved with the use of fertilizer drills. Good fertilizer application equipment helps to apply the fertilizer at the recommended rate, at the right time and at the right place. It also avoids wastage, reduces the cost of operation and improves fertilizer use efficiency and crop yields.

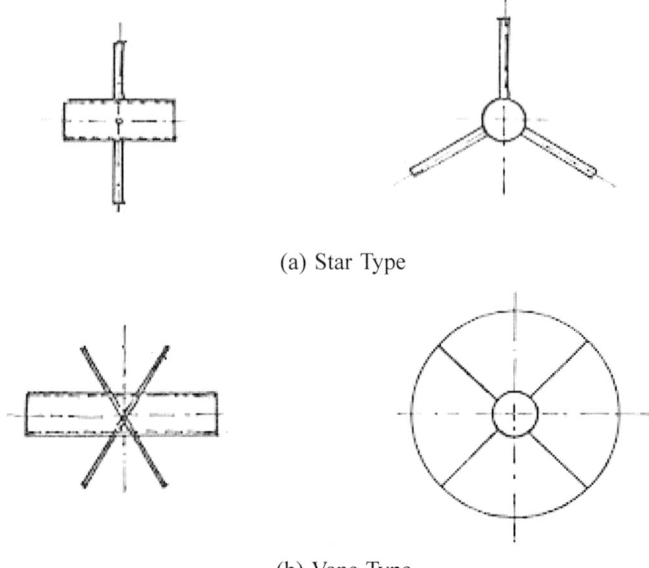

(a) Star Type

(b) Vane Type

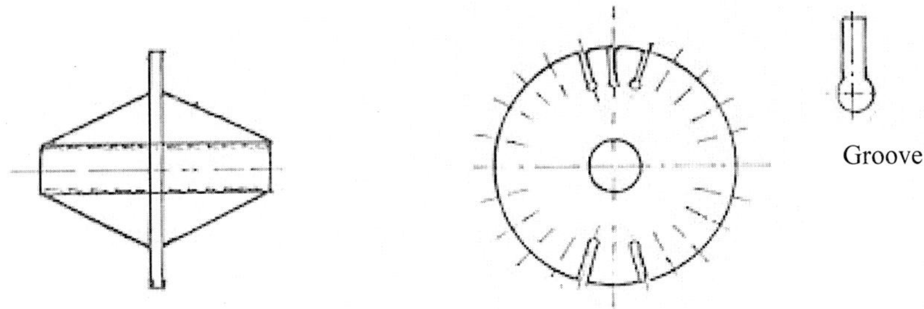

(a) Notch Type

Fig. 5.37 : Different types of agitators used in fertilizer drills.

Selection of fertilizer application equipment

While selecting fertilizer application equipment one should keep in mind that the machine is suited to the source of power available, easy to adjust, operate and maintain, have a positive and accurate metering device, have proper type of furrow openers, proper calibration arrangement and index plate to adjust the application rate, the hopper, metering device and delivery tube must be made of non-corrosive material, must have an instant closing orifice, and easy to clean left over fertilizer (Singh, 2007; Pandey *et al.*, 1997; Singh *et al.*, 2015). Dry fertilizer is applied practically to most of the crop in all types of field conditions. The rate of application varies according to the type of fertilizer, crop, soil and irrigation levels etc. Some of the application methods for dry fertilizer are i) broadcasting the fertilizer before ploughing or place it at ploughing depth; ii) broadcasting the fertilizer and mix into the soil before sowing; iii) deep placements of fertilizer with deep plough viz. chisel plough, sub soiler; iv) applying fertilizer during drilling or planting; and v) applying fertilizer through sides (side dressing) in row crops.

In general, distributors used for applying dry fertilizers may be classified as those that broadcast the material onto the surface of the ground and those designed for placing the fertilizer in rows or bands below the surface. Drop type broadcaster is suitable for spreading either fertilizer or lime. Furrow opener is also provided in some of such equipment for band placement below the soil surface and can be used for side dressing of row crops. These may be either pull type or mounted type implements. These are also available as an attachment to various implements. Centrifugal type of fertilizer broadcaster uses hopper that meters fertilizer and distributes it laterally over the field with the help of one or two horizontally rotating ribbed discs

(Garg and Singh, 2002). This type of broadcaster can also be used for broadcasting small seeds in the field. Most of these units employ centrifugal spreading primarily because of the compactness and simplicity in comparison to throw type distributors. In this type of broadcasters, non-uniform lateral distribution is a problem, which has been solved by taking proper left and right turnings to overlap the distribution pattern and make it as uniform as possible. These fertilizer broadcasters may be manually operated or tractor operated.

Manually operated fertilizer broadcaster

Majority of farmers still practices the hand broadcasting of fertilizer. The principal disadvantage of hand broadcasting is the non-uniformity of distribution causing uneven growth of crop, which ultimately results in poor yield. The problem of non-uniform distribution of fertilizer in the field can be easily overcome, if manually operated fertilizer broadcaster is used (Anonymous, 1985). It can be used for broadcasting granulated fertilizer viz. urea etc. The manually operated fertilizer broadcaster consists of a cylindrical hopper made of either sheet metal or special grade plastic with tapered bottom, fertilizer-metering mechanism, spreading disc, transmission unit, agitator, handle and strap (Fig. 5.38a). All these components are fitted over a lightweight rigid frame. The hopper bottom normally has a circular hole for metering fertilizer. This hole is having a sliding gate, which reduces or increases the notch opening area with the help of a lever. The ratio between the diameter and height of the hopper is kept in the range of 0.8 to 1.25. The hopper can be covered with a lid with a peeping hole of at least 75 mm in diameter; to see the quantity of fertilizer in the hopper during operation. The hopper should be sufficiently strong and should not buckle when filled fully with fertilizer. The capacity of hopper varies between 12 to 15 liters. The machine can cover about 1.2 ha in one hour at a forward speed of 2 km/h. The effective width of spread is about 6.4 m.

Spreading disc is made of aluminum alloy or MS sheet. It is mounted at the bottom of the hopper, which may have 6 to 8 equally spaced blades/fins. The spreading disc is provided with a vertical clearance of at least 30-mm from the hopper bottom. This is rotated with the help of transmission unit and it spreads the fertilizer falling over it from notch opening of hopper uniformly in the field. A suitable feed control mechanism with loading device is provided to control the flow of fertilizer through the aperture. This mechanism is controlled by a lever provided on the side of the hopper and can be moved easily to the left or right from mean position to have required notch-opening area. An index pointer is also provided on the hopper, which correspond to the notch opening position on the chart having marking ¼, ½, ¾ and full. In order to avoid the clogging of the aperture and for feeding the fertilizer to the aperture properly, a suitable agitator is provided near the orifice in the hopper. This agitator is provided just above the aperture and a suitable vertical clearance is provided between the agitator and the aperture. The basic frame is normally made of flat aluminum alloy. The hopper, transmission unit, handle, spinning disc, axle etc are mounted over to this frame. Transmission unit, which rotates the spreading disc, consists of set of gears and shafts with bearing and bushes. The transmission unit rotates the spreading disc at a peripheral speed of 450-550 cm/s. The set of gears are so arranged that it converts the rotation of handle into right angled rotation of spreading disc. Handle rotates the pinion which rotates a helical gear over which spreading disc is mounted. The speed ratio is kept in such a way that the disc rotates at much faster rate than the handle rotation. Normally, two straps of suitable length are provided in the fertilizer broadcaster to help easy carriage of broadcaster. The length of strap can be adjusted as per requirement of the operator. Hence, operator can have the broadcaster mounted at his front at a convenient height. Some of these straps are also provided with cushioning pads at shoulder position. This cushion is covered with cotton or leather. This cushioning in the strap reduces the fatigue of the operator.

The total weight of the machine is kept as low as possible, so that it can be easily carried in the field by the operator (Fig. 5.38b). Normally, the weight of empty machine is around 5 kg. Fertilizer is filled in the hopper and broadcaster is mounted at the front of the operator. By rotating the handle fertilizer from the notch-opening fall over to the rotating disc, which spreads the fertilizer evenly into the field. The operator is asked to move in the field at its own normal speed. The spread pattern of this type of broadcaster is generally skewed towards right of the line of travel. The skewness is a function of drop ratio. The skewness can be reduced to a minimum with the help of proper notch opening location, height of drop and speed. Else, taking proper overlapping can solve the problem of skewness. Under such condition, there will be a bottleneck to recommend proper left and right turning in the field to have proper overlap pattern to get maximum possible uniformity. The non-uniformity pattern of distribution is quite apparent for broadcasting seeds.

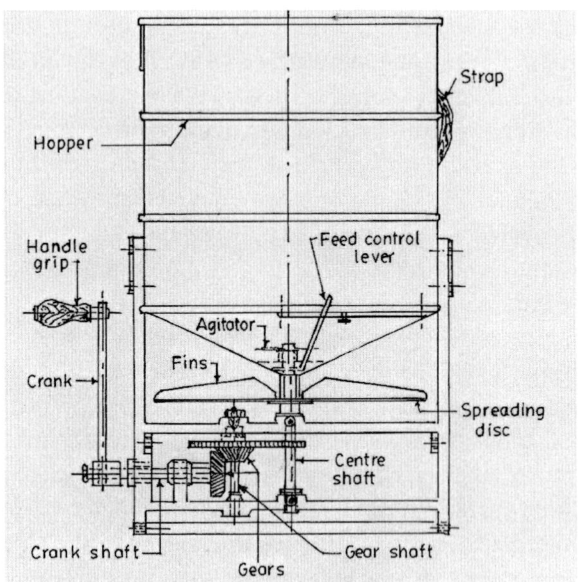

a) Components of manual fertilizer broadcaster

b) Manual fertilizer broadcaster
(Courtesy: National Agro Industries, Ludhiana)

Fig. 5.38 : Manually operated fertilizer broadcaster

Tractor operated fertilizer broadcaster

This type of fertilizer broadcaster also works on the similar principal, as in the manually operated broadcaster. The required rotations to the spinning disc are provided through tractor PTO shaft (Singh, 2007). It consists of an MS sheet hopper of conical shape. At the bottom of the conical hopper there is an adjustable gate, through which fertilizer is allowed to pass (Fig. 5.39). Amount of fertilizer dropped over to the spinning disc can be varied to obtain desired fertilizer rate. The disc is rotated with the help of shaft mounted at the centre of the disc. The other end of disc shaft is connected to the gearbox shaft. On the other side of the gearbox, there is a splined shaft that can be connected to tractor PTO. Gearbox is provided to reduce the speed and increase the torque of disc. The required rpm of disc is achieved by providing suitable reduction gear. Fertilizer broadcaster is mounted at the rear of the tractor (Fig. 5.40). As the tractor PTO moves the disc rotates and the fertilizer from the gate falls over to the disc and spread uniformly towards both left and right side of the tractor. The pattern of fertilizer distribution is not as skewed as in the case of manual broadcaster. A tractor mounted fertilizer broadcaster with or without oscillating spout is also used for top dressing of granular fertilizers (Fig. 5.41). It has a hopper capacity of 2.5 to 3.0 quintals of granular fertilizer. It spreads the fertilizer over a swath width of 6 to 8 m and requires 35 hp tractor. It maintains good uniformity of distribution and is used under soil conditions with adequate moisture and where irrigation facilities are available. It can also be used for top dressing application.

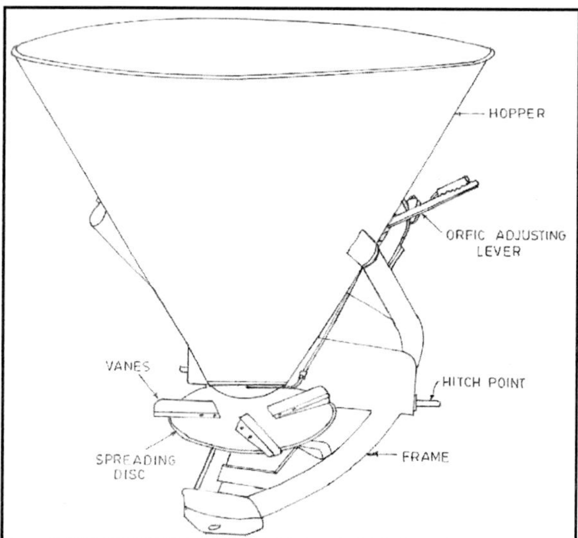

Fig. 5.39 : Components of tractor operated fertilizer broadcaster

Fig. 5.40 : Tractor operated fertilizer broadcaster
Courtesy: Tirth Agro Technology Pvt. Ltd. Rajkot (Gujarat)

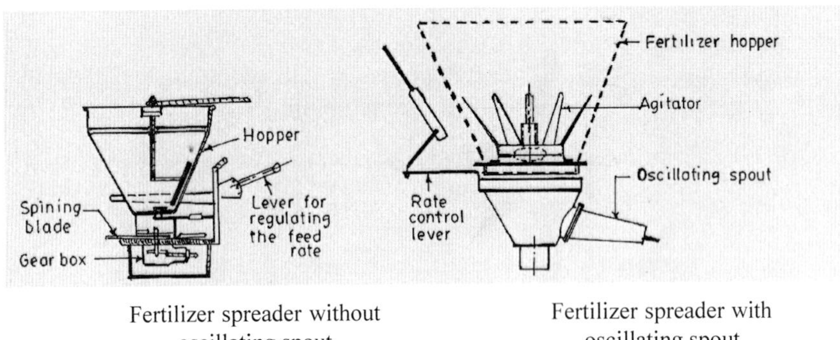

Fertilizer spreader without oscillating spout

Fertilizer spreader with oscillating spout

Fig. 5.41 : Tractor operated fertilizer broadcaster with and without oscillating spout

Fertilizer drills

Manually operated fertilizer drill: Manually operated seed-cum-fertilizer drills for small holdings have gravity type metering device with an adjustable orifice to control the fertilizer rate. An agitator is also provided to ensure uniform flow of fertilizer. Two persons are required to operate the drill; one pulls the drill and other guides it in the field. It can cover 0.3 to 0.4 ha/day.

Animal-drawn fertilizer drills: The simplest device for placement of seeds and fertilizer is an attachment to the country plough called 'Pora' method of sowing. It consists of a funnel with tube attached to a country plough. One man guides the plough and other drops the seed/fertilizer in the funnel. In some animal-drawn 'desi' plough provision for agitator in fertilizer box is made. The drill uses gravity or fluted feed metering devices to control fertilizer rate. It can cover 0.4 to 0.6 ha/day. Several types of 2 to 5 rows animal drawn drills with variation in metering devices and furrow openers are also available. The seed and fertilizer are placed in different boxes. Mostly fertilizer box is having gravity type metering device with an adjustable opening and an agitator. Agitators are of disk or spur wheel type. An agitator prevents fertilizer from bridging over the opening and ensures uniform flow. The fertilizer is conducted through transparent plastic tubes, which helps in detecting the clogging of fertilizer in tube and also avoids corrosion. Its field capacity varies from 1 to 1.5 ha/day.

Tractor mounted drills: Several types of tractor mounted drills and planters with fertilizer attachment are available in India (As discussed in Chapter 4). The most commonly used seed-cum-fertilizer drill uses vertical disc type metering device with grooves on the periphery for both seeds and fertilizers (Chauhan *et al.*, 1984). These machines are comparatively cheaper, more durable and easy to operate. It covers 2.5 to 3.0 ha/day at the forward speed of 4-5 km/h. A seed-cum-fertilizer drill using external fluted feed roller for metering fertilizer has also become popular. The drill has an average field capacity of 2.5-3.0 ha/day. The drills usually places fertilizer in band on one side of the seed at the same or slightly higher depth than the seed. Agronomic research have shown that the fertilizer placement in a band about 2.5 cm to the side and 2.5 cm deeper than seed lead to about 8 to 10% increase in yield of wheat and other cereal crops.

Fertilizer drill for dryland farming: The simplest device for line sowing of cereals and band placement of fertilizer used in dry farming areas is an attachment to country plow having a bamboo or metal pipe with a bowl. Such type of drills is available in one to three furrows. Seeds and fertilizer are metered and dropped by hand. The accuracy of metering depends on the operator's skill and experience. However, this method of placing seed and fertilizer is certainly superior to hand broadcasting and leads to better fertilizer use efficiency. For the light and medium soils, single-row animal drawn and two-row tractor mounted ridger type seeder with fertilizer attachment is also available. It has provision to sow the seeds in the furrow for the rabi crops when the moisture is scarce and on the ridges during kharif. The furrows during kharif season help to harvest the rainwater and promote better plant growth. It has been found that bed planting in low rainfall area help in establishing good crop stand and increases the crop yield significantly. This equipment has provision to drill the fertilizer in a band on one side of the seed.

Common type of fertilizer metering devices

Fertilizer drill uses mainly two types of metering devices namely: gravity feed type with agitators and positive feed type (Singh, 2007; Chauhan *et al.*, 1984). The variation in fertilizer placement in rows and between rows depends on the type of metering devices used in the machine. The positive metering devices with vertical wheel having grooves on the periphery and helical fluted rollers have low intra row variations in the rate of fertilizer dropped as compared to the gravity feed metering device with adjustable orifice and agitator. The metering device with vertical wheel having grooves on the periphery has the maximum intra row variation of 13-16 percent as against 63-66 percent in the gravity feed device. This variation is lowest (6-8%) in case of helical fluted roller type metering device. There are different types of metering devices used to meter the granular fertilizer to get uniform metering action under varying conditions (Fig. 5.42). These common metering devices are:

Star wheel: It is an old metering device and has been in use for metering fertilizer from years together. It is used in seed drills and also in some planters. There are star wheels, which carrying fertilizer through gate opening from hopper into delivery compartment. The star wheel is rotated in horizontal plane. The fertilizer trapped between the teeth of wheel falls into the delivery tube by gravity. The material carried on the top of the wheel is scrapped off and it does not come into delivery tube. Raising or lowering a gate just above the star wheel can control the rate of discharge or fertilizer application rate. Speed ratios can be altered through transmission unit. The star wheels are driven through transmission unit which consist of set of gears from feed shaft below the hopper and gets the drive through ground wheel.

Rotating bottom plates: In this type of fertilizer metering device there is a horizontal rotating bottom plate which is fitted over to a stationary bottom ring of hopper base. This type of attachment is used in row crop planter. An adjustable gate fitted over side outlet can control fertilizer rate.

Auger feed metering device: It uses an auger to meter the required amount of fertilizer. Auger is closely fitted into tube. A star wheel type agitator is also fitted in the hopper to avoid clogging of fertilizer in the auger. From the hopper auger carries the fertilizer into tube and drops in the delivery tube.

Edge cell feed type: It consists of a vertical rotor fitted in the hopper, having cells on its entire periphery. Number of such metering devices is installed over to entire length of hopper to meter granular fertilizer. Varying the rotor speed can vary the discharge rate.

Stationary opening metering devices: This type of metering device uses a stationary opening at the bottom of the hopper. In this opening a moving gate is provided to vary the size of opening to control the rate of fertilizer application. Sometimes an agitator is also provided to avoid any clogging of opening with fertilizer. In this type of fertilizer broadcaster, the bottom is made conical for easy flow of material.

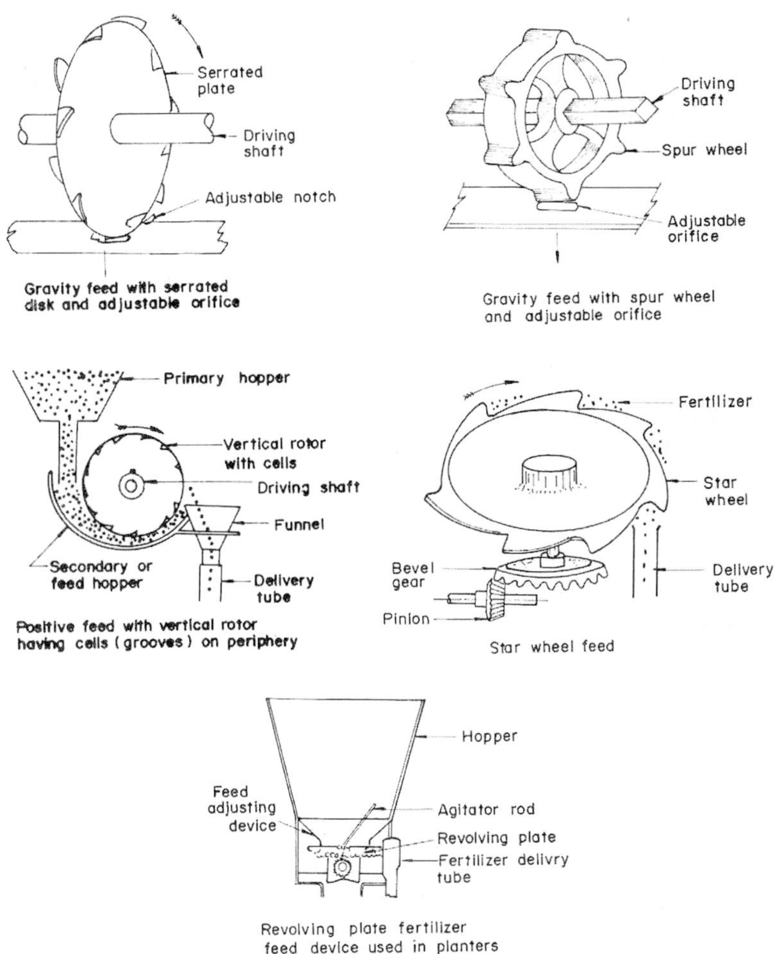

Fig. 5.42 : Common types of fertilizer metering devices used in fertilizer drills

Tractor trailer-cum-farmyard manure spreader

The farmyard manure application is a basic input operation in crop production. The application system includes equipment for loading of manure at storage facilities, transporting to the application sites, and applying the proper quantity of manure uniformly to soil-crop systems (Singh and Singh, 2013). The equipment for the application of solid, semi-solid manure and other organic materials discharge the products at varying rate depending upon ground and PTO speeds of tractors, equipment settings and manure moisture content. This equipment is calibrated during normal manure application times and recalibrated whenever a new source of manure with different moisture content is applied. Because of the equipment limitations, most solid and semi-solid manure are broadcasted on the surface of the land. Uniformity and application rate emerge as the most important technological challenges for the efficient use of solid and semi-solid manure for the field crops. Most equipment currently available are not capable of achieving manure application having coefficient of variation lower than 20%, which is generally considered as a upper acceptable boundary for application of organic fertilizers. In India farmyard manure is applied through manual broadcasting and bullock-carts/tractor trailers, resulting in human drudgery, more time per unit area and loss of nutrients with poor application uniformity and wide variation in the application rate. Tractor PTO operated trailer type manure spreaders have been developed and evaluated (Singh and Singh, 2013). It has ben reported that the variation in manure application rate and uniformity are highly significant at tractor forward speed of 2 - 4 km/h. Therefore, a farmyard manure spreader has been developed for two tonne capacity by modifying the design of commercial available two-wheel tractor-trailer at Central Institute of Agricultural Engineering, Bhopal (Fig. 5.43 and Fig. 5.44).

The spreader consists of main frame, manure spreading unit, feeding auger, manure box and sheet sliding mechanism (Fig. 5.43). A rotating type manure spreader, having a longitudinal axis is positioned in front of pneumatic wheels, below the chassis of tractor trailer. The beaters bars are provided over the periphery of drum in four rows. The spreader is designed to rotate at speed of 540 rpm to spread the organic manure efficiently. The concave made of 13 MS squire bars of 12 mm diameter, spaced at 50 mm apart is provided below the beater unit to reduce the size of manure clods and mitigate the problem of clod size distribution of existing manure spreader. An auger as anti-bridging device is provided over the manure spreading unit in the centre of body of the trailer. Spiral discs of 250 mm diameter and 250 mm pitch are mounted on a GI pipe (diameter 48 mm and thickness 2 mm) parallel to each other to convey the material directly to the spreading unit. Auger is designed for impact rotational speed of 0.5 m/s required to decompose and breaking of manure clods.

On the frame of trailer two number slanting platforms are provided to make trapezoidal shape manure box for sliding of manure to the feeding auger. Slanting platforms made of G I sheet are hinged on the body of the trailer with an angle of 40^0 for sliding the manure from the top of the box to the feeding auger on the basis of angle of friction of farmyard manure (moisture content, 36%) to the GI sheet. Because angle of friction is more important for sliding the manure over a sheet compared to angle of repose, as farmyard manure is a heterogeneous material and its moisture content changes with the depth of storage pit. The capacity of trapezoidal shaped manure box is calculated by using volume (length, width and height) and considering the average density of farmyard manure as 0.55 t/m^3. A sheet sliding mechanism has been developed to control the manure delivery rate to the spreading unit. It has a screw mechanism for rotating a handle to slide the GI sheet for allowing the estimated quantity of manure to the spreading unit. Thus by increasing/decreasing the opening width of sliding sheet the manure delivery rate is controlled for uniform application of manure. The manure spreader is calibrated for manure delivery rates/discharge rates by varying the opening space by adjusting the width of opening, keeping the length constant as 1.6 m at 540 and 70 rpm of manure spreading cylinder and feeding auger, respectively.

Average swath of manure spreading is 2.0 m and field capacities of machine is 0.23 0.38, 0.58 ha/h at tractor forward speeds of 0.41, 0.69 and 1.10 m/s, respectively (Fig. 5.44). The field capacity of machine increased with increase in forward speed and swath. The field efficiency of machine is about 76%. It is due to time loss in turning, because during the turning at headlines the speed of tractor is slowed down. Based on this principle animal drawn manure spreader has also been developed at CIAE Bhopal (Fig. 5.45) and SHIATS Allahabad (Fig. 5.46).

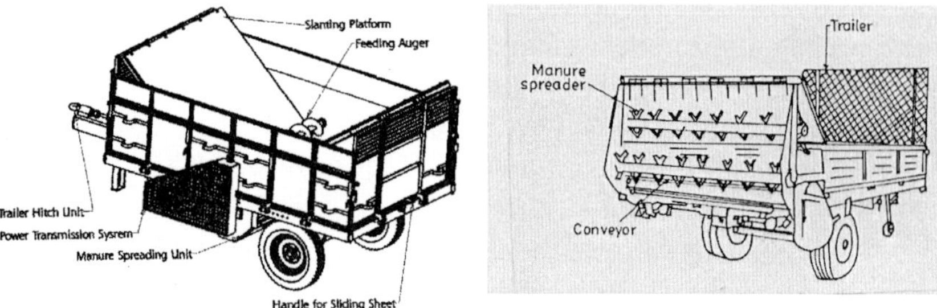

Fig. 5.43 : Tractor trailer-cum-farmyard manure spreader

Fig. 5.44 : Tractor trailer-cum-farmyard manure spreader

Fig. 5.45 : Animal drawn CIAE manure spreader

Fig. 5.46 : Animal drawn manure spreader SHIATS type

Tractor operated compost spreader

The machine is simple and spreads compost or farm yard manure evenly in the field. It is multi-purpose implement which is useful in spreading compost in farms, open field, green houses etc. The machine can be easily hitched with three-point hitch of the tractor. It consists of spreading unit, fertilizer box, hydraulic units and supporting wheels (Fig. 5.47). Width of spread and thickness of the spread can be controlled hydraulically by the operator. This allows operator to spread the compost only where it is required. While operating, a thick rubber strip shoots out any non-granule material such as small rocks and stone thus avoids possibility of spreading of such particles with compost. It covers 1.5 m width. It can be operated by 35 hp tractor and above. Thickness of compost layer can be 2-10 mm. The container has capacity of 750 kg.

Fig. 5.47 : Tractor operated compost spreader

Courtesy: Tirth Agro Technology Pvt. Ltd. Rajkot (Gujarat)

Power tiller mounted earthing-cum-fertilizer applicator

In the wide row (1.20 – 1.50 m) sugarcane cultivation, interculturing, earthing up operation and fertilizer applications are labour intensive (90-100 man-h/ha). Bullock drawn and tractor drawn implements due to restriction in their movement cannot perform inter-culture operation in wide row sugarcane cultivation. For such situations, power tiller operated unit suitable for earthing up-cum-fertilizer application in wide row sugarcane crop has been developed by MPKV, Rahuri (Anonymous, 2007). The earthing up unit consists of single ridger body, which also performs shaping of ridges in addition to earthing up operation (Fig. 5.48). The ridger body includes V shaped shovel of 380 mm length and 250 mm in height. The wings are fixed to the body by means of hinges and then connected to adjustable MS struts.

The normal setting provision for the wings are from 35 to 50° with reference to the direction of travel of power tiller. The unit is provided with 500 mm bar of 40 mm diameter for depth control attached to the power tiller rotary unit. The auxillary bar of 500mm height and 25 mm diameter are provided for fixing it to fertilizer box. The fertilizer application unit consists of long trapezoidal shape fertilizer box with 25 kg fertilizer capacity. The unit employs cup feed type metering mechanism having four cups (18 g capacity each) on fertilizer rotors. Provision of auxiliary chamber facilitates free fertilizer delivery (Fig. 5.49). The ground drive wheel of 380 mm diameter is employed for transmitting power to fertilizer agitation system. The working width is 60 cm. The average depth of fertilizer placement is 75 mm. The average weeding efficiency is 80%. At the forward speed of 1.2 km/h, its field capacity is 0.50 ha/h and field efficiency 90%.

Fig. 5.48 : Power tiller drawn earthing-cum-interculture and fertilizer applicator for wide row sugarcane crop

Fig. 5.49 : Power tiller mounted earthing up-cum-fertilizer applicator in operation in wide row sugarcane crop

Power tiller operated farm yard manure spreader

Uniform spreading of manure is possible only by mechanical spreaders. A power tiller operated farm yard manure spreader has been developed (Anonymous, 2006). It consists of main frame, manure tub, conveying mechanism, spreading mechanism and adjustable rear door (Fig. 5.50). A chain conveyor carries the farmyard manure from the front end to the rear end of the manure tub. The spreader assembly mounted on the rear end of the manure spreader consists of discs with eight straight vanes. The disc is driven by a hydraulic motor at 200 rpm. An adjustable door mounted in the rear end of the manure tub controls the amount of manure fed to the spreader. The field capacity of spreader is 0.90 ha/h and manure application rate 16.6 t/ha. It can cover 6 m in one go with field efficiency of 75%. It saves cost and time 85 and 94 per cent respectively as compared to conventional method of manual spreading.

Fig. 5.50 : Power tiller operated farm yard manure spreader

References

Anonymous. 1985. Biennial Report 1983-85. Dept. of Farm Power & Machinery, P.A.U. Ludhiana.

Anonymous. 2006. Research Highlight. AICRP on Farm Implements and Machinery, CIAE Bhopal. Technical Bulletin No.: CIAE/2006.

Anonymous. 2007. Annual report – 2006-07. AICRP on Farm Implements and Machinery, MPKV Rahuri .

Anonymous. 2008. Research Highlight. AICRP on Farm Implements and Machinery, CIAE Bhopal. Technical Bulletin No.: CIAE/2008/141.

Anonymous. 2010. Research Highlight. AICRP on Farm Implements and Machinery, CIAE Bhopal. Technical Bulletin No.: CIAE/2010/151.

Anonymous. 2012. Directory of Successful Farm Machinery in SAARC Countries. SAARC Agriculture Centre. BARC Complex, Farmgate, Dhaka – 1215 (Bangladesh)

Anonymous. 2013. Research Highlight. AICRP on Farm Implements and Machinery, CIAE Bhopal. Technical Bulletin No.: CIAE/2013/158.

Anonymous. 2013a. Success Stories. AICRP on Farm Implements and Machinery, CIAE Bhopal. Extension Bulletin No.: CIAE/FIM/2013/40.

Anonymous. 2013b. Annual Report (2012-13). Niche Area of excellence (NAE), Farm Mechanization in Rainfed agriculture. Department of Farm Machinery & Power, IGKV, Raipur.

Anonymous. 2014. Annual Report (2013-14). Niche Area of excellence (NAE), Farm Mechanization in Rainfed agriculture. Department of Farm Machinery & Power, IGKV, Raipur.

Anonymous. 2015. Research Highlight. AICRP on Utilization of Animal Energy with Enhanced System Efficiecy, CIAE Bhopal. Technical Bulletin No.: CIAE/UAE/2015/180.

Bhardwaj K C; Ganesan S; Pandey M M; Singh Gyanndra. 2004a. Directory of Agricultural Machinery & Manufacturers. . CIAE Bhopal. Technical Bulletin No.: CIAE/2004/116.

Bhardwaj K C; Pandey M M; Singh Gyanndra. 2004. Horticultural Tools and Equipment. CIAE Bhopal. Technical Bulletin No.: CIAE/2004/110.

Chaudhuri D; Singh R C. 2013. Improved Technology for Utilization of Draught animals. AICRP on Utilization of Animal Energy with Enhanced System Efficiency, CIAE Bhopal. Technical Bulletin No.: CIAE/UAE/2013/160.

Chauhan A M; Singh Surendra; Rawal G S. 1984. Role of seeding and planting equipment in farm mechanization. Paper presented at the all India seminar on, "Role of Agricultural Engineering in Rural Development" organized by the Institution of Engineers, Agril. Engg. Division, New Delhi.

Garg I K; Singh Surendra. 2002. Farm equipment for Punjab agriculture. Department of Farm Power & Machinery, Punjab Agricultural University, Ludhiana.

Goswami N M; Kamath M B. 1982. Fertilizer use in multiple cropping systems and measures to increase efficiency. Proceedings of FAI-Northern region Seminar on Accelerating the Pace of Fertilizer Consumption, New Delhi. Pages 17-34.

Khan A U. 1982-84. Deep Placement Fertilizer Applicator Project, Semi-annual Progress Reports No. 1-5. Agricultural Engineering department, IRRI, Manila, Phillippines.

Pandey M M; Ganesan S. 2005. Farm Mechanization Package for Dryland Agriculture, Technical Bulletin No: CIAE/FIM/2005/117. Central Institute of Agricultural Engineering, Bhopal.

Pandey M M; Majumdar K L; Singh Gyanendra; Singh Gajendra. 1997. Farm Machinery Research Digest. Technical Bulletin No. CIAE/97/69, Central Institute of Agricultural Engineering, Bhopal, 328 p.

Pandey M M; S Ganesan; R K Tiwari. 2006. Improved Farm Tools and Equipment for North Eastern Hills Region, Technical Bulletin No. CIAE/2006/121. Central Institute of Agricultural Engineering, Bhopal.

Sharma V K; Garg I K; Singh Surendra; Sodhi K S. 1987. Performance evaluation of weeding equipment on wheat crop. *J. Agric. Engg*. 24(2): 122-126.

Singh R C; Singh C D. 2013. Development and Performance Testing of a Tractor trailer-cum-farmyard manure spreader. *AET.* Vol. 37(2): 1-6.

Singh Surendra. 2007. Farm Machinery – Principles and Applications. Directorate of Information & Publication of Agriculture, Indian Council of Agricultural Research, Krishi Anusandhan Bhawan-I, Pusa Campus, New Delhi.

Singh Surendra; Manes G S; Dixit Anoop. 2015. Mechanization of cultivated crops. New India Publishing Agency, Pitam Pura, New Delhi.

Singh Surendra; Pandey M M. 2008. X Plan Achievements (2002-2007). AICRP on Farm Implements and Machinery, CIAE Bhopal. Technical Bulletin No.: CIAE/2008/137.

Singh Surendra; Verma S R. 2009. Farm Machinery Maintenance & Management. Directorate of Information & Publication of Agriculture, Indian Council of Agricultural Research, Krishi Anusandhan Bhawan-I, Pusa Campus, New Delhi.

❑❑❑

6

Plant Protection Equipment

Ever since man started domesticating plants and animals for food, feed and fiber, he is facing competition from insects, disease and other pests. The struggle is going to continue as long as man wants plants and animal. The need to save plants/crops from the ravages and depredations of pests and to reduce the losses caused by them became more urgent (Patel, 2004).

India is a predominantly agricultural country and will continue to be so. Neither individuals nor nations can live without agriculture. India is witnessing growth of its agriculture, industry and economy in general, at the same time the population is growing at an alarming pace.

The crop losses by pests in our country are enormous. Chemical control continues to play a major role in crop protection; at the same time it enhances the environmental pollution problem. One has to use chemical control in a precise way by adopting Integrated Pest Management (IPM). In applications of pesticide (which are poisonous), the technique used should be such that minimum quantity of pesticides is applied to cover the Target-Leaf-Foliage with less environment pollution. Water is usually used as a carrier to apply pesticides to the Target-Leaf-Foliage. Beside water, air stream is also used, to carry fine droplet-particles to reach the Target-Leaf-Foliage.

Classification of spraying technologies

The method, by which pesticides are applied on crops, weeds and soil, is known as Application Technique. It means applying appropriate dose and distribution of spray droplets. There are a number of application techniques to distribute pesticide evenly over the target. Chemicals are used in doses as low as 100 g to as high as a few kg/ha. Carriers or diluents are employed to secure an even distribution for greater efficacy. Since the pesticides are expensive, it is very important to apply them correctly on the target in required quantity. The importance of pesticide application technique is to cover the target with maximum efficiency and minimum effort to keep the pest under control. Spraying techniques are classified as: i) High volume (400 l/ha and above), ii) Low volume (5-400 l/ha; 100-200 l/ha in ground application and 15-75 l/ha in aerial application), iii) Ultra low volume (up to 5 l/ha), iv) Controlled droplet application, and v) Integrated Pest Management (IPM) according to the total volume of liquid applied per unit of ground area. Initially high volume spraying technique was used for pesticide application but with the advent of new pesticides the trend is to use least amount of carrier or diluents' liquid. In spraying, the optimum droplet size differs for different types of application. Fine droplets are required to control insects, pests or diseases and bigger size droplets for application of herbicides, etc. The greater the number of fine droplets produced by the device better is deposition on target area. The size of droplet is important as it affects drift and penetration distance of droplets towards the target. Hence a compromise is to be made to prevent drift, achieve wide coverage of plant or target area and more penetration.

The basic concept of pesticide application technique depends on the carrier, which is used for applying a particular pesticide. Application technique plays an important role in achieving efficacy. Its importance is as much as pesticide. It depends on various factors, such as Target - Pest and Pesticide, Pesticide Formulation Characteristic, Evaporation and Drift. The optimum droplet sizes required for different target groups are indicated in Table 6.1.

Table 6.1: Optimum droplet sizes for different targets groups

Target group	Droplet size (microns)
Flying insects (drift)	10-15
Crawling and sucking insect (drift)	30-50
Plant surfaces (limited drift)	60-150
Soil application (no drift) as in case of herbicide application	250-500

The Spray Spectrum produced by any nozzle has different size of droplet particles. All of them are not of same size. In most cases, droplets are in wide range. In sprayer, which is working on centrifugal principle, the droplet size produced is in narrow range. Usually fine droplet particles are more in number than bigger. The total volume of small droplet particles, which are more in number, does not exceed the total volume of less number of big droplets. A way or means has to be found so that fine droplet particles which are more in number and has less volume should reach not only the target but also on upper and lower side of Target-Leaf-Foliage.

In a spraying system, nozzles are the ultimate controllers, which are supposed to be kept clean, calibrated and regularly replaced. The cost of this is small relative to the value of the pesticide sprayed through them. Droplet size is most important. Reducing the droplet diameter improves the spreading efficiency. However it cannot be reduced indefinitely.

In plant protection, the application of pesticide depends on definition of the target to assure reaction of the pesticide on the pest. The target leaf area requiring treatment may be much greater than the land area. Some insects harbour on the lower side of leaves. The amount of leaf surface varies considerably with the crop growth and spacing. The geometry and orientation of leaves also affect deposition. Pesticide droplets must travel in and around the plant canopy in order to cover on both sides of leaves as far as possible. The target for a pesticide is seldom a flat piece of land, except perhaps for pre-emergent herbicides. Most recommendations simply state "so much" pesticide per hectare. This is particularly inappropriate for crops, which changes the target area with its growth over a period of time during its life cycle and have a considerable change in leaf area. Pesticide application should be carried out in accordance with the development stage of the plantation crop. Pests may be quietly localized within a plantation canopy. The pre-requisite in the evaluation of any pesticide application technique is to establish the biological target precisely. Although much may be known about the biology of a pest, the important factor is to define at what stage the pest is most vulnerable to a pesticide deposit.

In the orchard, the target is complex; it may be an insect, pathogen, mite, or it could be the leaves or fruit etc. The target location within the canopy varies, outside edges of canopy, or inside top centre. The tree size and shape changes during an individual growing season as well as during the span of tree age. A large proportion of the pesticide may never reach its intended target due to factors such as tree density, seasonal growth patterns, pruning and the differential retention properties of leaves due to "poor match" between standard set-up of sprayers and plant geometry. Although considerable progress has been made in developing and perfecting the pesticide application, yet there is much room for improvements to achieve high efficiency in pest control with reduced costs, minimum drift and residue hazards to mankind, domestic animals, wild life, beneficial insects and crops. However for further improvements in pesticides application, the design of pesticide application equipment, one should be concerned with; i) the possibility of placing the pesticide maximally where the entomologist wants it and minimally where he does not want it, ii) optimum or adequate distribution in the sense of uniformity within the target area, and iii) minimal escape from the target area, such escape producing drift, damage, general pollution and wastage which vary in importance with the chemicals and the conditions (Singh, 2007; Singh and Kandoria, 1999).

Pesticides, crop, target and operating conditions present difficult application problems. In such cases, the compromise is the only practical solution. Development and selection of specialised application appliances require a study of several factors such as the pesticide – its formulation, mode of action, dose, toxicity – mammalian and crop; effect of equipment materials; the target; the crop; the operator and the operating conditions. The aim is to produce an appliance which has efficiency in application and operation, reliability, simplicity and economical. With the advent of new pesticides and the use of it in small doses, the technique of its application has assumed great importance. The present day agro chemicals are not only highly poisonous but also costly and hence are to be used in small doses. The safe and effective use of pesticides is therefore, a must. Hence, one has to adopt – ***'AN INTEGRATED PEST MANAGEMENT'.***

Integrated Pest Management (IPM)

'IPM' is a system that, in the context of the associated environment and the population dynamics of the pest species, utilizes all suitable techniques and methods in as compatible manner as possible and maintains the pest populations at levels below those causing economically unacceptable damage or loss (Patel, 2004; Ahuja, 2004a). Use of pesticide cannot be avoided at the same time the drift has to be minimized and hence the only alternative left out is Integrated Pest Management (IPM). Farmers perceive chemical control to be a simple option – buy a pesticide, mix it and apply it. But environmental and safety concerns have increased as considerable numbers of farmers use highly toxic chemicals which adversely affect their health and non-target organisms. One alternative seen by many is organic crop production with a return to traditional pest control measures such as crop rotation, host plant resistance and biological controls. To many people this approach is synonymous with integrated pest management (IPM), because they have ignored the role of chemical control in IPM. In reality, relying purely on biological and cultural controls generally will not maintain the high levels of crop production needed to feed the growing world population. Chemical control will continue to play a major role in the future protection of crops. New chemistry is providing different pesticides, some of which are more selective than previous ones, and are active at very low dosages. Their effective use in IPM programs emphasizes the need for more precise integration of applications, both in space and time. Such improved application requires crop monitoring to provide relevant in-field information. Unfortunately it is the application aspects of chemical control that are largely ignored.

Sprays are still applied almost entirely by pressure through hydraulic nozzles. However, there have been major improvements in developed countries to reduce exposure of farmers to the chemical by improved packaging, closed transfer systems and changes in equipment to allow nozzle selection in terms of spray quality and drift reduction. Improved cultivars, including transgenic crops, can raise the potential yield, not only this but it is needed that insect pest, pathogen and natural enemy identification must be teamed up with better farming practices that include judicious use of pesticides.

Application equipment

The function of pesticide application equipment is to store, meter, atomize and distribute pesticides accurately on the target pests. Once the pesticides mixture is discharged, it is atomized by nozzles, which create various shapes of thin liquid sheets. These sheets of fluid break into droplets that move with the induced air produced by the liquid flowing from the nozzle, toward the target. The size of the droplets produced from liquid sheets depends on the natural frequencies within the liquid during formation of the sheet.

By considering the economy of plant protection technology it has been observed that the machine costs is only 5 to 10% of total plant protection costs. On the other hand, the pesticides cost is as high as 60 to 80%. Therefore, to save on capital investment on equipment is not practical, since the utilization of expensive pesticides will depend on it and on the application technique. The development costs of new pest control equipment is also significantly high but is less if compared with the development cost of pesticides.

Drift: Drift is the movement of a pesticide through air, during or after pesticides application, to a site other than the intended site of application such as ground. The affected area could be another crop, or any other vegetation within range of the pesticide movement. In most cases, this movement is limited to the edges of the field or fence. Problems occur when this movement affects a sensitive crop or another person's property. Many factors play an important role in drift creation viz. nozzle type and size, spray pressure, droplet size, nozzle orientation, spray height, chemical formulation, and evaporation due to weather, wind velocity and direction, humidity and temperature. The increase of pest control activities has unfortunately led to environmental pollution and residual effect on food and feed.

Therefore, sprayer should be designed so that it must have a provision not only to atomize spray solution in to fine droplet particles but also to provide extra force to carry the droplet on to the crop foliage. Air assistance spraying technique is a must for applying pesticide now a day for effective utilization.

Functions of sprayer

Sprayer is a machine to apply fluids in the form of droplets. The sprayer is used to apply herbicides to kill weeds, to apply fungicides to minimize fungus disease, to apply insecticides to control insect pests and to apply micro nutrients on the plants. The main function of a sprayer is to break the liquid into droplets of effective size and distribute them uniformly over the plants. It regulates the amount of insecticides to avoid excessive application that might prove harmful or wasteful on crops. The sprayer should produce a steady stream of spray materials in the desired fineness of the particles so that plants to be treated may be covered uniformly. It should deliver the liquid at sufficient pressure so that it reaches all the foliages and spreads entirely over the sprayed surface. It should be light, yet sufficiently strong easily workable and repairable. A sprayer that delivers droplets large enough to wet the surface readily should be used for proper application of surface residual sprays. Extremely fine droplets of less than 100 micron size tend to be diverted by air currents and get wasted. Crops should, as far as possible, be treated in regular swaths. By use of a boom, uniform application can be obtained with constant output of the machine and uniform forward travel.

Most sprayers in use today are hydraulic types in which spray pressure is built up by the direct action of pump on liquid spray material. The pressure thus developed forces the liquid through nozzles, which break the liquid into fine size droplets and disperse them in the desired spray pattern. Also sufficient energy is imparted to the spray droplets to carry them from nozzle to the surface to be treated. The main parts of hydraulic sprayer are; pump, tank containing an agitator, framework for mounting sprayer, pressure regulator and relief valve, pressure gauge, strainers and screens, control valves, pipes and fittings, distribution system and power source.

Types of spraying equipment

Different designs of spraying equipment have been developed for different types of applications and field and crop conditions. Manually operated hydraulic sprayers viz. knapsack sprayers, twin knapsack sprayers, foot sprayers, hand compression sprayers; air carrier sprayers such as motorized knapsack mist blower cum duster (LV) and centrifugal rotary disc type sprayers are specially suitable for spray applications in crops. The present trend is to apply concentrated pesticides by means of low and ultra low volume sprayers. This has been possible through the development of better formulations and special nozzles. New Controlled Droplet Application (CDA) atomizers require less than 15 l/ha of spray mixture and are easy to operate. Although new type of spray atomizers is available but the correct chemical formulations are commonly not produced. Hence their use is limited to particular crop, pest or disease. The instructions of the manufacturer should be carefully read whether a formulation is recommended for ultra low or low volume application. Conventional knapsack sprayers require 400-600 l/ha of the spray solution. This requirement puts considerable hardship on a farmer who has to fetch clean water from a long distance. Hence, a reduction in the volume of water needed for spraying mitigate some of the problems faced by the farmer. Low and ultra-low volume spraying techniques reduce the spray volume requirements considerably.

A farmer willing to buy a new sprayer is often faced with a problem of selecting a machine from a large choice of models of varying sizes and capacities available in the market. It is necessary that the farmers make the right choice of spraying equipment because the selection of proper type of equipment is very important if it is to be used economically and effectively. A choice must be made as to which of the four: duster, high-volume sprayer, low-volume sprayers or ultra-low volume (ULV) sprayers are required. The dusters are preferred in arid and semi-arid areas where water is scarce. If a sprayer is the choice, select one that could be used for applying spray liquid in the high-volume as well as low-volume. High volume sprayers are selected where more than 400 litres of spray liquid per hectare is to be sprayed. If spray volume is in between 5 to 400 litres per hectare, low volume sprayers are selected. Ultra-low sprayers are only selected where volume of spray liquid to be applied is less than 5 litres per hectare.

Foot and rocking sprayers give satisfactory performance for small trees and bushy crops (Fig. 6.1). Knapsack sprayers are good for spraying field crops. Motorized knapsack sprayers perform better for spraying field crops, bushy crops and small size fruit trees. In hilly regions and waterlogged areas, manual and power operated knapsack sprayers are preferred. The sprayer selected should be easy to operate, maintain and requires least number of operators and technical skills. The sprayer must be purchased from the reputed manufacturers using high quality nozzles. Easy availability and after-sales service should also be considered while buying the sprayer. Size and type of sprayer to be purchased depends on the crop, area to be covered and investment capability of the farmer. The operator must be trained in advance in safety precautions and effective application techniques. Multi-purpose farm sprayers are available to meet the many spraying needs on diversified farms. It has necessary pressure range to be used for weed spraying or for spraying fruit trees or livestock. These sprayers have free choice of spray materials because of type of pump and agitation provided. Wettable powders, cold water paints and other spray materials that are abrasive or difficult to keep in suspension can be used satisfactorily in this type of sprayer.

Power sprayers are used for those spraying jobs, which are just too large for coverage with hand sprayers (Fig. 6.2). These sprayers are popular on golf greens, large gardens or green houses. Commercial growers of fruits and vegetables use high pressure and high volume sprayers. The machine combines the high pressure and volume delivery desired to get complete spray coverage of high growing fruits and shade trees in full foliage as well as the dense growth of vines and other crops. Plunger-type multiple cylinder reciprocating pumps is used exclusively with large sprayers at maximum pressure of 28-69 kg/cm^2 delivering 30-227 litres per minute. Low-pressure, low volume sprayers have been primarily designed to meet the specific requirements for low-volume field spraying (Fig. 6.3). They are very popular for use in controlling weeds and insects in field crops. Because of their relatively low cost, their use is limited to sediment-free type of spray material due to type of pump and agitation provided. These are available either tractor-mounted type or trailed type. Sprayers are available for every use, in the home, garden, orchard and field. Garden sprayers are manually operated but field and orchard sprayers are manually as well as power operated. Sprayers are of four types viz. manually operated, engine operated, tractor operated and air plane sprayers.

Calibration of sprayer: The spray volume required to cover a unit area depends on discharge rate of nozzle, forward travel speed, and spray swath width. Therefore, it is essential to have trial run with plain water on a small area and note down the above parameters. Based on these parameters, the quantity required to spray per unit area can be worked out as:

$$\text{Spray volume required (l/ha)} = \frac{\text{Field area (ha)} \times \text{spray volume (l) used on trial plot area}}{\text{Trial plot area (ha)}}$$

Manually operated equipment

Hand sprayer: The hand sprayer is a small capacity pneumatic sprayer. It consists of chromium plated brass tank having a capacity of 0.5 to 3 litres (one litre is more common) which is pressurized by a plunger pump (Fig. 6.4). The air pump remains inside the tank. The sprayer has a short delivery tube to which a cone nozzle is attached. In some models, the nozzle is attached at the top of the tank with flow spring actuated lever, which regulates the flow of the spray liquid. For spraying, the tank is usually filled to three-fourths capacity and pressurized by air pump. The compressed air causes the agitation of the spray liquid and forces it out, on operation of the trigger or shut off type valve. Usually the chemicals with suspension characteristics cannot be effectively sprayed with this type of sprayer. For spraying wettable powders the sprayer is shaken frequently to prevent settling of the chemical. For operation, the spray nozzle is directed to the target after charging. It is fitted with mist spray nozzle with gooseneck bend. The pump assembly is made of brass and operated by one person. It is ideal for small nurseries, rose plants, kitchen gardens and spraying wettable insecticides and fungicides.

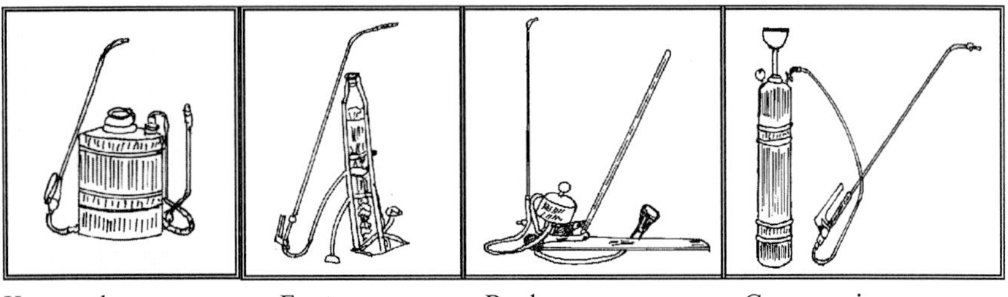

Knapsack sprayer Foot sprayer Rocker sprayer Compression sprayer

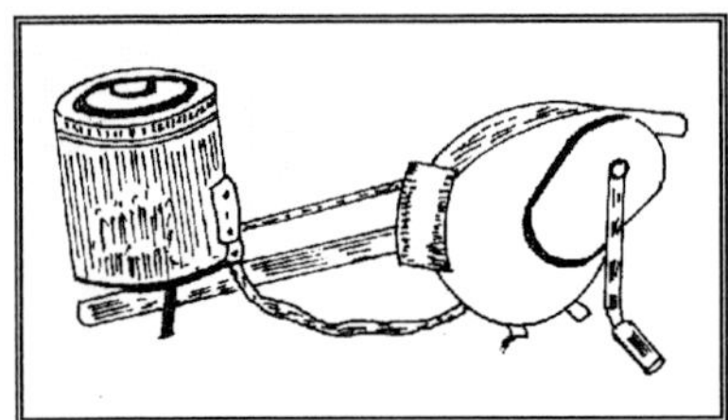

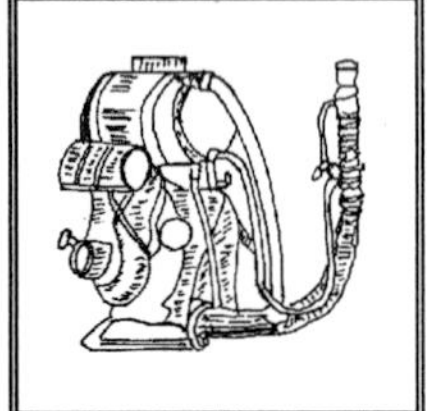

Hand rotary duster Knapsack power sprayer

Fig. 6.1 : Manually operated spraying equipment

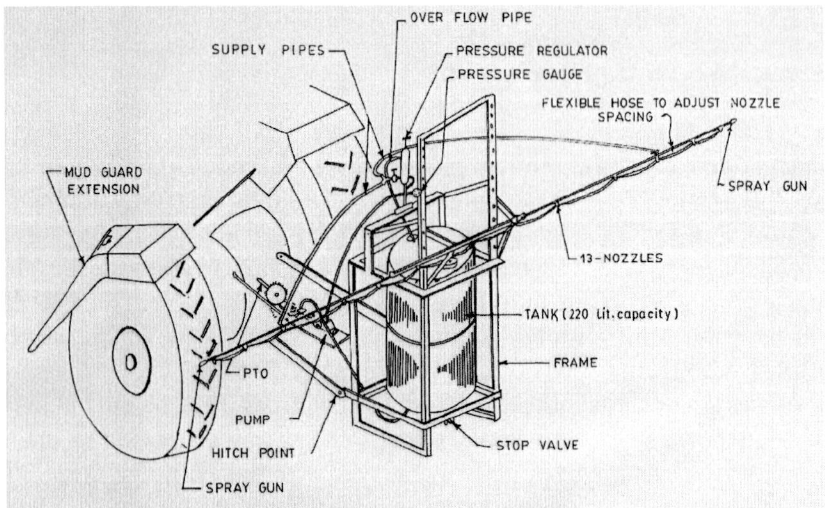

Fig. 6.2 : Tractor operated sprayer

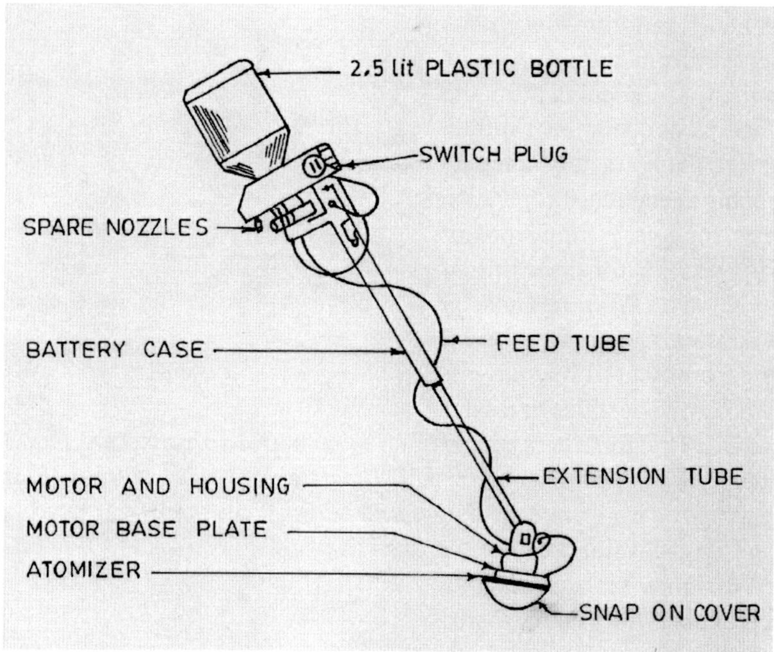

Fig. 6.3 : Ultra-low volume sprayer

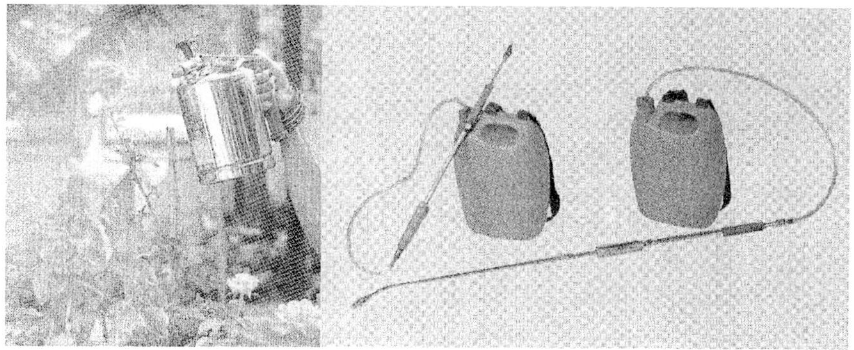

Fig. 6.4 : Hand sprayers

Knapsack sprayer

It is used for spraying insecticides and pesticides on small trees, shrubs and row crops. It is manually operated (Pandey *et al.*, 2006). It consists of a pump and an air chamber permanently installed in a 9 to 22.5 liters tank (Fig. 6.5). The handle of the pump extending over the shoulder or under the arm of operator is operated with short up and down strokes to pump with one hand and spray with the other. Uniform pressure can be maintained by keeping the pump in continuous operation. Uniform pressure maintained by keeping the pump in continuous operation forces the spray liquid out of spray nozzle on opening of control valve. Working capacity of knapsack sprayer is 0.4 ha/h. It is generally used for spraying low crops, vegetables and trees up to 2.5m height. The frame of sprayer is shaped so as to conveniently fit on the back of operator. The capacity of container is 10-14 litres. It is made of galvanized iron, brass, or stainless steel. Translucent plastic materials for visual fluid-level indication are used for spray tank. Sometimes it is provided with a double-barrel pump of piston type with a lever to operate the pump. In some cases, the tank is provided with a single pump and a pressure barrel having a piston pump and a mechanical agitator. Some models are equipped with externally mounted pumps and such pumps have more life because it is not immersed in fluid. A pressure of 3-5 kg/cm^2 is maintained in the pressure chamber. Now-a-days most of the spray pumps are made of plastic body which is light in weight and easy to carry by operator (Fig. 6.6). This sprayer consists of a spray tank, mini pump, spraying system, rubber tubes, spraying rod, nozzles (s) and shoulder trap. Tank capacity is 16 lit. It works in the pressure range of 0.25 to 0.45 MPA. Lance is of fibre material.

Fig. 6.5 : Knapsack sprayer during field operation

Fig. 6.6 : Manually operated spray pump
Courtesy: Khedut Agro Engineering Pvt. Ltd., Rajkot (Gujarat)

Compression knapsack sprayer: It consists of a tank for holding spray fluid with compressed air, vertical air pump with handle, filler hole, spray lance with nozzle and a cut-off device (Fig. 6.7), Singh (2007) and Singh & Kandoria (1999). The capacity of tank varies from 10-20 litres. It can develop 4-5 kg/cm^2 pressure in the tank filled with fluid to ¾ of its capacity. It is mostly used in gardens, small farms, and livestock and poultry houses. The major drawback is that pressure in the tank drops progressively as spraying proceeds.

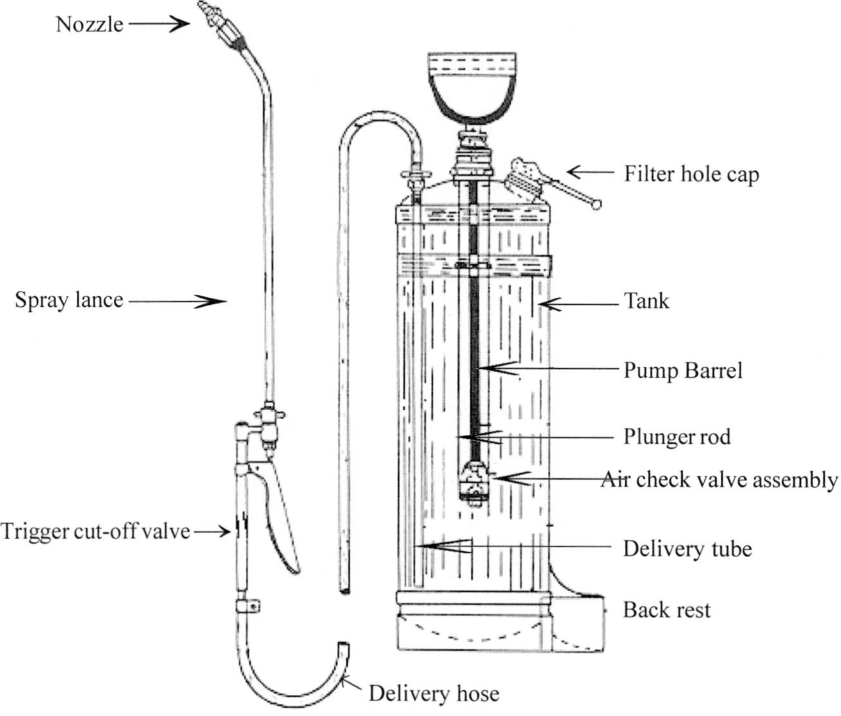

Fig. 6.7 : Components of compression knapsack sprayer

Stirrup pump sprayer : It consists of a brass pump, a footrest (stirrup), hose, lance and nozzle (Fig. 6.8). It may have a single barrel or double barrel pump (Singh, 2007 and Singh & Kandoria, 1999). In this type of sprayer continuous pumping is necessary for getting uninterrupted discharge of spray fluid. It develops a pressure of 4-10 kg/cm^2. It is suitable for small scale field spraying. It does not give uniform spray, as single barrel pump has no mechanism to retain pressure. Two persons are required for operating this sprayer; one to operate the pump and other to direct spray from the nozzle at the end of spray lance. The length of nozzle varies depending upon the requirement in the field but it is generally kept as 5 m.

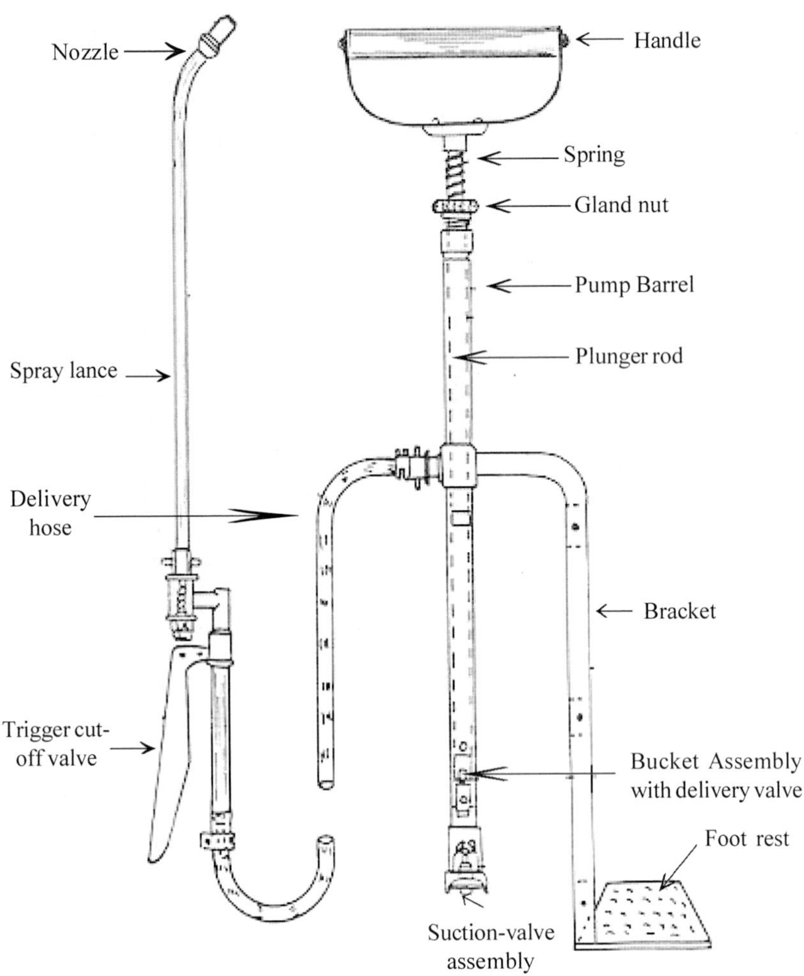

Fig. 6.8 : Components of stirrup pump sprayer

Motorized knapsack sprayer: This is also called power operated gaseous-energy knapsack sprayer (Fig. 6.9a). The hopper having capacity of 7-12 litres, is made of high density polyethylene (Singh, 2007). Another small tank of 0.75-2.25 litres capacity is also provided for fuel. It is powered by 1-3 hp engines. The fuel consumption varies from 0.6-2.0 l/ha. The frame is provided with shock- proof cushion that comfortably fits on the back of operator and eliminates the transfer of vibrations of the engine to the operator (Fig. 6.9b). It is provided with an on-off control switch and a mechanism for adjusting the discharge rate from 0.5-5.0 l/min. The spray liquid is blown-out by an air current generated in the tank. A part of air generated is diverted into the hopper to form an air cushion over the liquid in the tank. This ensures a uniform delivery of liquid. The liquid from tank passes through a tube to the nozzle on the spray lance by gravity. Effective swath width is 4-5 m horizontally and 3-4 m vertically. It can cover approximately 3 ha a day. The machine fitted with a rotary pump and tree-spray lance can spray trees up to 6 m high. Many of these sprayers can be converted for dusting and ULV application. It is used for spraying in orchards, coffee estates and tall crops. The centrifugal fan is usually mounted vertically. The fan produces a high velocity air stream, which is diverted through a 90-degree elbow to a flexible (plastic) discharge hose, which has a divergent outlet. The spray tank that has also a compartment for fuel and engine-fan unit is mounted on a common frame, which fits to the back of operator. The spray liquid flows due to gravity and suction created at the tip of nozzle, thus remains in the air stream. Some models of the sprayers have roller pump for pumping the spray liquid into the discharge hose and have tall tree spraying attachment. For operation, the tank is filled with the spray liquid and cranking with the rope starts the engine. The control valve for the spray liquid is opened gradually and adjusted for the desired flow rate. The operator directs the discharge hose to the target.

a) A view of motorized knapsack sprayer

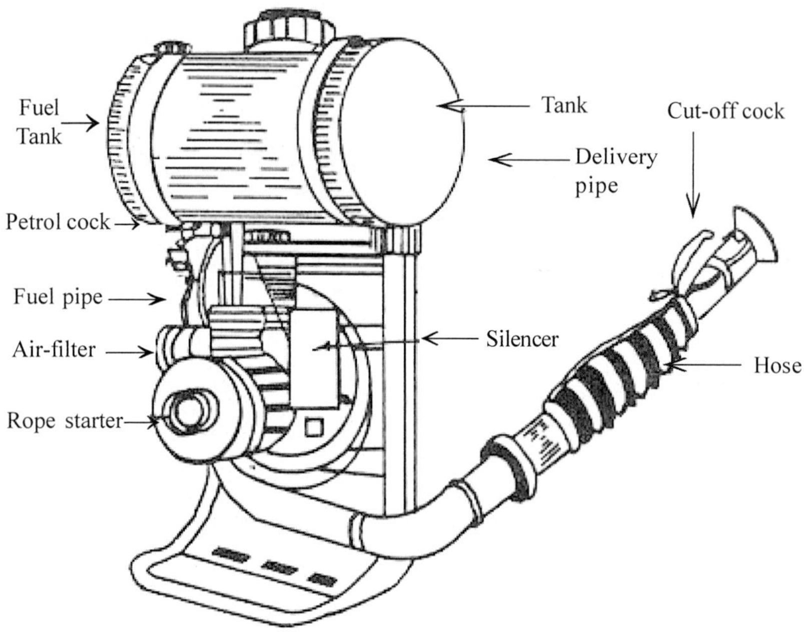

b) Components of motorized knapsack sprayer

Fig. 6.9 : Motorized knapsack sprayer.

Foot operated sprayer: It is also called pedal pump (Singh & Kandoria, 1999; Pandey *et al.*, 2006). It consists of a plunger assembly, stand, suction hose, delivery hose and an extension rod with nozzle (Fig. 6.10). One end of suction hose is fitted with a strainer and other end with a flexible coupling. Similarly, one end of delivery hose is fitted with a cut-off valve and other end with a flexible coupling. An additional container is required to hold spray fluid, as this sprayer does not have a built-in tank. Continuous pedalling is required for uniform spray. It can develop a pressure of 17-21 kg/cm^2. It is easy to operate and can be used for spraying tall crops and fruit trees up to 4 m height. Sprayer can be used to spray trees up to 6 m height with additional hose.

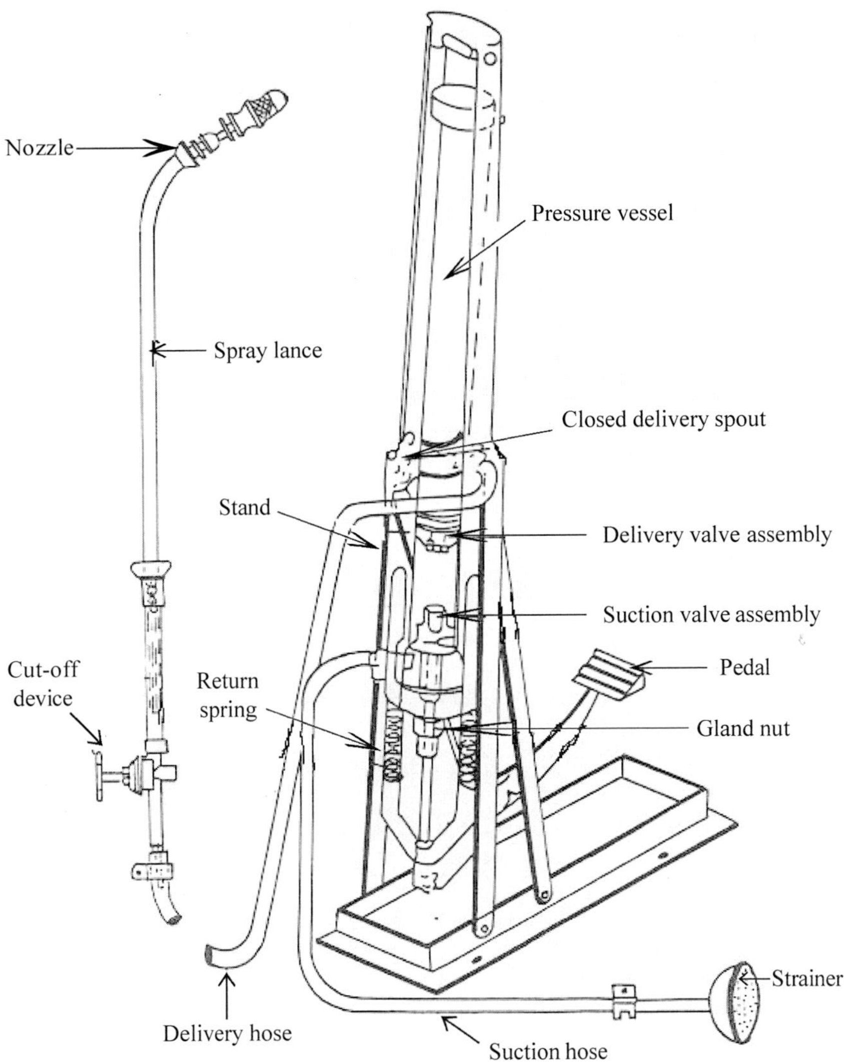

Fig. 6.10 : Foot-operated sprayer

Rocking sprayer: It consists of a pump assembly, platform, operating lever, pressure chamber, and suction hose with strainer, delivery hose and an extension rod with spray nozzle (Fig. 6.11), Singh & Kandoria (1999). Rocking movement of handle operates the pump resulting in built up pressure in pressure chamber. This also requires an additional container to hold fluid. It can develop a pressure of 14-18 kg/cm^2, but in some cases it can develop up to 36 kg/cm^2. It can be used for spraying coconut and palm trees.

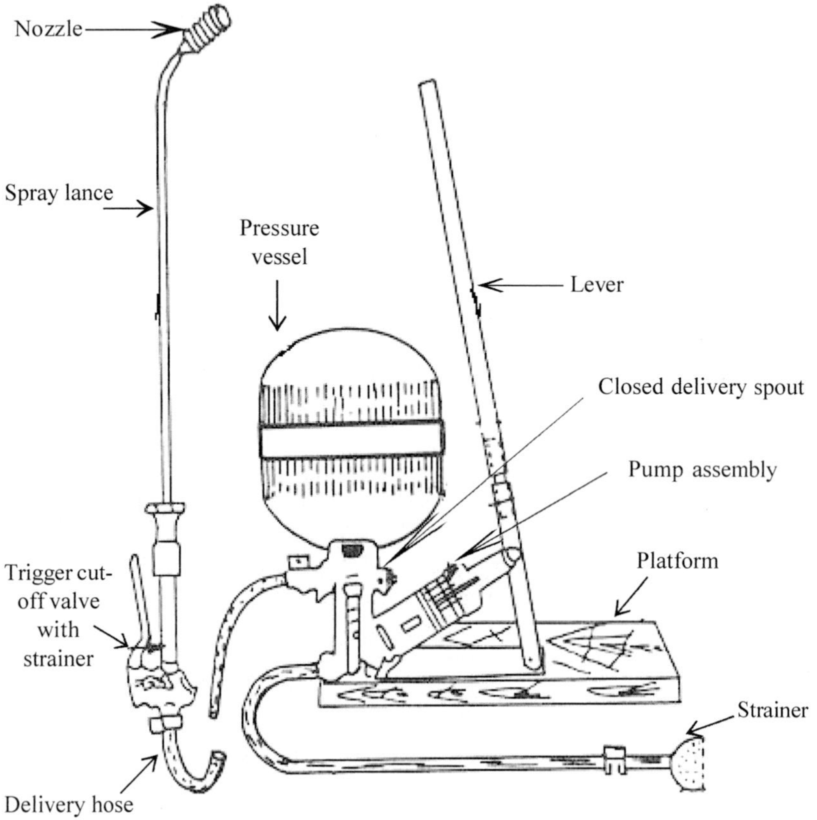

Fig. 6.11 : Rocking sprayer

Battery powered low volume knapsack sprayer

The sprayer consists of plastic tank, grooved spinning disc, a fractional horsepower DC motor to spin the disc at very high speed, 12 volt power supply source which can be a dry battery or lead acid battery and a light weight handle on which the spray head is mounted (Fig. 6.12a). The handle has provision to adjust the angle of the spray head (Dixit *et al.*, 2006). The spinning disc is also made of fine quality plastic, which has very fine radial grooves (Fig. 6.12b). For operation, the spray liquid in minimum dilution is filled in the tank, which flows in the form of drops at the centre of spinning disc. The disc is attached to the motor, which rotates it at a very high speed. The spray liquid travels in the grooves of the disc and gets fragmented into very fine droplets when it leaves the periphery of the disc. The droplets are thrown outwards due to centrifugal force. It is suitable for spraying in crops like paddy, cotton, groundnut, pulses and vegetable. It saves 30 per cent labour and operating time and 15 per cent on cost of operation compared to manual spraying. It also results in 27 per cent increase in yield compared to spraying by manual sprayer.

a) Manually controlled battery operated spray pump
Courtesy : Khedut Agro Engineering Pvt. Ltd., Rajkot (Gujarat)

b) Manually controlled battery operated spray pump with spinning disc
Fig. 6.12 : Manually controlled battery operated knapsack spray pump

Rotary duster

Duster is a machine to apply chemicals in dust form. Dusters make use of air streams to carry pesticides in finely divided dry form on the plants. The hand rotary duster is available in two models, shoulder mounted and belly mounted (Singh & Kandoria, 1999). It is a common type of duster being used by the farmers. The duster consists of a hopper, fan/blower, rigid/flexible discharge pipe, reduction gearbox, rotating handle, shoulder straps, and metering mechanism (Fig. 6.13). The hopper is either made of plastic or aluminium. The hopper made from mild steel sheet is coated with anti-corrosive material for longer life. The duster has mechanical agitator connected to the gearbox placed in the hopper, which churns the chemical and prevent clogging of the outlet. The adjustable orifice plate mounted below the hopper outlet controls the application rate. The fan/blower is enclosed in the casing and is rotated with the handle through gearbox. This is mounted on the shoulder/belly with the help of adjustable straps. The discharge pipe fitted with spoon type deflector is directed towards the target continuously rotating the handle. The chemical in dust/powder form drops from the hopper in the discharge pipe having an air stream created by the blower. These dust particles emerging in the form of cloud from the discharge pipe are carried to the plant where these settle on the leaves, stems and other parts. This is used for control of pests and diseases by use of chemicals in the dust forms in nursery, vegetable gardens, field crops, tea and coffee plantations, green houses, glasshouses and godowns. These dusters are extensively used for dusting field crops, vegetables, bushes, small trees etc. They can cover 0.8-1.2 ha/day. The belly-type is more convenient to operate in tall crop like sugarcane as it does not get entangle in plants during dusting. The capacity of hopper is 5 kg and should not be filled more than ¾ of its capacity. There are different types of dusters such as plunger type; knapsack type, rotary type and power operated dusters. Dusters can also be set for given feed rate of chemicals using the same formula as used for sprayers.

Field dusters or power operated duster can be mounted upon a platform bolted to the rear of a tractor and operated by PTO. Operating the tractor in high gear makes it possible to dust a large area. It mainly consists of a power driven fan, a hopper and a delivery spout. The fan creates strong air, which causes the dust to blow off from hopper to a considerable distance either vertically or horizontally. A movable delivery spout suitably fitted with the unit regulates the direction of dust. An agitator is provided in the bottom of hopper to keep the dust in fluffy condition to meter out uniformly through the feed system into the air stream. This type of dusters is used for large areas. Tractor-drawn trailing dusters with PTO drive and auxiliary-engine drive are also available. An auxiliary engine operates most orchard dusters. They have only one large flexible metal hose, which can be turned to direct dust in any direction.

Plunger type duster is a simple duster with a small piston. The piston moves current of air over the dust in the hopper. The dust is carried away through a delivery spout. This is suitable for small areas where vegetable or flowers are grown. Knapsack type duster consists of powder container mounted on the back of operator. It has a hopper through which a current of air is blown to pick up the dust. The air current is produced by lever operated leather bellows. These dusters are suitable for small areas only. Rotary duster having a hand operated rotor is mounted in front of the operator.

Rotary dusters are preferred where more force of delivery is required for spraying tall crops. The dust is fed from a hopper into a current of air produced by a rotary fan and is blown out through a delivery pipe. It has a stirring device actuated by the fan crank to ensure a steady flow of dust. A valve below the hopper can regulate the rate of delivery. Controlling the speed of fan controls the delivery force.

Rotary pump is commonly used for low-pressure sprayers. They are compact, lightweight and cheap. These pumps are commonly mounted on and powered by tractor PTO shaft. Oversize units are usually specified to provide sufficient by-pass liquid for hydraulic agitation in the tank. Most common type of rotary pumps is gear type (internal and external) and roller pumps. Gear pumps are classified as positive displacement pump and available in various sizes. Their service life is greatly reduced if used to pump abrasive type of spray material. Rollers in roller pump are commonly made of nylon, but sometimes rubber and steel are also used. Pressures above 600 kPa are generally not recommended for rotary pumps when pumping non-lubricating liquids.

Centrifugal pump works on high speed having column discharge. The discharge rate varies directly with speed. It is not a positive displacement pump. It is extensively used on sprayers. This type of pump can handle any type of spray material. They have long life and have wide range of capacity. They are directly mounted on PTO shaft of tractor. They are relatively unaffected by abrasive-type spray material. These pumps are well suited for airblast sprayers and aircraft sprayers, for which high flow rates are needed at relatively low pressure. Diaphragm-type pumps are also used on sprayers where flow rate is limited to 19-23 l/min and pressure does not exceed 550 kPa. Since the valves and diaphragm are the only moving parts in contact with spray material, these pumps can readily handle abrasive materials.

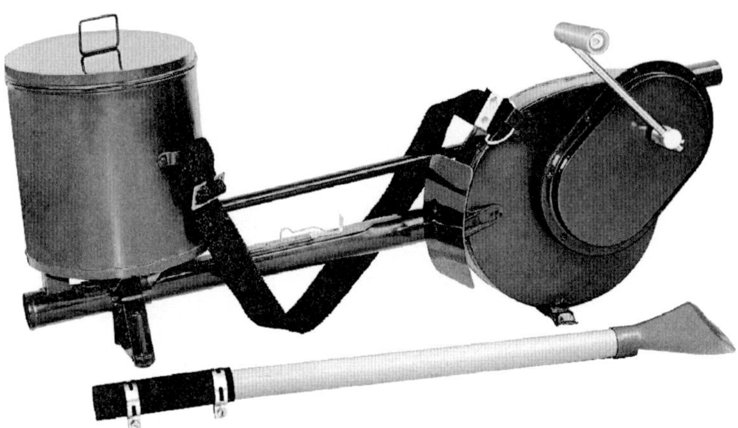

Fig. 6.13 : Shoulder mounted hand rotary duster
Courtesy: Almighty Agrotech Pvt. Ltd., Metoda, Dist. Rajkot (Gujarat)

Animal-drawn sprayer

Animal-drawn sprayer developed commercially has been modified at CIAE, Bhopal for hitching system to uniformly distribute the load, alignment of nozzles in one plane to increase ground clearance and the control valve to regulate the flow of liquid (Anonymous, 2008b). The sprayer works well on soybean crop and on cotton crop (Fig. 6.14). The conventional practice is spraying of pesticides by use of knapsack sprayer or foot sprayer. It can cover about 0.5 ha/h. The annual use of this machine is 50-100 h.

a) Animal drawn sprayer in soybean field

b) Animal drawn sprayer in cotton field

Fig. 6.14 : Animal drawn sprayer being used in field

Power sprayers

It is suitable for spraying in orchards, tea and coffee plantations, rubber plantations, vineyards and field crops. Tall trees up to a height of 15 meters can be sprayed with these types of sprayers (Pandey and Ganesan, 2005; Singh, 2007; Singh *et al.*, 2010). The pump draws the spray liquid from the tank, imparts pressure energy and sends it to the delivery line/lines (Fig. 6.15). Adjusting the nozzle or selecting the appropriate nozzle, adjusts the spray pattern. These sprayers are either operated by auxiliary engines or electric motors. For operation, the shut off trigger valve of the lance is closed and the engine/ electric motor is started to actuate the pump. The operator directs the lance towards the target and operates the trigger/shut off valve. For delivering the spray liquid to large distances/ height a bamboo lance can also be used. Number of spray lance is up to 6 and it can develop pressure up to 40 kg/cm^2. Discharge of the sprayer is up to 25 lit/min and working capacity 0.2-0.3 ha/h.

Fig. 6.15 : A view of power sprayers

Power tiller mounted/small tractor operated orchard sprayer

It is suitable for spraying of foliage, in orchard crops like pomegranate, orange, sweet lime and grapes (Singh and Kandoria, 1999). It consists of an HTP (horizontal triplex piston) pump, trailed type main chassis with transport wheels, chemical tank with hydraulic agitation system, cut off device and boom equipped with turbo nozzles (Fig. 6.16). It is fitted with turbo nozzles with operating pressure of 9-18 kg/cm^2. It generates droplets of 100-150 micron sizes. Depending upon the plant size and their row spacing, the orientation of booms can be adjusted. The spray booms are mounted behind the operator. The sprayer is mounted on power tiller driven by power tiller power take-off pulley (Fig. 6.16a). The power tiller is steered to move in between two rows of crops at recommended speed to maintain the desired spraying rate. Working capacity of machine is 0.40 to 0.70 ha/h at travel speed of 1.20-1.50 km/h and turbo nozzle discharge rate of 2.20-4.85 l/min. Orchard sprayer can also be operated by small tractor of 18-20 hp (Fig. 6.16b).

a) Power tiller operated orchard sprayer

b) Small tractor operated orchard sprayer

Fig. 6.16 : Power tiller mounted orchard sprayer

Tractor operated air assisted sprayers for fruits, vineyards and garden

The sprayer is basically used to sprinkle the pesticides over the fruit trees, and vineyards. The main aim of sprayer is to sprinkle fewer amounts of costly pesticides to most of the areas of tree (Fig. 6.17). The machine covers the plants from bottom to top. It gives best penetration and coverage all over plants and protects from all insects, pests, residual due to effectiveness on underside of the leaves. It gives continuous air pressure, prevents the chemical from drifting and moreover the chemical can be sprayed at low rate, which reduces the quantity of chemicals and maximum coverage of area. The main advantage is that it is operated by small tractors of 18-20 hp which can move easily in vineyards of row-to-row spacing 240 cm, plant-to-plant spacing 150 cm, and height of plant minimum 1.2 m and maximum 2.25 m. The sprayer can be operated at forward speed of 2-4 km/h. It has field capacity of 0.4 ha/h with field efficiency of 88% and discharge per nozzle 1150-1200 ml/min. Water required per ha is 594 l/ha.

Fig. 6.17 : Tractor operated air assisted sprayers for fruits, vineyards and garden
Courtesy: Birar Equipments, Nashik (Maharashtra)

Blower sprayers: They are also called mist sprayers. They are used to apply pesticides in concentrated form (Pandey *et al.*, 1997). It requires 20-80% less water as diluents and saves labour cost substantially. Further savings can be made in the quantity of chemicals used by reducing runoff from foliage when the equipment is operated properly. These sprayers are used for spraying chemicals on fruit trees, large shade trees, vegetable and certain other crops. Large fruit trees may require thinning and pruning to permit proper penetration of small airborne droplets to inner and top-most branches. These machines are similar to power dusters and employ liquid chemical instead of dry powder. This sprayer also has low-pressure, low-volume pump, which forces the spray material under low pressure to the fan where it is discharged into the air stream in small spray droplets by a group of nozzles.

Power sprayers or tractor operated sprayer

Power sprayers are operated usually with internal combustion engines of 1-5-hp capacities or by a tractor. A small engine operated power unit ensuring a constant steady pressure operates the pressure pump. They are operated at pressure from 20-55 kg/cm^2. The pump is operated by tractor PTO also (Shukla *et al.*, 1987). Power sprayer consists of a pump, one or more drums, control valves, pressure gauge, pressure regulator, relief valve and a spray boom fitted with nozzles (Fig. 6.18). The boom may be of flexible hosepipe on which nozzles are mounted to meet different row spacing. The boom is tied with a rigid beam by clamps. Inlet liquid supply to boom is provided at two points for even distribution of liquid. Holes are provided on the frame to lower or raise the beam to suit the crop height. Normally sprayer boom has 13-15 triple action nozzles and can cover 7-8 m width. The capacity of tractor-mounted sprayer is about one hectare per hour while operating at a speed of 3.5 km/h. The tractor is operated in the same pathways and in the same direction for all sprays. Sprayers are used in row and tall crops such as cotton, maize, sugarcane etc (Fig. 6.19). The tractor-operated sprayer can be used for spraying all the 6-7 sprays on the cotton crop if same path and direction is used during all the sprayings. This reduces crop damage under the tractor. The pest control is better in case the spraying is done by tractor-operated sprayer as compared to knapsack sprayer (Shukla *et al.*, 1987). There is a net increase of about 29% yield in the field sprayed by tractor-mounted sprayer as compared to that sprayed by knapsack sprayer.

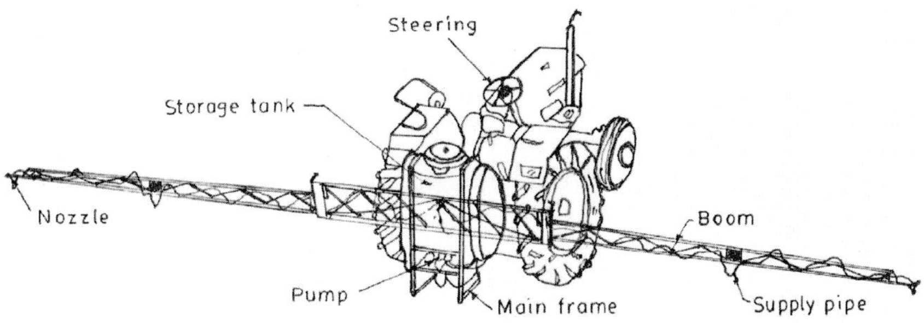

Fig. 6.18 : An isometric view of tractor mounted high clearance sprayer for tall crops

Fig. 6.19 : Tractor-mounted high clearance sprayer for tall crops

Tractor operated sprayer for tall crops

It is used for spraying in vegetable gardens, flower crops, vineyards and for tall field crops like sugarcane, maize, cotton, sorghum, millets etc (Fig. 6.20a). It is operated by PTO of 35 hp tractor (Pandey *et al*., 1997). These are hydraulic energy sprayers. They utilize PTO power of the tractor to operate the pump of the sprayer. The spray boom can be arranged in two ways; ground spray boom and overhead spray boom. The overhead spray boom is designed for tall field crops and the planting is done in such a way that it leaves an unplanted strip of about 2.5 m width for operation of the tractor. Therefore a planted strip may be 18-20 m wide and after every planted strip a fallow strip has to be left for tractor operation. For ground spray boom the planting has to be done in rows keeping in view track width of the tractor. It is suitable for use when the crop is small. The sprayer essentially consists of a tank which is made of fibre glass or plastic, pump assembly, suction pipe with strainer, pressure gauges, pressure regulators, air chamber, delivery pipe, spray boom fitted with nozzles (Fig. 6.20b). It uses high pressure and high discharge pump as the number of nozzles may be up to 20 depending upon the crop and make of the sprayer. Working capacity is 1 ha/h with 14 nozzles.

a) A view of tractor operated sprayer

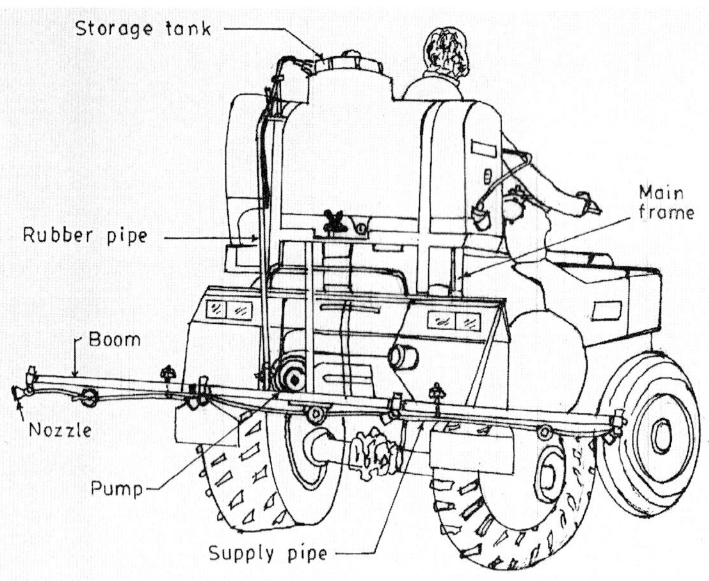

b) Components of tractor operated sprayer

Fig. 6.20 : Tractor operated sprayer for tall crops

Self-propelled light weight boom sprayer

This is used for chemical application on wheat, vegetable and other crops. It is operated by 3.73 kW diesel engines and is controlled by the operator from the handle (Singh and Garg, 2002; Ahuja *et al.*, 2004). It consists of a lightweight power tiller unit and a spraying unit. Spray pump and two narrow pneumatic wheels get power from the engine through gears, chains and sprockets (Fig. 6.21). The ground clearance of the machine is 50 cm. The boom height can be adjusted from 60 cm to 130 cm to suit different crops. The machine has two narrow rubber wheels which are powered from the engine through gears and chain. The third wheel is also provided at the back which acts as supporting wheel. The sprayer consists of a tank of 120 liters capacity, roller type spray pump and a boom with 12 nozzles. The sprayer covers a width of 7 m in one pass. The nozzles spacing can be adjusted to suit different corps. A provision has also been made to adjust the tract width from 90 to 105 cm. The spray boom is mounted on the power unit through a canopy frame. Working capacity of machine is 1.0 ha/h. Saving in labour for this sprayer in comparison to knapsack sprayer is about 70-80% and the saving in cost is about 40-50%.

Fig. 6.21 : Self-propelled light weight boom sprayer

Self-propelled high clearance sprayer

The high clearance sprayer is most suitable for spraying on tall crop like cotton. It can also be used for spraying on sunflower, wheat and other crops (Singh and Garg, 2002; Ahuja *et al.*, 2004). It is a self-propelled unit and operated by 14.9 kW diesel engines (Fig. 6.22a). It had a chassis with

120 cm ground clearance, four wheels, and 20 hp diesel engine, gearbox, water tank, seat for the operator, spray pump and boom with 18 nozzles (Fig. 6.22b). The nozzles are spaced at 675 mm and total boom width is 10.80 m (Fig. 6.22d). The boom height can be adjusted from 31.5 cm to 168.5 cm to suit different crops and can be folded during transport. The wheel track is 135 cm and during operation 2 rows of cotton crop comes under the machine chassis. Machine has four forward speeds and one reverse speed. The 1st and 2nd gears are for field operation while 3rd and 4th gears are for road transport. The field speed is up to 5 km/h and the road speed is up to 25 km/h. The front two wheels are narrow in width (20 cm) and are given drive, while the rear wheels are steering wheels. Fenders have been provided in front of the drive wheels to deflect the crop branches away from the wheels for reducing mechanical damage. These machines usually have low-pressure, low-volume sprayer. The spray boom is mounted on the rear of the machine (Fig. 6.22c). The height of boom can be varied mechanically/hydraulically to permit spraying either in short or tall crops (Anonymous, 1997). The power from the engine is transmitted to two front lugged wheels through chain from gearbox. A clutch is provided to engage or disengage the engine power to drive wheels. The gearbox provides 4 forward and one reverse speed. Drive to pump is given directly from engine through V belt. The spraying pressure can also be adjusted by relief valve. The machine can cover about 1.5-2.0 ha/h. Liquid application rate can vary from 150-175 litres/ha and fuel consumption is 1.5 l/h.

(a) A view of self-propelled high clearance sprayer

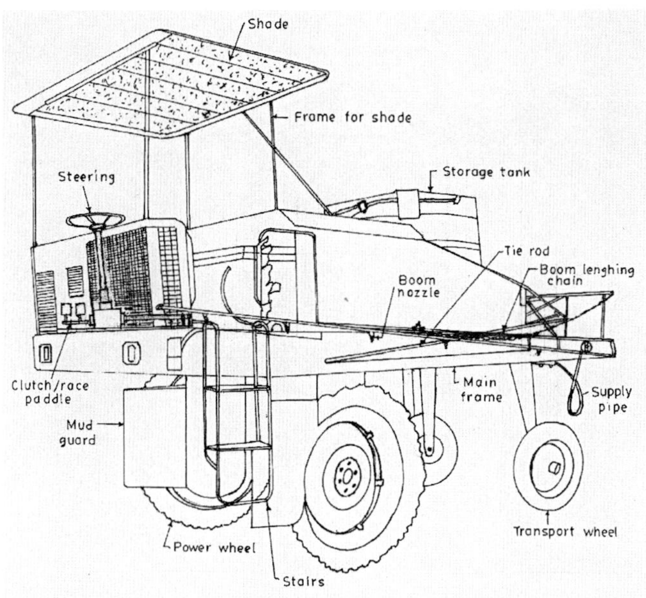

(b) Isometric view high clearance sprayer

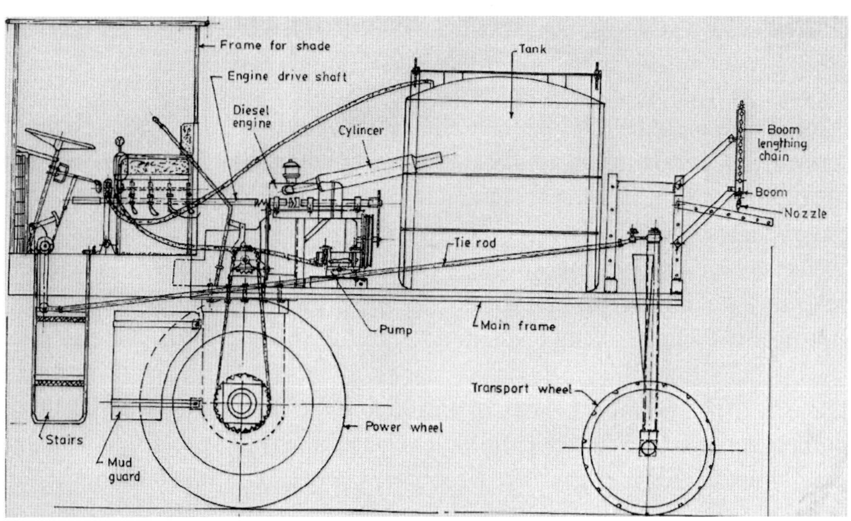

(c) Side view of high clearance sprayer

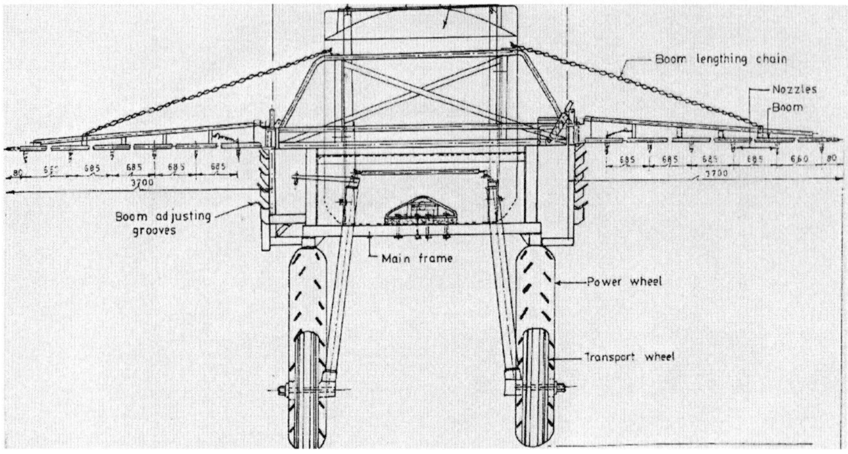

d) Front view of high clearance sprayer

Fig. 6.22 : Self-propelled high clearance sprayer for tall crops

Tractor operated aero blast sprayer

Cotton is cultivated in three distinct agro-climatic zones. North zone comprising Punjab, Haryana and Northern Rajasthan. Central zone comprising Maharashtra, Gujarat and Madhya Pradesh and South Zone comprising Tamil Nadu, Karnataka and Andhra Pradesh. Central Zone contributs 57.5% of total cotton production in the country. Cotton is sown in India during March to September and harvested during September to April. Now India grows all types of cotton i.e. Asiatic, American, Egyptian types and hybrids. The farm machine has great potential for its introduction in the cotton growing area, because as many as 6-7 sprays are required for complete control of insects. Tractor operated sprayers damage the crop, which comes under chassis especially when the crop is tall. On the other hand, aero-blast sprayer results in low crop damage and hence contribute towards increase in yield (Garg and Singh, 2002; Singh and Pandey, 2008; Anonymous, 2008, 2008c, 2008d, 2010 and 2012). Also, by introduction of this sprayer the farmer's having large orchards are benefited. The problem of spraying on cotton, sunflower & horticultural trees can be solved by the introduction of this machine (Fig. 6.23a). The area under these crops may also increase with introduction of this machine. Because of high capacity of this machine, this may also promote custom hiring.

Aero blast sprayer consists of tank of 400 litres capacity, pump, fan, control valve, filling unit, spout adjustable handle and spraying nozzles to release the pesticide solution in to stream of air blast produced by the centrifugal blower (Fig. 6.23b). The machine is comprised of spraying nozzles to release the pesticide solution into stream of air blast produced by the centrifugal blower. The air blast distributes the chemical in the form of very fine particles throughout its swath width, which is on one side of tractor. Other parts of the machine are pump, fan, control value, filling unit and spout adjustment handle. The pump mounted on the sprayer has high flow rate ranging from 110 to 150 litres/min. The sprayer is equipped with a high capacity centrifugal blower to produce air blast that helps to distribute the chemical in form of very fine particles throughout its swath width. Flow rate regulation valve helps to regulate the flow rate ranging from 8.0 to 25.0 l/min at different lever settings of 0.5 to 4.0. The sprayer is equipped with two in-line filters, one for main spout and other for the auxiliary rotary atomizer. These filters have 80 mesh-size capacities and optimize the filtering of the chemical mixture that flow to the nozzles. Spout is easily adjusted both vertically and horizontally to suit the crop/ tree height. As the adjustment handle is located in a convenient position to the tractor drive, it can be adjusted even during the operation. The job of the nozzles is to generate droplets and distribute them uniformly over the area being sprayed. It is operated by 26.1 kW tractors' PTO. The major portion of swath is taken care of by the main blast through the main spout and the auxiliary nozzles cover the swath area near the tractor. Working capacity of aero blast sprayer is 1.5 to 2.0 ha/h.

Air blast sprayers also known as air carrier sprayers have been developed for applying spray liquid to tall trees. Air blast sprayer uses air stream to carry the droplets and utilize smaller droplet size to obtain adequate coverage with low amount of spray material per unit area. The effectiveness of sprayer depends upon its ability to disperse air in all parts of trees with spray-laden air. Deposits on leaf surfaces decrease in proportion to air velocity. Drift problems with this machine are comparable with those of aircraft spraying. Most of the large orchard-type air blast sprayers have axial flow fans with guide vanes to direct the air radially outward through a partial circumferential slot. Some sprayers have two opposite axial flow fans blowing towards each other from either side of the slot. The machine can cover one side of one row or the adjacent side of two rows (Fig. 6.23c). The included angle of delivery on each side is adjustable to accommodate different size of trees.

a) Tractor-operated aero-blast sprayer being used in cotton field and tree

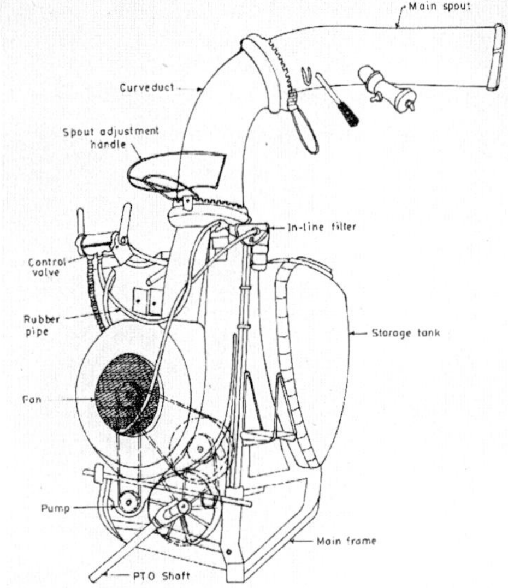

b) Components of aero blast sprayer

c) Aero blast sprayer covers one side of row or the adjacent side of two rows
Courtesy: Gursukh Agro Works, Samrala, Ludhiana (Punjab)

Fig. 6.23 : Tractor operated aero blast orchard sprayer

Tractor operated air sleeve boom sprayer

In pesticide application technique the most popular and commonly used sprayers are based on hydraulic principle. In this technique the deciding factor in the effectiveness of a spraying system is the behaviour of each spray droplet from the moment it emerges from the nozzle until it reaches its target. From the moment the drop leaves the nozzles, it is affected by outside temperature, atmosphere, wind, humidity and their own gravity force applied to the droplet particle by nozzle (Ahuja, 2004 and 2004a; Singh and Kandoria, 1999). In a conventional sprayer, the pressure in the spray line forces the drops out of the nozzle at a force, which throws them at maximum distance of 30 cm. After that, the drops are on their own as they begin the long descending under the pull of gravity. Strong winds blow the droplets off from the intended path to follow and in turn get vaporized and lose momentum if they remain in hotter atmosphere for longer period. Due to this, the droplet size reduces and affects the coverage of the target.

Besides this, if droplet particles have to reach lower side of leaves then it is not possible to reach there as it requires extra force for tilting the crop foliage up and down. To cover such target the droplet particle size should be bigger to have its own momentum.

Air assisted spraying system has been found more efficient way for application of pesticides for control of insects and pest (Singh *et al.*, 2007; Singh *et al.*, 2010; Anonymous, 2008, 2008a, 2010, 2013). It is useful for spraying on tall crops like cotton, sunflower, pigeon pea etc. It is operated by 26.1 kW tractors' PTO. An air-assisting system is used to produce a continuous airflow, which consists of an axial fan, its casing and ducts (Fig. 6.24a). Axial fan is designed and fabricated to provide air at a required flow rate and pressure. There is provision to change the angle of the blade. On the air outlets of the fan casing, sleeves (ducts) of 46 cm diameter are fixed. Throughout the length of the sleeve (duct) 40 mm holes 80 mm apart are made on bottom side. The total length of the sleeves (ducts) plus casing width is 600 cm. A horizontal triplex pump (HTP) is used to generate pressure for the application of spray formulation. A boom of GI pipe having diameter of 35 mm is used for nozzle mounting. Spacing between nozzles, its working pressure and height are selected as 45 cm, 3.5 kg/cm^2 and 50 cm respectively for uniform spray distribution using patternator but can be changed as per need. Nozzle boom is placed just below but slightly behind of the holes of the sleeves. Working capacity of air assisted sprayer is 1.5 to 2.0 ha/h. With the air-assistance facility, the sprayer is able to put an effective number of drops on underside of leaves at any section of plants (Fig. 6.24b). The reduction in number of whitefly adults is 29.5 to 69.7% more for air-assisted sprayer as compared to conventional sprayer. The reduction in freshly shed damaged fruiting bodies is almost double in case of air-assisted sprayer as compared to conventional sprayer. It can be used for spraying chemicals on cotton, pigeon pea and other pulse and oil seed crops to control insects/ pests. The machine creates swirling action in leaves throughout the plant canopy before chemical is being sprayed that makes sure that chemical is being sprayed on both sides of leaves. Tractor operated air assisted boom sprayer also used in vertically folded mode to spray on both side and/or one side in orchard (Fig. 6.25a). The components of this sprayer are shown in Fig. 6.25b).

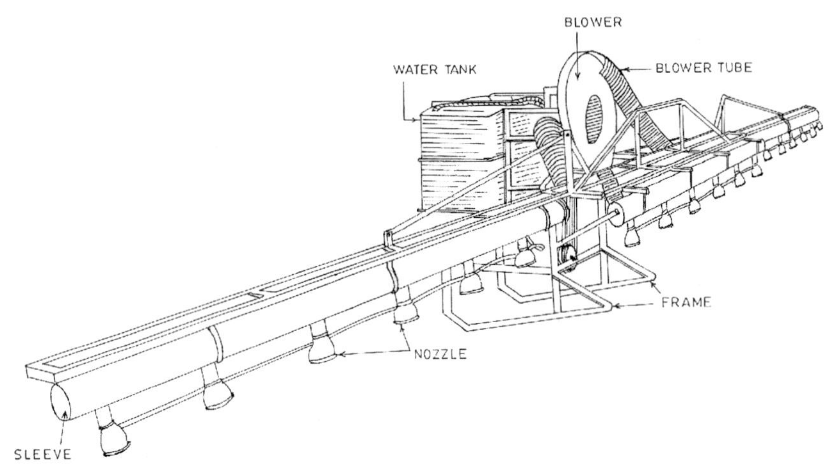

a) Components of air assisted sprayer

b) Different types of air Assisted boom type sprayer being used in cotton field
Courtesy: Gursukh Agro Works, Samrala, Ludhiana (Punjab)
Fig. 6.24 : Tractor operated air assisted boom sprayer

a) Tractor operated air assisted boom sprayer used in vertically folded mode to spray on both side or one side

Courtesy: Gursukh Agro Works, Samrala, Ludhiana (Punjab)

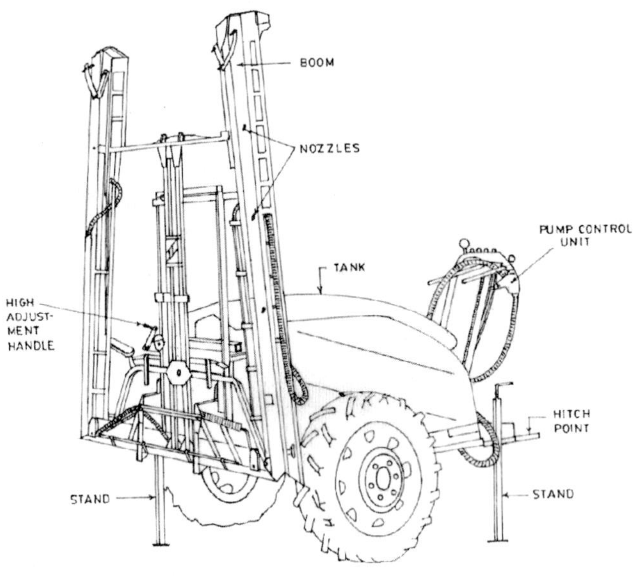

b) Components of air assisted folded type sprayer

Fig. 6.25 : Tractor operated air assisted boom sprayer used in vertically folded mode to spray on both side or one side

Tree duster

Tree duster is used for dusting of the orchards and tall trees. The tree duster consists of an engine which may be petrol or petrol/ kerosene run, blower/ fan directly coupled to engine that produces a high volume and high velocity air stream, hopper for the storage of the chemical, discharge chute and other control attachments. All the subassemblies are joined together and mounted on the stretcher frame. The equipment can be conveniently carried by two persons to the dusting place and moved around the periphery of the tree. For operation, the hopper is filled with the chemical and the metering mechanism is adjusted for the desired application rate. The equipment is placed under the tree to be dusted and cranking starts the engine. The chemical drops in the discharge chute where it comes in contact with high velocity air stream and is thus carried to the target. The air velocity imparts enough energy to the dust particles for their penetration in the canopy of the tree. The chemical usually comes out from the discharge chute in the form of cloud. The equipment is moved around the tree for completion of the dusting with the help of stretcher frame and moved to the next tree. The air velocity produced with this duster is about 70 m/s. The duster can throw the powder to the height of about 25 m. The capacity of the duster is 3 to 4 ha/day depending upon the plant distance and density. The fan output is 15-20 m^3/min. It can be used for dusting in orchards and tall trees.

Tractor operated orchard sprayer

Tractor operated orchard sprayer working on the principle of air blasting has been developed (Anonymous, 2008). The sprayer consists of one plastic tank of 300 litre capacity, ASPEE triplex pump, one centrifugal blower of 1 m^3/ s capacity: one main by pass valve and two flow rate control valves to regulate the flow rate of chemical in two spouts of the sprayer, two pressure vessels, one pressure gauge fitted on the outlet of the pump, two funnel type nozzles fitted in two spouts and pipes and fittings (Fig. 6.26). The sprayer has provision to change the orientation of two spouts with help of two ratchets and pawls to adjust the spouts for maximum coverage of plant canopy. The power to the blower and pump is transmitted by V - belts and pulleys from PTO shaft of the tractor. Effective width of coverage is 10 m. The field capacity of sprayer is 0.87 ha/h and field efficiency 80%.

Fig. 6.26 : Tractor operated orchard sprayer

Ultra low volume spraying equipment

Field crop and Orchard production system in the country is currently dependent on low capacity knapsack or tractor-mounted or recently introduced air-carrier orchard sprayers using large volume of water to apply plant protectants to combat pests (Ahuja, 2004). This not only reduces the field capacity and field efficiency but also drastically cuts down the application efficacy of the pesticide. Thus the overall objective of research and development is to improve field crop and orchard spraying technology. That means concurrently to reduce drift, maintain crop and fruit quality, reduce quantity of needed pesticides, increase deposition efficiency and uniformity on the target, and maintain economic viability for the grower. Air carrier sprayers use water and air to carry agricultural chemicals to the target canopy. In an effort to reduce operating cost and loading time many growers have reduced the volume of water in the spray solution. This trend towards "low volume" spraying has tempted growers to purchase sprayers with too small delivery system. The result is poor pesticide use, loss of quality, and/or higher operating cost.

Ultra low volume spraying technology is being used extensively in many western countries without or with very small dose of water. In ULV spraying droplet size have been further reduced and air has been mainly used as carrier in place of water. Manually operated ULV is a simple but robust hand-held spinning disc Controlled Droplet Application (CDA) sprayer powered by torch (D-cell/R20) batteries, with one set of good quality batteries giving up to twenty hours spraying time (Fig. 6.23). It is designed primarily for the foliar application of insecticides and fungicides – both water-based mixtures (e.g. ECs, WPs) at 10 to 20 litres/ha of total spray volume and in volatile Ultra-Low Volume (ULV) formulations at 1 to 3 litres/ha. An electric motor pins the atomiser disc to produce uniform spray droplet size ranges (the actual size of droplets produced depends on the atomiser disc speed which is determined by the number of batteries fitted). Liquid is fed by gravity through colour coded feed nozzles. The ULV can apply insecticides and fungicides to various row crops.

Air assisted electrostatic sprayer

Air-assisted electrostatic sprayer produces electrically charged spray droplets that are carried to the targets at low pressure (Anonymous, 2015). Air and liquid enter separately at the rear of the nozzle. Just before leaving the nozzle, the air hits the liquid stream to make many thousands of tiny spray droplets that pass through the charging ring. An electrical charge is applied to the spray droplets by the charging electrode. Then the charged spray droplets are blown out of the nozzle and move into the plant canopy where they attract plant material by electrostatic forces. The electrostatic charge induced by electrostatic nozzle is strong enough to allow the droplets to move in any direction to cover all plant surfaces i.e. the undersides of leaves, back side of fruits, vegetables and objects. Air assisted electrostatic sprayers give more than twice the deposition efficiency of hydraulic sprayers and non-electrostatic type of air assisted sprayers. The user benefits in terms of significant reduction in cost of application and significant increase in insect and disease control. The spray gun is powered by 9 Volt battery. This consists of nozzle, power on/off switch, charging indicator, battery, battery housing, air and liquid supply, air filter, liquid filter, trigger and locking system (Fig. 6.27), Anonymous (2015). This gives uniform and better coverage on leaves and hidden areas, reduction in soil deposition, reduction in air pollution, less liquid consumption and higher efficiency.

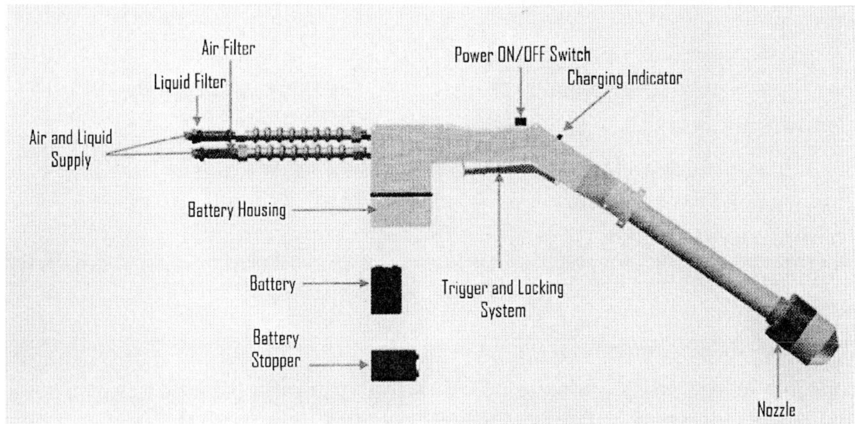

Fig. 6.27 : Components of air assisted electrostatic sprayer

Nozzles

Nozzle is most important component of the spraying technology. Efficacy and control of any of the pesticides will depend upon the type of nozzles being used, its operating pressure and other environmental considerations (Ahuja, 2004a; Singh *et al.*, 1988; Singh and Kandoria, 1999; Singh *et al.*, 2006). The nozzle helps to control the rate, uniformity, thoroughness and safety of pesticide application. Four basic types of nozzles are most commonly used in pest control sprayer's equipment.

Hollow-cone nozzle: They are commonly used for spraying insecticides and fungicides. The pattern of spray is circular with tapered edges to form a hollow cone or ring type spray pattern (Fig. 6.28). This type of nozzle provides adequate coverage of pesticides from all angles. The liquid is fed into a whirl chamber through tangential entry or through a fixed spiral passage to give a rotating motion. The liquid comes out in the form of a hollow conical sheet, which then breaks into small drops.

Solid-cone nozzle: These nozzles produce a solid-cone spray pattern and are used for spot spraying of herbicides in lawns (Fig. 6.28). They cover entire circular area evenly. The construction is similar to hollow cone nozzle with the addition of an internal jet, which strikes the rotating liquid just within the orifice of discharge. The breaking of spray droplet is mainly due to impact action.

Flat-fan nozzle: The flat-fan nozzle produces a spray pattern in the form of a flat sheet and is used for spraying herbicides (Fig. 6.28). The spray pattern has been designed in such a way that 30-50 per cent overlap can be achieved for even distribution of spray liquids. The nozzle tip is provided with an outlet that is elliptical. This nozzle is mostly used for low pressure spraying.

Flooding nozzle: It makes a wide-angle flat spray pattern (Fig. 6.28). It works at lower pressure than other nozzles. The spray pattern is fairly uniform across its width. It is used for spraying herbicides and fertilizer solutions. The flooding flat nozzles result in less drift than the other type of nozzles.

Triple-action nozzle: These nozzles are very popular because they can produce a jet, a solid-cone and a hollow-cone, three types of spray pattern by adjustment. They are used for high-volume as well as low-volume spray applications and are mostly used with tractor-drawn sprayers. The triple action nozzle consists of a swirl plate, swirl chamber, a nozzle disc and a cap. Nozzles are usually made of brass, stainless steel or plastic. Pressure has a considerable effect on nozzle performance. For a given size and design of nozzle and depth of swirl chamber, the higher the pressure, the higher the throughput, the wider the spray angle and the finer the spray droplets.

Basic functions of nozzle is to atomizes liquid into droplets, disperses the droplets in a specific pattern, meters liquid at a certain flow rate, and provides hydraulic momentum. A nozzle has 3 or 4 components; body, cap, strainer and the tip (Fig. 6.29). Nozzle tip is one of the most important functional components of the whole system. Performance of the nozzle is very much dependent upon the nozzle tip. Different types of nozzles are being used. However, different nozzles perform different functions. The nozzle helps to control the rate, uniformity, and thoroughness of pesticide application. As far as spray penetrations are concerned it is wider in flood jet and narrower for the hallow cone but narrow to minimum is the flat type. The shape and size of the nozzle tip is to control the spray angle, discharge rate, spray pattern whereas spray angle influence the swath of the sprayer. A comparative performance of different nozzle spray is given in Table 6.2.

Table 6.2 : Comparative performance of different types of nozzle

S. No.	Parameter	Type of spray pattern		
		Hollow cone	**Floodjet**	**Flat fan**
1.	Droplet size	Finer	Coarse	Medium
2.	Spray penetration	Better in foliage	Wider coverage	Even & uniform
3.	Suitability of spray	Insecticide	Herbicide	Herbicide
4.	Spray angle	Narrow to medium	Wider	Narrow to medium

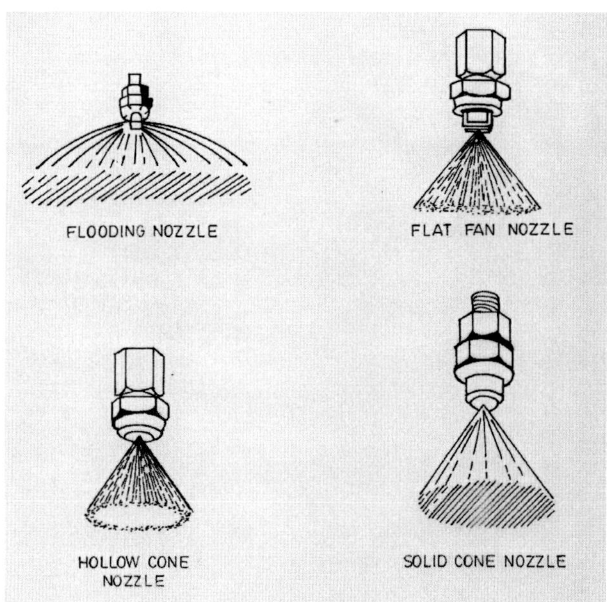

Fig. 6.28 : Various types of nozzles

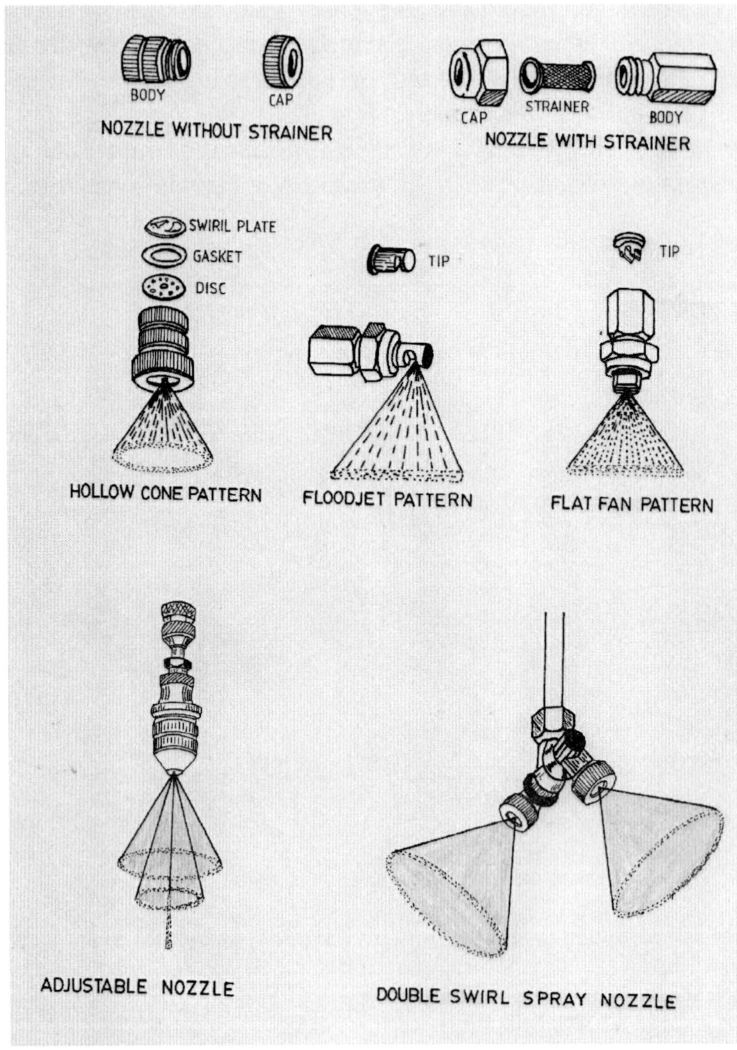

Fig. 6.29 : Different types of nozzles used plant protection equipment and their sectional view

Adjustable nozzle is most suitable for spraying targets which are not within the reach of a man whereas double swirl spray nozzle gives two directional spaying. Adjustable nozzles are difficult to calibrate whereas double swirl spray nozzle can spray higher volumes. Swath width also depends upon the high spray narrow angle, narrow swath, wide angle and wide swath. Pressure also controls the droplet size. Droplet size increases as orifice size increases. Droplet size decreases with an increase in fan angle. Normally tractors have more than one nozzle. In such situation correct overlap is used because more overlap will result in loss of pesticides whereas inadequate overlap leaves untreated gaps & area of poor biological control efficiency (Fig. 6.30). Adjustable nozzle is most suitable for spraying targets which are not within the reach of a man, gives a wide angle hollow cone to a straight solid stream that is, it gives a jet to a cone type of spray pattern, and difficult to calibrate as the flow and droplet sizes vary widely with the nozzle angle. Double swirl spray nozzle is used for spraying in two different directions simultaneously, used in row crops to cover both sides, nozzles can be fitted with different types of tips like hollow cone, solid cone or flat fan, and suitable for high volume applications.

Optimum droplet size of application of pesticides is generally specified within a range of the droplet diameter. Besides this, fate of droplet from the time of their formation by a nozzle until their deposition onto a target is influenced by several factors such as velocity of droplet ejection, gravitational force, and wind velocity, air turbulence caused by thermal movement, volatility of the spray liquid and characteristics of target surface. Droplet size is an important parameter for efficient application and for minimum environment contamination. To reduce wastage, narrow range of droplet spectrum is essential. Coarse droplets are largely influenced by gravitational force though remain relatively unaffected by the turbulence. Fine droplets are influenced by wind and turbulence and have a tendency to drift. Depending upon the target object droplet size also vary. Droplet size more than 300 microns are lost by drip whereas less than 100 microns are lost by drift. Loss of spray material by drip or drift is more prominent in high volume or low volume spray techniques respectively. In case of insecticides, for satisfactory biological efficacy minimum of 20 droplets are required per sq/cm. To achieve efficient deposition a narrower droplet spectrum is needed. In fact larger droplets cause most of the wastage and even if they fall on the target these are off. Thus there is considerable wastage if the density of lager droplet is more.

Nozzle is an important mechanism, which breaks the spray liquid into the desired size of droplets for application to the surface to be sprayed. The nozzles help to control the rate, uniformity, thoroughness and safety of pesticide application. The shut-off may be provided in the supply line or in the nozzle itself. The valve is used to impart the whirling motion to the spray liquid as it flows at high velocity through valve into eddy chamber of nozzle body. The whirl mechanism is usually a spiral on the valve that fits into nozzle body. Since no single nozzle can meet various spray requirements, they are now commonly manufactured with inexpensive replaceable nozzle tips or discs, which can be selected to give, desired spray characteristics and volume for the specified job. The hole at centre of nozzle disc results in a hollow cone of spray. If there is a hole through the centre of valve, a solid cone of spray is produced. Built into some nozzles is a removable strainer with slightly smaller openings than the nozzle orifice to prevent clogging. Most of these nozzles require a minimum hydraulic pressure of 75 kPa whereas for full development of spray pattern a hydraulic pressure of 300 kPa is required.

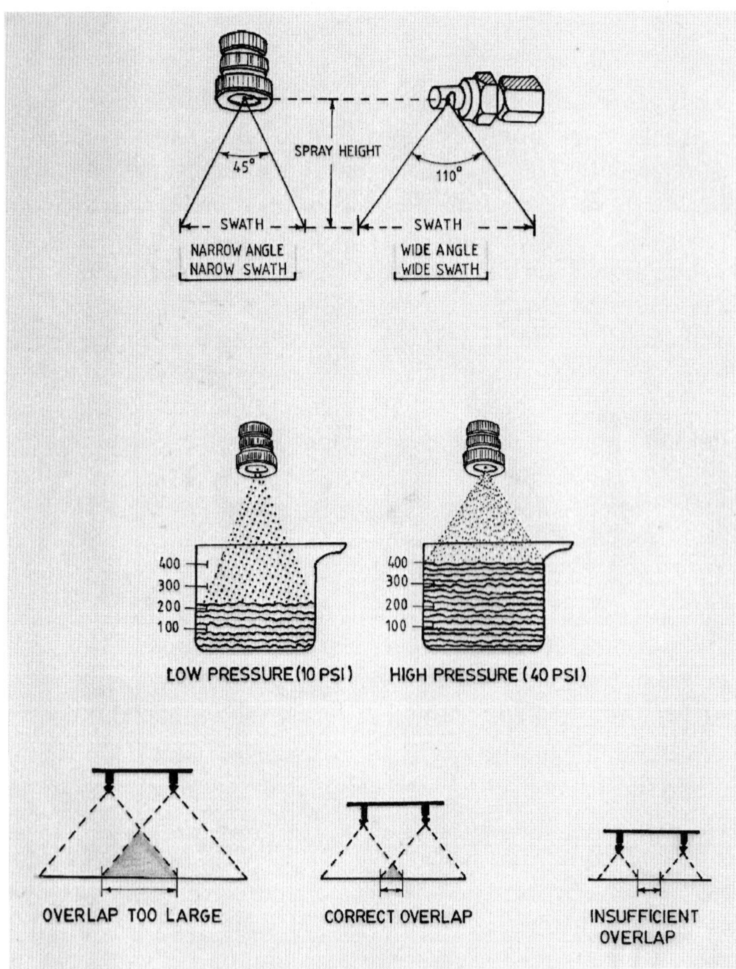

Fig. 6.30 : Proper use of nozzles

Rotary nozzle for tractor operated sprayer for orchard

At present the nozzles used for spraying on trees results into insufficient coverage due to less impinging velocity, droplet size and less droplet density. It is observed that using the most commonly available tractor mounted orchard sprayer with centrifugal blower, having 5 to 9 hallow cone nozzles results into insufficient deposition. Rotary nozzles are used for cotton crop and weed control in western countries. As rotary nozzle gives effective spraying, reduce loss of pesticide, cover large area of plant canopy it is decided to develop rotary nozzle for mango orchard. Rotary nozzles have been developed after considering crop and machine parameters (Anonymous, 2008; 2008b). The performance of the rotary nozzles for spraying in orchards is found satisfactory (Fig. 6.31).

Fig. 6.31 : Tractor operated tycoon sprayer with rotary nozzles during operation

Selection of nozzles: Most of the nozzle manufacturers give discharge of various nozzles manufactured by them at different pressures. This information can be used for selection of correct nozzle for spray job to be done. Before selecting the right nozzle, the points should be considered are: i) type of spray job i.e. spraying of weedicides, insecticides etc; ii) total amount of spray solution to be applied; iii) row spacing; iv) number of nozzles to be used per row and nozzle spacing; v) type of spray pattern desired i.e. fan or cone type; vi) speed of travel, and vii) approximate pressure to be used in spraying. It has been observed that the use of single hollow cone nozzle directly over the crop row gives better insect control and results in higher yields. The fan type spray nozzles do not give as good results as hollow nozzles. The higher yields are obtained from the rows nearest to the nozzle and where more complete coverage is achieved with smaller droplets.

Agitators for chemicals

Positive agitation of spray material in the tank is essential for proper mixing of insecticides and to permit using full range of spray materials including wettable powders, emulsions, fungicides, cold water paints or any other spray material. A propeller or paddle type mechanical agitator in sprayers is equipped with reciprocating type pumps. Hydraulic agitation is used in sprayers provided with pumps mounted on PTO shaft of tractor. Hydraulic agitation usually is not as thorough as mechanical agitation. Mechanical agitation is obtained by means of either flat blades or propellers on a shaft running length-wise in the tank near the bottom. The speed of agitator is kept in the range of 100-200 rpm. The peripheral speed of paddles and shaft input at any peripheral speed can be obtained for round-bottom tanks with flat, I-shaped paddles sweeping close to the bottom (Singh, 2007). Hydraulic agitation is obtained by diverting a portion of pump output back into the tank through a jet nozzle(s). The turbulence from jet provides the mixing action. The jet nozzles are located in a pipe along the bottom of tank.

Safety measures

All chemicals used for dusting and spraying plants are very strong poisons and should be handled with utmost care (Singh and Verma, 2009; Singh and Kandoria, 1999). The safety precautions be taken while handling and spraying the insecticides and pesticides i.e. i) do not allow youngsters, pregnant women and nursing mothers to take part in seed dressing, dusting and spraying of plants; ii) make sure that all workers are familiar with safety rules governing the handling of chemicals and are supplied with respirators, goggles and overalls; iii) always keep a supply of clean water, soap, towels and first-aid-kit with necessary antidotes on the working site; iv) do not use leaky sprayers; v) during spraying avoid smoking and do not take food; vi) at the end of working day wash the face and hands; vii) wash the clothes after use; and viii) store the left over chemical in safe place under lock and key and destroy the container properly. For dry dressing of seeds, choose an open place at a distance from living quarters, cattle sheds, storehouse, and water storage pools and pastures grounds. Transport the chemicals in tight and well-closed containers. Supply all workers with overalls and protective devices while handling and spraying chemicals. Should there be a case of poisoning, send the victim immediately to nearest hospital. If the cause of poisoning is arsenic, fluorine, mercuric dichloride or phenol, give the victim some warm milk with egg. If the cause of poisoning is copper compounds, do not give milk. Remember the following points as safety precautions before using sprayers:

i) always keep pesticides under lock and key,

ii) do not store leaky containers,

iii) do not keep pesticides near the food and fodder storage,

iv) read the label and directions for use before opening a pesticide container,

v) do not use cooking utensils for preparing the solutions,

vi) ensure that sprayer is not leaking,

vii) always use safety wares like apron, goggles, facemask, gumboots and gloves while spraying,

viii) remove nozzle clogging by thin pin and not by mouth,

ix) smoking, eating, drinking etc during pesticide application is strictly prohibited without washing hands, face, body etc,

x) do not store spray solution over night,

xi) clean the spraying equipment thoroughly to decontaminate it,

xii) change the clothes and wash them after spray,

xiii) have a thorough bath with soap and water, and

xiv) store the unused contents of container in the original container.

Maintenance of pesticides application equipment

Pesticides are expensive and so they must be applied efficiently. Proper maintenance of spraying equipment is, therefore, essential. Points should be kept in mind to minimize the labour problems and prolong the useful life of pesticide application equipment are: i) use clean water; ii) keep screens in place; iii) never use a metal object to clear the nozzles; iv) flush sprayers before using them; v) clean sprayers thoroughly after use; vi) remove and clean all screens and nozzle tips in kerosene using a soft brush; vii) when pump is not in use, fill it with a light oil and store it in a dry place; and viii) lubricate all the points required. Common troubles, causes and remedies of pesticide equipment are listed in Tables 6.3, 6.4 and 6.5.

Table 6.3: Common troubles, causes and remedies of pesticide equipment

Trouble	Cause and Remedy
1) Single-barrel stirrup pump	
Spray liquid leaks along the sides of the piston rod during working.	This occurs when the packing twine below the cap of the barrel is worn-out, replace it.
2) Double-barrel stirrup pump	
i) Plunger rod of the pump gets pushed up automatically after the downward stroke	This is due to leakage in the delivery valve. Clean the valve and its seat. If the valve is worn-out, replace it.
ii) Spray liquid is properly discharging only during the pressure stroke	This is due to the rupturing of the delivery tube at some place. Repair or change the tube.
3) Rocking sprayer	
i) Water leaks along the sides of the plunger spraying.	Tighten the locknut of the piston.
ii) PVC piston does not move freely in the pump barrel.	Loosen the locknut.
iii) Pressure chamber does not retain pressure required for efficient spraying	Replace the rubber washer below the pressure chamber
4) Foot sprayer	
i) Spray fluid leaks along the plunger rod during operations.	Tighten the gland nut that touches the plunger rod. If necessary, replace the packing twine of the gland nut.
ii) The pedal does not move upward automatically after the downward stroke.	Change the pedal-return spring.
5) Knapsack sprayer	
Spray fluid is properly discharging only during the pressure stroke.	Repair or change the delivery tube.
6) Compression knapsack sprayer	
i) Plunger rod of the pump gets pushed up automatically after the downward stroke.	This is due to leakage of the air-check valve. Clean it. If defect persists, replace the valve.

ii) On the downward stroke, strong resistance is felt.	A faulty valve lets the spray liquid into the pump barrel. Since liquid cannot be compressed, resistance is encountered. Replace the valve.
7) Rotary duster	
i) Dust not discharged on operating the duster.	It may be due blockage of suction pipe or stoppage of movement of the feeding-brush. Clean suction pipe and tighten the feeding brush on the shaft by tightening the nut.

Table 6.4 : Common faults and remedies of different types of manually operated sprayers

Defects/Possible Causes	Remedies
1) Pressure drop	
i) Clogged suction screens	Clean them
ii) Loose connection or missing gaskets in suction line or partially opened suction valve; broken or leaky suction hose, drain cocks not closed	Find out the specific cause and rectify the fault
iii) Imperfect sealing due to deposition of grains of sand under the ball valve or certain solids on the surface of the ball or wearing out of ball valve and its seat	Cleaning the ball and its seat usually improves the performance of pump. If necessary, replace the valve seat or valve or both
iv) Worn out nozzle disc	Replace the disc
v) Pump speed slow	Increase the pump speed to regain pressure
2) Leakage	
Loose nuts and clamps or worn out gaskets	Tighten the nuts and replace the gaskets
3) Lack of suction	
Cup leather (plunger bucket) is either wrinkled or worn out or it is completely dried	Clean the suction assembly set. Replace the cup leather. If cup leather is hard and dry, lubricate with vegetable oil, viz., castor oil, groundnut oil, coconut oil or palm oil.
ii) Suction valves temporarily stuck	Put water through the inlet so that it reaches the suction valve to loosen it. If necessary, open the valve

4) Failure to retain pressure	
i) Lid of the tank may be loose or the gaskets may be spoiled	Tighten the lid and check the gasket and replace it, if necessary
5) Difficult working of the plunger	
i) Plunger may be bent through the full length of the barrel	Straighten the plunger rod, if found bent
6) Uneven discharge of spray	
i) Nozzle obstructed	Open the nozzle and clean its various parts, especially the orifice

Table 6.5 : Common faults and remedies of engine driven and motorized knapsack sprayers and dusters

Defects/Possible Causes	Remedies
A) Engine does not start/runs irregularly/misfires	
i) Fuel cock closed	Open the fuel cock
ii) Lack of fuel in the tank	Fill the fuel tank
iii) Air vent in tank cap clogged	Open the air vent
iv) Main jet blocked	Remove it and clean it by blowing with air and not with needle or a wire
v) Engine flooded with fuel	When the engine is running, close the fuel cock. Open the throttle lever completely and turn the engine over a few times. Unscrew the spark plug, clean and dry it
vi) Spark plug damaged	Clean the spark plug and see whether the gap is proper. If the lead gives a spark but the plug does not, change the spark plug. If the plug is loose, tighten it
vii) Carburetor dirty	Clean the carburetor
viii) Short circuit	The lead meant for stopping the engine may be in contact with the frame. Brake this contact
ix) Ignition wire loose or damaged	Fasten it or replace it, as need be
x) Contact-breaker points dirty	Clean the points and make the contact-breaker point opening equal to 0.3-0.4 mm, i.e. equal to thickness of a post card
xi) Air cleaner dirty	Clean it with petrol, dry and refit

xii) Starting difficulty in winter	Remove the air cleaner and close the air passage in the carburetor with hand to stop the intake of air. Start the engine with rope and immediately fit the air cleaner
xiii) Dirty fuel pipe screens	Clean the fuel pipe filter at the fuel-cock or in the carburetor supply nipple, or both
xiv) Silencer faulty	Replace the silencer
xv) Water in the fuel	Change the fuel
B) Engine does not gain momentum	
i) Air filter partially choked	Clean the air filter element
ii) Choke closed	Open the choke
iii) Carburetor partly blocked	Clean the carburetor
iv) Exhaust port or silencer choked	Remove the carbon deposit
v) Oil seal of the crank case leaking	Change the oil seal
vi) Cylinder bore and rings worn-out	Change the barrel and the rings
vii) Ignition timing incorrect	Correct the ignition timing
C) Engine overheating	
i) Engine over speeding	Run at the speed recommended by manufacturer
ii) Fuel incorrect	Correct the proportion of fuel and oil in the mixture
iii) Mixture lean	Check the dirt in the jet and set the carburetor needle
iv) Ignition setting incorrect	Correct the ignition setting
v) Air-flow to engine obstructed	Clean the surface of fins and fan blades
vi) Exhaust system choked	Remove the carbon deposit
D) High fuel consumption with a strong smell or unburnt fuel	
i) Float needle damaged	Replace the defective parts
ii) Needle seating damaged.	
iii) Float punctured	
iv) Main jet too big	
v) Air cleaner dirty	Clean the air cleaner

Terminology in spray application: Some of the terms related to the effectiveness of application of chemical are listed here (Singh, 2007; Singh and Verma, 2009; Singh and Kandoria, 1999; Ahuja, 2004):

Coverage: is defined as the extent to which a pesticide spray has been distributed on a target surface. The biological implication is that good coverage increases the probability that a pest will encounter a pesticide.

Retention: is the amount of spray liquid retained on crop plants (mostly on the leaves). In other words, it is the remaining proportion of pesticide that has not "run off". There is an interaction between formulation effects on the tenacity of a deposit and the surface of the leaf to which it adheres. Droplets often bounce on leaves that are waxy and poor retention may occur with water-based formulations, especially those with high dynamic surface tensions. On the other hand, absorption of active ingredient may occur with oil-based formulations. Leaf exudates (*e.g.,* in apples and broad beans) may also contribute to the redistribution of a pesticide.

Volume application rate (VAR): is the amount of formulation applied per hectare. This gives a general classification of volume application rates (in l/ha) for field and tree/bush crops.

Volume median diameter (VMD): This parameter is usually employed to describe the droplet spectrum. It is droplet diameter that satisfies the condition that half the spray volume is of droplets smaller than a droplet whose diameter is VMD and other half of spray volume contains larger droplets. This can be obtained from a volume cumulative curve of droplet size spectrum. It is the diameter corresponding to 50% cumulative. Droplets in the range of 120-300 micron VMD are found efficacious.

Number median diameter (NMD): It is an average diameter of the droplets without any reference to their volume. The diameter corresponding to 50% cumulative number curve gives NMD.

Uniformity coefficient (UC): The ratio of VMD and NMD is called uniformity coefficient. It indicates the range of size of droplets. The more uniform the size, the nearer the ratio is to unity.

Drift: Drift is of particular concern when deposits of harmful chemicals must be avoided or minimized in the fields nearby the area where chemical is used. The parameters affecting drift are rate of fall of particles, initial height of fall, wind velocity and direction, atmospheric stability and other meteorological factors. Evaporation of water or other volatile materials from droplets reduces the droplet size and thus adversely affects the deposition efficiency and drift. Employing atomizing devices that produce spray having large volume mean diameters can minimize drift. Drift is more serious with dust than with sprays because of smaller particle size.

Droplet density (DD): The number of droplets per unit area of leaf surface is called droplet density. Droplet density of 20-50 droplets of pesticide per square centimetre is most effective against cotton pests.

Leaf area index (LAI): It is the area of one side of the leaf divided by the corresponding ground area. It is considered to be a gauge to assess the crop growth.

Spread factor: The ratio of the diameter after incidence to the diameter before incidence is called spread factor for a particular object and liquid spread. It is greater than unity because after impingement the droplet spreads. Size of droplet, surface of object and formation of the spray liquid affect the spread factor.

Droplet size spectra: It is generally refers to the measurement of and statistics for describing spray droplet spectra. At least two important measures of spray distributions are usually required, and the statistics used here are: i) **size**: the *volume median diameter* (VMD) or MMD - mass median diameter): where half of the volume of spray contains droplets larger than the VMD (in µm) and the other half is in smaller droplets. The number median diameter (NMD) is the drop diameter such that 50% of the drops by number are of smaller diameter; ii) **quality**: the most reliable statistic is probably *relative span*, which is a dimensionless parameter calculated from the volume distribution only.

Descriptions of droplet size: especially with respect to standard hydraulic nozzles i.e very fine, fine, medium, coarse etc.

Factors for improving spray efficiency

The three primary factors to improve application technology are: timing, targeting and towers. **Timing** consists of two components: First, full implementation of Integrated Pest Management (IPM) strategies to assure that pesticides are applied only "as needed". It is no longer efficient to apply protect ant sprays following the traditional "Spray Calendar". Second, availability of high capacity sprayers to cover the critical production area within a narrow window of concurrent "as needed" and "acceptable weather" conditions (typically 24 to 48 hours). Less pesticides are required to provide control if they are effectively applied at the optimum time. **Targeting** consists of three operations: first, manually close all the nozzles that direct spray either above or below the target canopy, second, adjust the application rate of each nozzle to match the density of the canopy it targets, and third, install an automated canopy sensing system, "electric eyes," that turns a nozzle or group of nozzles "on" when it senses a target canopy. **Tower sprayers** have a radically different air distribution system that produces three significant benefits.

i) A conventional air-carrier orchard sprayer directs the air radially out from one central source. Many small droplets are carried through the canopy and released as "drift" in the atmosphere above the trees. An ideal tower sprayer is taller than the target canopy and produces a continuous curtain of "straight stream", uniformly spray laden air. The top of the tower focuses the air slightly downward toward the center of the canopy as well as providing a boundary of clean air above the canopy. This boundary of clean air traps the spray laden air into the target canopy, thus minimizing drift.

ii) A tower sprayer moves the spray laden air horizontally, parallel to the ground. The horizontal air movement provides each spray droplet the maximum opportunity to deposit on a target surface before it evaporates or lands on the soil.

iii) A tower sprayer typically produces smaller size spray droplets. Because of the greatly expanded length of the air outlet on a tower sprayer, the manufacturer needs to install many more, smaller nozzles to atomize the spray. Smaller nozzles produce smaller drops which result in improved deposition uniformity. Pest problems start in the areas where it is the most difficult to deposit spray droplets (i.e. back side of leaves and fruit). Careful observation of the target canopy reveals that big drops tend to deposit on the front side of the first layer of leaves and only small drops land in the hard to reach areas. Spraying smaller drops decrease deposition on the outside edge of the tree where there is excess and increase deposition in the hard to reach areas of the fruit canopy. Some tower sprayers use mechanical rotary atomizers with individual peristaltic metering pumps to minimize nozzle plugging, maximize the number and uniformity of small spray drops, and provide a wide range of continuously variable application rates.

References

Ahuja S S. 2004. Equipment for ultra low volume spraying. Proceedings of Summer School on 'Advances in Interculture and Plant Protection Equipment. Department of Farm Power & Machinery, PAU Ludhiana. Aug. 17, 2004 to Sept. 6.

Ahuja S S. 2004a. Procedure for effective and Safe Utilization of Sprayers for different farm operation. Proceedings of Summer School on 'Advances in Interculture and Plant Protection Equipment. Department of Farm Power & Machinery, PAU Ludhiana. Aug. 17, 2004 to Sept. 6.

Anonymous. 1997. Biennial Report 1995-97. Dept. of Farm Power & Machinery, P.A.U. Ludhiana.

Anonymous. 2008. Research Highlight. AICRP on Farm Implements and Machinery, CIAE Bhopal. Technical Bulletin No.: CIAE/2008/141.

Anonymous. 2008a. Annual Report (2006-08). AICRP on Farm Implements and Machinery, MPUAT Udaipur.

Anonymous. 2008b. Annual Report (2008-10). AICRP on Farm Implements and Machinery, MPKV Rahuri.

Anonymous. 2008c. Success Stories. AICRP on Farm Implements and Machinery, CIAE Bhopal. Extension Bulletin No.: CIAE/FIM/2008/80.

Anonymous. 2008d. NAAS Report on the evaluation of Plan Scheme Central Institute of Agricultural Engineering, Bhopal for the X Five Year Plan (2002-2007). NAAS New Delhi

Anonymous. 2010. Research Highlight. AICRP on Farm Implements and Machinery, CIAE Bhopal. Technical Bulletin No.: CIAE/2010/151.

Anonymous. 2012. Directory of Successful Farm Machinery in SAARC Countries. SAARC Agriculture Centre. BARC Complex, Farmgate, Dhaka – 1215 (Bangladesh.

Anonymous. 2013. Research Highlight. AICRP on Farm Implements and Machinery, CIAE Bhopal. Technical Bulletin No.: CIAE/2013/158.

Anonymous. 2015. Operation & maintenance manual – Air assisted electrostatic sprayer. CSIR Organization Chandigarh.

Dixit Anoop; Singh Surendra; Singh Santokh. 2006. Ergonomics studies on power operated knapsack sprayer. *J Res Punjab Agric Univ*, 43(4): 323-27.

Garg I K; Singh Surendra. 2002. Farm equipment for Punjab agriculture. Department of Farm Power & Machinery, Punjab Agricultural University, Ludhiana.

Pandey M M; Ganesan S. 2005. Farm Mechanization Package for Dryland Agriculture, Technical Bulletin No: CIAE/FIM/2005/117. Central Institute of Agricultural Engineering, Bhopal.

Pandey M M; Ganesan S; R K Tiwari. 2006. Improved Farm Tools and Equipment for North Eastern Hills Region, Technical Bulletin No. CIAE/2006/121. Central Institute of Agricultural Engineering, Bhopal.

Pandey M M; Majumdar K L; Singh Gyanendra; Singh Gajendra. 1997. Farm Machinery Research Digest. Technical Bulletin No. CIAE/97/69, Central Institute of Agricultural Engineering, Bhopal, 328 p.

Patel Sharad L. 2004. Role of Agricultural Engineering in Integrated Pest Management. Proceedings of Summer School on 'Advances in Interculture and Plant Protection Equipment. Department of Farm Power & Machinery, PAU Ludhiana. Aug. 17, 2004 to Sept. 6.

Shukla L N; Sandhar N S; Singh Surendra; Singh Joginder. 1987. Development and evaluation of wide-swath, tractor mounted sprayer for cotton crop. *Agricultural Mechanisation in Asia Africa & Latin America (AMA)*. 18(2): 33-36.

Singh S K; Singh Surendra; Dixit Anoop; Khurana R. 2010. Development and field evaluation of tractor mounted air assisted sprayer for cotton. *Agricultural Mechanisation in Asia Africa & Latin America (AMA)*. 41(4):49-54.

Singh S K; Singh Surendra; Sharda Vaishali. 2007. Effect of air-assistance on spray deposition under laboratory conditions. *Journal of Institution of Engineers (Agril. Engg.)*, Vol. 88 (December): 3-8.

Singh S K; Singh Surendra; Sharda Vaishali; Singh Nirmal. 2006. Performance of different nozzles for tractor mounted sprayers. *J Res Punjab Agric Univ*, 43(1): 44-49.

Singh Surendra. 2007. Farm Machinery – Principles and Applications. Directorate of Information & Publication of Agriculture, Indian Council of Agricultural Research, Krishi Anusandhan Bhawan-I, Pusa Campus, New Delhi.

Singh Surendra; Kandoria J L. 1999. Selection, Use and Maintenance of Pesticide Application Equipment. *Directorate of Information & Publication of Agriculture, Indian Council of Agricultural Research*, Krishi Anusandhan Bhawan-I, Pusa Campus, New Delhi.

Singh Surendra; Pandey M M. 2008. X Plan Achievements (2002-2007). AICRP on Farm Implements and Machinery, CIAE Bhopal. Technical Bulletin No.: CIAE/2008/137.

Singh Surendra; Shukla L N; Sandhar N S. 1988. Comparative study of critical dimensions and performance of tripple action hydraulic spray nozzles. *J. Agril. Engg.* 25(1): 6-11.

Singh Surendra; Verma S R. 2009. Farm Machinery Maintenance & Management. Directorate of Information & Publication of Agriculture, Indian Council of Agricultural Research, Krishi Anusandhan Bhawan-I, Pusa Campus, New Delhi.

7

Harvesting and Threshing Equipment

7.1 Harvesting and threshing equipment for cereal crops

Harvesting and threshing is an important operation in the farming. This can be done either manually or with the help of power operated machinery. Harvesting of field crops continues to be one of the most labour intensive operations in agricultural production system. Harvesting of crop is generally done by hand using sickle in India. Power operated harvesting machinery has not caught up with Indian farmers because of limited size of land holdings, economic constraints and non-uniformity of field conditions. Manual harvesting with different hand tools continues to be dominant, which takes 170-200 man-hours to harvest one hectare of paddy crop. Due to high labour demand at the time of harvesting, the operation continues for weeks together, resulting in over drying of crops in the field causes grain losses to the extent of 5 to 15 percent and even results in loss of crop due to untimely rain during harvesting. Mechanical devices to harvest wheat and paddy crops have become popular especially in northern India to minimize the field losses resulting from shattering, unfavourable weather conditions and pilferage.

Higher shattering losses occur while harvesting crops like wheat, barley and gram in over dried conditions. It has been noticed that loss of moisture from the crops in the months of April and May takes place at a much faster rate. As a result of this, the moisture content drops down far below the optimum values of these crops. Harvesting of crops like maize, sorghum and pigeon pea is done at relatively low moisture content. The crop stacks are thick and woody and the hand sickle is adequate tool used for harvesting. For other conditions, a hand chopper is sometimes also employed. Now mechanized harvesting devices such as reapers, combine harvesters etc are being used for harvesting these crops.

Indian farmers having low income and low investing capacity are reluctant to change the traditional methods of harvesting and threshing. Small farm holdings, small plot size, poor weed control and lack of proper approach roads to the field, limits the mechanization of harvesting operation. There are some crop constraints such as excessive moisture content at the time of harvesting, uneven ripening, severe lodging and low straw-grain ratio, which also limits the mechanization of harvesting and threshing operations.

Harvesting of wheat and rice crops is traditionally done by using local sickle. Improved serrated blade sickles are also in use. The machines available for efficient harvesting of these crops are; self-propelled walking type reaper, reaper binders, tractor front and rear mounted reapers and combine harvesters. Sorghum is harvested by local sickle and is the traditional practice followed by farmers. Suitable machines are not available for harvesting this crop. However, combine harvesters are in use in advanced countries. In case of maize the matured cobs are collected manually and threshed by multi crop threshers where grains are separated from cobs after removing the cover. Maize combines are also available with corn-head snapping unit which collects cobs and feeds into the threshing unit. Harvesting of crops like soybean has to be done carefully as the matured grains easily detach from the ear heads/pods and, therefore, cannot be harvested by fast working tools or machines. Bengal gram, green gram, lentil is to be harvested at ground level. Oilseed crops pose different type of problems to engineers for mechanization of their harvesting. Safflower is a spiny crop and difficult to harvest even manually. In case of sunflower, harvesting is simpler as only flower heads are to be collected. In sesamum crops, pods containing seeds are attached to the main stem and they are mostly raised by broadcasting. This also needs gentle handling. Farmers follow different methods for harvesting of rapeseed/mustard and pigeon pea. Mostly, farmers harvest these crops at branch level, but small farmers harvest these crops at ground level. Harvesting of root crops involves digging, shaking to remove adhering soil, windrowing or stacking and picking. A good root crop harvester should give maximum recovery and cause minimum damage to pods or tubers.

Crops are harvested after normal maturity with the objective to take out grain, straw, tubers etc. without much loss. It involves cutting / digging / picking, laying, gathering, curing, transport and stacking of the crop. In case of cereals like wheat and paddy the plants are straight and smooth and ears containing grains are at the top whereas most of oilseed and pulse crops have branches, which create problems in harvesting by manual or mechanical means. As per Bureau of Indian Standards the cutting and conveying losses should not be more than 2 per cent. There are various designs of tools and equipment used for harvesting the crops and threshing it separately. Sickles, hand tools and reapers for grain crops and diggers for tuber crops and rhizomes, operated with different power sources are used. Combine harvesters, both tractor mounted and self-propelled, are being very widely used for different grain crops.

The harvesting of crops is traditionally done by manual methods. Harvesting of major cereals, pulse and oilseed crops are done by using sickle (Fig. 7.1) whereas tuber crops are harvested by country plough or spade. All these traditional methods involve drudgery and consume long time. Timeliness of harvest is of prime importance. During harvesting season, often rains and storms occurs causing considerable damage to standing crops. Rapid harvest facilitates extra days for land preparation and earlier planting of the next crop. The use of machines can help to harvest at proper stage of crop maturity and reduce drudgery and operation time. Considering these, improved harvesting tools, equipment, combines are being accepted by the farmers.

Fig. 7.1 : Harvesting by sickle

Different type of mechanical harvesting tools / equipment

Serrated blade sickle: It has a serrated curved blade and a wooden handle. The handle of improved sickle has a bend at the rear for better grip and to avoid hand injury during operation. Serrated blade sickles cut the crop by principle of friction cutting like in saw blade. The crop is held in one hand and the sickle is pulled along an arc for cutting. Cutting of crop close to the ground is possible with modified handle. Energy requirement is 80-110 man-h/ha. It can be used effectively for harvesting of wheat, rice and grasses.

Reaper: Reapers are used for harvesting of crops mostly at ground level. It consists of crop-row-divider, cutter bar assembly, feeding and conveying devices. Reapers are classified on the basis of conveying of crops as given below:

Vertical conveying reaper windrower: It consists of crop row divider, star wheel, cutter bar, and a pair of lugged canvas conveyor belts. This type of machines cut the crops and conveys vertically to one end and windrows the crops on the ground uniformly. Collection of crop for making bundles is easy and it is done manually. Self-propelled walking type, self-propelled riding type and tractor mounted type reaper-windrowers is available. These types of reapers are suitable for crops like wheat and rice. The field capacities of these machines vary from 0.20-0.40 ha/h.

Horizontal conveying reapers: This type of reapers is provided with crop dividers at the end, crop gathering reel, cutter bar and horizontal conveyor belt. They cut the crop, convey the crop horizontally to one end and drop it to the ground in head-tail fashion. Collection of crop for making bundles is difficult. This type of reapers is tractor mounted and suitable for wheat, rice, soybean, and gram. Performance of reapers with narrow-pitch cutter bar is better for soybean and gram crops.

Bunch conveying reapers: This type of reapers are similar to horizontal conveying reapers except that the cut crop is collected on a platform and is being released occasionally to the ground in the form of a bunch by actuating a hand lever. Here, collection of crops for making bundles is difficult. Bullock drawn and tractor-operated models are available and they are suitable for harvesting wheat, rice and soybean crops.

Reaper binders: The cutting unit of this type of reapers may be disc type or cutter bar type. After cutting, the crop is conveyed vertically to the binding mechanism and released to the ground in the form of bundles. Self-propelled walking type models are available but these are not popular due to high cost of twine. Reaper binders are suitable for rice and wheat.

Strippers: The design of a tractor front mounted stripper is available for collection of matured grass seeds from the seed crops. It consists of a reel having helical rubber bats which beat the grass over a sweeping surface where the ripened seeds get detached and the seeds are collected in the seed box.

Diggers: The design of groundnut and potato diggers of animal drawn and tractor operated types are available. The digging units consist of V-shaped or straight blade and lifter rods are attached behind the share. These lifter rods are spaced to allow the clods and residual material to drop while operating the implement. The plant along with pods/tubers is collected manually.

Combines: Various designs of combine harvester having 2 to 6 m long cutter bar are commercially available. The function of a combine harvester is to cut, thresh, winnow and clean grain/seed. It consists of header unit, threshing unit, separation unit, cleaning unit and grain collection unit. The function of the header is to cut and gather the crop and deliver it to the threshing cylinder. The reel pushes the straw back on to the platform while the cutter bar cuts it. The crops are threshed between cylinder and concave due to impact and rubbing action. The threshed material is shaken and tossed back by the straw rack so that the grain moves and falls through the openings in the rack onto the cleaning shoe while the straw is discharged at the rear. The cleaning mechanism consists of two sieves and a fan. The grain is conveyed with a conveyor and collected in a grain tank.

The points to be kept in mind for successful use of mechanical harvesting equipment are: i) field must be fairly level without undulations to facilitate smooth operation and uniform stubble length, ii) for small reapers and binders, plants must be grown in rows, iii) field efficiency of harvesting machines is high in large fields, and iv) water control in rice field is essential to ensure that the fields are drained and are relatively dry at harvest time. For harvesting tall varieties, plants in rows should not be entangled with each other. Factors affecting performance of harvesting machines are: i) Crop factors such as crop variety, ambient temperature, maturity of crop, crop moisture, crop condition, crop density; ii) Machine factors such as shape and size of crop divider, reel position and speed, cutting blade shape and speed, conveyor speed, machine vibrations and machine settings; and iii) Operational factors such as height of cut and operation speed.

Basic principles of harvesting: The operation of cutting plant is carried out by four different actions i.e. slicing action with a sharp smooth edge, tearing action with a rough serrated edge, high velocities single element impact with a sharp or dull edge, and a two-element scissors type action or shearing type cutting. Forces used while operating grain harvesting equipment in the field are:

i) Shearing force - e.g. cutting, straw chopping
ii) Impact force - e.g. threshing, cutting
iii) Rubbing force - e.g. threshing, shelling
iv) Pneumatic force - e.g. cleaning
v) Gravitational force - e.g. separating, cleaning and sieving (one 'g')
vi) More than one 'g' - e.g. axial flow combine separation
vii) Combination forces - e.g. pneumatic and gravitational in cleaning
viii) Frictional and kinetic - e.g. conveying, rolling and feeding
ix) Centrifugal force - e.g. threshing, separation
x) Compression force (squeezing) - e.g. threshing

Generally manual harvesting involves slicing and tearing actions those results in plant structure failure due to compression, tension or shear. The serrated sickle combines a slicing and sawing action. It does not require repeated sharpening as in the case of smooth edge sickle. A single element impact cutting may be either moving or stationary type. An impact cutter has a single high speed cutting element and cuts mainly due to inertia. This cutting method is an economical method widely used in rotary lawn mowers, forage choppers and in some tractor mounted cutter bar. Usually a single element, sharp edged blade requires a velocity of about 10 m/s for impact cutting. A dull edged single element blade requires a velocity of about 45 m/s. In the rotary cutter the knives rotate in a horizontal plane as in the rotary mowers, whereas, in flail shredder the knives rotate in a vertical plane parallel to the direction of travel. In shearing type cutting, cutting takes place due to shear. A system of forces acts upon the material in such a manner as to cause it to fail in shear. Shear failure is invariably accompanied by some deformation in bending and compression, which increases the energy required for cutting. A common way of applying the cutting force is by means of two opposite shearing elements, which meet and pass each other with little or no clearance between them. Both or one of the elements may be moving with a linear uniform, reciprocating or rotary motion. This type of cutting mechanism is most widely used for harvesting agricultural crops. The reciprocating cutter bars that are commonly used for harvesting wheat or paddy crops use this principle. The inclined angle between the cutting edge is about 30 degrees. The serrated blades cannot easily slip between the two cutting edges. Reciprocating cutter bars do an excellent job of harvesting but are characterized by high-energy losses, short dynamic imbalance and limited operating speeds.

Harvesting Equipment

Sickle

It is suitable for harvesting wheat, rice, soybean and cutting of grasses. It is a manually operated tool used in squatting position (Fig. 7.2). It is a serrated blade sickle suitable for harvesting wheat, rice and grasses. The blade is made from medium carbon steel or alloy steel, hardened and tempered to suitable hardness. The back of serrated edge is ground to bevel profile to expose the cutting teeth. Upon wearing of teeth, the bevel profile is ground and teeth are exposed again. The wooden handle has a bend at the rear for better grip and to avoid hand injury during operation. It saves 26 per cent labour and operating time and 27 per cent on cost of operation compared to harvesting by local sickles. Working capacity of sickle is 61 m^2/h for paddy, 151 m^2/h for wheat and 184 m^2/h for soybean. Weight of the machine is 0.28 kg. Blade thickness is 1.5-4.0 mm, tapered to the cutting edge and radius of curvature 260 mm.

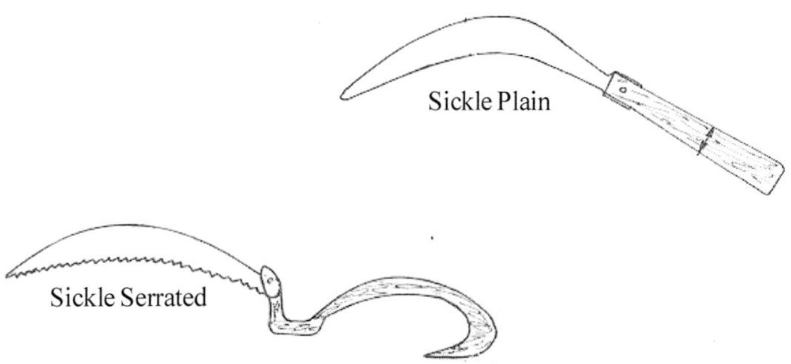

Fig. 7.2 : Sickles used in harvesting of crops

Manually operated long handle scythe

It consists of a curved blade and a long pipe handle (Fig. 7.3a). The blade is made from medium carbon steel or low alloy steel and forged to shape. The cutting edge is sharpened for smooth cutting of fodder crop. Handle and blade are nearly at right angle to each other. A person operates the scythe in standing posture by gripping the handle at suitable positions and the blade is swung in a curvilinear motion. With the scythe it is possible to cut an area of about 1.2 m wide and 0.6 m long in one stroke. Cut crop is swept and windrowed in second stroke of the blade. During harvesting blade is kept close to the ground. Length of handle is 170 cm, length of blade 65 cm and width of blade 8 cm. Scythes are also used for harvesting barseem fodder. Power operated hand controlled blades are also used for harvesting standing crops such as wheat, rice, barley etc (Fig. 7.3b).

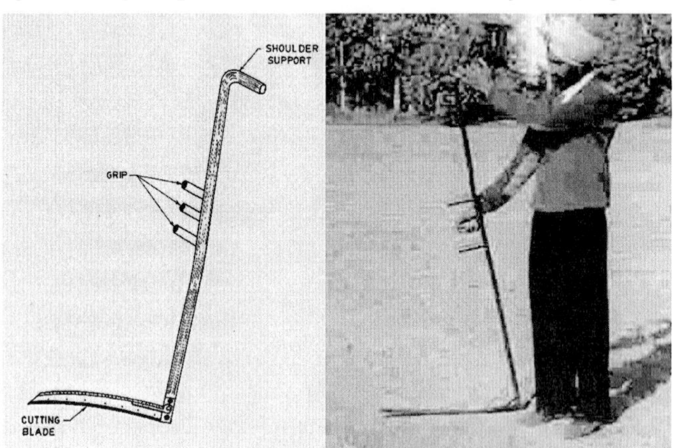

a) Manually operated long handle scythe

b) Manually controlled power operated long handle scythe

Fig. 7.3 : Long handle scythe

Reaper

The most appropriate method of mechanization of grain crop is cutting, binding and stacking. As such, harvesting the grain by a vertical conveying reaper-windrower, manual collection of windrows, and stationary threshing be the most appropriate system from the socio-economic point of view (Gupta, 2001). The operations performed by this reaper windrower include guiding of the standing crop towards cutter bar with the help of row dividers and star wheels, cutting of crop, conveying of cut crop under control to the right side of machine, and laying it on ground in the form of a windrow with ears on one side. A vertical conveyor of reaper windrower has many features, which enable it to have an edge over other harvesters. The plants are conveyed in the vertical position and the gathering and collecting of standing crop towards the machine is achieved without the panicles coming into contact with the moving parts of the machine. This helps in minimizing the shattering loss. The crop is conveyed in vertical position and is under control and is laid on the ground in a clear windrow maintaining the direction of tillers perpendicular to direction of travel. This also provides ease in manual collection and tying of crop. The machine is small in size due to compact design. This results in better manoeuvrability in the field. The weight of machine per unit length of machine is less and therefore the cost of unit is low and thus the cost of harvesting operation is low. For paddy crop where soil conditions are poor for traffic ability, the lightweight power tiller unit can be operated easily.

Different sizes of the vertical conveyor reaper-windrower have been developed for mounting in front of a power tiller as well as a tractor. It cuts the crops and lays it in a windrow, which can be easily picked up by the farm labourers. The cutter bar cuts the crop. The cut plants are held in vertical position by means of pressure springs and are guided by the star wheels to the lugged conveyor belt. The cut plants are transferred in the vertical position until discharged at the outer end into a windrow at right angle to the direction of motion. The main parts of the reapers are lifting and gathering mechanism, lugged belt conveyor and cutter bar. The lifting and gathering mechanism consists of a row crop divider fitted with a lugged V-belt, pressure springs and the star wheel. The star wheels for each row are driven by the lugs attached to the vertical conveyor belt thus guiding the plants to the cutter bar. The speed of star wheels or lugged V-belt depends upon the forward speed of machine as they handle the plants to be cut. The lugged V-belt and the star wheel do the job similar to that of a reel in case of conventional header unit. The lugged V-belt is particularly helpful when the crop is in lodged condition. The star wheel is inclined at an angle with ground and therefore the velocity component of the star wheel, which is responsible for guiding the crop, will be Vs Cos α where Vs is the average peripheral velocity of star wheel along its plane. The angle α depends upon the length of the row divider. In order to have a compact machine, α should be as large as possible and it may be kept = 20 degree. The effective velocity of star wheel i.e. Vs Cos α and this should be greater than the forward velocity of machine in order to have the gathering action on crop towards the cutter bar by the star wheel. The speed ratio of 1.026 between Vs Cos α and forward speed is recommended for the proper operation. The high velocity of star wheel results in shattering loss. Therefore, the star wheel speed should take into account the crop varieties to be handled by the machine.

The vertical flat belt conveyor with lugs drives the star wheel. The speed of conveyor belt for machine speed of 1.0 m/s and normal crop yield levels should be around 1.5 - 1.6 m/s. Taking speed of vertical conveyor belt as 1.5 m/s the speed of drive pulley and the pitch of lugs and number of arms on the star wheels are determined. The lug should be strong and rigid enough otherwise it bends during operation and the lug spacing does not remain uniform. It affects the performance in a big way. The bolts used to fix lugs on the belt should preferably be properly imbedded in the backside of the belt or there should be sufficient clearance between the belt and the back vertical metal sheet so that wearing of the metal sheet does not occur.

The windrower power requirement is the total power consumed in cutting and conveying of the crop in the form of windrow, in propelling the machine and the power transmission losses. The power for cutting and conveying of crop is determined to be 0.513 hp/m of cutter bar length for a knife speed of 1.52 m/s. The power required for the conveyor belt for conveying crop is assessed to be 50% of cutting power. The power required to propel the machine is estimated by rolling resistance x speed. Similarly for tractor also, the total power requirement is quite below the available tractor power. The size of tractor operated reaper is limited because of some other problems. Firstly, very high labour requirements for collecting the cut and laid crop are the major limitation. The present size of tractor operated reaper requires about 16 persons for an output of about 2.5-3.0 ha/day. Secondly, the whole of the cut crop gets accumulated at the end before it is laid in windrows. But there is a limitation to the size of end clearance, thus limiting the amount of crop to be cut and passed through the end clearance and then laid on the ground. Thirdly, the reaper is front operated, thus adding a large amount of weight on the front side of the tractor. This can cause steering problems.

Self- propelled walk behind vertical conveyor reaper

It is used for harvesting and windrowing cereals and oilseed crops. It is a self-propelled reaper and operated by a 3.7 kW diesel engine (Fig. 7.4a). It is an engine operated, walk behind type harvester suitable for harvesting and windrowing cereals & oilseed crops. It consists of engine, power transmission box, lugged wheels, cutter bar, crop row dividers, conveyor belts with lugs, star wheels, operating controls and a sturdy frame (Fig. 7.4b & 7.4bc). The engine power is transmitted to cutter bar and conveyor belts through belt-pulleys. During forward motion of the reaper, crop row dividers divide the crop, which come in contact with cutter bar, where shearing of crop stems takes place. The cut crop is conveyed to one side of the machine by the conveyor belt fitted with lugs and is windrowed in the field. The crop is bundled manually and transported to threshing yard. Working capacity of reaper is 0.20 ha/h for rice and wheat; 0.21 ha/h for soybean; 0.23 ha/h for safflower; and 0.21 ha/h for sesamum. Width of cut is 1.6 m and number of crop row divider is 5. Reaper is also available with 3 and 4 crop row dividers i.e. 1 m and 1.3 m width (Fig. 7.5). Type of pickup and gathering mechanism is star wheel type. Angle of star wheel w.r.t. horizontal is 20 degrees. Ratio between conveyer and forward speed is 1.6. Recommended forward speed is 2.5 - 3.5 km/h and cutter bar speed 600 rpm at full throttle. Type of crop conveying mechanism is conveyor flat belts fitted with lugs which convey the cut crops towards the right side of the machine keeping the crop in vertical orientation. At the end, the crop is discharged and laid on the ground in the form of windrow. The power is transmitted from the engine to the cutter bar and conveyor belts through V-belts and pulley. The top conveyor belt drives the star wheels on the crop row dividers. Pressure springs are fitted below the star wheels and between the conveying platforms to keep the cut crop in upright position while it is being conveyed out of the machine. Total labour required for machine operation and collection of crop is 25.0 - 40.0 man-h/ha. Saving in labour requirement is 60% and cost of operation 40% in comparison to traditional method. The reaper can be used for harvesting of soybean crop (Fig. 7.6). The average area covered, fuel consumption, stubble height and total machine losses are 0.42 ha/h, 0.74 l/h, 85 mm and 4.66%, respectively.

a) Self- propelled walk behind vertical conveyor reaper
Courtesy: Khedut Agro Engineering, Rajkot (Gujarat)

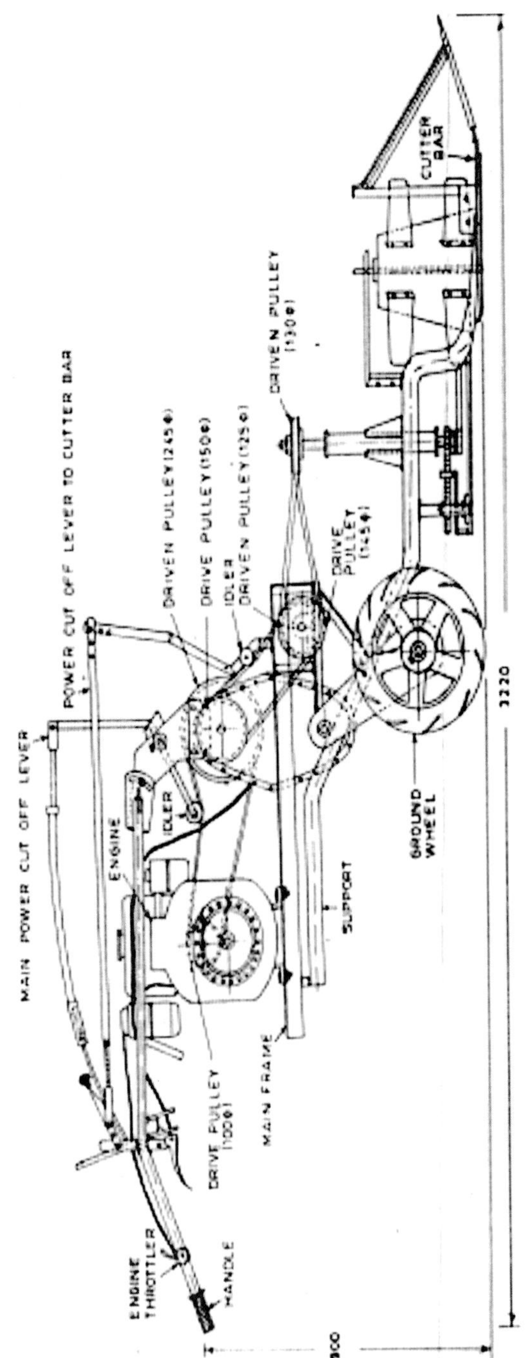

b) Side view of reaper

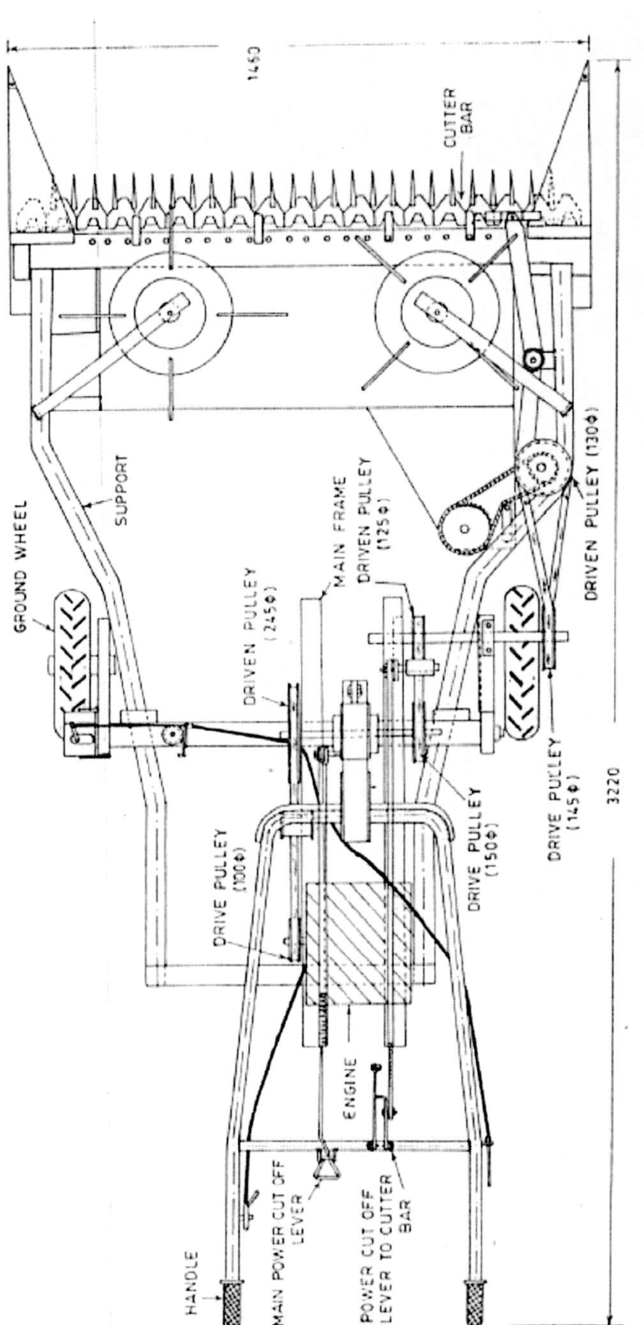

c) Top view of reaper

Fig.7.4 : Self-walk behind reaper

Fig. 7.5 : Self-propelled walk behind reaper
Courtesy: BCS India Pvt. Ltd. Ludhiana (Punjab)

Fig. 7.6 : Harvesting of soybean by self-propelled walk-behind type reaper

Self-propelled riding type vertical conveyor reaper

A riding type self-propelled vertical conveyor reaper windrower has been developed to avoid human drudgery of walking behind the reaper (Fig. 7.7). The reaper is powered by a 6 hp diesel engine and the effective cutter bar width is 153 cm. It is provided with four pneumatic wheels, two driving wheels in the front and two steering wheels on the rear. Simple systems of clutch, brakes, steering and hydraulics have been incorporated and an operator's seat is available to make the machines riding type. It has two forward and one reverse speed. In case of wheat the machine can be operated at an average speed of 3.31 km/h and effective field capacity, fuel (diesel) consumption and total machine losses are 0.260 ha/h, 0.99 l/h and 0.47%, respectively (Fig. 7.7). The overall mean values of forward speed, effective field capacity, fuel consumption and total machine losses for soybean are 3.25 km/h, 0.288 ha/h, 0.96 l/h and 5.89%, respectively (Fig. 7.8).

Fig. 7.7 : Harvesting of wheat by self-propelled riding type reaper windrower

Fig. 7.8 : Harvesting of soybean crop by self-propelled riding type reaper

Tractor front mounted vertical conveyor reaper

A machine called vertical conveyor reaper–cum-windrower can cut the crop and lay it in the form of windrow for easy picking (Anonymous, 1984). It consists of a conventional cutter bar assembly, crop row dividers with star wheels, covers, pressure springs and vertical conveyor belts (Fig. 7.9a). Cutter bar is given reciprocating motion by crank wheel. Crop row dividers with star wheels enter the standing crop, help in lifting, gathering and guiding the crop towards the cutter bar. After the crop is cut, held in a vertical position during its passage by means of pressure springs and star wheels. Vertically held crop is then delivered towards right side of the machine in a windrow perpendicular to the direction of movement of machine with the help of lugged conveyor belt. The gearbox and windrower is coupled to the drive shaft of the prime mower. The output shaft transmits power to the shaft driving the lugged flat conveyor belts and a crank is attached at the lower end of the output shaft to operate the cutter bar through connecting rod. The serrated blade cutter bar with standard knife guards is fitted. The top lugged conveyor belt drives the star wheels on the crop row dividers. Pressure springs are fitted below the star wheels between the conveying platforms to keep the cut crop in upright position while it is being conveyed out of the machine. The power in case of power tiller units is transmitted through an intermediate shaft to the gearbox on windrower either by belt or by shaft drive. In tractor mounted models, the power to the gearbox is transmitted from PTO shaft through gearbox by a long shaft running beneath the tractor body to the front and with the help of universal joint and telescopic shaft which is connected to the gearbox. Lowering and raising of reaper is carried out with the help of hydraulic system of a tractor. In case of power tiller the machine pivots on the power tiller wheels. By pushing the handle, the cutter bar can be raised.

A front mounted vertical conveyor reaper is the most common reaper, to harvest wheat and paddy crops. It can also be used for harvesting of soybean and other similar crops. Power tiller operated reaper (Fig. 7.10) can be operated with a 5-6 hp engine, whereas, tractor operated reapers (Fig. 7.9b) can be operated with 25-35 hp tractor. Width of cut is about 1.6 m in power tiller reaper, and about 2.05 m in tractor operated reapers. Stroke per min of cutter bar is 1225 and 1550 in case of power tiller and tractor operated reapers, respectively. Power tiller and tractor-front mounted vertical conveyer reaper windrower can cover about 0.2 ha/h and 0.4 ha/h, respectively. Labour required for crop collection is 40-45 man-h/ha. Saving in labour requirement is 60-70% and cost of operation 40-50% in comparison to traditional method.

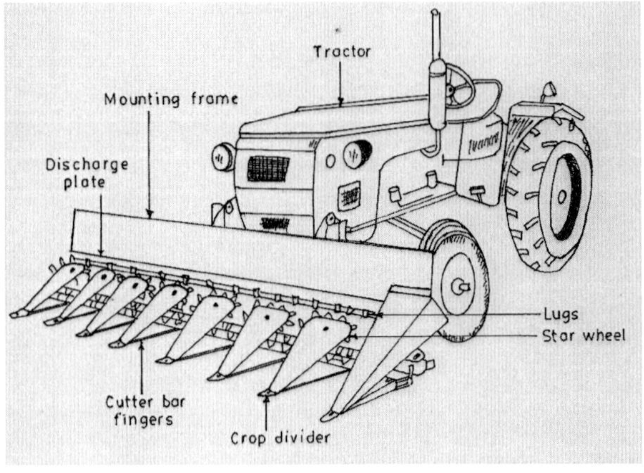

a) Components of reaper

b) Wheat crop being cut and windrow by reaper

Fig. 7.9 : Tractor front mounted vertical conveyor reaper

a) Power tiller operated reaper working in field

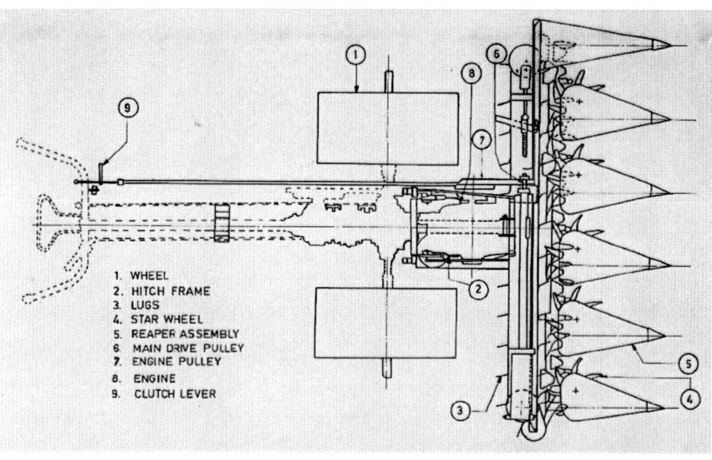

b) Top view of power tiller operated reaper

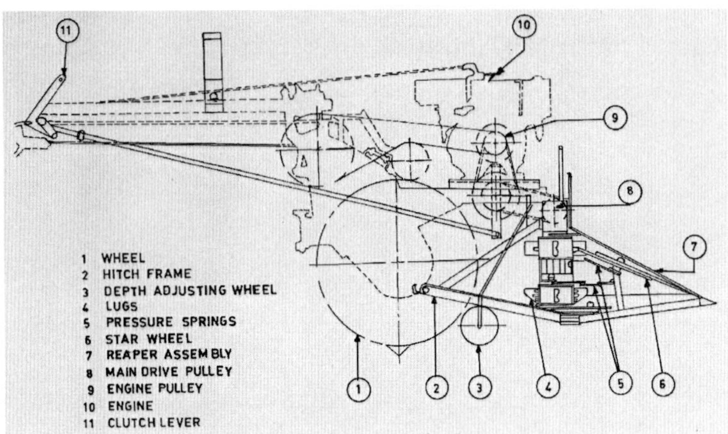

c) Side view of power tiller operated reaper

Fig. 7.10 : Power tiller-operated vertical conveyor reaper windrower

Animal-drawn reaper

It consists of a cutter bar of 1.05 m length. The power to drive the knife bar is given from the ground wheel by means of gear box, crank and connecting rod mechanism. As the machine is pull forward by a pair of bullocks, a reciprocating motion is imparted to the knife bar with a peak cutting velocity of about 100 m/min. The crop is cut due to shearing action. The effective field capacity of machine varies between 0.2-0.3 ha/h.

Animal-drawn engine operated reaper

It has a cutter bar of 1.35 m length. A 2-hp 4-stroke petrol engine operates the cutter bar. The drive to the cutter bar from engine is given through V belt and gearbox. It can cover 0.2-0.4 ha/h with field efficiency of about 80%.

Tractor rear mounted PTO operated self-raking reaper

The machine carries a cutter bar of 1.5 m length, the drive to which is given from PTO of a tractor (Anon. 1965-79). It is a side delivery machine in which crop is collected over a platform and is delivered on one side in the form of bound bunches of desired size (Fig. 7.11). The raking and sweeping of harvested crop is done mechanically. A profile cam controls raking motion. An index lever regulates the movement of cam rollers in such a way that either of first, second, third or fourth rake sweeps out the cut crop laid on the platform. The crop is tied into bundles of desired size manually. It can cover about 3 ha/day with field efficiency of 85%. The grain loss varies between 0.2-3.1percent.

Fig. 7.11 : Tractor side mounted reaper

Self-propelled reaper binder

It is suitable for harvesting and making bundle of wheat, paddy, oats, barley and other grain crops of height from 85 to 110 cm. The riding type self–propelled vertical conveyor reaper windrower is powered by a 9 kW, single cylinder, water cooled diesel engine having rated engine speed of 3000 rpm (Fig. 7.12). It is provided with four pneumatic wheels; two driving wheels in the front having agricultural tread pattern tyres and two steering wheels at the rear having automotive tyres. Other systems include clutch, brakes, steering, hydraulics, and power transmission and an operator's seat is available to make the machine riding type. The harvesting system include crop row dividers, star wheels, standard cutter bar having 76.2 mm pitch of knife section, vertical conveyor belts and wire springs. The effective cutter bar width is 1.2 m. The crop row dividers enter the standing crop and the star wheels guide the crop towards the cutter bar and help in slightly lifting the crop after it is cut, and in turning it at right angle, prior to its conveying by the lugged conveyor belt. The two lugged flat belts convey the cut crop towards the centre of the machine and moves back on a platform where it makes a bundle of about 5 kg each. At the end, the crop is discharged on the ground in the rear. Working capacity of reaper binder is 0.3-0.4 ha/h. Weight of the machine is about 450 kg. Height of cut is 5-7 cm.

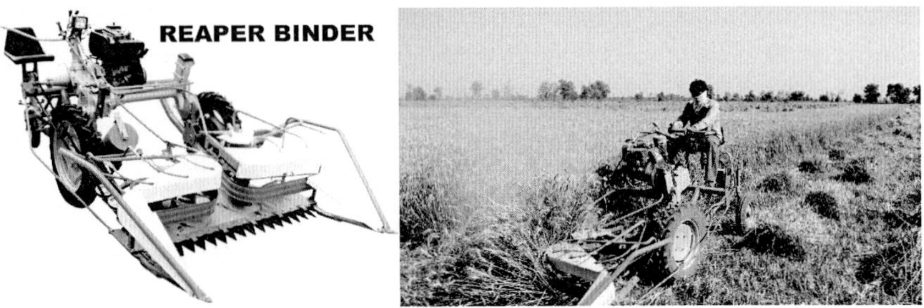

Fig. 7.12 : Self-propelled reaper binder

Courtesy: Courtesy: BCS India Pvt. Ltd., Ludhiana (Punjab)

Tractor-rear mounted reaper binder

The machine consists of a cutting, gathering knotting mechanism mounted on a high pressure pipe frame with a 3-point linkage arrangement for hitching at the rear of a tractor (Fig. 7.13a). The machine can cover 1.5-2.0 ha/day at a forward speed of 2 km/h (Fig. 7.13b). Machine can be used for harvesting wheat and paddy both. The major components of the machine includes standard cutter bar assembly, crop board, reel and power transmission system. The drive to the cutter bar is given from the tractor PTO shaft. The cutter bar is of conventional reciprocating type 2 m long with 76.2 mm knife sections. A 2 m long and 0.5 m wide mild steel sheet board is hinged behind the cutter bar assembly. The two lugged flat belts convey the cut crop towards the centre of the machine and moves back on a platform where it makes a bundle. At the end, the crop is discharged on the ground in the rear. Working capacity of reaper binder is 0.5-0.6 ha/h. Height of cut is 5-7 cm.

a) A view of tractor side mounted reaper binder

b) Tractor side mounted reaper binder in field operation
Fig. 7.13 : Tractor side mounted reaper binder
Courtesy: BCS India Pvt. Ltd., Ludhiana (Punjab)

Grain combine harvesters

The grain harvesting equipment is available in various forms and sizes (Saijpaul, 2001; Garg *et al.*, 2003). Their varied nature is due to different size of land holdings, crop intensities, crop diversification etc. Food grains are very important all over the world. Therefore, any advances in grain harvesting equipment has significant impact on the timeliness of operation to aid crop intensification. The unit operations necessary in all the grain harvesting equipment consist of one or more of the following:

i) Gathering e.g. stripper gatherer, reel, and vertical flat belt; lugged conveyor; pick-up reels; windrower (Fig. 7.14),

ii) Cutting e.g. cutter bar (Fig. 7.14),

iii) Threshing e.g. rasp bar thresher, peg type thresher, thresher with partial or full concave (Fig. 7.15),

iv) Separation e.g. straw rack, straw walker, rotary separation,

v) Cleaning e.g. pneumatic, mechanical or combination of two in chaffer sieves (Fig. 7.16),

vi) Size reduction e.g. chopping or bruising of straw,

vii) Spreading e.g. straw spreader,

viii) Handling e.g. feeding mechanism to thresher, augers to convey cut crop, gain, tailings, delivering, and bagging of clean grain; storing grain in grain tank; grain pan and agitation and shifting on cleaning shoe, and

ix) Rubbing and rolling-e.g. corn picker, sheller etc.

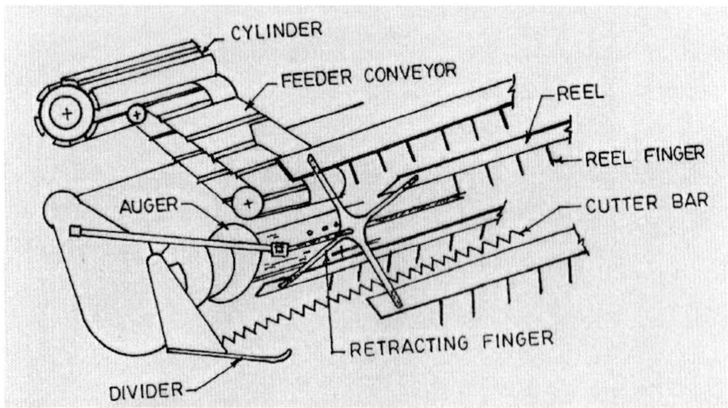

Fig. 7.14 : Cutter bar assembly in combine harbesters

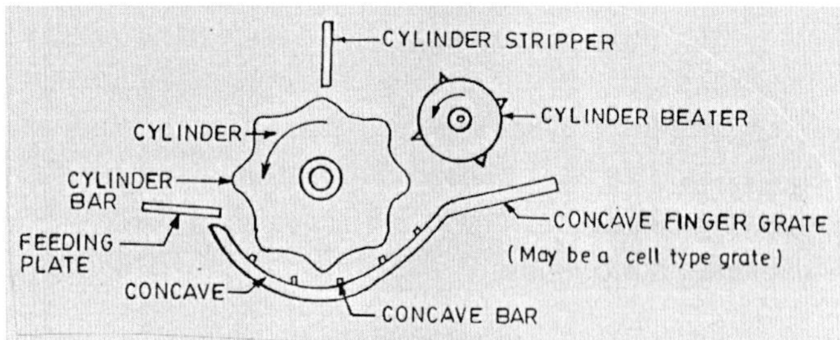

Fig. 7.15 : Threshing unit in combine harvester

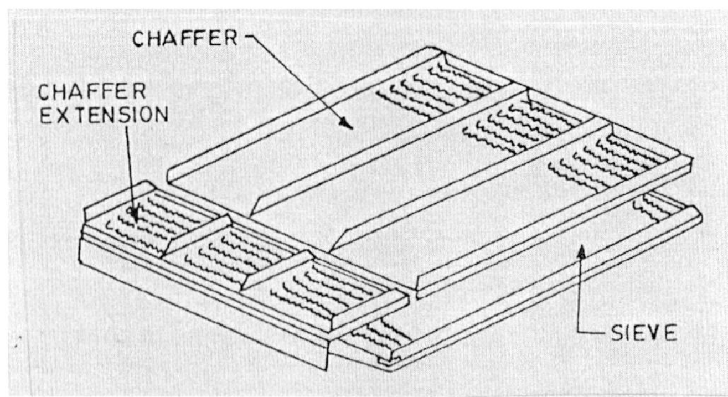

Fig. 7.16 : Cleaning unit in combine harvester

The grain harvesting equipment can be classified on the basis of:

i) Source of power (manual with or without power, animal drawn-ground operated or power operated, tractor mounted - PTO operated, tractor-trailed-PTO operated, self-propelled,

ii) Mode of hitching-trailed, mounted, side mounted, self-propelled,

iii) Width of cut,

iv) Power consumed,

v) Drive used - engine driven, PTO driven, self-propelled,

vi) Mode of delivery - windrow, swath, binder, bagging of grain, storing grain in tank,

vii) Mode of threshing and separation - conventional combine harvester (with straw-walkers), axial-flow combines (without straw walkers),

viii) Field terrain - hill side combine and prairie combine (level land combines), and

ix) Mode of transport i.e. wheel type or track type

Since different types of new grain harvesting equipment are being designed, developed, modified and manufactured, it is necessary to define certain typical terms associated with them. Some of them are:

i) Combine-harvester-thresher - A machine designed for harvesting, threshing, separating, cleaning and collecting grain while moving through standing crop. It may be used for handling crop that has been swathed.

ii) Header - It comprises the mechanism for gathering, cutting, stripping or picking up the crop and delivers it to cylinder.

iii) Straw spreader-cum-cutter - A component used to cut the straw in pieces and to spread in the field.

iv) Cutter bar effective width - The distance between points at which the tips of the knife sections meet the last effective shearing edges of the guards (fingers) or shoes at the extremities of the cutter bar.

v) Maize header - effective width - The average distance between the centre lines of adjacent picking units multiplied by the number of units. If header width is adjustable maximum and minimum dimension shall be stated.

vi) Cylinder or drum - A balanced rotating assembly comprising rasp, beater bars or spikes on its periphery and their support for threshing the crop.

vii) Turning radius - The distance from the turning centre to the centre of ground contact of the wheel describing the largest circle while the combine is taking the shortest turn for the respective steering brake condition.

viii) Harvesting - The operation of detaching, picking or cutting the crop from the undesired portion of the same rooted to the ground.

ix) Threshing - The operation of detaching the grains from the ear heads cob or pod.

x) Separating - The operation of isolating the detached grain, small debris and incompletely threshed or completely un-threshed grain from the bulk of straw stem or stalk.

xi) Cleaning - The operation of isolating the desired grain from chaff, small debris and incompletely threshed and completely un-threshed grains.

xii) Screening - The operation of isolating the desired grains by a mechanical device where the desired grain is carried over the device and the undesired material penetrates the device.

xiii) Sieving - The operation of separating the desired grains by a mechanical device where the desired grain penetrates the device and the undesired material is carried over the device.

xiv) Chaffing - The process of pneumatic cleaning of grains.

xv) Combine capacity - The maximum sustained total feed rate measured in kilograms per second or tonnes per hour at which rack losses and shoe losses shall be within acceptable limits while the combine is operating at rated speed on level ground without choking of threshing, separating, cleaning and grain conveying mechanisms and without stalling the prime-mover. Combine capacity is also defined as the total feed rate at a 3% processing loss (cylinder + walker + shoe).

xvi) Total feed rate - The sum of grain feed rate and material other than grain. Feed rate expressed in kg/s or tonnes/h.

xvii) Clean grain - The threshed grain free from damaged grain and foreign matter.

xviii) Collectable loss - The un-threshed ear heads in main grain outlet or grain tank; threshed, un-threshed and damaged grains from the secondary cleaning or grading unit.

xix) Non-collectable loss - The header loss, shoe loss, rack loss and secondary blower loss.

xx) Processing loss - The damaging of the grains, un-threshing of grains, loss of threshed (damaged or undamaged) grain and un-threshed grains after completion of threshing, separating and cleaning operations. It is expressed as percentage of grain feed rate.

xxi) Cleaning efficiency - Clean grain present in the total grain obtained from the main grain outlet expressed in percentage by mass.

xxii) Threshing efficiency - Threshed grains from all the outlets of the combine with respect to grain output in tank expressed in percentage by mass.

xxiii) Field efficiency - The quotient of effective field capacity and theoretical field capacity expressed in percent.

Feeder convener: The feed conveyor or feed rake is designed to feed the crop in a steady even flow into the threshing unit. Feeder beater takes the crop from the feed conveyer and feed it uniformly to the threshing unit.

Threshing unit: The function of this section of combine is to thresh the gain from the heads. This is done by passing the grain between a rapidly revolving cylinder and a stationery surface underneath which is called the concave. The grains are separated from the pods by impact, rubbing or squeezing actions between cylinder and concave. The threshing cylinder may be of different types: Rasp bar type – having corrugated bars; Angle bar type – right angle bar with rubber facing; and Spike tooth type – pegs or spikes on it.

Concave: The concave is the stationary part that the cylinder works against in the threshing action. The concave is a grate composed of rods and bars or wires. It is at the concave grate and finger grate that as much as 90% grain is separated from the crop. The separated grain falls through the grate on to the shoe pan where it is delivered to the cleaning unit. The straw and the remaining grain pass on to the separation unit.

Cylinder beater: The beater behind the cylinder slows down the material coming from the cylinder tears apart the straw and delivers the material to the straw rake or walker. The beater helps in cleaning the straw from the cylinder thus preventing cylinder wrapping and feedback.

Separating unit: The separating unit agitates the straw after it comes from the threshing unit. This shakes out the loose grain remaining in the straw and delivers it to the cleaning unit. The straw is carried out of the combine by the rack.

One piece straw rack: The straw rack is a one piece unit with risers pointed toward the rear of the combine. The straw rack is mounted on cranks located at the front and rear which give it an oscillating motion. As the crank moves rearward and upward the straw is tossed up and to the rear. As the rack returns forward and downward the straw stays in mid air for a short time and then falls in to the section of the rack nearer the end of combine. In this way the straw moves step by step out of the combine.

Walker type straw rack: Some large combine may use a walker type straw rack which operates on the same principle as the rack. The straw walker has three or more narrow sections placed side by side. Each section is mounted on multiple throw cranks located at the front and rear. The crank throws for section are equally spaced around the circle of rotation thus the sections do not operate as a unit as one piece rack does.

Grain return pan: It is located under the straw rack. It catches the grain as it falls through the rack and moves forward to the grain pan.

Grain return conveyor: In place of grain return pan some combines uses a conveyor to catch the grain and move it forward.

Grain pan: The grain pan is located under the forward part of straw rake behind and below the cylinder. Its function is to catch the grain from the concave and cylinder grates and from grain return pan or conveyor for delivery to the cleaning unit.

Cleaning unit: The function of cleaning unit is to separate the clean grain and deliver it to the grain tank, return tailings to the cylinder for rethreshing and move the remaining material out of the combine. This is accomplished by means of gravity and air blast.

Adjustable chaffer: The adjustable chaffer act as a sieve. It is made up of a series of cross pieces mounted on rods and fastened together so they can be moved at the same time to adjust the size of the openings.

Chaffer extension: This is an extension of chaffer having adjustable lips. The un-threshed portion of gain heads fall through the chaffer extension into the tailing anger.

Sieve: The sieve is like chaffer except that the lips and openings are smaller. The final job of cleaning is done here.

Cleaning fan: The fan furnishes a blast of air. The strength of air blast is controlled by wind board. The function of the air blast is to keep the material alive on the chaffer and sieve, but not strong enough to blow grain out of the combine.

Tailboard: The tailboard keeps the un-threshed material from being carried out of the rear of the machine while still allowing the chaff to blow. It may be raised or lowered as needed.

Material handling: The grain auger collects the grain and augers it to the clean grain elevator which delivers the clean grain to the grain tank.

Combine adjustments: The losses could be minimized by running the combine at proper adjustment. The adjustment should be made in the following order to reduce losses:

i) **Cutting and conveying:** The height of cut can be adjusted from 5 cm to 75 cm in most of the combines. The rate of feeding can be adjusted by manipulating height of cut and forward speed of machine. Forward speed range of 2.5-4.5 km/h for standing crop and 1.0-1.5 km/h for lodged crop has been recommended by BIS. The speed of cutter bar varies from 400 to 550 rpm.

ii) **Reel adjustment:** The horizontal positioning should be such that real bats have a distance of 50 to 100 mm in front of the cutter bar. The optimum value of reel index should be 1.10 to 1.25.

iii) **Machine speed:** Check machine speed to see that combine is running at recommended speed. Adjust the engine governor so that the speed is 3-5% above normal when combine is running empty.

iv) Adjust the cylinder concave clearance and cylinder speed. Small grain – high cylinder speeds and narrow concave spacing and large grain or seeds – low cylinder speeds and wide concaves spacing.

v) **Over threshing:** Cracked grain, broken or chewed straw which overloads the shoe, grain losses over shoes.

vi) **Under threshing:** Un-threshed heads, excessive tailings, overload straw walker, grain losses over straw walker

vii) **Proper threshing:** No cracked grain, no un-threshed heads or excessive tailing, no broken or chewed straw, no grain losses over straw walker & shoe.

To determine proper threshing action, straw discharged from rear of the machine should not be broken & chewed; grain tank should have very few cracked grains; tailing returns should be very few; straw walker is overloaded if under threshing is occurring at high speed; cleaning shoe is overloaded with excessive chewed and broken straw when over threshing.

i) **Adjust cutter bar height:** If it is too high heads or pods left in the field and if it is too low too much material will run in the combine causing overload the rack.

ii) **Reel adjustment:** Reel may be adjusted for speed, height and position forward and reward. Speed too fast means more shattering loss, and grain carried over top of the reel. Speed too slow means cut grain may fall on the ground, cut grain may fall on cutter bar and choke it, and short plant may not move to the platform. Reel too low means the crop may tend to wrap around the reel bats and ride over the reel. Reel too high means the crop will not move to platform. Reel too forward means the crop will not move to platform. Reel too far back means the crop may wrap on the reel bats and ride over on the reel. The adjustment of reel will depend on the condition of crop.

iii) Forward speed of travel of combine be set according to manufacturer recommendation.

iv) **Cleaning sieves and fan blast:** Losses in cleaning area are mainly due to over threshing, overloading (by low height of cut or high forward speed)

v) **Chaffer opening adjustment:** Chaffer opening too large allows too much chaff to fall on the sieve. Chaffer opening too small allows grain to be carried over the chaffer. The chaffer opening should be large enough to allow the grain to work through the chaffer before it passes over two thirds of its length.

Sources of grain loss from the combine: Each of the four separate areas of the combine-cutting and feeding, threshing, separating and cleaning – can be a source of loss (Fig. 7.17). The losses in these areas are usually known as cutter bar, cylinder, rack and shoe losses.

Cutter bar loss: This is also called header loss. Any or all the following items may cause cutter bar losses:

a) Heads of grain misses by the cutter bar,

b) Grain shattered out of the head as the knife cuts the straw,

c) Grain cut and dropped to the ground before reaching the feeding platform,

d) Grain shattered out when the reel strikes the standing grain, and

e) Heads of grain thrown out by the reel.

It is determined on those portions of ground, which are protected from combine afflux by the use of rolls of cloth. The loose grains and complete and incomplete ear heads fallen on the marked area, where pre-harvest losses are determined, shall be collected manually. This gives the header loss.

$$\text{Header loss (\%)} = \frac{\text{Grain collected from 1 m}^2\text{ area after harvest} - \text{grain collected from same area before harvest}}{\text{Gross yield}} \times 100$$

Cutter bar loss should be in the range of 0.5 to 2%. The grain that has been shattered into the ground ahead of the combine should not be including with the cutter bar loss.

Cylinder loss: The cylinder can cause loss in two ways i.e. un-threshed grain left in the heads and carried to the rear of the machine by rack, and cracked grain in the grain tank caused by the cylinder running too fast or concave cylinder clearance being too close. Cylinder loss should be in the range of 0.5 to 1%.

$$\text{Cylinder loss (\%)} = \frac{\text{Un-threshed grain collected from straw rack \& sieve}}{\text{Gross yield}} \times 100$$

Rack & shoe loss: The rack loss is the loose grain which has not been separated from the straw as it passes over the rack and is carried out of the machine with the straw. Rack loss should be in the range of 0.2 to 0.4%. The shoe loss is the grain that is carried over the rear of the sieves with the chaff or blow out of the combine with the fan. Shoe loss should be in the range of 0.2 to 0.4%. For determining the rack and shoe loss, the straw and chaff afflux is collected separately. To collect these, two rolls of cloth 30 m in length and one and half times the width of straw/chaff outlet is suspended on especially attached fittings beneath the rear of machine. As the sheets of cloth unroll, one sheet retains the afflux from straw walker and other from sieve for 20 m run length. Unrolling operation starts 5 m in advance and terminates 5 m ahead of end point.

$$\text{Rack loss (\%)} = \frac{\text{Free grain collected from straw rack sample}}{\text{Gross yield}} \times 100$$

$$\text{Sieve loss (\%)} = \frac{\text{Free grain collected from sieve sample}}{\text{Gross yield}} \times 100$$

The four losses taken together shows how good a job combine harvester is doing.

Pre-harvest loss: In addition to above machine losses, there may be a pre-harvest loss. This is shown by the grain found on the ground before harvesting. It is determined at minimum of three places randomly selected in the field where combine harvester is to be operated. The sample should be collected from the area having one-meter length in the direction of travel and full or half width of cutter bar of machine depending upon its size. All the loose grains, complete and incomplete ear heads fallen in the marked area have to be picked up manually without vibrating the plants before the machine is to be operated. This will give pre-harvest loss.

Grain crackage: It is determined from the samples taken from grain tank. Only visible damaged grains are separated and expressed in percentage of sample taken. At least three samples should be taken.

$$\text{Grain crackage (\%)} = \frac{\text{Damaged grain wholly or partially collected from sample}}{\text{Gross yield}} \times 100$$

Net yield = Grain collected in the bag from combine test area

Gross yield = Net yield + header loss + cylinder loss + rack loss + sieve loss.

Total combine loss = Cutter bar loss + cylinder loss + rack loss + sieve loss.

$$\text{Performance efficiency, \%} = \frac{\text{Net yield}}{\text{Gross yield}} \times 100$$

$$\text{Un-threshed (\%)} = \frac{\text{Un-threshed grain in tank + cylinder loss}}{\text{Gross yield}} \times 100$$

$$\text{Cleaning efficiency, \%} = \frac{\text{Clean grain}}{\text{Total grain collected from main outlet}} \times 100$$

$$\text{Threshing efficiency, \%} = \frac{\text{Threshed grain from all outlets}}{\text{Grain output in tank}} \times 100$$

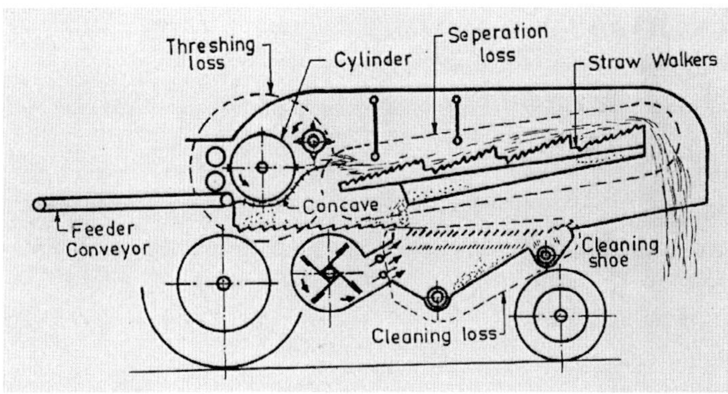

Fig. 7.17 : Various loss sections in combine harvesters

Combine harvester

A combine is farm machine that combines the reaper and thresher to harvest the standing crop, thresh it and clean the grain from straw in one operation. Combine harvesters are suitable for harvesting wheat and paddy, however; other crops can also be harvested with combines like, sunflower, maize, soybean, pulses etc with slight changes in the combine. Combine harvesters are of two types namely: self-propelled combine harvesters and tractor operated combines. Self-propelled and tractor operated combines use rasp-bar threshing cylinder (Fig. 7.18) for wheat and spike tooth threshing cylinder (Fig. 7.19) for paddy. Combines are used for cutting, threshing, cleaning and loading in one operation. It is operated by 4-Stroke water cooled 37 to 45 kW tractor (Fig. 7.20). It consists of cutting unit, threshing unit, cleaning and grain handling units (Fig. 7.21). The cutting section includes reel, cutter bar, an auger and a feeder conveyer. Threshing section has threshing cylinder, concave and cylinder beater. The cleaning section mainly consists of straw walker, chaffer sieve, grain collection pan and blower. The grain handling section consists of a grain elevator and a discharge auger. The crop after being cut is delivered to the cylinder and concave assembly through feeder conveyor where it is threshed and grains & straw is separated in different sections. Working capacity of combine harvester is 0.4-0.5 ha/h. Spiral type auger is used for conveying the grain to the tank. Grain loss in the combine, if properly adjusted, is 1.08 - 2.04%; grain damage, 0.57 - 2.3% and fuel consumption, 7.0 - 12.0 lit/h. Saving in labour requirement is 80 - 90% and cost of operation, 33% as compared to traditional method. Self-propelled combine harvesters has 82 kW engine (Fig. 7.22). It has all other components as tractor operated combines except the power source. Width of cut varies from 2 m to 4.5m. Working capacity of combine harvester is 0.7-0.8 ha/h.

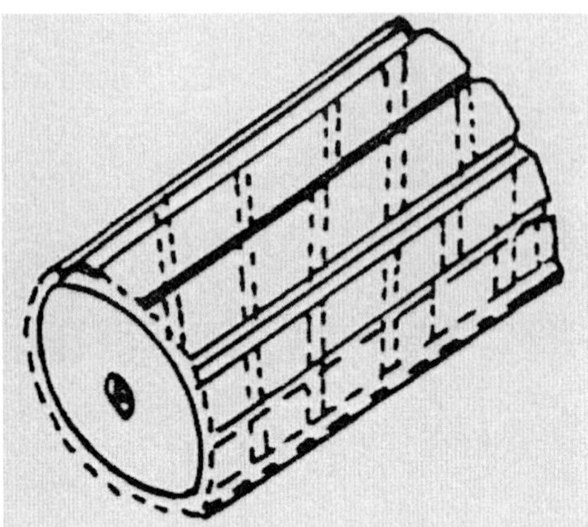

Fig. 7.18 : Rasp bar threshing cylinder

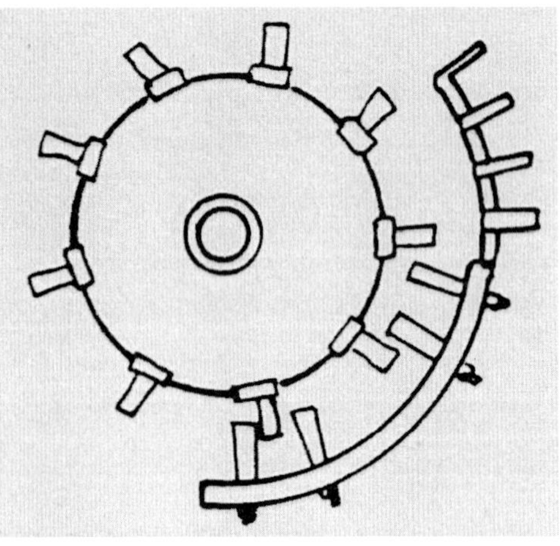

Fig. 7.19 : Spike tooth threshing cylinder

Fig. 7.20 : A grain combine harvester with 55-hp tractor mounted on it

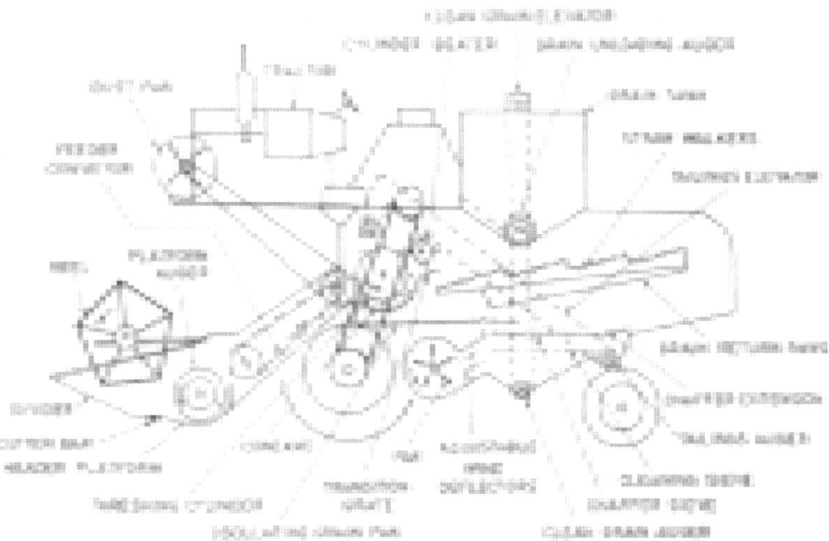

Fig. 7.21 : Details of tractor operated combine harvester

Fig. 7.22 : Self-propelled combine harvester

Whole straw rice combine harvesters have been imported from Japan, Korea and China and are under field operation on large scale (Fig. 7.23). The whole straw combine harvester for rice is useful for obtaining the grain and straw (Anonymous, 2008b). Due to the design of track, the damage to rice fallow crop is reduced. Since the straw is obtained without damage the fodder value is not reduced and hence the machine is suitable for harvesting whole crop. The self-propelled combine is fitted with about 18 to 65 hp diesel engine. It is either track type or wheel type. If field is dry wheel type is used and if field wet tract type combines is used. It consists of crop lifting mechanism with retractable fingers, reciprocating cutter bar, crop conveying system, ear head thresher and cleaning and bagging unit. The crop is held between two chain conveyors, threshed and the remaining straw is released in the field in the form of windrows. It has a provision also to chop the straw and spread it uniformly in the field for subsequent sun drying and incorporation into the soil. The field capacity of the machine is 0.20 ha/h with field efficiency of 90-95%.

Rice is cultivated in entirely different and difficult situations in Kerala. Kuttanad, Kole, Pokkali, and Kaipad regions in Kerala constitute major problematic rice growing area. Because of soil conditions and labour scarcity, farmers are facing problems in cultivating paddy. Heavy combines experience sinkage problems in this area. A light weight combines is beneficial for the farmers in this area. A float type light weight combine has been development under the scheme "Development of Innovative Farm Mechanization Package for Kerala" for Pokkali field at Kerala Agricultural University funded by Govt. of Kerala (Fig. 7.24). The Kukje DKC 515 50 hp hopper combines and Kukje DKC 685 64 hp grain tank combines have been recently introduced in Southern India. These combines have the features like crawler track, finger fed with ear head thresher, 50 hp diesel engine, no damage to grain and straw, full straw recovery, and field coverage is 0.36 ha/h (Fig. 7.23). The Champion TRF 15 is a newly introduced whole straw combine harvester in Kerala. It has the features like crawler track, finger feed with ear head thresher, 65 hp diesel engine, CVT for control, manoeuvrability and stability, header can be adjusted for fallen crops, multiple threshing drums for quality grain and reduced grain loss, and field coverage is 0.33 ha/h. The latest combine introduced in Kerala is Dae-Dong (Anonymous, 2008b). The combine harvester is having the features such as crawler track, finger fed with ear head thresher, 48 hp diesel engine, HST for control, manoeuvrability and stability, axial flow threshing drums, and field coverage 0.32 ha/h. Recently Redlands Ashlyn Motors (P) Ltd Thrisure has also introduced small tract type combine in southern India. The Dae-dong ear head combine is found to perform better compared to other combines in the highly sinking fields with heavy weed infestation. All the combine harvesters are found to have savings in the cost of harvesting, labour use and time of operation.

Fig. 7.23 : Self-propelled combine harvesters with ear head thresher of different makes

Fig. 7.24 : A float type light weight combine for harvesting rice under water

Axial flow combines

In a conventional grain harvesting combines the separation and cleaning units become limiting factors to increasing its capacity. Grain losses (separation losses) increase rapidly as feed rate increases. The separation process in straw walkers is dependent on gravitational forces necessitating large cross-sectional area and making it slop dependent. These problems have been overcome in axial flow combines where the transverse threshing cylinder, rear beater and straw-walkers have been replaced by a longitudinal rotor. Its front portion does the threshing and rear portion the separation. Whereas conventional threshing is single pass process the axial flow thresher separator is multi-pass process. The total dwell time of crop in threshing and separation process is less in axial flow combines than in conventional combines. Axial flow combines reduce number of moving parts from 36 to 6 in the thresher separator. Since axial flow concept results in compact design, a larger grain tank can be provided. The centrifugal separator is self-emptying and relatively maintenance frees due to high "g" force fields (Anonymous, 2010b). Rotor power required for harvesting in axial flow combines is function of mainly helix angle of transport vanes (11°, 22°, 33°), feed rates (10 to 15 t/h) and moisture content (10% to 18%) whereas the losses are affected mainly by rotor speed (800 to 1000 rpm), vane angle and feed rate and to looser extent by moisture content and the concave clearance (7 mm to 15 mm).

Tangential axial flow (TAF) combines has a threshing cylinder which has a provision of moving crop in a tangential axial flow manner that eliminates threshing losses and ensures clean and unbroken grain even in wet and green conditions. It has hexagonal reel with curved arm support that ensures easy pick up of lodged and small crops especially basmati, soybean, gram etc (Fig. 7.25). It consists of different size of diesel engine depending upon the width of combine, tangential axial flow threshing cylinder (Fig. 7.26), cleaning system, transmission system, universal joint type unloading system (Fig. 7.27). The cleaning system consists of forced air-cleaning fan with shutter control and speed from 1200 to 1500 rpm. Tangential axial flow combines are also available in different sizes (with different width of cut) and pneumatic wheels and track type. It can be used for variety of crops such as wheat, rice, maize, basmati, sunflower, soybean, gram etc. The field capacity of the combine varies from 0.45 to 0.60 ha/h. The total grain loss varies from 0.67 to 2.03% and grain breakage from 0.27 to 0.51%. Threshing efficiency varies from 97 to 99% for paddy and 99% for wheat. Cleaning efficiency is 96% for paddy and 99% for wheat. The labour and cost saving in combine harvesting is 93 and 41 percent for paddy and 96 and 59 percent for wheat respectively when compared with manual harvesting and mechanical threshing. Tangential axial flow (TAF) combine like CLAAS Crop Tiger combines has the features like crawler track, reel fed, full crop beater thresher with grain tank, maximum cutter bar lifting height, 54 hp diesel engine, and field coverage is 0.38 ha/h.

Fig. 7.25 : Hexagonal reel with curved arm support along with cutter bar in TAF combine
Courtesy: CLAAS India Pvt. Ltd.

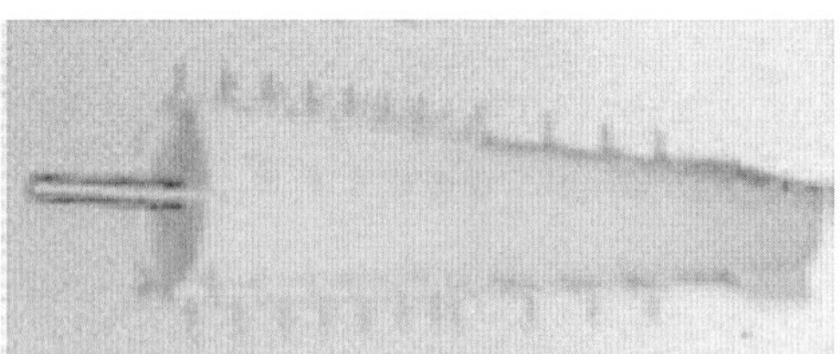

Fig. 7.26 : Tangential axial flow threshing system
Courtesy: CLAAS India Pvt. Ltd.

Fig. 7.27 : Self-propelled tangential axial flow combine harvesting wheat crop
Courtesy: CLAAS India Pvt. Ltd.

Threshers

Threshing is recognized to be a critical farm operation. Although, grain combines have become popular in some parts of the country to harvest crops like wheat, paddy and soybean, stationary threshers remain very popular for major crops like wheat, barley, sorghum, millets, soybean, etc (Sharma, 2001; Sharma *et al.*, 1986a). An important requirement of these threshers is their ability to bruise the straw in addition to threshing, separation and cleaning. Threshing and straw bruising operations require maximum energy (55-82%) as compared to other systems (20%), Sharma *et al.,* 1986. Moreover, grain damage, threshing efficiency and straw quality are also dependent upon the design of threshing/straw bruising system. Thresher is a machine to separate grains from the harvested crop and provide clean grain without much loss and damage. The threshing can be achieved by three methods: Rubbing action, Impact and Stripping. During threshing, grain loss in terms of broken grain, un-threshed grain, blown grain, spilled grain etc should be minimum. Bureau of Indian Standards has specified that the total grain loss should not be more than 5 per cent, in which broken grain should be less than 2 per cent. Clean un-bruised grain fetch good price in the market as well as it has long storage life.

The traditional method of threshing using manual labours requires 150-230 man-h/ha. Threshing is normally done after the grain moisture content is reduced to 15 to 17%. Trampling of paddy under feet, beating shelves of rice or wheat crop on hard slant surface (Fig. 7.28), beating crop with a flail, treading a layer of 15 to 20 cm thick harvested crop by a team of animals (Fig. 7.29) are traditional methods followed by farmers depending upon capacity, lot size and situation. Threshing by bullock treading is practised on large scale in the country but it is also time consuming and involves drudgery. Tractor in many places is now used in place of animals for treading. Introduction of animal drawn olpad thresher reduced the drudgery of the operator and gave comparatively higher output per unit time (Fig. 7.30). In all above methods the threshed materials are subjected to winnowing either in natural wind flow (Fig. 7.31) or blast from winnowing fan for separation of grain from straw (Fig. 7.32). Threshing wheat by traditional method involves drudgery and takes more time to obtain required quality of *bhusa*. Due to these, mechanical threshers are widely accepted by the farmers.

Olpad thresher

'Olpad' threshers (Fig. 7.30) are used mostly for threshing wheat and paddy crops. A pair of bullocks pulls it around over the dried crop spread in a circular form on the threshing ground. Threshing is continued till the entire material becomes a homogeneous mixture of grain and 'bhusa' (chaff). It consists of about 20 circular grooved discs each of 45-cm diameter and 3-mm thickness placed 15 cm apart in three rows. An operator's seat is provided on the frame to control the movement of animals. All discs are mounted staggered to give more effective cutting of the straw. It has 3 or 4 wheels to facilitate its movement from one place to other. Threshing by this thresher is fairly efficient and cheap but is quite slow with low output capacity. This machine can be used for threshing wheat, barley, gram etc.

Fig. 7.28 : Beating shelves of rice or wheat crop on hard slant surface

Fig. 7.29 : Treading a layer of 15 to 20 cm thick harvested crop by a team of animals

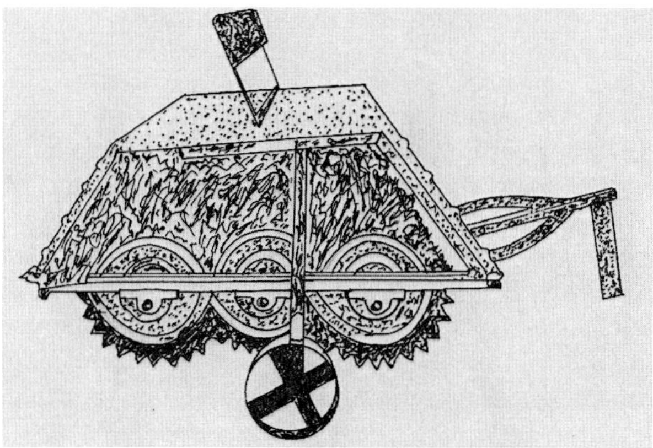

Fig. 7.30 : Olpad thresher

Fig. 7.31 : Manual winnowing

Fig. 7.32 : Winnowing fans – hand and power operated
Courtesy: Mecfa Enterprises Varanasi (U.P.)

Different type of thresher and their suitability for crops

The type of thresher is generally designated according to the type of threshing cylinder fitted with the machine (Fig. 7.33). Drummy type, hammer mill type and syndicator type threshers are suitable for threshing wheat crops only and they can produce fine quality of '*bhusa*'; rasp-bar type, wire-loop type and axial flow type threshers are suitable for paddy and they do not make fine straw. Rasp-bar type threshers can be used for threshing other crops but farmers do not prefer this machine because it does not make fine 'bhusa'; and cost is very high due to its bulky size. Though the hammer mill type threshers can produce fine quality '*bhusa*' its use is decreasing day by day due to high power requirement. Portable wire loop type paddle operated threshers are widely used by farmers in paddy growing areas. Spike tooth type thresher can thresh wheat crop and can produce fine quality of '*bhusa*'. This thresher can be used for threshing other crops if the blower is mounted on a separate shaft so that the cylinder speed can be varied independently. Majority of farmers prefer spike tooth type threshers because of their simplicity in design, low cost and their ability to make fine '*bhusa*'. The major type of threshers commercially available are:

Spike tooth type cylinder: In this type of threshing drum, there is a hollow cylinder, made out of MS flat over to its entire periphery, a number of spikes/pegs of square /round bars or flat iron pieces are welded or bolted (Fig. 7.34). Now days, in most of threshers, round peg with adjustable length are used. These spikes are staggered on the periphery of the drum for uniform threshing. The crop is fed along with the direction of motion of the rotating drum. Direction of the threshing cylinder is such that the crop materials after entry into the threshing system impacts against the upper concave where these are threshed/bruised. Subsequently, these materials are passed through the concave opening in lower concave after being reduced sufficiently. In the process, grains are threshed and separated and the straw is reduced suitably to pass through the concave, because the crop materials receive repeated impacts by the threshing cylinder against the concave. The spike tooth cylinders are available in various sizes. These variations for the same range of horsepower include cylinder diameter, number of spike rows, number of threshing spikes per row, diameter of spikes, shape of spikes etc. Besides these variations, the effect of operational variables like feed rate, peripheral speed and cylinder-concave clearance are not exactly the same as for the conventional rasp bar/spike tooth threshing system as the two designs are basically different. A spike tooth cylinder with spikes of flat front and streamlined back has lower energy consumption. Spikes are mounted on the periphery of a cylinder that rotates inside a closed casing and concave. Increasing number of rows of the spikes results in decreasing the power requirements. When the number of rows are more, straw is bruised relatively faster which escapes through the concave. Thus, straw bruising is more efficient, if the number of rows is more. Increasing the number of rows, however, does not result in any effect on the threshing efficiency, grain crackage and straw size index. Length of spikes for the same diameter of the cylinder does not affect power requirements, threshing efficiency and straw size index. However, it affects grain crackage i.e. lower values of grain crackage are observed when length of spikes is higher. With increase in diameter of the spikes, grain crackage increases whereas straw size index decreases. Increase in axial distance between the spikes (i.e. lesser number of spikes per row) results in lower values of grain crackage and higher values of straw size index (i.e. coarser straw). When the number of spikes per row is high, power requirements are high because of the greater resistance offered to the movement of the threshing cylinder, whereas at lower values of number of spikes per row, crop retention time within the threshing system perhaps increases. It is provided with cleaning sieves and aspirator type blower. Spikes tooth type threshers can handle a variety of crops.

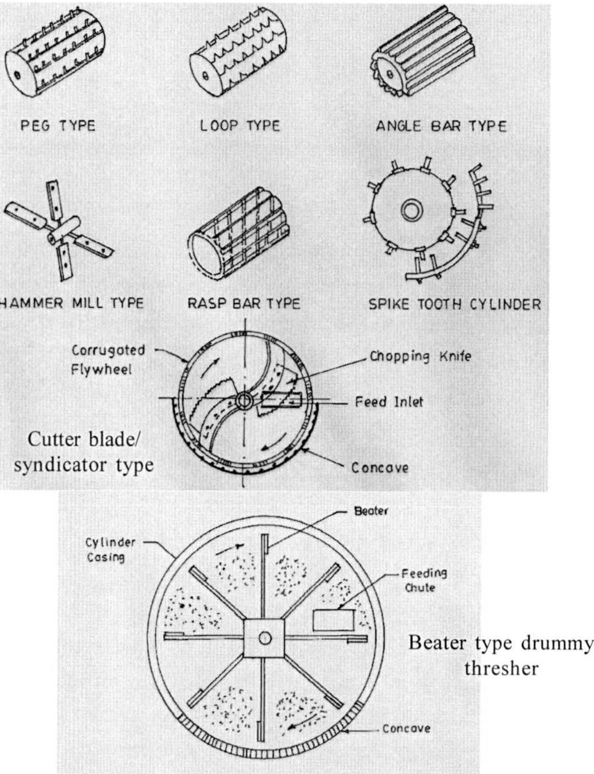

Fig. 7.33 : Different types of threshing systems

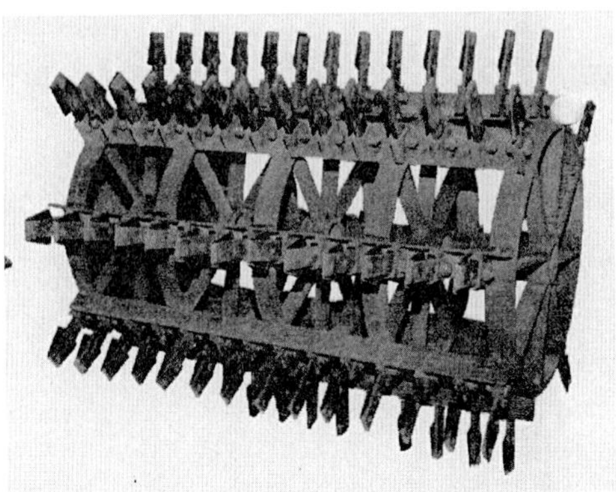

Fig. 7.34 : Commercial spike tooth cylinder

Courtesy: Balaji Agricultural Industries (P) Ltd. Allahabad (Uttar Pradesh)

Rasp bar type cylinder: In this type of cylinder, there are slotted plates, which are fitted over to the cylinder rings, in such a way that the direction of slot of one plate is opposite to another plate. This type of cylinder is commonly used in threshers. It gives better quality of bhusa and it can be used for a wide variety of crops viz. wheat, paddy, maize, soybean etc. Corrugated bars are mounted axially on the periphery of the cylinder. It is fitted with an upper casing and an open type concave at the bottom of the cylinder. The cleaning system is provided with blower fan and straw walker.

Wire loop type cylinder: In this type of threshing drum, there is hallow cylinder, over which a number of wooden or MS plates are fitted. On these plates, number of wire loops is fixed for threshing purposes. This type of cylinder is common in the manually operated paddy threshers. Holding the bundle against the loops of revolving cylinder does threshing of paddy crop. Wire-loops are fitted on the periphery of a closed type cylinder and woven wire mesh type concave is provided at the bottom.

Cutter blade type cylinder or **Syndicator type**: The cylinder consists of a flywheel with corrugation on its periphery and sides, which rotates inside a closed casing and concave (Fig. 7.35). The rims of the flywheel are fitted with chopping blades. It consists of 2-4 chaff cutter blades mounted on the radial arms of a flywheel. The rim of flywheel is corrugated to provide additional rubbing action. This type of cylinder performs satisfactorily in moist crops. Moist crops, in case of spike tooth or hammer mill type cylinders results in frequent choking, over loading etc. This type of thresher is generally provided with position feed roller and endless conveyor belt. A chaff cutter type thresher can handle crop with grain moisture content up to 24% wet basis with reduced output (43-52%), high grain crackage (about 10%) and relatively inferior quality of 'bhusa'. In view of the advantages of lower energy requirements in case of chaff-cutter type threshers, high capacity threshers for wheat based on this system called 'Harambha' have become very popular in northern Indian states particularly Punjab and Haryana. In the existing design of chaff-cutter type threshers, materials after cutting, which contain grains, un-threshed heads, nodes and 'bhusa' have to pass through the threshing mechanism. The concave used in the chaff-cutter type thresher has concave openings of 12 mm as compared to 8 mm in spike tooth threshers. This factor also contributes towards lower energy requirements in case of chaff cutter type threshers due to higher percentage of open area for passage of straw. Higher concave openings can be adopted in the case of chaff-cutter type threshers because the straw is already chopped to the desired length by the chaff cutters. On the other hand, the straw has to be beaten and bruised in the spike tooth type threshing system to achieve the desired quality of straw, which does not permit higher size of the concave openings.

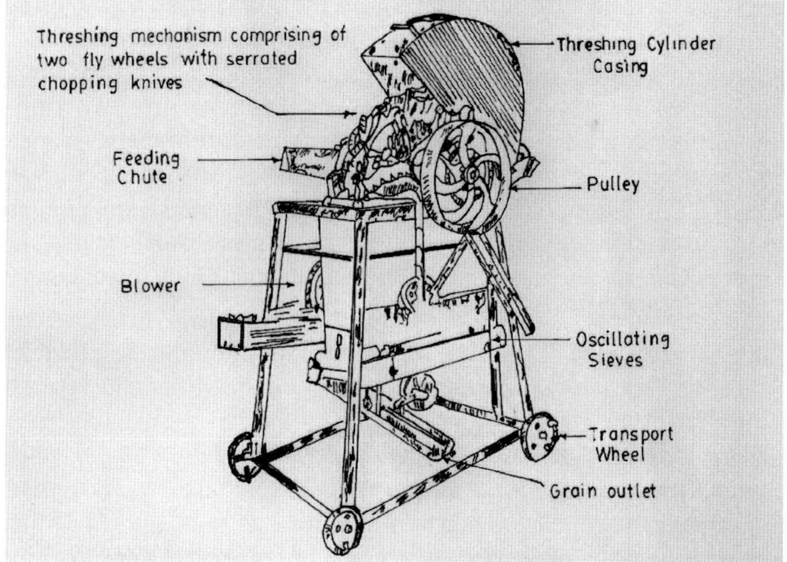

Fig. 7.35 : Details of chaff cutter types of thresher

Hammer mill type cylinder: It uses beaters to do the required job of threshing. The shape of this type of cylinder is different from the above-discussed cylinder. The beaters are made of flat iron pieces and are fixed radially on the rotor shaft. Generally feeding chutes are used with hammer mill type threshing cylinder. The cut crop is fed perpendicular to the direction of motion of rotating beaters. This type of thresher requires more power as compared to spike tooth type of thresher. It is similar to dummy type but it is provided with aspirator type blower and sieve shaker assembly for cleaning grains.

Drummy type: It consists of beaters mounted on a shaft which rotates inside a closed casing and concave.

Axial flow type: It consists of spike tooth cylinder, woven-wire mesh concave and upper casing provided with helical louvers.

Working principle of a thresher

During operation, the crop material is slightly pushed into the threshing cylinder through the feeding chute, which gets into the working slit created between the circumference of the revolving drum having attached spikes and the upper casing. The speed of the spikes is greater than the plant mass due to which they strike the latter which results in part of the grain being separated from straw. Simultaneously, the drum pulls the mass through the gap between the spikes and the upper casing with a varying speed. The angle iron ribs on the other hand, restrain the speed of the travelling of stalks clamped by the spikes. Due to this the spikes move in the working slit with a varying speed in relation to the shifting mass of material, which is simultaneously shifted, with a varying speed with respect to the upper casing. As a result, the material layer is struck several times by the spikes against the ribs, causing threshing of the major amount of grains and breaking stalks into pieces, and also accelerating them into the inlet of the lower concave.

As the material layer shifts towards the progressively converging slit of lower concave, its size reduces. The vibration amplitudes, therefore, decrease, whereas the speed of the layer increases. This causes mutual rubbing of the ear stalks, as well as rubbing of the ears against the edges of the concave bars and causes breaking of stalks depending on the concave clearance. Since the system is closed, the thicker stalk, which cannot be sieved through the concave, again joins the fresh stalk and the same process is repeated until the stalk size is reduced to the extent that it can pass through the concave apertures. Thus fine bruised straw is produced. The effective threshing process means that the loss of un-threshed kernels ejected with the straw through the concave and the loss of grain damage is low and the amount of the material passed through the concave high.

Factors affecting thresher performance: The factors that affect the quality and efficiency of threshing are broadly classified in three groups viz. i) Crop factors i.e. variety of crop and moisture in crop material; ii) Machine factors i.e. feeding chute angle, cylinder type, cylinder diameter, spike shape, size, number, and concave size, shape and clearance; and iii) Operational factors i.e. cylinder speed, feed rate, method of feeding, and machine adjustments (Ahuja and Sharma, 2001; Singh *et al*;., 1986).

Beans are more susceptible to damage due to impact and the variety of grain has much influence on grain loss during threshing. Damage of large beans is more than smaller beans at same impact velocity and orientation. The amount of damage increases rapidly below ambient temperature of 10°C. So, handling of dry beans at low temperature should be avoided. Moisture content of grains is a major factor in controlling grain damage. Decrease in moisture content greatly increases the brittleness of grains. Un-threshed grains are more at high pod moisture content whereas grain damage decreases with increase in grain moisture content. More threshing effort is required for threshing high moisture crop, which causes more internal grain damage and thus affects viability. Soybean moisture content between 8 and 12 per cent (w.b.) is optimum for low mechanical damage.

The base angle of feeding chute affects the feed rate. It should be tangential to cylinder drum for maximum feed rate and minimum physical effort. The threshing cylinder requires power as high as 60-75 per cent of total power input. Hammer mill type threshers bruise the straw very fine but the specific energy requirement is the highest among all types of threshers. Rasp-bar cylinder design can thresh most of the crop except groundnut but these machines do not provide bruised straw. The concept of a straw bruising attachment to rasp-bar thresher is not economically viable. Spike tooth type threshers having independent drive to cylinder and blower can thresh major crops effectively but the cylinder speed is to be adjusted according to the crop conditions. Larger cylinder diameter has lower power requirements than smaller ones at higher feed rates. Higher rib spacing in upper concave increases un-threshed grain but reduces power consumption. The performance with flat spikes is better than round and square spikes. Larger spike spacing in a row reduces power consumption and broken grains whereas power increases and broken grains reduce with the increase in number of rows of spikes. However, uniformity of spike distribution over cylinder periphery is more important for better performance. Power consumption and grain damage increases with the increase in spike length and thickness. The grain damage decreases and un-threshed grains increase with the increase in concave gap. Higher concave clearance reduces power consumption whereas straw bruising is more at low concave clearance.

Different parts of a thresher and their functions

A mechanical thresher consists of the following parts:

a) Feeding device (chute/tray/trough/hopper/conveyor)

b) Threshing cylinder (hammers/spikes/rasp-bars/wire-loops/syndicator)

c) Concave (woven-wire mesh/punched sheet/welded square bars)

d) Blower/aspirator

e) Sieve-shaker/straw-walker.

The crop is fed from the feeding tray into the threshing cylinder. The threshing cylinder is fitted with spikes/bars/hammers or wire-loops around its periphery according to the type of thresher. Below the cylinder there is a concave and it covers lower portion of the cylinder. The cylinder rotates at high speed and thus the crop is threshed and the entire or a portion of threshed material falls from the concave on to top sieve of cleaning system. Due to reciprocating motion of top sieve lighter material accumulate at the top and grain falls on to the bottom sieve. In case of spike-tooth thresher, an aspirator blower sucks out the lighter material from the top sieve and throws it out from blower outlet. The sieves help in further cleaning of the grain by allowing heavier straw to overflow. Various adjustments are required before starting threshing operation. The machine is to be installed on clean level ground and is to be set according to crop and crop conditions. The adjustments necessary to get best performance from the machine are (i) concave clearance, (ii) sieve clearance, (iii) sieve slope, (iv) stroke length and (v) blower suction opening. Besides these, cylinder concave grate, top sieve hole size and cylinder speeds for threshing different crops are important for a multi crop thresher.

Functional components of threshing unit: A power thresher essentially consists of feeding unit, threshing unit, cleaning unit, power transmission unit, main frame and transport unit (Fig. 7.36). The operation of conveying the cut crop into threshing unit is known has feeding. Normally, one of the two types of feeding units 'throw-in-type' or 'hold-on-type' is used in power threshers. In 'throw-in-type' feeding unit, the cut crop is pushed into threshing cylinder, where as in 'hold-on-type' the heads are only pushed into the cylinder and straw is manually or mechanically held. Throw-in-type feeding device is quite common in the threshers, which may be a feeding hopper or feeding chute (Fig. 7.37).

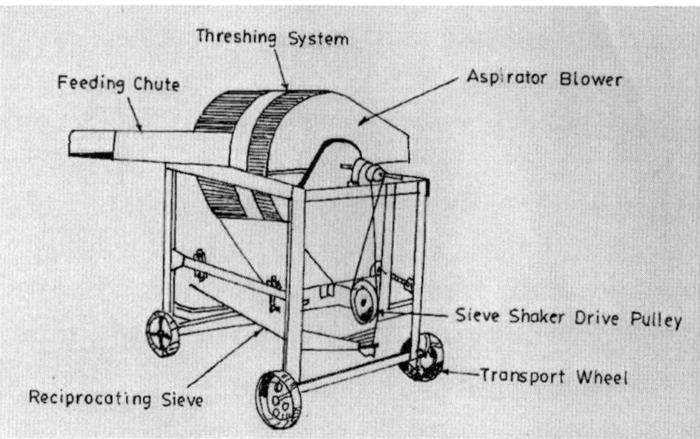

Fig. 7.36 : Components of wheat thresher

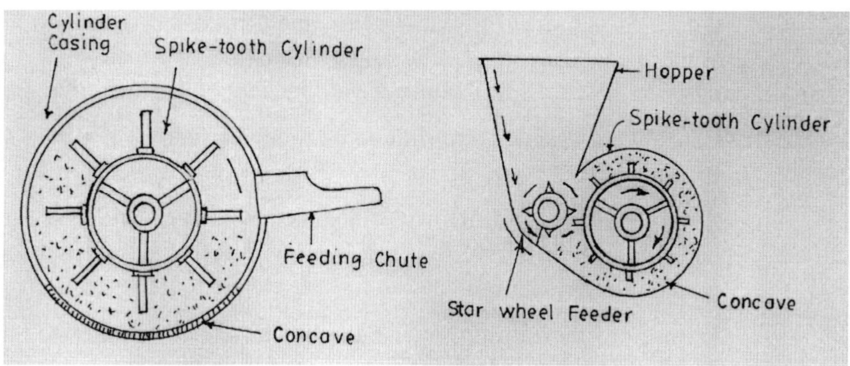

Fig. 7.37 : Throw-in-type of crop feeding systems

Feeding hopper: In this type of feeding device there is a hopper, placed on the top of the threshing cylinder. Generally hopper type of feeding units have a rotating star wheel mechanism between the hopper and threshing drum to facilitate the uniform feeding of crop to the drum. The initial cost of this system is high, hence is mostly used on a large thresher e.g. axial flow thresher of large capacity.

Feeding chute: Introduction of power wheat threshers has greatly reduced the time required for threshing as well as physical burden and drudgery of work for human beings. However, these machines have lead to the problem of involvement of the operators in accidents while using the thresher. The threshers are generally of spike-tooth cylinder type, chaff-cutter type and hammer mill or beater type. The crop is fed manually into these machines through a feeding chute. It has been observed that human factors such as inattentiveness, un-skilfulness, overwork, and physical incapability, wearing of loose clothes, hand-wears and use of intoxicants are mainly responsible for about 73% of accidents (Verma *et al.*, 1978). The machine factors such as improper design of feeding systems, substandard material and defective design contribute to about 13% of accidents. Crop factors such as feeding of ear-heads, short crop stalks and wet crop contributs about 9% of accidents whereas inadequate light, crowded surroundings and slipping on the threshing yard contributs to about 5% of accidents. The threshing accidents can be minimized provided the farmers adopt the following measures:

i) The farmer should buy only those threshers, which are fitted with safe feeding chute as per B.I.S. standards (Fig. 7.38). For safety, the minimum length of feeding chute should be kept 90 cm, covered upto a minimum of 45 cm and inclined to the horizontal at an angle of 5-10 degrees. The angle of covered portion with the base length of feeding chute should be kept equal to 5 degrees,

ii) Employ only skilled and trained workers for feeding the crop to the thresher,

iii) Avoid feeding ear-heads without stalks as it may lead to serious hand injuries, Similarly, feeding of wet crop should also be avoided which otherwise might lead to fire accidents,

iv) Ensure proper lighting in case the machine is to be operated at night, otherwise poor visibility may lead to accidents,

v) Keep the work place and surroundings of thresher free of all kinds of obstructions,

vi) Do not smoke or light a fire near the threshing yard,

vii) Keep a first aid box handy for use in the event of need,

viii) Provide warning and caution plates on each machine,

ix) Provide covers, shields, guards and fence on all revolving parts,

x) Educate farmers about safe use of threshers,

xi) Do not work on threshers for more than 8-10 hours. Fatigue and drowsiness during operation result in serious accidents,

xii) Do not talk while working on the thresher,

xiii) Do not work on the thresher under the influence of alcohol or any other intoxicants,

xiv) Do not wear loose garments, wristwatch etc while working on the thresher,

xv) Do not cross over the driving belt while the thresher is in operation,

xvi) Read the Operators' Manual before using the thresher,

xvii) Bearing and other working parts should always be properly greased and oiled,

xviii) The sieves should be inspected frequently and cleaned from time to time to avoid clogging,

xix) While feeding the crop or ear heads, operator should not insert the hand(s) deep in the feeding trough, and

xx) The thresher should be operated at the speed recommended by the manufacturer or Bureau of Indian Standards (BIS). The cylinder peripheral speed for some of the important and common crops is given in Table 7.1.

Table 7.1 : Recommended speeds of threshing drum for different crops

Crop	**Cylinder peripheral speed**	
	rpm	m/s
Paddy	675-1000	16-25
Wheat	550-1100	20-30
Barley	740-1080	20-26
Gram	400-750	12-22
Jawar	400-675	12-20
Bajra	400-550	12-16
Peas	430-750	12-22

Source : IS 9019: 1979

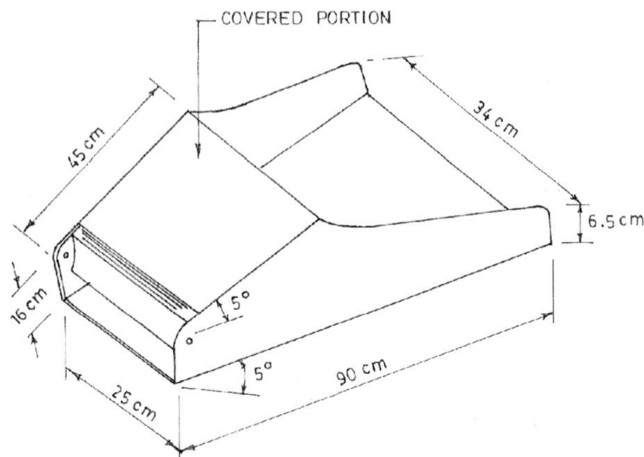

Fig. 7.38 : Specifications of BIS recommended safe feeding chute

Threshing unit: The threshing is accomplished by the impact of the rotating pegs mounted on the cylinder, over to the ear heads, which force out the grain from the sheath holding it. In the threshing of wheat crop, the straw is also bruised and broken up by the impact, thus converting it into 'bhusa' (straw). Threshing unit is mainly consists of a cylinder and concave. There are different types of threshing cylinders (Fig. 7.33) such as spike tooth/peg type cylinder, rasp bar type cylinder, angled bar type cylinder, wire loop type cylinder, cutter blade or syndicator type cylinder, and hammer mill type cylinder.

Concave: Cylinder and concave together makes the threshing unit. It separates the grain from the crop and removes grain from the straw. Concave is provided in the thresher to hold the fed crop inside the threshing chamber and allows only grain and small amount of chaff to pass through it. The threshing takes place only in this space. It is a curved unit, made of iron steel or iron bar, fitted near the threshing cylinder. The clearance between cylinder and concave is adjustable, depending upon the size and type of grain. The concave clearance for wheat is 5 to 13 mm and for paddy is 5 to 10 mm. As the concave clearance is reduced, the threshing efficiency increases but losses increase and vice versa. The concave clearance at the inlet is less as compared to outlet. There are different types of concave, which are used in thresher. Screen type concave is made of MS rod. It is semicircular in shape and sometimes made with wire also. The screen allows the material after threshing to pass through its perforation. In perforated concave, perforations are made in a mild steel sheet. The concave is closed from both the ends by iron sheet. The size of perforation is made as per the size of grain of a crop.

Cleaning unit: This unit is provided to separate the grain from chaff. It further uses sub units, like aspirators or blowers, sieves and sieve shaking mechanisms to separate out grains from chaff. The thresher that is provided with aspirator unit is usually called aspirator type thresher. Those threshers fitted with blower which blows air in horizontal direction is called drummy threshers.

Blower or aspirator: After threshing unit carries out threshing, the cleaning and separation of straw from grain is required. The fan is generally installed on the main shaft over which cylinders, flywheel and driven pulley are mounted. Fan lifts/sucks the lighter material chaff and other plant portion and throw away from the out let. Rest of the separation-cum-cleaning is done by screen with its oscillating motion.

Screens: Most of the power threshers are equipped with two screens. Top screen is provided so as to pass the grain to second screen and chaff etc is taken out from it. Other screen sieves out the smaller grain or weed seeds and delivers the cleaned grain towards outlet. The size of screen hole is selected on the basis of grain size. These screens are effective when kept under oscillation.

Shaking mechanism: The screens are oscillated or shaken with a crank attached to the screen. This crank is powered from main axle either by belt or by rod. The circular motion of the main shaft is converted into oscillating motion of screen, which shakes it and separates the grain from other foreign material and chaff. The separating effectiveness depends on the frequency of strokes of crank, which is adjustable.

Power transmission unit: Threshers are usually powered with tractors and sometimes with electric motors or diesel engine also. After installing the thresher into the threshing floor in the field, tractor PTO shaft is coupled with a flat pulley. A corresponding matching pulley of appropriate size is provided over to the thresher main shaft. These pulleys are connected with a proper rating of flat belt and thresher is operated. Blower fan is provided into the main shaft of the thresher, which rotates and does the required job. The screens are oscillated with the help of a V-belt and a crank wheel, powered with main shaft of thresher. A heavy flywheel is also provided on the main axle of the thresher. It is very important part of any thresher. It is provided to store the energy to supply continuously and equally to the entire threshing cylinder. It is made up of cost iron, and fitted on one end of the main shaft of thresher.

A very strong frame is provided in the thresher on which all the functional parts are attached. The frame is made usually of heavy angle iron sections. It should be strong enough to sustain vibrations of machine, during its operation in the field. Thresher is provided with wheels at its legs, so that transportation can be done easily. These wheels are made mostly with cost iron but new and large capacity threshers are equipped with pneumatic wheels for better performance during transportation.

Thresher adjustments: The following adjustments can be done on a stationary power thresher:

Cylinder and concave clearance: In order to get cleaned grains and proper threshing, it is very important to set the proper clearance between tip of cylinder and concave. On an average, concave clearance is kept about 25 mm at the mouth, 10 mm at the middle and 15 mm at the rear end. Start operating the thresher, by keeping proper recommended speed, and check if any grain is left in the ears. If it is so, reduce the concave clearance gradually, until drum is threshing cleanly. Too close concave setting is likely to crack some of the grains.

Cylinder speed: The drum of the thresher should be rotated at proper speed for better threshing and cleaning efficiency (Table 7.1). Normally, manufacturers specify the cylinder speed for different crops. The cylinder speed can be checked using tachometer. Operator should check the speed occasionally under load for proper functioning of thresher.

Fan adjustment: Fan(s) fitted on thresher must provide the proper amount of blast. The shutter(s) at each end of fan should be adjusted properly so that it could provide blast sufficient enough to remove chaff and light materials without grain. Watching the sample and adjusting the blast can help in getting the desired results.

Performance of threshing system: The performance of a threshing system is the percent of seed detached from the non-grain parts of the plant and the percent of seed damaged. Two other parameters are performance of cleaning and separating units. The performance of threshing system as defined by Bureau of Indian Standards in IS: 6284-1971 is:

Total grain input: It is the feed rate multiplied by grain content as obtained from straw-grain ratio.

$$\text{Un-threshed grain (\%)} = \frac{\text{Quantity of un-threshed grain obtained from all outlets (kg)}}{\text{Total grain input (kg)}} \times 100$$

Threshing efficiency: It can be defined as the percentage of grain threshed, collected from all the outlets of a thresher with respect to grain input. It depends upon cylinder speed, concave clearance, spike shape, moisture content of crop, crop mass velocity, crop mat thickness etc. It can be calculated from un-threshed grain.

$$\text{Threshing efficiency (\%)} = \frac{\text{Total grain input (kg)} - \text{un-threshed grain from all outlets (kg)}}{\text{Total grain input (kg)}} \times 100$$

$$\text{Blown grain (\%)} = \frac{\text{Quantity of threshed grain obtained at `Bhusa' outlet (kg)}}{\text{Total grain input (kg)}} \times 100$$

$$\text{Damaged grain (\%)} = \frac{\text{Quantity of damaged grain from all outlets (kg)}}{\text{Total grain input (kg)}} \times 100$$

$$\text{Sieve loss (\%)} = \frac{\text{Healthy grain obtained at sieve overflow + sieve underflow + struck grain (kg)}}{\text{Total grain input (kg)}} \times 100$$

Cleaning efficiency: It is the percentage clean grain in the total grain obtained from the main grain outlet. It is affected by various factors such as crop mat thickness, crop mat velocity, cylinder speed, concave clearance, spike shape etc.

$$\text{Cleaning efficiency (\%)} = \frac{\text{Total grain received at main grain outlet (kg)} - \text{refraction at main grain outlets (kg)}}{\text{Total grain received at main outlets (kg)}} \times 100$$

Total Loss = Un-threshed grain + blown grain + cracked grain + sieve loss

Power wheat thresher

Power wheat thresher (Fig. 7.36) is a machine, which thresh the wheat crop and performs several other functions such as feed the harvest crop to the threshing cylinder, thresh the grain out of the ear head, separate the grain from the straw, clean the grain, and make 'bhusa' suitable of animal feeding. During the 50's and 60's in the country, power threshers have become quite popular. The famous Ludhiana thresher was first introduced in India during 1956-57. The thresher was tractor operated type and used mainly for wheat. It threshed, cleaned and bagged the grain, at the same time it made the quality straw (bhusa). Further development work took place during the period from 1965 onwards for low horsepower threshers. The most widely used design; spike tooth cylinder thresher was commercially marketed in the country around 1970. This simple design has been able to maintain the cost of machine low as the total weight of machine greatly reduced. The output capacity also improved. These threshers are available in various sizes operated by 3-40 hp power source. The grain output is 20-25 kg/hp-h.

Pedal operated thresher

It is used for threshing rice crop. It is pedal operated by human being (Fig. 7.39a). It consisted of wire-loop type threshing cylinder, power transmission system, mild steel sheet body and foot pedal (Fig. 7.39b). The threshing cylinder consists of wire-loops of 'U' shape embedded in wooden or metallic strips joined to two discs. A shaft carries the threshing cylinder and is connected to the transmission system. The transmission system consists of meshed gears or sprocket-chain mechanism. The larger gear or sprocket is connected to foot pedal/bar with links. The foot pedal/bar is always in raised position. On pressing the pedal the threshing cylinder starts rotating. For continuous rotation of the cylinder, the pedal is lowered and raised repeatedly. For threshing, paddy bundle is held in hands and ear head portion of the crop is placed on the rotating cylinder. The wire-loops hit the ear heads and grain get detached from the rest of the crop. Working capacity of pedal operated thresher is 44 kg/h. Weight of the machine is about 36.0 kg.

a) A view of machine

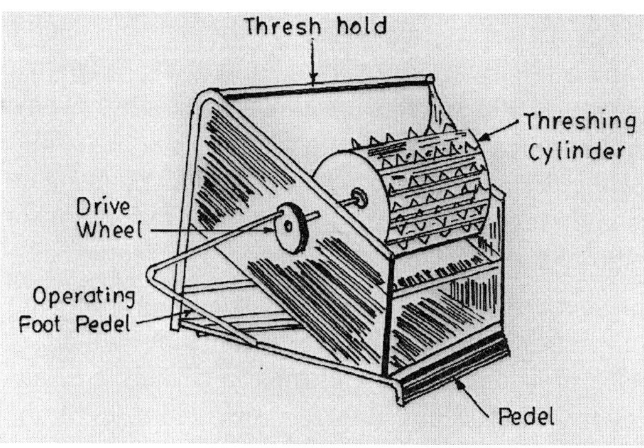

b) Components of thresher

Fig. 7.39 : Pedal operated thresher

Drummy thresher

These threshers were very popular in the beginning when threshers were introduced because of its simplicity and low cost. The radially arranged arms known as beaters are mounted on the shaft (Fig. 7.33). These are made of mild steel square section with mild steel flat welded or bolted at the top. The beaters revolve inside an enclosed casing. Ribs are provided inside of upper half of the cover in order to have better threshing. The lower half (known as concave) has rectangular openings made out from square bars. The crop is fed through feeding chute. Crop receives impacts from the rotating beaters till size is reduced to pass through concave. The clearance between beater and concave is kept about 18-20 mm. The crop should be well dried before feeding in the thresher. A wet crop raps around the beater shaft and machine becomes overloaded. These threshers do not have provision for separation and cleaning of grains. The threshed material is later separated and cleaned by small pedal type blower (Fig. 7.32).

Multi-crop threshers

Since, the Indian farmers raise variety of crops as per the suitability of particular region, climate and soil conditions, there was need to thresh all these crops for timelines of operation. Developing a multi crop thresher has solved this problem. It can thresh crops like wheat, moong, paddy, grain, soybean etc. For these crop requirements are different, as in the case of wheat bruised straw (bhusa) is the main requirement; for paddy farmers need long straw; for pulses, seed damage should be minimal; as damaged seeds lower the quality and causes spoilage in storage (Singh and Pandey, 2008; Anonymous, 2008 & 2008c; Anonymous, 2010c) . The crop factors such as moisture content, grain size, grain-straw ratio, condition of straw etc influence the design consideration of main components of threshers. The suitable multi crop threshers for cereals and pulses are commercially available in the country (Fig. 7.40).

The thresher consists of a feed tray, spiked cylinder, straw thrower, blower and cleaning sieves (Fig. 7.41). The threshing cylinder is fitted with rectangular shaped flats as beaters, welded on eight bars in staggered fashion. On the rear end of the threshing cylinder, impeller blades for throwing long straw in case of paddy are provided. The concave is made out of square bars with fixed opening and clearance. Three concave grates are provided for threshing various crops. A semi-hexagonal top cover with spiral louvers is provided for threshing paddy. The feed tray, hinged for folding during transport, is of sufficient size to hold crop material for continuous feeding. Crop is fed at the feed opening and is taken inside the machine by the cylinder, which rotates on ball bearings. The beaters on the threshing cylinder hit the material, separating the grain from the straw and at the same time accelerating them around the cylinder. The spiral louvers in the top cover move the material axially from the feed end to discharge end. The long straw is discharged from the machine by the paddles at the discharge end of the cylinder. For threshing crops other than paddy, the semi-hexagonal top cover is to be replaced by a semi-circular cover and a semi-circular disc to be inserted in between cylinder and straw thrower. Threshed material is passed through the openings between the concave bars and falls on the upper oscillating screen. An aspirator blower is mounted behind the cylinder with two suction openings, one at the separating chamber and another at main grain outlet. The blower sucks out the lighter material and blows out of the machine from blower outlet. Cleaning sieves are hinged at the bottom of cylinder concave on adjustable hangers. Further separation of grain from straw mixtures takes place due to oscillating motion of shaker assembly, the grains fall through the holes of top sieve over bottom sieve and heavier straw and un-threshed tailings overflow from top sieve. Three screens are provided for threshing various crops. The secondary inlet of blower does final cleaning and thus clean grain is obtained at the main grain outlet. The cylinder and shaker assembly get power from blower shaft through V-belt and pulley drive. The concave clearance, sieve clearance, screen slope and cylinder speed are adjustable. The salient features of multi crop thresher over commercial spike tooth threshers are: i) it can be used effectively for threshing variety of crops without much loss of grains; ii) the cylinder and blower are mounted on separate shafts and the speeds can be varied independently as per crop requirement. This ensures minimum grain damage, blower loss and better cleaning; iii) different accessories are provided for threshing different crops; iv) axial flow threshing principle is used for threshing rice crop; and v) though the cost of the machine is slightly more (about 12%) but it is quite economical to use, as the machine use is 400 h per year against 200 h for single crop thresher.

A multi-crop thresher attains the axial movement of the crop while handling paddy and all crop material is made to move through the concave in case of wheat (Sharma *et al.*, 1986, 1986a; Sharma, 2001). The axial flow of material can be accomplished by providing seven louvers with spacing of 150 mm in the hexagonal casing. The clearance between louvers and tip of cylinder spikes is 20 mm. For wheat threshing, the first three louvers are placed with ribbed casing and side plates are fixed with top casing and concave to prevent material flow in the second portion. The direction of rotation of threshing cylinder is opposite for wheat than paddy. That is why; straw outlet of aspirator blower is repositioned. The top sieve has holes of

9-mm diameter for wheat and 5 mm for paddy grains. The lower sieve has holes of 1.5 mm diameter common for both the crops. The upper sieve can be changed easily depending upon crop to be threshed. The cylinder-concave clearance in the first section of threshing system (i.e. facing the feeding chute) has to be more while handling paddy than wheat. Broken grain and blown grain losses are found to be in the range of 0.3-0.8% and 0.5-0.9% respectively, and threshing and cleaning efficiencies 99% and 95.7% respectively. There is a saving in labour and cost of operation by 85-90% and 65% respectively over conventional method. This thresher can be operated by 10 to 25 hp power source i.e. electric motors/ diesel engines/ tractors. The thresher can thresh wheat, soybean, barley, bajra, Urd, masur, gram, paddy etc and has the capacity ranging from 10 to 25 q/h.

Fig. 7.40 : High capacity multi crop thresher
Courtesy: Gobind Industries, Barabanki (U.P.)

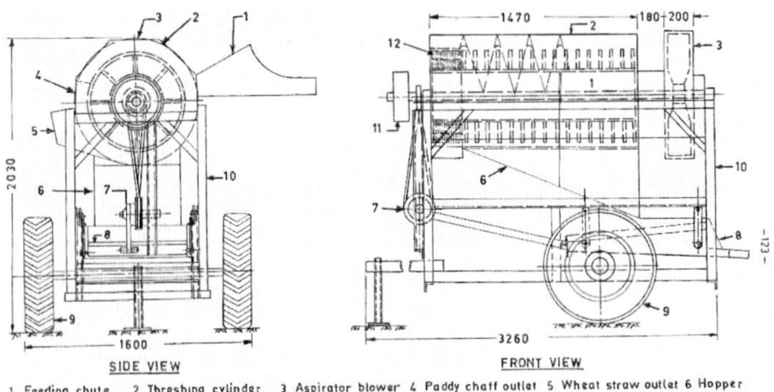

1 Feeding chute 2 Threshing cylinder 3 Aspirator blower 4 Paddy chaff outlet 5 Wheat straw outlet 6 Hopper 7 Cam for oscillating sieves 8 Oscillating sieves 9 Transport wheels 10 Frame 11 Main pulley 12 Louvers

Fig. 7.41 : Components of multi crop thresher

High capacity (Hadamba) wheat thresher

It is a basically a chaff-cutter type thresher. It consists of a threshing cylinder, concave, two aspirator blowers, reciprocating sieves, feeding chute, feeding conveyor, feed rollers, safety lever in the feeding chute and flywheel (Fig. 7.42). A platform is attached to the main frame of thresher, on which a person stands and feeds the crop into thresher. All the crop materials are fed through the conveyor of feeding chute and feed rollers move the crop into threshing cylinder. A safety lever provided in feeding chute prevents the entrapping of hands by the feed rollers. Threshing cylinder has two chaff-cutter type blades and beaters. Chaff-cutter blades cut the crop into pieces and beater helps to detach grain from crop. All the threshed materials pass through the concave where it is subjected to aspiration action of blower. Light materials like chopped straw (Bhusa) are blown away with aspirator blower while the heavier materials like grains, nodes etc fall on a set of reciprocating sieves. The sieves clean the grain and an auger elevates the grains and conveys directly into a trolley. It can be used to thresh the crop having high moisture content also. It is used for threshing wheat crop. It is basically a combination of chaff-cutter and beater type thresher and has high capacity (Fig. 7.43). The machine is operated by PTO of a 35-hp tractor and is mounted on two pneumatic tyres for easy transportation. Working capacity of thresher is 15-20 q/h. Weight of the machine is about 700-800 kg. Threshing efficiency is about 100%, total grain loss, 0.5 - 1.5% and cleaning efficiency about 99%.

Fig. 7.42 : High capacity (Hadamba) wheat thresher

Courtesy: Dasmesh Mechanical Works, Amargarh (Punjab)

Fig. 7.43 : High capacity (Hadamba) wheat thresher
Courtesy: Amar Agricultural Implements Works, Ludhiana (Punjab)

Axial-flow high capacity paddy thresher

Axial flow drum is provided with spikes/bolts arranged in spiral manner to allow axial movement of threshing material (Ahuja and Sharma, 2001; Garg and Singh, 2002). In this process different layer of the matter move across the lengths of cylinder causing rubbing and squeezing action. This intern results in separation of grain/kernel from the ear head. This method is quite useful for threshing of maize, sunflower and paddy crop. Eight rows of threshing/separation spikes are equally spaced on the four rings of 340-mm diameter (Fig. 7.44). The rings are made of 50 x 15 mm flat. The first 11 spikes have a spacing of 62.5 mm while the rest have a spacing of 125 mm. The last portion of length 150-mm have 4 straw throwing paddles of 465-mm tip to tip diameter. The cylinder spikes are arranged in staggered rows. The second and third portion of the cylinder is used for paddy threshing along with the first portion. Machine is suitable for threshing paddy and it works on the principle of axial flow. It consists of a threshing cylinder, concave, cylinder casing, cleaning system and feeding chute (Fig. 7.44). In axial flow concept, the crop is fed from one end and the straw is taken out from the other end after completing threshing of crop. During this period the crop is rotated three and half times and all the grains are separated. The threshing cylinder is of peg type and has a diameter of 77 cm and length 150 cm. The first part of cylinder of length 133 cm has spikes for crop threshing and the 2nd portion of cylinder of length 17 cm has straw throwing blades. The concave is made of 8 mm round bar and cylinder-concave clearance is 4 cm. The casing of the thresher is provided with 7 louvers for moving the crop axially. For cleaning purposes, it is provided with 2 aspirator blowers and 2 sieves. The opening of the top sieve is 9 mm and of lower sieve 1.5 mm. A pulley and a cam drive the sieves, which in turn get the drive from the threshing cylinder shaft through a V-belt. It is used for threshing of paddy crop (Fig. 7.45). It is operated by a PTO of

26.1 kW tractors. Working capacity of thresher is 12-20 q/h. Cleaning efficiency is 99%, threshing efficiency 100%, and total grain losses less than 2.5%. Savings in labour requirement is 70% and cost of operation 20% in comparison to traditional method. Axial flow threshers as shown in Fig. 7.46 have a special feature of filtering unit and router fitted into it does not allow breakage of grain during threshing. The grains and husk are entirely separated and blown in opposite direction. It is fitted with two large wheels for easy movement from one place to other. Threshing drum is bar type with spikes. It has one blower and three sets of cleaning sieves. It is operated at cylinder speed of 700 rpm.

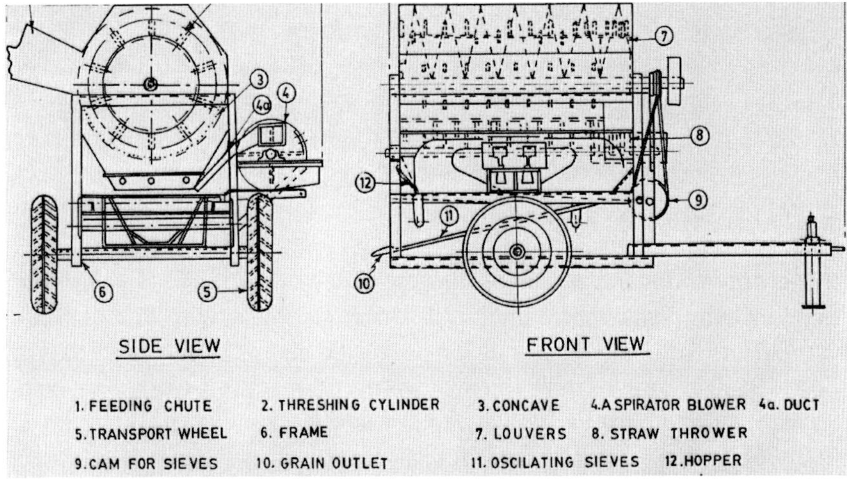

Fig. 7.44 : Detailed view of axial-flow high capacity paddy thresher

Fig. 7.45 : Axial-flow high capacity paddy thresher during operation

Courtesy: Madho Agro Industries, Moga (Punjab)

Fig. 7.46 : Axial flow paddy thresher
Courtesy: Dasmesh Mechanical Works, Amargarh (Punjab)

Paddy thresher flow through

In conventional practice (beating the bunch of crop against a platform) 8 labours can thresh approximately 1 to 1.2 q/h, while with flow through paddy thresher the threshing rate can be raised up to 13 times with saving of 85 to 92% in labour - hour and 65% in cost of operation in comparison to conventional method (Fig. 7.47); Pandey and Ganesan (2005); Pandey *et al.* (2006). In this machine crop is fed from one end and straw is thrown out from the other end after completing threshing. During threshing crop rotates inside along the cylinder and all the grain get detached. Threshing cylinder is peg type. Aspirator blowers and sieves are provided for cleaning. Threshing and cleaning efficiencies are 99% and 95.7% respectively.

Fig. 7.47 : Flow through paddy thresher in working

Portable rice thresher

The thresher consists of a peg tooth cylinder with straw throwing paddles on one end enclosed by a cover with spiral louvers, wire mesh concave at the bottom and a blower (Fig. 7.48); Pandey *et al.* (1997). It can be shifted from one field to another by two persons. It is suitable for threshing rice. The rice crop is fed through the feed inlet. The grain separation and axial movement of the straw takes place in threshing cylinder. The bruised straw is thrown from the straw outlet at the end of the threshing cylinder. The louvers provided on the casing help in axial movement of the straw. Threshing efficiency is 98%, cleaning efficiency 90%, output capacity 100 kg/h and labour requirement 2 man-h/q. It can be operated by 5 hp engine or electric motor. It is used for threshing crops like rice, sorghum, pearl millet and safflower.

Fig. 7.48 : Portable rice thresher

7.2 Harvesting and threshing equipment for oilseeds and pulses

Harvesting of crops like oilseeds and pulses has to be done carefully as the matured grains easily detach from the ear heads/pods and, therefore, cannot be harvested by fast working tools or machines (Singh, 2007). Bengal gram, green gram, lentil is to be harvested at ground level. Oilseed crops pose different type of problems to engineers for mechanization of their harvesting. Safflower is a spiny crop and difficult to harvest even manually. In case of sunflower, harvesting is simpler as only flower heads are to be collected. In sesamum crops, pods containing seeds are attached to the main stem and they are mostly raised by broadcasting. This also needs gentle handling. Farmers follow different methods for harvesting of rapeseed/ mustard and pigeon pea. Mostly, farmers harvest these crops at branch level, but small farmers harvest these crops at ground level.

Different harvesting and threshing methods are followed by farmers for various oilseeds and pulse crops. Bengal gram is harvested by local sickle (Pandey *et al.*, 1997). Improved serrated blade sickles are also in use. The performance of narrow pitch cutter bar with horizontal conveyor is better than other types of available reapers. Combines with floating cutter-bar are also in use. Pigeon pea is traditionally harvested at ground level by using a chopper or local sickle. No suitable machine for harvesting this crop is available in the country. Crop stems are being used by farmers for domestic use. Urad, moong and cowpea are traditionally harvested by using local sickle. Improved serrated blade sickles are also in use. Digging of groundnut crop with country plough and blade hoe at proper soil moisture level and manual pulling and gathering of pods using hand hoe is common practice. Animal drawn and tractor operated diggers and digger windrowers are improved implements developed for groundnut harvesting. The blade harrow is widely used for digging of groundnut crop in Gujarat. PAU Ludhiana, TNAU Coimbatore, CIAE Bhopal designs are some of the improved animal drawn groundnut diggers. Tractor operated groundnut diggers have wide blade, which cover 1.25 to 2.0 m width and operate at 10 to 15 cm depth.

The traditional practice of harvesting rapeseed and mustard is to harvest manually using sickles. In tall varieties, farmers cut the plants above ground level and leave long stubbles in field, which are subsequently ploughed in. In some areas, where plants are used as fuel or thatch material, harvesting with serrated blade sickles close to ground level, is practised by farmers. Soybean harvesting by local sickle is the traditional practice followed by farmers. However, modified serrated blade sickles are recommended, as plant stem is 8 to 12 mm thick. Self-propelled vertical conveyor reaper windrower, tractor front mounted reaper and combine harvester have been found suitable for soybean harvesting. When the available harvesters are to be used for soybean, these are required to be modified and adjusted to reduce field losses and suit crop and soil conditions. Cutting of crop close to ground with low stubble height and crowding and stripping effect are the main requirements. Combine harvesters with floating cutter bars are recommended for low harvesting losses. Narrow pitch cutter bar has been reported to give lower harvesting losses as compared to conventional cutter bar.

The traditional practice of harvesting sunflower and castor is to manually harvest the flower heads of sunflower and castor plants. These are stacked and sun dried for threshing. Suitable machines are not available for harvesting of sunflower and castor crops. Harvesting of whole plant would require separation of flower heads for threshing and thus the time saved by harvesting the whole plants would not reduce the labour requirement. The combine harvesters are used for harvesting of above crops using specially designed header. These are in use in advanced countries. Safflower is harvested manually using sickles. Because of thorny and spiny nature of crop, harvesting and handling of safflower plants is a problem. Use of hand gloves and covers on legs and arms is recommended during harvesting. Hayforks are used for gathering and stacking the plants in field or on trailers. Self propelled (1 metre wide) vertical conveyor reaper and combine harvester are also used for harvesting of safflower.

Engine operated walk behind vertical conveyor reaper

It is used for harvesting and windrowing cereals and oilseed crops (Pandey and Ganesan, 2005). It is operated by a 3.7 kW diesel engine (Fig. 7.49). It is an engine operated, walk behind type harvester suitable for harvesting and windrowing cereals & oilseed crops. It consists of engine, power transmission box, lugged wheels, cutter bar, crop row dividers, conveyor belts with lugs, star wheels, operating controls and a sturdy frame. The engine power is transmitted to cutter bar and conveyor belts through belt-pulleys. During forward motion of the reaper, crop row dividers divide the crop, which come in contact with cutter bar, where shearing of crop stems takes place. The cut crop is conveyed to one side of the machine by the conveyor belt fitted with lugs and is windrowed in the field. The crop is bundled manually and transported to threshing yard. Working capacity of reaper is 0.21 ha/h for soybean; 0.23 ha/h for safflower; and 0.21 ha/h for sesamum. The weight of the machine is about 85 kg.

(a) Stationery view (b) Machine harvesting soybean field

Fig. 7.49 : Engine operated walk behind vertical conveyor reaper

Self-propelled riding type reaper

It is suitable for harvesting soybean and other oilseed crops. The riding type vertical conveyor reaper is a self-propelled unit in which the operator rides on the machine (Pandey *et al.*, 1997). Drive is by means of two large pneumatic wheels and steering is by rear idlers. The prime mover is a 6 hp diesel engine. Convenient clutch, break, steering, hydraulic system and simple power transmission are provided for ease of operation (Fig. 7.50). It consists of crop row divider, star wheel cutter bar, conveyor belt and wire spring. This reaper has two forward and one reverse speed. Working capacity of reaper is 0.25 to 0.30 ha/h.

Fig. 7.50 : Self-propelled reaper harvesting soybean

Groundnut harvesting and threshing equipment

Groundnut is a root crop and an important oil seed crop. It is grown during Kharif season. However summer groundnut variety has also been developed. It is grown in light to medium to heavy soil. Main groundnut growing states in India are Gujarat, Tamil Nadu, Andhra Pradesh, Madhya Pradesh, and Maharashtra. The crop is planted on flat land as well as ridges & beds. Row spacing is kept 30 cm with a plant to plant spacing of 20 cm. Requirement for groundnut harvesting is i) to cut the tap roots slightly below the pod zone; ii) to loosen the soil up to the pod zone such that it is easy to pull out the vines (plants) along with the pods manually or lift the vines by an elevator shaker conveyor and place them in a windrow such that the maximum number of pods get exposed to the sun for rapid sun drying; iii) to minimize the pod loss; iv) to shake-off the soil; and v) to lay the crop in a windrow for subsequent combining. Common type of groundnut harvesting equipment used are: i) full or half-sweep type blades; ii) sweep-type blade with a pressing roller; iii) digger shaker windrower; and iv) groundnut combine.

In the traditional method of groundnut harvesting, the vines are uprooted with a hand tool along with the entire root system. Groundnut is a legume and tap-rooted crop. The groundnut pods are located up to a depth of 5-10 cm usually referred to as pod zone. The manual harvesting leads to avoidable depletion of soil fertility due to removal of the complete root-system bearing the nitrogenous nodules. It requires 120-150 man-h/ha to harvest the crop. Mechanical harvesting of groundnut has the advantage of reducing the cost and labour requirements and is conductive to better soil fertility as the blade of the digging machine cuts the roots below the pod zone and leaves the remaining root system in the soil itself. Harvesting by machines, however, require that the fields should have sufficient moisture to enable the blade to penetrate to the desired depth. Also it requires the crop to be harvested without letting it over mature. There are three types of groundnut harvesting equipment commonly being used by the farmers in Punjab (Verma, 2001; Singh, 2007). In all these equipment, a single-piece blade slightly curved in the centre is used. Further, disc coulters are provided on both sides of the blade to make a pre-cut on the crop being loosened by the blade. In the absence of coulters, the vines extending outside the blade width are dragged which leads to detachment of pods from the vines. The action of a groundnut-digging blade also results in thorough cultivation of the soil.

Animal drawn groundnut digger

Animal drawn groundnut digger developed at CIAE, Bhopal is used for effective digging of groundnut at faster rate compared the manual digging by use of hand hoe/narrow spade (Pandey *et al*, 1997). By use of animal drawn digger the recovery of groundnut is 5% more compared to the manual digging. Besides one uniform tillage operation for the next crop is the added advantage compared to the non-uniform tillage by manual digging to uproot the groundnut crop (Fig. 7.51). Losses are well within 3%.

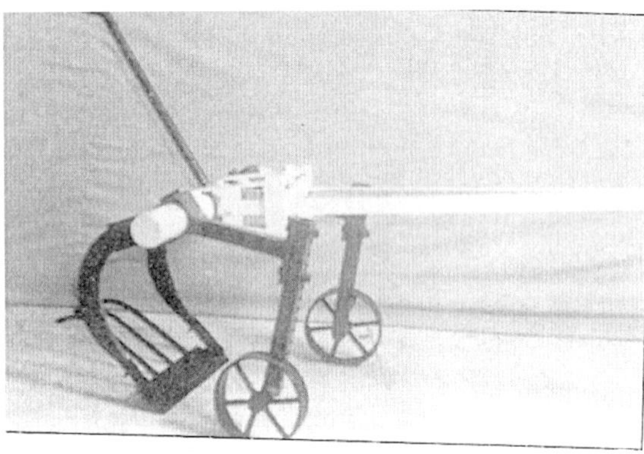

Fig. 7.51 : Anumal drawn groundnut digger

Tractor mounted digging blades

The digging blade works on the principle of a sweep. The depth of working is adjusted with the help of the gauge wheels and tractor hydraulic lift. The vines are cut below the pod-zone and are left in their places without disturbing them. The vines are later collected manually and stacked in bunches in the field for 3-4 days for sun drying before transporting them to the threshing floor. The tractor mounted digging blade (Fig. 7.52) is suitable for harvesting groundnut from light soil where adhering of the soil to the pods and plant roots is not a problem (Singh, 2007). Loose soil falls through the extension rods behind the blade and groundnut tubers are left over the soil for easy picking and gathering. In heavier and harder soils, it does not work satisfactorily due to clod formation for which there is no provision to shake off the soil. Groundnut digger blades with corrugated roller at rear (Fig. 7.53) are also used in medium and heavy soils where the corrugated roller helps in crushing the clods. This helps in easy picking of groundnut tubers. A hollow corrugated roller measuring about 40 cm in diameter, made of angle-iron sections is attached behind the blade. The length of the blade is kept about 4 cm per hp. The depth of working is adjusted with the help of the top link of the tractor and by adjusting the corrugated roller up or down. The roller in this way serves the purpose of the gauge wheel and helps in removal of the soil adhered to the roots and also minimizes chances of clogging as the vines get pressed after digging. It is not suitable for wet soil conditions as the soil has a tendency to adhere to the vines while being pressed under the roller. As in the case of the digging blade, gathering of the vines into small bunches is carried out manually.

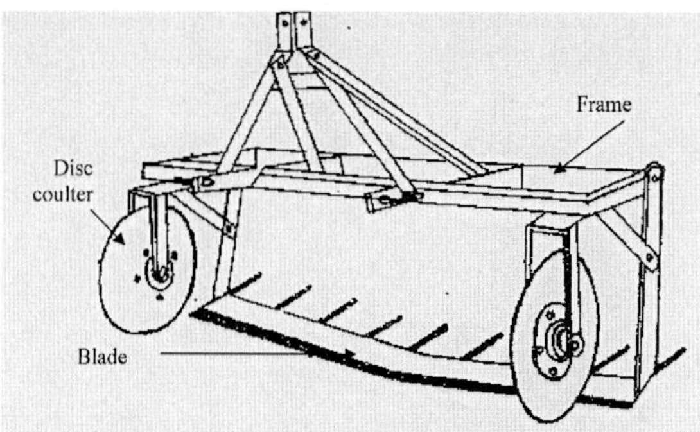

Fig. 7.52 : Tractor operated groundnut digging blade

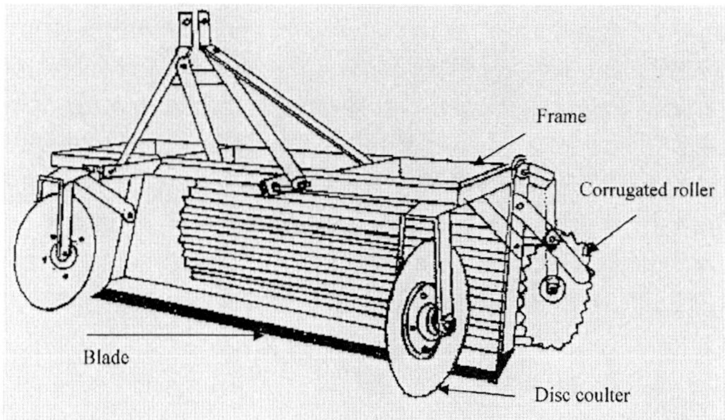

Fig. 7.53 : Tractor operated groundnut digging blade with corrugated roller

Tractor mounted groundnut digger elevator

The tractor mounted groundnut digger shaker comprises of a blade, elevator-cum-pickup reel, fenders, gauge wheel, coulters and power transmission system (Fig. 7.54); Anonymous (2008, 2008e & 2010). The front end of the pickup-cum-elevator reel is adjustable in accordance with the depth of working of blade. The frame, made up of angle iron, has provision of mounting blade, coulters, gearbox and three-point linkage in the front. Its rear portion is inclined at an angle and pick up-cum-elevator assembly is mounted on it, which consists of 15 cross members bolted on the endless chains running on sprockets. The power is transmitted to the machine from tractor PTO. The power is transmitted to the rear drive shaft, which in turn rotates the chain driving sprockets. Two MS shanks one on each side are mounted on the frame. A single piece blade made up of high carbon steel is bolted with the lower end of these shanks. The working depth of the machine is adjusted with the help of two gauge wheels and the top link of the tractor. The front end of the pickup rod is so adjusted that the spikes comb about 3 cm of the topsoil to lift vines gently from the loosened soil. The groundnut vines are cut below the pod zone and simultaneously lifted by a shaker-conveyor. The soil attached to the vines is shaken off in the process and the crop is dropped at the rear in the form of a narrow fluffy windrow (Fig. 7.54). The vines are dropped in such a manner that the pods get exposed to the sun for speedy drying. It requires a tractor operator and one attendant for its working. The effective width of coverage is less than 1 m, which is one of the constraints. The machine can uproot and invert 0.16-0.21 ha/h. The machine saves 65% labour and 32% cost of operation. It digs out the plant along with groundnut. The material is moved over the conveyor; the soil is loosened and removed. It has been designed to suit a tractor of 30 hp or more and is being operated by PTO shaft of the tractor. The digging shovel may be either of pointed type or conveying cutting edge type. The forward speed of digger is kept between 2.4-3.0 km/h. In order to pick and gather the tubers dropped in a windrow behind the digger, nearly 10-12 workers are required for continuous working of the digger. Depth of operation is 15.4-18.2 cm; field efficiency, 74-78%; and digging efficiency, 98.02%. Another type of tractor operated groundnut digger has also been developed (Fig. 7.55); Anonymous (2010 & 2010a). It has field capacity of 0.11 ha/h with field efficiency of 67%.

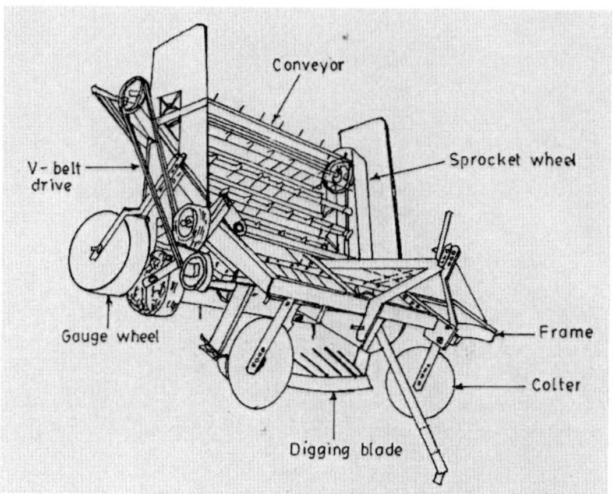

a) Details of tractor operated groundnut digger shaker

b) A view of tractor operated groundnut digger shaker

c) Tractor operated groundnut digger elevator working in field

Fig. 7.54 : Tractor operated groundnut digger elevator shaker

Fig. 7.55 : Tractor operated groundnut digger-cum-elevator used in Andhra Pradesh

Power tiller mounted groundnut digger

The common method of harvesting of groundnut is manual uprooting and labour requirement is very high. In dry clay loam soil and lateritic clay soils, pods remaining inside soil could be as high as 50%. Digging soil by hand hoe is a common practice, which is labour intensive and the pods are collected manually. In such cases power operated diggers can reduce the cost involved in groundnut harvesting. Bullock drawn groundnut digger could not function efficiently in dry field condition due to heavy draft requirement. A power tiller operated groundnut digger has been developed by OUAT Bhubneswar (Anonymous, 2008a). The groundnut digger (Fig. 7.56) designed has two bottom unit operated by a power tiller of 7.5 – 9.0 kW. The bottoms are placed at a horizontal distance of 140 mm along the direction of travel to avoid clogging of uprooted plants. Each bottom with a shank and a tyne is clamped on a square bar by means of 'U' bolts. The height of the shanks can be increased or decreased by loosening the clamps. The spacing between the two bottoms is changed by loosening the clamps and adjusted according to the prevailing row spacing. The tyne consists of a 520 mm long flat and a 'V' blade of 200 mm horizontal width with an internal angle of 120°. A shield is provided behind the cutting blade for better conveyance of uprooted plants and to avoid clogging of plants in front of the shank. The angle of inclination of the cutting blade to the ground level is fixed at 38°.

Fig. 7.56 : Power tiller mounted groundnut digger

High capacity groundnut thresher

The axial flow groundnut thresher has been developed (Garg and Singh, 2002; Singh,2007; Anonymous, 2008). It is a spike tooth type thresher. It is used to remove the groundnut pods from the vines. The crop is fed to the threshing cylinder, which removes the pods from the vines. It consists of an axial flow-threshing cylinder, concave, cleaning system, a blower and feeding platform (Fig. 7.57). It has a threshing unit, separation and cleaning unit. The threshing unit includes a threshing cylinder and a concave. The separation and cleaning unit has two sieves and a blower. The threshing cylinder has two units namely threshing and straw or vines throwing. In the first unit, the threshing spikes has been mounted whereas the second unit (straw/vines ejecting section), comprising of blades fitted on the cylinder to throw away the straw/vines. Two blowers are provided to increase the intensity of air and to provide uniform airflow to both sieves to increase cleaning efficiency. The cylinder is of closed type and has pegs of 130-cm length in the 1st portion. The 2nd portion of 22-cm length has 5 throwers. The cylinder concave opening is 3.0 cm. The casing is of hexagonal shape with 7 louvers inside for axial movement of the crop materials. It has 2 sieves. The upper sieve has 15 x 45-mm holes where as the lower sieve is made of 8-mm MS round bar having 15-mm spacing. Machine is operated through a pulley of the tractor with flat belt. The harvested crop at moisture content varying from low to high is fed to the threshing cylinder through feeding platform (Fig. 7.58). The pods after threshing fall on the upper sieve through concave and straw is thrown out from the outer end. It is suitable for threshing of groundnut crop. It is operated by a 7.5 kW engine/tractor. Total losses are in the range of 1.0 - 2.25%; pod damage less than 0.5%; cleaning efficiency, 92 - 98%; threshing efficiency, 99.0 - 99.5%; saving in labour requirement, 70%; and cost of operation, 40% in comparison to traditional method. A view of commercial groundnut and caster thresher is shown in Fig. 7.59. Threshers have been developed for threshing wet groundnut crop also (Fig. 7.60).

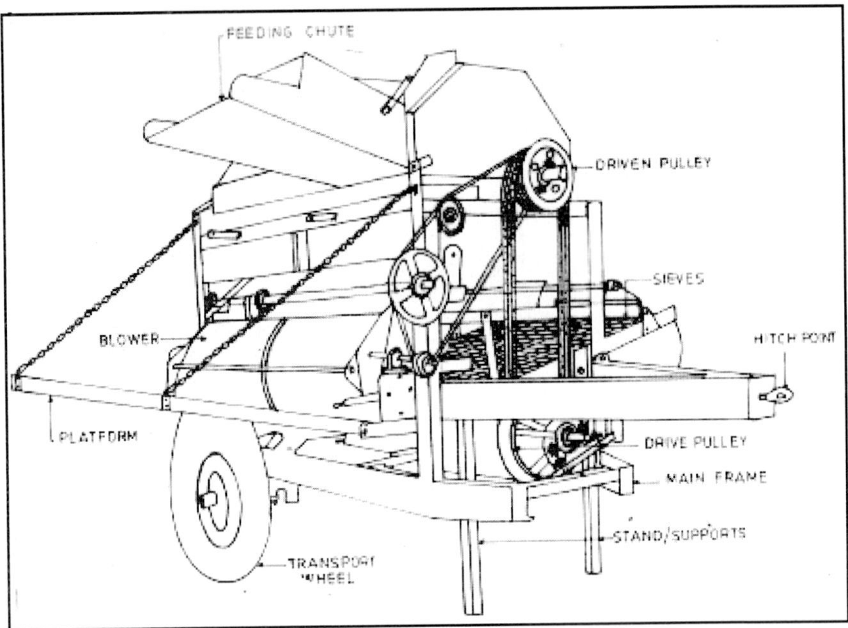

Fig. 7.57 : Details of groundnut thresher

Fig. 7.58 : Axial flow groundnut thresher (Stationary & during operation)

Fig. 7.59 : Caster and groundnut thresher
(Courtesy: Umiya Agriculture Industries, Visnagar (Gujarat)

Fig. 7.60 : Groundnut wet pod thresher

Groundnut combine

China make self-propelled combine harvester for groundnut suitable for harvesting and threshing green pods have been imported by TNAU Coimbatore (Anonymous, 2010 & 2013). It consists of 30 hp engine, chassis, power transmission system, digging assembly, gathering system, chain conveying system, stripping system, pod collecting system and cleaning system (Fig. 7.61). The digging efficiency is around 98%, picking and conveying efficiency 94%, stripping efficiency 88% and damage to the pods 12 per cent.

The field capacity of groundnut combine is 0.12 ha/h. The machine is suitable for harvesting groundnut crop raised on beds. Growing crop on raised bed system with one or two irrigation gives higher yield as compared to their conventional practice of raising groundnut under purely rain fed conditions.

Fig. 7.61 : Self-propelled groundnut harvester

Groundnut pod stripper

It is suitable for stripping of groundnut pod from freshly harvested crop (Anonymous, 2008 & 2008f). It is operated by a 1.5 kW electric motor. It consists of a wire spike type cylinder (Fig. 7.62). Stripping is done by holding the pod portion of a bunch manually over spiked cylinder. There are 120 beaters with 60 mm height. There is a blower of throwing type. Three persons can work at a time. It saves 66 per cent labour, 80 per cent operating time and 30 per cent on cost of operation and it also results in 10 per cent reduction in losses compared to conventional method of manual stripping. Working capacity of stripper is 100-120 kg/h, threshing efficiency, 100% and cleaning efficiency, 96%.

Fig. 7.62 : A view of groundnut pod stripper during operation

Groundnut-cum-caster decorticator

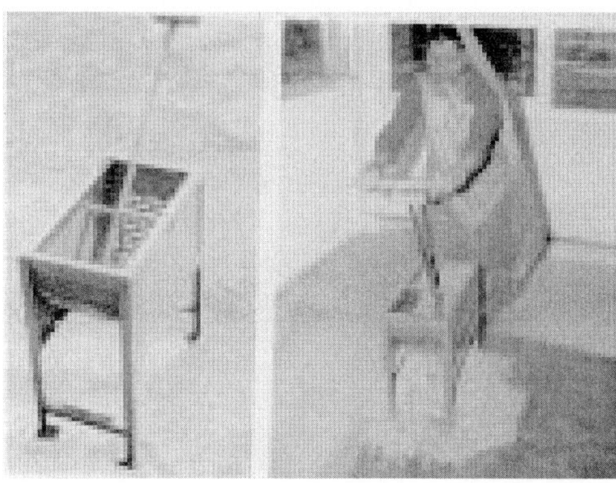

Fig. 7.63 : Groundnut-cum-caster decorticator

It is suitable for decorticating groundnut pods to separate kernels. It is manually operated. Groundnut decorticator is a sitting and standing type (Pandey and Ganesan, 2005). It consists of an angle iron frame the two sides of which are covered with mild steel sheet, a perforated mild steel sieve the openings of which are designed to suit a groundnut or caster, decorticator shoes made of cast iron or nylon, a handle made of steel mild flat iron, hand grips attached to the handle made of wood and a mild steel rod which acts a fulcrum for providing reciprocating motion to handle (Fig. 7.63). The pods are fed in batches of nearly 1.5 kg i.e. up to half of its hopper capacity so that oscillating arm can easily be operated. Pods are decorticated by swinging the lever forward and backward by hand. For proper decortications, the shoes, which are mounted on oscillating arm need to be adjusted. Working capacity of decorticator is 60-65 kg of pods decorticated/h. Broken kernels are 1-2%; total kernel losses, 1-2% and shelling efficiency, 98.0%. The groundnut decorticator is either manually driven or power driven. The power driven decorticator consists of two main parts for shelling the groundnut and crushing plates and grates. The material is fed through the hopper provided on the machine, which moves to shelling unit. The grain like structure provided at an angle in the shelling parts breaks the shells. There after nuts and shells fall on the pan. Shells are blown out with the help of blower and clean nuts are collected separately.

Sunflower thresher

It consists of a threshing cylinder, concave, casing fitted with louvers, cleaning system, feeding hopper and frame (Fig. 7.64a); Garg and Singh (2002); Anonymous (2008). The cylinder concave clearance is 40 mm and is uniform throughout its length. The diameter of cylinder is 65 cm and length 150 cm. The first part of cylinder of length 133 cm has flat bars for crop threshing and the 2nd portion of length 17-cm has straw throwing blades. The cylinder casing is of hexagonal shape and is fitted with 7 louvers. The louvers help the crop to move axially and the crop is rotated three and half times for complete separation of grains. The cleaning system has a blower and two sieves. The opening of top sieve is 16 mm and of lower sieve 6 mm. A tractor or 7.5 hp motor can operate machine. It is a used for separating seeds from sunflower heads. It works on axial flow principle and suitable for threshing sunflower crop. Recommended cylinder and blower speeds are 300-350 rpm and 1200-1400 rpm respectively. It can thresh the harvested crop directly at high moisture content (Fig. 7.64b). Threshing efficiency of the machine is 100 percent whereas cleaning efficiency is about 90 percent. Grain losses vary from 0.65-2.94 % depending upon the moisture content. Capacity of the machine is 600-900 kg/ha of clean grain. There is a saving of 70-80 percent of labour, saving in time 70- 80% and saving in cost, 40-50%; besides reduction in drudgery and reduction in losses.

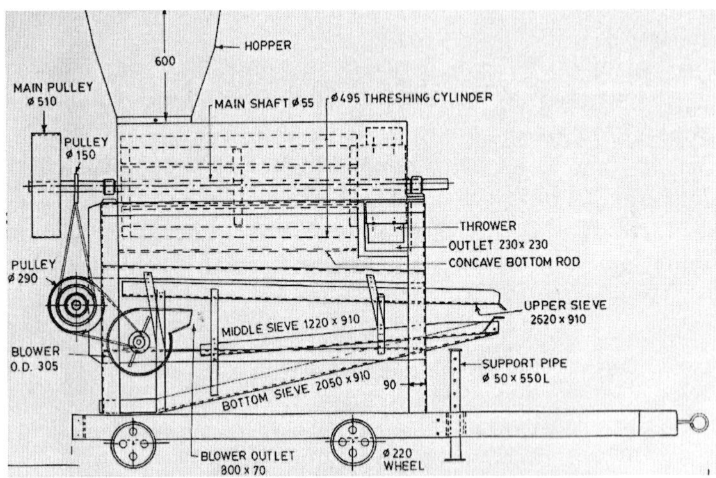

a) Side view of sunflower thresher (Dimensions in mm)

b) Sunflower thresher during operation

Fig. 7.64 : Tractor operated sunflower thresher

Multi-crop thresher

Pulses are important and integral ingredients of rich protein diet for human being. However, availability of pulses per person in the country is dropping and prices are sky-rocketing. One of the main hurdles in promotion and adoption of pulse crops by farmers is lack of mechanization of various farming operations. At present farmers thresh it manually, tread the crop below tractor tyres or use stationery combine harvester for threshing of harvested crop. Another option is to use spike-tooth type multi-crop thresher for threshing by making minor adjustments viz. reducing number of spikes on threshing cylinder and arranging them axially on cylinder, increasing concave clearance and reducing peripheral speed. This eliminates the need for a special thresher for pulse crops thus saves cost. Commercially available spike tooth type thresher has been modified (Fig. 7.65); Anonymous (2008, 2010, 2010b).

Fig. 7.65 : Spike-tooth multi-crop thresher during operation on Arhar & Masoor crops

The threshing cylinder has thirty six spikes placed six in each row. For threshing pulses six spikes are retained on cylinder in 6 rows i.e. 1 in each row. These spikes are 16 mm studs. The arrangement of spikes on cylinder periphery is axial. Aspirator I & II are centrifugal type and having 4 & 3 blades respectively. Aspirator I separate the major chunk of chaff from sieve falling directly below the concave. Aspirator II accomplishes the stage II cleaning. It picks chaff from screen just prior to discharge of grain from main outlet. Reciprocating sieve system consists of replaceable set of sieve and screen mounted on oscillating frame. The amplitude of oscillations could also be varied. Upper sieve separates chaff bigger than grain while lower screen let the dust, smaller particles than grain pass through thus help in delivering clean grain at main outlet. The thresher is provided with four wheels made of cast iron for transportation and motor stand to fit the motor on it. It has provision for attaching universal shaft for operating the thresher by tractor PTO. It also has provision to hitch it with three point linkage of tractor. The thresher can also be operated by a 10 hp 3-phase electric motor. The crop is dried to suitable moisture content before threshing. Mash is a hardy crop could sustain 720 rpm (~22 m/s). To attain threshing efficiency of 99.5% concave with bar to bar spacing of 5 mm had to be used. The output of thresher is higher for mash touching 250 kg/h compared to 100-150 kg/h for other pulse crops. Sieve overflow is more and even after re-feeding few very small pods remains un-threshed. The machine saves labour and cost of operation as compared to traditional method. The bar to bar spacing of concave used for pea is about 12 mm. The concave clearance is kept 25 mm. The moisture content of the crop and grains may vary from 13.2 to 16.14% and 19.2 to 40.0 percent on wet basis respectively. The percent grain loss ranges from 0.026 to 0.080 and the percent grain damage from 1.2 to 3.24. The threshing efficiency is 98-100% and cleaning efficiency 96-98%.

Generally 36 spikes are provided on threshing drum for threshing wheat. However only six spikes are provided on the threshing cylinder in 6 rows i.e. 1 in each row for threshing **Arhar** (Fig. 7.66). The arrangement of spikes on cylinder periphery is helical. Similarly, concave clearance is increased from 15 to 25 mm by reducing length of spikes. The machine with similar settings of threshing cylinder that of Arhar is used for threshing **Masoor** and broken percentage is reduced below 2%. The bar to bar spacing of concave used is 8 mm. The concave clearance is kept 25 mm. The peripheral speed is 20 m/s, feed rate 10.5 q/h, grain damage 0.57%, non-collectable losses 0.4% and cleaning efficiency 99%. Similarly, only 6 spikes are provided on the threshing cylinder in 6 rows i.e. only one spike in each row for threshing **Mash and Mungbean**. The arrangement of spikes on cylinder periphery is helical. One spike is mounted in each row on the cylinder periphery and arranged helically because mash is a pulse crop and it requires less intensive threshing than the wheat crop for which the thresher developed. Only 15 spikes are provided on the threshing cylinder in 6 rows i.e. 2 & 3 alternatively in each row for threshing **radish**. The arrangement of spikes on cylinder periphery is helical. Similarly, concave clearance is increased from 15 to 25 mm by reducing length of spikes. For threshing of **Raya** the cylinder concave clearance is kept 25 mm and number of spikes reduced to 12 (two/row in helical pattern) instead of 36 spikes used in wheat thresher. Sieve size is kept as 7.5 mm. Only six spikes are provided on the threshing cylinder in 6 rows i.e. 1 in each row for threshing **lentil**. The arrangement of spikes on cylinder periphery is helical. Similarly, concave clearance is increased from 15 to 25 mm by reducing length of spikes. At this combination the grain damage is 0.98%, non-collectable losses 0.60%, cleaning efficiency 99% and threshing efficiency 99%.

Fig. 7.66 : A view of arhar, red gram and pea thresher during operation

High capacity multi crop thresher

High capacity multi crop thresher is used for threshing of wheat, gram, mustard, green gram, urad, guar, lentil etc. It is operated by a PTO of 26.1 kW tractors (Anonymous, 2008, 2010, 2010b, 2010c). The thresher consists of a feed tray, spiked cylinder, straw thrower, blower and cleaning sieves (Fig. 7.67). The threshing cylinder is fitted with rectangular shaped flats as beaters, welded on eight bars in staggered fashion. The concave is made out of square bars with fixed opening and clearance. Three concave grates are provided for threshing various crops. A semi-hexagonal top cover with spiral louvers is provided for threshing paddy. The feed tray, hinged for folding during transport, is of sufficient size to hold crop material for continuous feeding. Crop is fed at the feed opening and is taken inside the machine by the cylinder which rotates on ball bearings. The beaters on the threshing cylinder hit the material, separating the grain from the straw and at the same time accelerating them around the cylinder. The spiral louvers in the top cover move the material axially from the feed end to discharge end. For threshing crops other than paddy, the semi-hexagonal top cover is to be replaced by a semi-circular cover and a semi-circular disc to be inserted in between cylinder and straw thrower. Threshed material is passed through the openings between the concave bars and falls on the upper oscillating screen. An aspirator blower is mounted behind the cylinder with two suction openings, one at the separating chamber and another at main grain outlet. The blower sucks out the lighter material and blows out of the machine from blower outlet. Cleaning sieves are hinged at the bottom of cylinder concave on adjustable hangers. Further separation of grain from straw mixtures takes place due to oscillating motion of shaker assembly, the grains fall through the holes to top sieve over bottom sieve and heavier straw and un-threshed tailings overflow from top sieve. Three screens are provided for threshing various crops. Final cleaning is done by the secondary inlet of blower and thus clean grain is obtained at the main grain outlet. Weight of the machine is in the range of 600-800 kg. Working capacity is 15-20 q/h for wheat, 3-4 q/h for raya; 4-5 q/h for gram, moong, urd, arhar; 5-6 q/h for gaur.

Fig. 7.67 : High capacity multi crop thresher threshing different crops

High capacity pigeonpea thresher

Physical nature of the pigeonpea stalk is different than other pulse and grain crops. Threshing of the crop is mainly done by beating with sticks since the stalks of the pigeonpea crop is valuable by-product for farmers. There is no thresher in which threshed stalk could be saved for domestic use. The machine therefore has been developed to thresh the pigeon pea and saves the stalk intact. A high capacity pigeonpea thresher (Fig. 7.68) has been developed powered by 7.5 kW (10 hp) electric motor (Anonymous, 2010b). The design consists of an automatic feeding mechanism. Crop bundles are fed one by one from feed end of the cylinder through chain conveyor and the bundles are discharged through other end after threshing. The triangular spikes are arranged in 15 rows comprising five sets of helix so that crop is released after one and half rotation. The cylinder concave is made of 15x15mm wire mesh supported on a separate concave made of 6 mm square bar covering 120 degree of cylinder. The upper concave covers 170 degrees of cylinder. Two aspirator blowers are provided with blower speed within 950-980 rpm. Threshing efficiency is found as 96%.

Fig. 7.68 : High capacity Pigeonpea thresher

Multi crop thresher for seed spices

The traditional practice of threshing of seed spices is either by beating on drums or wooden logs or treading by tractor wheels on mud floors. This method requires lots of labour and lot of dust and stone comes in during the threshing process which are removed manually by hand sieves or cleaning machine in factories reducing the quality of seed spices thus it necessitates the mechanization of seed spices threshing (Anonymous, 2008 and 2008d). Therefore, a tractor/electric motor operated thresher has been developed at MPUAT Udaipur in collaboration with M/S Makewell Industries, Unjha (Gujarat). The threshing cylinder has flat steel spikes with size of 30 x 120 x 6 mm (Fig. 7.69). The throat opening of thresher is kept as 580 x 240 mm. The diameter of threshing cylinder is kept as 660 mm and length as 550 mm. The thresher is mounted on rigid frame with covered power transmission system for threshing cylinder, three blowers and sieve shaking mechanism. The rpm of threshing cylinder is kept as 540, blower 480 and drive pulley 440. The drive to power transmission is given by telescoping shaft connected to PTO of tractor or by electric motor of 7.5 hp mounted on frame. Output of thresher is 240 -260 kg/h with threshing efficiency, 100%; cleaning efficiency, 99%; saving in cost over manual threshing, 55%; and saving in time over manual threshing, 72%.

Fig. 7.69 : Operation of seed spices thresher for cumin threshing

Castor sheller

Castor oil is an important export product used by pharmaceutical industry and cosmetic industry. Traditionally this crop is being shelled manually after drying in sunlight which is very labour intensive and low output method and need winnowing after shelling (Anonymous, 2010 and 2010d). Small hand operated castor sheller are also available but suitable to very small group of farmers and more labour requirement for shelling and then corresponding winnowing operation. Looking to large area of cultivation a tractor operated castor sheller is found more feasible proposition for timely shelling and marketing of castor oil seeds. Therefore, a castor sheller has been developed at MPUAT Udaipur in collaboration with M/S Ganesh Raj Industries,

Mehsana (Gujarat). It consists of feeding hopper, beater shelling unit, concave, sieve, blower and PTO coupling unit (Fig. 7.70). Three people are required to operate this machine. The threshing efficiency is 85.2% and cleaning efficiency 97.6%. The broken grains are in the range of 6-7%. The output capacity is 1040 kg/h. The shelling efficiency of sheller is low due to wide variation in size of fruit which causes problems in adjustment of clearance between rotor and concave. The unshelled fruits are collected at outlet in basket and again fed in hopper for full threshing. The broken grains are only marginal. The husk grain ratio of castor fruit is 1:4.

Fig. 7.70 : Castor sheller operated on farmer's field
Courtesy: M/S Ganesh Raj Industries, Mehsana (Gujarat)

7.3 Harvesting and threshing equipment for maize

Maize is an important crop grown for grain, fodder and silage. In India it was introduced in the seventeenth century and the seed was brought from USA (Mittal, 2001). Harvesting and shelling methods depend upon the final use of the crop. In case it is used as fodder/silage, it is harvested at the time of appearance of silky hairs and plant-stalks are green. In case the crop is used as grain crop, it is allowed to ripen till the cobs become yellowish and hard. Then these are cut from the plant stalks and dried in the sun. The grains are threshed from the cobs by beating them with sticks and cleaned and manually removed. This traditional method is followed by a majority of farmers throughout India. However medium and large category farmers have started using modern innovations and methods. The tools/equipment are being used for maize harvesting and shelling in the light of advancement that has taken place in agricultural technology.

Sickle harvesting of maize is still the most prevalent method of harvesting. Shelling is done manually by beating. It consumes 120 to 140 man-h at an average yield of 30-40 q/h. Maize shellers both manually as well as power operated were introduced in the early 60's. Power operated shellers have become popular because of their high shelling capacity, low cost and machine simplicity. Power operated sheller however, requires maize crop to be dehusked. Dehusking itself is very labour intensive operation. The progressive farmers who cultivate the crop on a large scale need efficient dehusking and shelling equipment. The combine owning farmers can use their combines for threshing of maize with husk. The adjustments for maize shelling needed are change of cylinder speed, concave clearance and adjustment of chaffer. The tractor-operated machine can thrash about 20 q/h of clean grain. The labour involved in dehusking is completely saved by the use of this machine.

Maize sheller

Maize Sheller is used to separate grain from cobs. Before shelling, the foliage is removed manually. Maize shellers are either manually operated or power operated. A tubular maize sheller removes grains from cobs, about 5 kg of grain can be removed in one hour. Octagonal hand maize sheller (*fin type*) is used for shelling of dehusked maize cobs, especially for seed purposes. It is manually operated (Fig. 7.71). It is simple device to remove maize grains from the dehusked cobs (Pandey *et al.*, 1997). The sheller consists of 4 mild steel fins tapered along their length, one edge of the fin is taper. In each fin, two holes are provided for riveting. Each fin is bent at two places in a manner for assembling in octagonal shape. The corners of the fins are rounded in order to avoid injury to the operator during shelling operation. The fins are joined together with rivets. The assembled sheller has thus four tapered projections inside the sheller body that accomplishes removal of the grain from the maize cob. In order to avoid corrosion, the sheller is powder coated which also increases its working life. For operation, the sheller is held in left hand and the dehusked maize cob in right hand (for right hand person). The cob is inserted in the sheller and is given forward and backward twist or given clockwise and anticlockwise strokes repeatedly. The tapered edges of the fins dig into the space between the rows of the grains in the cob and with the forward or backward stroke the grains are released. After grains are separated from one end of cob, the other end is inserted in the sheller to complete the removal of grains from cob. Due to the taper edges of the fins, which are projected inside the sheller body, one end of the sheller has larger opening and the other smaller. Therefore, for shelling the larger end of the cob it is inserted in the larger opening of the sheller and the smaller opening of the sheller is used for smaller end of the cob. Working capacity of maize sheller is 15 to 20 kg/h and weight of the machine is 0.22 kg.

Metal maize shellers are available in two types of models i.e. RAU Model and CIAE-Model designed and developed at Rajendra Agricultural University, Pusa and Central Institute of Agric. Engg. Bhopal respectively. The RAU model of hand operated maize sheller is made of 20 gauge G.I. sheet of 20 cm x 10 cm size. The sheet is folded into a cylindrical shape of 6-cm diameter and 14 cm length. The joining ends are fixed together with the help of two rivets. Four rows of 'V' shaped notches with four notches in each row are cut at both ends of pipe (Fig. 7.71). All the notches are inverted inside the pipe at an angle of 90 degree. The lengths of notches are respectively 12.5, 15.0, 17.0 and 19.0 mm for the first, second, third and fourth row respectively. The differences in the length of the notches make the device suitable for various sizes of maize cobs. CIAE model of hand operated metal maize sheller is made from a short length of 62.5 mm dia MS pipe of 16 SWG (Fig. 7.71). Four fins are made from an 18-gauge MS sheet and bent at 70°. These fins are tapered and fitted inside the tube. The tapered ends provide two variable openings - 39 mm at one side and 26.5 mm to the other side. Different sizes of cobs can be handled by this arrangement.

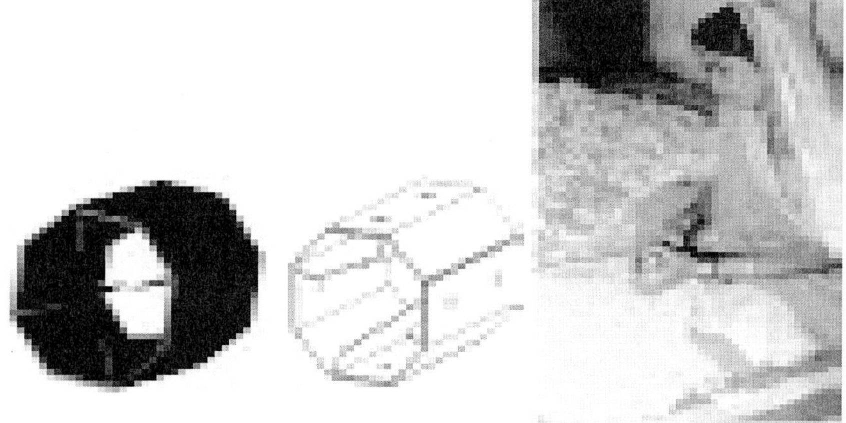

Fig. 7.71 : Octagonal hand maize sheller (Fin type)

Hand operated maize sheller

The maize sheller is operated manually by a driving handle, which rotates the flywheel (Fig. 7.72); Pandey *et al.*, 1997. The flywheel in turn rotates the driven pulley on the shaft on which is mounted the axial flow cylinder. The threshing mechanism separates the grain from stalk or cob through revolving cylinder and concave. The grain is rubbed against drum and is shelled as the cob moves to thresher end. Adjustments are provided for varying concave clearance. Separate outlets for grain and cobs are provided. The capacity varies from 30 to 40 kg of kernels per hour. The domestic sheller is a small hand operated tool suitable for shelling small quantities of maize. It can easily be mounted on a box. The feed opening is provided with spring to take different sizes of cobs. The picker gear, after shelling, directs the cobs to the outlet.

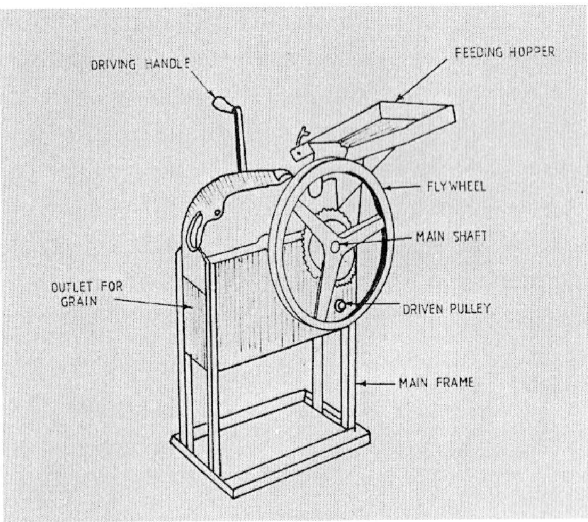

Fig. 7.72 : Hand operated maize sheller

Hand operated maize dehusker-sheller

The axial-flow maize dehusker-sheller consists of threshing cylinder (4 beaters with 3 solid lugs on each beater as threshing elements), louvers, main shaft, bottom sieve with equi-spaced square MS bars, top and bottom cylinder cover, hopper, trapezium shaped grain outlet, husk outlet, trapezium shaped frame, chain-sprocket system for power transmission etc (Fig. 7.73); Singh *et al.* (2011 & 2012); Singh (2010). The concave clearance is 35 mm and the diameter of cylinder 380 mm. The cylinder fitted in the maize-dehusker-sheller is operated by farm women in standing posture for dehusking-shelling the un-dehusked maize cobs (Fig. 7.73). The average output is 60 kg/h and variation in the range of 48 to 72 kg/h. The variation in output might be due to grain/un-spent cob ratio. The dehusking efficiency is

100 per cent but shelling efficiency 98 per cent. The grain breakage is 0.16 to 1.46 per cent. Some of the parameters can be calculated as below:

Moisture content: The moisture content on dry basis (M), % is calculated using formula

$$M = \frac{\text{Weight of wet sample including leaves} - \text{Weight of dry sample including leaves at } 103^{0}\text{C for 72 h}}{\text{Weight of dry sample including leaves}} \times 100$$

Maize grain/ cob ratio: After determining the weight of the dry samples as received from oven for moisture content, the cobs and maize grains are manually separated and weighed. The grains/ cob including outer sheath (leaves) ratio (K) is calculated as,

$$K = \frac{\text{Wet of dry maize grain}}{\text{Wet of dry kernels and leaves}}$$

Dehusking efficiency: Un-dehusked maize cobs received from outlet end or in cylinder after termination of threshing is weighed and dehusking efficiency is calculated as,

$$\text{Dehusking efficiency (\%)} = \frac{\text{Weight of un-dehusked cob received from outlet of machine, kg}}{\text{Total un-dehusked cob fed, kg}} \times 100$$

Shelling efficiency: The grain remained in the dehusked cobs received from outlet end or in cylinder after termination of threshing is weighed and shelling efficiency is calculated for cob-wise and grain-wise as,

$$\text{Shelling efficiency}_{(\text{cob})} = \frac{\text{Weight of dehusked cob having grain from outlet of machine, kg}}{\text{Total un-dehusked cob fed, kg}} \times 100$$

$$\text{Shelling efficiency}_{(\text{grain})} = \frac{\text{Weight of grain from un-shelled dehusked cob, kg}}{\text{Total grain received from fed un-dehusked cob, kg}} \times 100$$

Damaged grains: Damaged grains received from total grain received (grain outlet, cylinder and dehusked cob outlet) and calculated as,

$$\text{Total damaged grains (\%)} = \frac{\text{Weight of shelled damaged grains}}{\text{Weight of total grain}} \times 100$$

Throughput capacity: The ratio of grain obtained during dehusking-shelling of un-dehusked maize cobs in total time of run of machine and throughput capacity of machine is calculated as,

$$\text{Throughput capacity, kg/h} = \frac{\text{Weight of grain obtained, kg}}{\text{Total time of run of machine, min}} \times 100$$

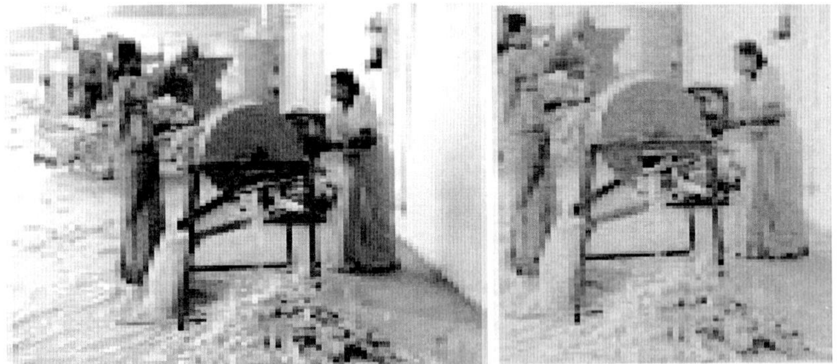

Fig. 7.73 : Farm woman operating the maize dehusker-sheller

Power operated maize shellers

The functionary principles are similar to those of hand operated maize sheller. The additional adjustments are provided for varying concave clearance and cylinder speed. A maize sheller, who requires 3 to 5 hp, usually has drum diameter of 30-36 cm, drum length of 80-100 cm with a cylinder speed of 500 to 600 rpm. On the periphery of the cylinder, there are pegs that remove the grain from cobs using axial flow movement. The cylinder speed is maintained in between 500-600 rpm. The cob moves toward the end of sheller from feeding side and during this process grains are rubbed against drum and posses through the concave. Blower is provided to remove lighter material. Concave clearance and cylinder speed can vary and adjusted as per recommendation. There are two types of maize shellers i.e. Spring type and cylinder type.

Spring type sheller: In the spring type sheller, the cobs are fed into an opening leading to a rotating fluted cylinder, a rotating disc and a spring pressure plate. The kernels are removed from the cobs as they move in between the cylinder and the disc. It may have a blower situated at the bottom to deliver clean grain. There are various sizes of this type available in the market e.g. Single hole sheller and Double hole sheller. The single hole sheller is generally provided with a ratchet type hand crank to avoid the danger of injury to the operator. A fan is provided to clean the grain from dirt and foreign material. Power operated maize shellers of this type have capacity of 220 kg/h. The double hole sheller is similar in construction as the single hole sheller except that rotating disc in this case has teeth on both sides. The cobs being fed from the two holes are shelled simultaneously. The feed openings are provided with spring plates to take different sizes of cobs. The double hole sheller when operated through a 1.5 to 2.0 hp motor gives an output of 400-450 kg/h.

Cylinder type sheller: This type of sheller consists of a cylinder with lugs, concave assembly and a blower unit. The lugs on the cylinder are placed spirally to advance the cob through shelling mechanism. The lugs are assisted by two spiral ribs to accelerate the movement. The cobs are fed between the concave and the cylinder and kernels are removed through the action of lugs. The blower with a strong blast of air removes small pieces of cobs and other material from the kernels. Cylinder type shellers are available with different sizes operated by 5 to 10 hp with output range of 8 to 18 q/h.

Power operated maize shellers are used for shelling of dehusked maize cobs and the chaff is removed by winnowing. It is operated by a 1.5 to 2.24 kW electric motor (Fig. 7.74). It consists of a threshing cylinder, concave and centrifugal blower mounted on a frame (Fig. 7.75); Pandey *et al.* (1997). Crop feeding is manual. The threshing cylinder is of spike tooth type. Round bars are used as spikes, which are fitted on circular rings. The head comes out through the opening at the far end of threshing drum. A blower is used for cleaning the grains. Working capacity of sheller is 200 kg/h at feed rate of 250 kg/h. Grain losses are less than 1%; cleaning efficiency more than 99%; threshing efficiency more than 99%; and grain crack age, 0.5-1.0%.

Fig. 7.74 : Power operated maize sheller

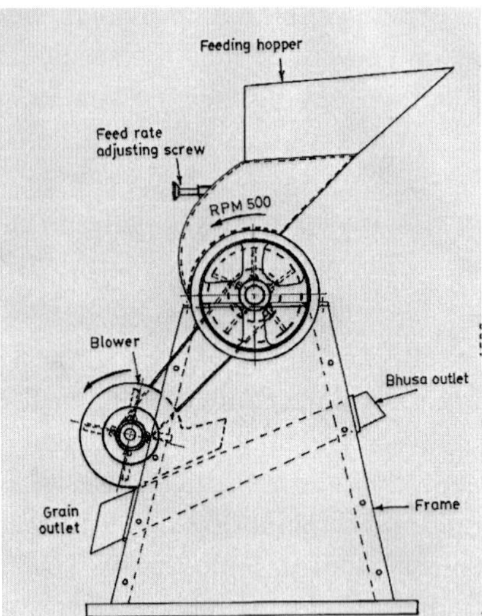

Fig. 7.75 : Components of power operated maize sheller

Maize dehusker-sheller

The Punjab Agricultural University, Ludhiana has developed two types of maize dehusker-cum-threshers namely spike tooth type (modified version of wheat thresher) and axial flow type (modified version of sunflower thresher) for threshing the maize along with the husk (Anonymous, 2008, 2010, 2010b). The capacity of the axial flow type machine varies from 1500 to 2500 kg/h at different moisture content and that of peg type 450-650 kg/h. Shelling efficiency is about 100 percent in both the cases and broken grains are maximum up to 2.0 percent. De-husking cum shelling saves lot of labour in comparison to traditional system. It is used for de-husking and shelling of maize cobs simultaneously. It is operated by a PTO of 26.1 kW tractors (Fig. 7.76a). In the spike tooth type sheller, pegs are staggered at varying heights for better shelling efficiency. The spikes are placed in 6 rows with 6 pikes in each row. The sieves have 1.25 cm diameter opening to separate the shelled maize from husk. In axial flow type threshers, pegs are provided on the cylinder and louvers are provided on the upper periphery of the drum to convey the crop to the outlet. Losses take place at various stages of de-husking and shelling of maize as shown in Fig. 7.77. Working capacity of sheller is 15-20 q/h. At moisture content of grain from 11.8 - 24.8%; cylinder speed, 700 – 750 rpm; threshing efficiency is 96 - 98%; cleaning efficiency 94 - 99%; broken grain 1.0 - 3.5%; sieve loss 0.40 - 0.76%; blown kernel 0.25 - 0.66%, and total grain loss 0.65 - 1.42%. High capacity maize dehusker-cum-sheller with axial flow technology and conveyor system (Fig. 7.76b) for feed and grain outlet has output capacity of 40-100 q/h, threshing efficiency 98-99% and cleaning efficiency 95-100%.

a) Maize dehusker-sheller during operation

b) High capacity maize dehusker-sheller during operation

Fig. 7.76 : Tractor operated maize dehusker-sheller

Courtesy : South East Farm Equipments Pvt. Ltd., Salem (Tamil Nadu)

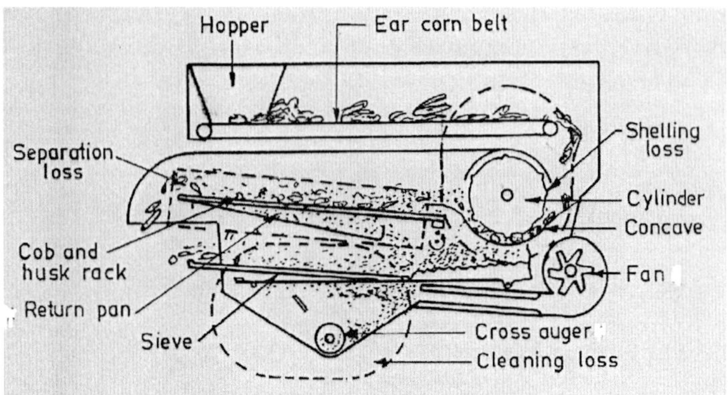

Fig. 7.77 : Losses at various stages of de-husking and shelling of maize

Whole crop maize thresher

Maize is a very predominant crop of southern Rajasthan and grown by farmers of all category of land holdings. It is also grown in rabi season though it is basically a kharif crop. Maize is also a crop for establishment of agro industries because many agro industrial products can be prepared from maize. The harvesting and threshing of maize is time and labour intensive operation spread from Sept to November every year. Since the land holding pattern in this region is mostly from small to medium category of farmers and mechanisation is of mixed pattern type mainly animal operated plus tractor operated system. The regular occurrence of draught in the region creates shortage of fodder requirement for cattle's. During scarcity the fodders are supplied by neighbouring states like Punjab and Haryana through trains & trucks. In some areas of southern Rajasthan farmers have started threshing of whole plant of maize in wheat threshers which are commonly available. These results in breakage of grain from 25-30 per cent reducing the quality of grain produced and also the output of grain from thresher is reduced substantially in the range of 1.5 to 2 q/h. But the practice is followed to fulfil the requirement of chaff (bhusa). In view of this there was a need to develop thresher for threshing of whole plant maize without the breakage of grains so that the farmer's demand of grain and chaff from maize crop can be met out. This will save the time & labour of female workers mainly employed for dehusking and transport operation.

The whole plant maize thresher (Fig. 7.78) has been developed at MPUAT Udaipur operated by electricity and tractor with modification in existing multi crop thresher (Anonymous, 2008, 2008d, 2010, 2010d). The developed thresher gives satisfactory performance for threshing of maize crop and stalks reduced to chaff in the size of 20 -50 mm size. Broken grain is in the range of 1.5-2.0 percent with output of 210 and 650 kg of grain per hour in motor and tractor operated thresher respectively. Substantial saving on monetary and labour aspect has been observed. A tractor operated multi crop thresher was modified with arrangement that spikes on threshing cylinder are made of 8 mm square bar with 19 mm spacing. The output of grain was observed as 710 kg/h with chaff size of 16 to 63 mm. This chaff fed to the animals and 85% material is consumed. The threshing efficiency is 99% and cleaning efficiency 96.4%.

Fig. 7.78 : The whole crop maize thresher during operation

Maize/Corn harvesting machines

The first picker-sheller was developed in middle 1930s. Then attempts were made to adapt the grain combine for harvesting the corn which includes the feeding the entire corn plant into the machine. The early investigators found that corn can be shelled and cleaned with a combine but introduction of the stalks into the machine was the major problem (Sandhar and Goyal, 2001). In 1950s rasp bar cylinder was evaluated as the corn husker or the sheller, same problem of introduction of the stalks into the machine existed. The next step in the evolutionary process was to equip the combines with corn snapping units in order to leave the stalks in the field and introduce only the ears into the machine. Following the development of acceptable corn heads, the popularity of grain combines for harvesting corn increased rapidly during 1960s. In 1970s approximately 70% of the corn produced was harvested by combines equipped with corn heads.

Four kinds of corn harvesting machines are in use i.e. corn snapper, corn picker, corn picker sheller and corn combine. Corn snapper removes or snaps the ear from the standing stalk but does not remove husks. Corn picker also called snapper husker snaps the ear from standing stalk and removes the husks from the rear. Corn picker sheller snaps off the ear, removes the husks and shells the kernels from the cobs all in one operation. A picker sheller attachment is available for the conventional corn picker. Corn combines have two, three, four, six or eight row heads as attachments. These attachments replace the grain cutter bar. The corn heads snap the ears from the standing stalks and feed the snapped ears into combines. Combine cylinder shells the kernels from the cobs and the separate in cleaning units of combine as they do in harvesting small grain. Standard corn heads harvest rows from 90 to 110 cm width. The 50 cm rows corn heads are available to harvest 4, 6 or 8 rows at a time.

Both mounted and pull type pickers and snappers are available. Mounted units are mostly 2-row machines, whereas 1-row, 2-row and 3-row sizes are common with pull type machines. Picker-shellers are usually two row machines. Most 2-row machines have snapping-unit centre distances of about 102 cm, but a narrow row models for 76 cm row spacing are also available. Corn heads for combines are mostly 2, 3, 4 or 6 units designed to handle row spacing of 92 to 102 cm or 71 to 81 cm. Five row and 7 row heads for these row spacing, as well as 8-row heads for 51 cm or 76 cm spacing are also available.

Self-propelled maize combine harvester

This machine consists of 101 hp 6 cylinder engine, heavy duty cutter bar, rasp bar type threshing drum with 8 bars, cleaning and conveying system (Fig. 7.79). Cutting height can be adjusted with minimum height 20 cm and maximum 80 cm. Width of cut is 3.65 m. It has 6 rows with row spacing varying from 460 to 610 mm. Diameter of threshing drum is 60 cm and length 126 cm. Rotational speed of threshing drum can be varied from 535 to 1210 rpm. It can cover 1 ha/h.

Fig. 7.79 : Self-propelled maize combine harvester

Courtesy: Kartar Agro Industries Pvt. Ltd. Bhadson (Punjab)

Shelling and cleaning in a grain combine: The conventional rasp bar cylinder in a grain combine is an effective device for shelling corn. The clearance between the cylinder and the concave is adjusted to about 25 to 32 mm at the font and 16 to 22 mm at rear. Cylinder peripheral speeds range from 12.7 to 20.3 m/s, often being about 15.2 m/s. Filler plates may be attached to the cylinder between the rasp bars to prevent ears from collecting within the cylinder. Separation of the shelled corn from the cobs, husks, and other trash is a relatively simple matter in a combine, as compared with separating small grains and other seed crops from straw and chaff. Except under extremely adverse crop or weather conditions, separating and cleaning capacity seldom is a limiting factor when harvesting corn with a combine. The chaffer and cleaning sieves are adjusted to larger openings than for the cereal grains, and a greater air blast is permissible through the cleaning shoe. Aerodynamic suspension velocities for corn kernels are considerably greater than for pieces of corn cobs and other lighter debris.

Shelling and cleaning in a picker-sheller: Although at least one model of picker-sheller has had a rasp-bar cylinder and concave as the shelling unit, most picker-shellers have axial-flow cage type shellers. A cage-type sheller has a rotating cylinder with lugs, spiral flutes, or paddles around the periphery. The cylinder operates within a stationary cage that is 1020 to 1420 mm long and 280 to 380 mm in diameter. The cage periphery is made of perforated metal or parallel round bars and has openings large enough to permit easy passage of shelled kernels but not the cobs. The cylinder is operated at 700 to 800 rpm and has a peripheral speed of 6.1 to 10.2 m/s. Ears from snapping units are fed radially into an opening at one end of the cage and pass circumferentially and longitudinally along the cylinder during the shelling process. Shelling is accomplished primarily by rubbing and rolling of ears and cobs against each other and against the cage (concave) and the rotating cylinder. The length of time that the ears remain in the shelling unit can be varied, by means of an adjustable cob gate at the discharge end of the cage, to accommodate different crop conditions. The shelled corn falls through the openings in the cage and is conveyed to the tank or wagon. Light trash is removed from the shelled corn by means of an air blast or with a suction-type fan. On some shelling units the material discharged from cage passes over a cleaning shoe for final separation of shelled corn from the cobs, husks, and other trash.

Functional components of corn harvester: The basic components for all types of maize harvester include a gathering unit to guide the stalks into the machine, snapping rolls to remove the ears form the stalks, and lugged gathering chains to assist in feeding the stalks into rolls and moving the stalks and snapped ears rearward through the snapping zone. Different units of maize harvester are gathering unit, snapping unit, trash removal, husking unit, and cleaning unit.

Gathering units: It is relatively simple matter for gathering device to control stalks and feed them into the snapping units when the stalks are standing. Unfortunately, however, many stalks become lodged (leaning or broken) during the season, due to adverse weather conditions, and high plant populations. The gathering assembly must be able to lift lodged stalks and guide them into the snapping unit with a minimum number of ears lost during the process. This requires having pick up devices close to the ground and handling the stalks gently to avoid excessive accelerations and consequent detachment of ears. Main parts of this unit are the snouts. These are hinged and can be set to float along the ground to slip under down stalks and then lift them into the gathering chains. Steel shoes beneath the snouts prevent them from digging into the ground.

Snapping units: Two general types of snapping rolls are employed. These may be described as spiral-ribbed or spiral-lugged rolls and straight-fluted rolls. Both types have tapered, spiral-ribbed points to facilitate stalk entry, and pull the stalks downward between the two rolls. With spiral-ribbed rolls, the ears are snapped off of the stalk when the ears contact the closely spaced rolls. Straight-fluted rolls pull the stalks down between two stripper plates and the ears are snapped off when they contact the plates. Ears are snapped from the stalks by a pair of long, closely spaced rolls. Because of their close spacing with upward plants and rotation towards each other, the rolls pull the stalks through quickly. The close spacing does not allow the ears to pass through instead they are passed on to husking rolls. Coarse trash and leaves are removed by ejectors and to husking rolls.

Trash removal: Special trash rolls are often provided on corn pickers to remove trash and broken stalks not expelled by spiral-ribbed snapping rolls. Fluted sections may be incorporated on the upper ends of the snapping rolls. Transverse, fluted rolls are sometimes mounted at the discharge end of a snapper-ear conveyor. The principal function of the teraservers rolls is to remove relatively long sections of stalk. Small blowers may also be used to assist in removing trash.

Husking units: The husking unit on a picker has pairs of rolls that grasp the husks and pull them downward between the rolls. Usually, all the husks are taken at once when one is caught. Most husking beds have either 2 to 3 pairs of rolls per corn row. The rolls may be in a separate husking bed at the discharge end of a conveyor from the snapping rolls and fed by snapping roll conveyor. With separate husking beds, there may be one per row or one per machine. Husking rolls are generally 64 to 76 mm in diameter and 760 to 1270 mm long. They operate at about 500 rpm. Husking units are available in two different types viz. continuous or combination rolls having the husking rolls as continuation of snapping rolls and separate husking unit has separate driven husking rolls independent of in snapping rolls. In the both types one or more pairs of husking rolls are provided for each row of corn. Various roll designs are used in different models.

Cleaning unit: Mechanical trash ejector and air blast accomplish the cleaning. Ejecting rolls and augers remove trash and husks from the husking rolls and discharge it to the ground. The air blast from the fan strikes the ears as they leave the husking rolls. The strength of the air can be adjusted as need be, to blow of loose husks, leaves and trash etc.

Shelled corn saver: Some kernels may be shelled from the cob by the snapping rolls and some by the husking rolls. Kernels shelled off by snapping rolls fall to the ground and are lost, the kernels shelled off by the husking rolls are saved because they fall into the shelled corn-saver where they are cleaned and then elevated to the wagon. It consists of a screen and a chain career.

Shelling attachment: Some models of corn pickers have shelling attachments. Revolving lugs on a screw feed cylinder rub the ears against one another and against the round bars of the cylinder cage to give the shelling action.

Wagon elevator and hopper: The hopper at the rear end of the machine receives the husked ears and also the shelled corn from the corn saver. The elevator can be controlled/adjusted for loading and unloading the wagon.

7.4 Harvesting/Digging Equipment for Root Crops

Potato is an important food and vegetable crop, which is also used as an industrial raw material for making starch, alcohol etc (Verma, 2001; Verma *et al.*, 1992). Potato has been grown in India for centuries by traditional methods by using hand tools and animal drawn implements. Modern practices of production demand precision in the placement & utilization of costlier inputs like seed, fertilizer, pesticides etc. Mechanization has played a pivotal role in timeliness of farm operations, reduction of drudgery as well as in cutting down the labour and cost of production for the crop. Root crop harvesting is different from the grain harvesting of cereal crops. A mechanical harvester needs to perform digging, separation of loose soil and small clods and stones, removal of vines and weeds and at least partial separation of tubers from similar size stones and clods (Verma *et al.*, 1992; Verma, 2001; Singh, 2009). In root crop harvesting equipment, digging blade and rod chain conveyor performs all above said jobs. Harvesting of root crops involves digging, shaking to remove adhering soil, windrowing or stacking and picking. A good root crop harvester should give maximum recovery and cause minimum damage to pods or tubers.

Potato harvesting equipment

Manual harvesting of potato requires 600-700 man-h/ha and accounts for nearly one third of the total labour requirement for production of the crop in India. In hand harvesting the ridges are split with the help of a digging tool, viz. spade, hoe, fork or "Ramba" (Khurpa). The tubers are exposed and picked up by hand. Apart from being slow and tedious, manual-harvesting leads to sizeable percentage of cut and bruised tubers. Root crop harvester is a machine used for lifting root crops out of the ground. The root harvesting equipment can be classified in several different ways. According to the source of power: i) manual operated, viz. hand tools and digging aids such as spade, khurpa, digging and picking fork etc, ii) animal-drawn equipment such as potato-plow, animal drawn groundnut digger etc, iii) tractor operated equipment such as spinner-digger, elevator-digger, potato-digger-shaker-windrower, sugarbeet-harvester, carrot-harvester, radish-harvester etc, and iv) self-propelled machine such as potato-harvester/combine, sugarbeet-harvester etc. According to the mode of hitching: i) trailed type, ii) mounted type, iii) semi-mounted type, and iv) self-propelled type. According to the unit operations performed: i) simple digger, ii) digger-shaker, iii) digger –elevator-windrower, and iv) digger-shaker-conveyor/digger-shaker-windrower.

Design considerations for potato digger/harvester

Potato is grown on ridges or beds. The tubers lie buried under ground while the leaves and stems remain outside. The unit operations involved in potato harvesting are (a) cutting or destroying of the potato haulms before or during harvesting operations, (b) splitting or breaking of soil-potato-ridge and exposing the tubers with a view to collecting them easily and promptly, (c) cutting, lifting and/or conveying of the soil-potato mass with a view to discarding the soil, stems, etc, and collecting the tubers with minimum injuries. Alternatively, soil-potato mass may be scattered with a view to attaining maximum exposure of the tubers thereby enhancing the efficiency of picking of tubers and reducing the labour and cost requirements for the operation. In the latter case, the separation of the potatoes is done by manual picking of the exposed tubers and leaving the soil and clods, roots, etc., in the field itself. For designing a root crop harvesting equipment, the following points should be kept in view:

i) **Variety of the crop and its cultural practices:** In case the existing varieties and cultural practices pose some problems in mechanical harvesting attempt should be made to make necessary modifications.

ii) **Soil type:** The soil type and moisture content at the time of harvesting are important factors influencing the machine performance. Suitable soil-amending substances may help to develop the proper soil-texture to eliminate problem of excessive blasé clogging and clod-formation. The soil for root crops should be friable for easy separation.

iii) **Method of irrigation:** One must carefully consider the method of irrigation being followed. Depending upon where the crop is sown on ridges/beds/flat surface, proper method of irrigation should be selected. The availability of right moisture content at the time of harvesting is very essential for proper penetration of blade and efficient soil-fragmentation.

Therefore general design requirements for a potato harvesting equipment can be outlined as follows:

i) The digger should serve the desired purpose, namely, to split the ridge and expose the tubers or split the ridge and pick and lift the material and separate the soil in a single pass, or split the ridge, pick, lift and convey the soil-potato mass, discard the soil and other extraneous material in the process through manual, mechanical, electronic or any other means or by a combination of these all, also separate the tubers into different size groups and collect them in sacks or barrels or any accompanying truck or trailer.

ii) Depending upon the intended purpose, the draft and power requirements of the digger/harvester should be well within the capacity of the source of power to be employed for its operation.

iii) The shape of the digging shovel/share should be such that the draft is low and the flow of the soil-potato mass is smooth.

iv) The digger/harvester should be suitable for a wide range of soil, crop and field conditions.

v) It should be suitable for harvesting the tubers under the prevailing or modified agronomic practices. This is important. For instance, an animal drawn digger can be used without much problems even when the crop is planted at row spacing of 45 cm. However, the minimum row spacing for a tractor-operated digger is 60 cm or more to allow the tractor tyres to move freely in the furrows and not to ride over the ridges.

vi) The working principles and mechanisms of digger/harvester should be such that the potato skinning, cutting and bruising losses during splitting of the ridge and picking, transporting and separation of the material are the minimum.

vii) In a digger requiring manual picking of the tubers from the field, the exposure of tubers should be maximum so that the time and labour requirement for picking and collecting of the tubers gets minimized.

viii) For a potato digger/harvester, the percentage of unexposed/ un-dug/ un-lifted tubers should be minimum.

ix) The tubers, in a digger requiring manual picking from the field, should not be scattered over too wide a swath to reduce the picking time and labour requirements.

x) Depending upon power source and the type of the digger, it should dig one or two ridges at a time.

xi) The digger should be suitable for different potato varieties. This is important especially for countries like India where large-tuber varieties are not yet prevalent, unlike the western countries.

xii) It should have necessary provision to adjust and maintain the desired depth of digging to minimize tuber injury and un-dug tubers.

xiii) The initial cost of the digger/harvester should be affordable i.e. within the purchasing capacity of the farmer

xiv) The field adjustments should be simple and quick.

xv) The fast-wearing components requiring rebuilding/replacement should be the minimum.

xvi) The design should be simple and it should be possible to manufacture its components out of available materials and with available manufacturing technology and expertise.

xvii) The overall size of the implement should be such that it does not pose any problem in its transportation and turning while in operation.

xviii) The design should be such that it should require minimum attention by way of greasing, oiling etc for its day to day maintenance.

xix) The weight of the digger if it is to be mounted behind a tractor and its rearward projection should not be excessive to be within the permissible hydraulic lift capacity of the tractor.

Following are some of the general considerations for efficient utilization of the root crop harvesting equipment:

i) Proper field layout,

ii) Proper planting method,

iii) Proper moisture content at the harvest time,

iv) Pre-harvest treatment such as haulm-cutting, vine-killing etc,

v) Proper adjustment of the depth of the digging,

vi) Proper hitching of the machine,

vii) Proper adjustment of the speeds of the picking, conveying and shaking system,

viii) Selection of desired forward speed in keeping with the field and crop conditions,

ix) Number of workers in the crew to pick and collect the material simultaneously,

x) Matching of the transport and handling equipment to the capacity of the digger/ harvester,

xi) Proper harvesting, maintenance and repairs of the equipment,

xii) Conveying and separation system must be efficient to handle and separate the loose material with minimum injuries,

xiii) Avoid excessive height of fall of the roots/ tubers to lesson injuries. If required use inside rubber padding at the components to minimize external/internal injuries,

xiv) The speeds of the components used for the translocation and separation of the material should be such that these do not lead to excessive agitation, damage and loss of the products,

xv) The various controls for raising and lowering of the blade, varying speeds, fletching etc should be carefully designed. Provision should be made for adjusting the speed of the components in accordance with the field conditions, and

xvi) Use of multi-speed gear box and vary-speed drives is desirable on root harvesting machines.

Potato harvesting equipment

Since potato is a root crop generally grown in ridges of the size of 60 cm x 15 cm. The tubers when well matured are laying in the ridge covered while leaves stems etc are outside the ridge. For this crop harvesting and digging are two synonymous words. Our ultimate aim is to get tubers from the ridge free from all other undesirable products/materials. The various operations involved are:

i) **Splitting of the ridge:** Since tubers and soil are forming a compact mass, the separation or getting the tubers free soil is only possible if the ridge is split up first or the material both soil and tubes are loosened. Any device which gives impact or stir the ridge can be harvesting equipment.

ii) **Scattering/exposing of tuber:** Once the ridge is loosened, some short of translocation between soil clods/particles and tubers take place, but tubers cannot be taken out until and unless these are exposed. This can be done by scattering the mass over a surface so that tubers are exposed. In case of manually operated tools, this exposure is achieved just by playing with the soil and tubers but in case of animals and power operated implements the mass is scattered over a surface in such a way that exposure is maximum and labour requirement in picking up the exposed tubers is minimum. Tractor operated potato spinners' work on the principle of scattering the soil potato mass for getting the tubers exposed.

iii) **Separation of the product:** After exposing the tubers, these are separated from the soil mass. This may involve simply picking of tubers in case of manually and animal operated harvesting equipment. In other equipment, the soil clods rather than tubers is separated from the tubers. After declaring the soil potato mass, the material is picked up on an agitating conveyor, where soil gets separated and falls down as the material is moving over a conveyor. This object is basically employed in the power operated potato digger elevator.

In mechanical harvesting destroying of potato vines/haulm is an essential pre-harvest operation. It offers the advantages such as: i) destroying of haulm 10 to 12 days before harvesting helps in curing the tubers. In turn, it minimizes the skinning and bruising damage during harvesting; ii) transmission of virus through leaves and stems, under favourable conditions, is checked; and iii) soil-potato separation during harvesting is improved significantly. In advanced countries, where potato cultivation is completely mechanized both chemical and mechanical methods are used for destroying of potato haulm. In the chemical method, aero-cynamide dust, ammonium sulphate, aero-sodium cynamide or copper sulphate-sodium chloride is sprayed about two weeks before harvesting to kill the vines. As regards the mechanical methods, potato haulm pulverisers (cutters) are in common use in overseas countries. They are used for cutting or failing of vines ahead of harvesting. The common practice in India is to cut the wines with the help of sickles, 10-12 days before harvesting. It requires 50-75 man-h/ha. Efforts have been made towards development of haulm-cutting equipment in India (Verma, 2001; Verma *et al.*, 1992) such as rotary haulm cutter and flail type haulm cutter.

Rotary haulm cutter: It is a tractor-mounted equipment consisting of two vertical rotors with discs of 25 cm diameter mounted at the lower ends. Two blades of 36-cm length are rigidly bolted to each disc. Power from tractor PTO is transmitted to the vertical rotors through a gearbox. The blades rotate at a speed of 1500-2000 m/minute corresponding to PTO speed of 540 rpm. A fence is provided on the rear and sides to prevent scattering of the cut vines as well as for safety. Gauge wheels are provided to adjust and maintain the desired height of cut. Haulm from four ridges are cut in each pass. The rotor shafts are aligned with the centre of the furrows such that the blades cut the vines from two adjacent ridges. Tests indicated that haulms from an area of 0.2-0.5 ha/h can be cut with this equipment while operating at forward speeds between 1.5 and 3.5 km/h. Depending upon potato variety and field conditions, the haulm-cutting efficiency varies from 80-90 percent. The main deficiency observed in the rotary haulm cutter is its inability to cope with the vines being on the sloping sides of the ridges. It also cuts some tubers projecting over the ridge if the field is not properly levelled. This implement could reduce the labour requirement by 70-80 percent.

Flail-type haulm cutter: This type of haulm-cutter has been in common use in the overseas countries. It employs beaters attached to a horizontal shaft. The length of the beaters is contoured to match the ridge profile. The rotary beaters consist of blades, chain or rubber dampers. The blade rotates at peripheral speeds of 2500-3500 m/min. Prototype of a flail-type haulm cutter using link chains has been developed at PAU, Ludhiana (Verma, 2001). It is mounted on the three-point linkage of a 35-hp tractor and driven by the PTO by means of a gearbox, V-belts and pulleys. The link chains of varying lengths attached to the rotor shaft revolved with a tip speed of 1500-2000 m/min. The height of cut is maintained with the help of the gauge wheels. Haulms from four rows is cut in each pass with an average field capacity of 0.5 ha/h. The uncut vines are found to be fewer than 5 percent. To minimize tuber damage, it is essential to plant the crop with a planter to ensure uniform size of ridges and furrows. Swinging blades of varying lengths are used instead of the link chains.

Animal drawn single row potato digger

This is a simple implement suitable for digging and exposing potato tubers from one row (Garg and Singh, 2002). It is operated by pair of bullocks (Fig. 7.80a). It consists of a V-shaped blade having round bars at the rear of blade, tyne, hitch, frame, handle and beam (Fig. 7.80b). The round bars are of 12-mm diameter welded at a spacing of 40-mm intervals to form a screen for separation of tubers from the soil. The length of these bars from the cutting point of the blade is kept 40 cm and the ends are bent slightly downward to facilitate the sliding of the soil over the bars. The length of the blade is 54 cm and is made from a one cm thick and 10 cm wide carbon steel flat. The height of the beam can be adjusted from the hitch frame. Working capacity is 0.12 ha/h, width of cut, 60 cm; field efficiency, 80%; total tubers loss, 1-2%; and tubers exposed, 90%. Saving in labour requirement is 40-50% and cost of operation 35-40% in comparison to traditional method.

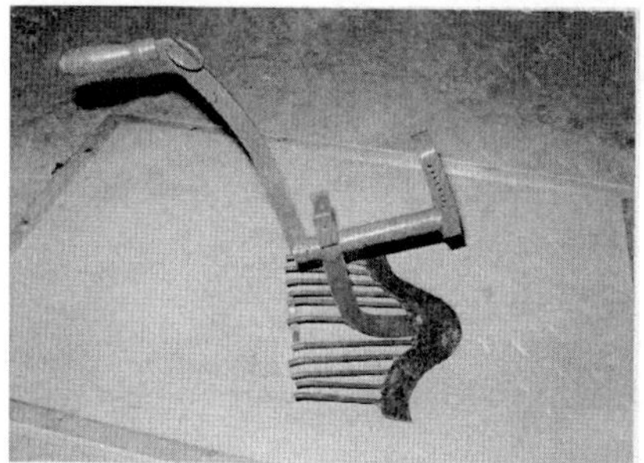

a) A view of bullock drawn single row potato digger

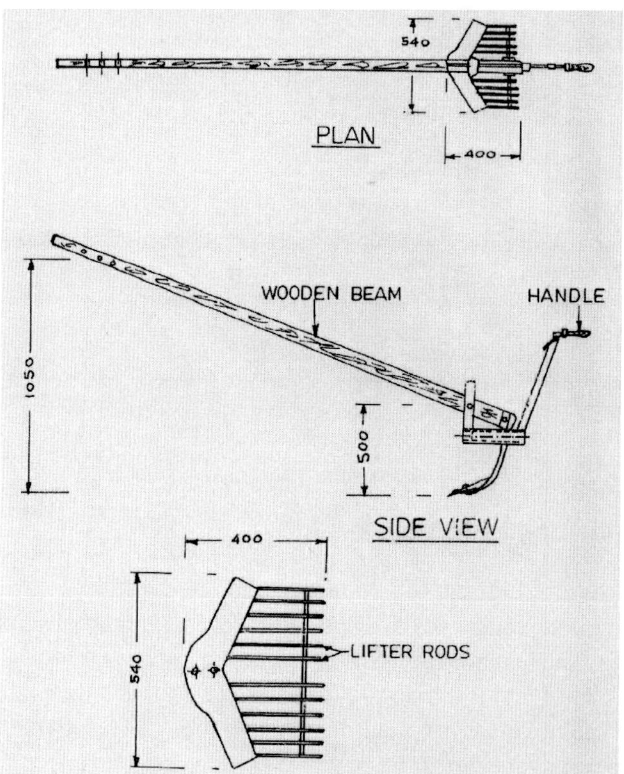

b) Details of bullock drawn single row potato digger (All dimensions in mm)

Fig. 7.80 : Bullock-drawn single row potato digger

Birsa animal drawn potato digger

It is an animal drawn digger consisting of a ridger shaped bottom with welded extension rods on the wings (Fig. 7.81); Pandey *et al.* (1997). These rods help in separation of soil and dirt from the potato tubers. It is suitable for digging potato tubers after removal of vines from the field. This implement was developed at BAU Ranchi. It saves 40 per cent labour and operating time and 18 per cent on cost of operation compared to conventional method of digging with spade. It also results in reduction of 11.3 per cent losses compared to conventional method of digging with spade. The width of cut is 35 cm and field capacity 0.03 ha/h. Damage to potato is 3.8%. Labour requirement is 230 man-h/ha (including picking of tubers).

Fig. 7.81 : Birsa animal drawn potato digger

Power tiller operated potato digger

It is used for digging of potato. It is operated by 8 kW power tiller. A power tiller potato digger is suitable equipment for digging potatoes (Pandey *et al.*, 1997). The machine consists of frame, beam, shoe type digging share and blade, wheel and hitch unit (Fig. 7.82). The share enters into soil first and the soil potato mass is cut with the help of blade. The whole soil potato mass is separated at rear of the blade where separation rods are provided. The depth of cut can be adjusted by the lever. It is simple in construction, light in weight and easy to operate on terraces. Working capacity of machine is 0.04 ha/h. Weight of the machine is about 23 kg.

Fig. 7.82 : Power tiller operated potato digger

Tractor operated two row potato digger

Fig. 7.83 : Tractor operated two row potato digger

Potato diggers are used for harvesting potato tubers (Garg and Singh, 2002; Pandey *et al.*, 1997). It is operated by a 26.1 kW tractors (Fig. 7.83). The two row potato digger has 2 blades fitted on the tynes. These are mounted on the cultivator frame having holes for adjustment of row-to-row spacing. The blades are of 540 mm length and have round bars of 12 mm diameter welded at the back at a spacing of 40-mm interval. The angle of penetration of the blades is adjusted from the top link of the hydraulic system of the tractor. The V-shaped blades with 145 degrees sweep angle are made from 100×10 mm carbon steel flat. After harvesting most of the tubers are exposed but some of them are covered by loose soil. The mounting of the blades on the frame is made in such a way that 2 rows are left un-dug between the blades. However, at the time of harvesting, a system is adopted so that only one un-dug ridge remains in between the blades (Fig. 7.84). Digging alternate rows give sufficient time to the labours for picking harvested tubers and clearing the strips for subsequent run of the digger. Working capacity of potato digger is in the range of 0.31 to 0.37 ha/h. Potato digger with width of cut 1.2 m has also been developed (Fig. 7.85) with field capacity of 0.32 ha/h, field efficiency, 70%; and total tubers loss, 1 - 2%. Saving in labour requirement is 40-50%; and cost of operation, 35% in comparison to traditional method.

a) A view of tractor operated two row potato digger

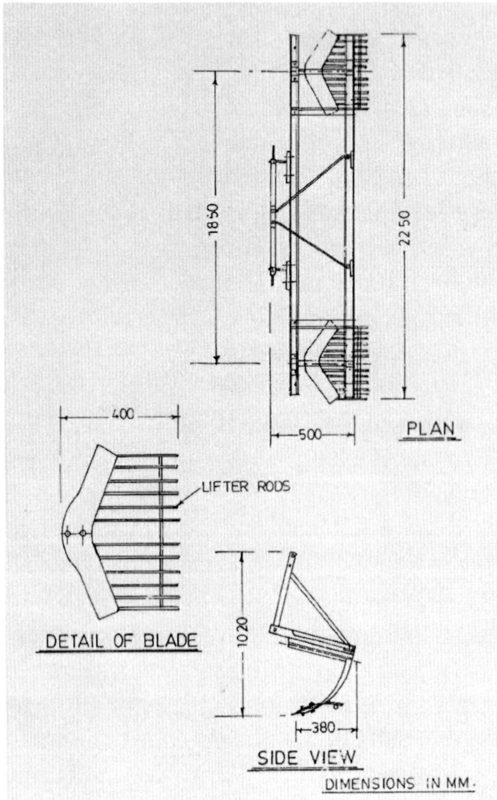

b) Details of tractor operated two row potato digger

Fig. 7.84 : Tractor operated two row potato digger spaced apart

Fig. 7.85 : Tractor operated potato digger

Tractor operated elevator type potato digger

It comprises of a frame, shovel type-digging blade of 550 mm width, endless rod chain conveyor, gearbox, two gauge wheels, idlers and driving sprockets (Fig. 7.86); Garg and Singh (2002); Singh and Pandey (2008); Anonymous (2008, 2010). The elevator conveyor is made of mild steel rods of 12 mm diameter, which are riveted/bolted to two endless flat belts. The pitch of the conveyor links is 25 mm. The length of the conveyor is 1500 mm and it makes an angle of 20 degrees with the horizontal. The machine is operated by a tractor of 35 hp. Power to the gearbox is transmitted by a telescoping shaft from the tractor PTO. Soil-potato mass is picked up and lifted by the chain conveyor. Two agitator sprockets oscillate the conveyor chain rod, which helps to separate the soil. Potato tubers with no or very little soil/clods are dropped on the ground behind the digger. Thus, the tubers are completely exposed which helps in speedy manual picking. Working capacity of machine is 0.16 to 0.25 ha/h, field efficiency, 60 - 65% and total tubers loss less than 1%. Saving in labour requirement is 75% and cost of operation 50% in comparison to traditional method.

Courtesy : Droli Mechanical Works, Moga (Punjab)

Fig. 7.86 : Tractor operated elevator type potato digger (Different makes and views)

Tractor operated root crop harvester-cum-elevator

A root crop harvester has been developed at Punjab Agricultural University, Ludhiana which has a digger blade and an elevator-conveyor for soil separation (Fig. 7.87); (Anonymous 2008, 2010, 2010b, 2012, 2013a). The digger blade is made from high carbon wear resistant steel. The width and thickness of the blade is 1144 mm and 16 mm. The blade is mounted on the machine at an angle of 20° with the horizontal. This helps in better lifting of the soil along with the onion bulbs. The elevator chain conveyor is attached behind the digging blade for cleaning, separating the soil from the bulbs and subsequent easy collection of onion bulbs. The spacing between the MS rods used for the fabrication of the elevator conveyor is 20 mm. The slope of the elevator conveyor is kept at 18°. Two oval agitators are provided in the conveying system for separation of soil particles from the crop bulbs or tubers. The power to the elevator conveyor has been provided through a gear box. Two coulter discs are provided in front of the blade at the outer ends which helps in easy slicing and lifting of soil by the blade. A roller behind the conveyor is provided to help in compaction of the soil as it falls through the grate of elevator chain. The crop bulbs and tubers fall over this compacted soil for easy collection.

The machine is used for digging root crops like onion, carrot, potato and garlic crop sown on beds in controlled traffic conditions (Fig. 7.88, Fig. 7.89; Fig. 7.90; and Fig. 7.91 respectively). The machine is operated by a PTO of 26.1 kW tractors. The field capacity of the machine is 0.28, 0.24, 0.21 and 0.21 ha/h for digging carrot, potato, garlic and onion crop respectively when operated at a speed of 2.78, 2.41, 2.10 and 2.10 km/h whereas respective damage is 1.98, 1.92 1.22 and less than 1.0 percent respectively. The saving in labour ranged from 62 to 71 percent and saving in cost of operation 46-52% over traditional method.

Fig. 7.87 : A view of tractor operated root crop harvester

Fig. 7.88 : Tractor operated root crop harvester harvesting onion

Fig. 7.89 : Tractor operated root crop harvester harvesting carrot

Fig. 7.90 : Tractor operated root crop harvester harvesting potato

Fig. 7.91 : Tractor operated root crop harvester harvesting garlic

Onion/Garlic digger

The onion digger imported was operated after de-topping operation (Anonymous, 2010e). The machine is operated by 55 hp tractor (Fig. 7.92). The operating speed is 2.33 km/h. The width of operation is 1.2 m and depth of operation 15 cm. The digger can be operated with minimum damage at the soil moisture content of 22.83%. If the moisture content decreases the clods are dug out along with the onion and causes damage to the bulb while conveying. The onion bulbs are spread behind the conveying unit and collected manually. The effective field capacity is 0.192 ha/h. The digging efficiency is about 98%. The damage during the de-topping operation is 4 to 5%. Another garlic digger-cum-elevator operated by tractor has been developed at MPUAT Udaipur with the support from manufacturer (Fig. 7.93). This has capacity of 0.2 ha/h.

Fig. 7.92 : Tractor operated onion harvester-cum-elevator

Fig. 7.93 : Tractor operated garlic harvester

Functional components of tractor-mounted root crop digger elevator

The root crop digger elevator has two main advantages over the spinner; it usually leaves more potatoes clear of the soil and it delivers them in a much narrower row. Picking in a spinner and the digger can usually be worked continuously with fewer pickers. The potato digger comprises of a frame, a digging shovel, an elevator conveyor, power transmission system, gauge wheels and fenders (Fig. 7.94). The elevator digger digs the potatoes from the ridge, lifts them along with the soil, separates the soil and drops the tubers behind the machine in rows or piles. The tubers are thereafter picked up by hand and placed in the baskets or racks. Following is a brief description of the digger elevator.

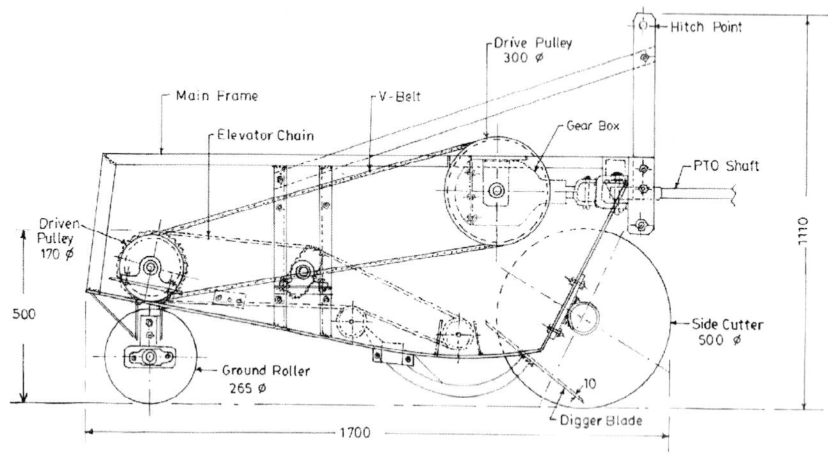

Fig. 7.94 : Side view of the root crop harvester cum elevator

Frame: The frame is made of angle iron sections of 50 x 50 x 6 mm. It consists of a rectangular tool bar on which the gear box is mounted. To the tool bar two baskets are welded in which a cross beam can be moved side wise for offset hitching of the digger with respect to the tractor. All other members, viz. hitch links, drive shafts, elevator conveyor, gauge wheels, blade and deflecting tines are attached directly or indirectly with the frame.

Elevator conveyor: Elevator conveyor is made up of interlocking cross bars of 10 mm diameter at a regular spacing of 30 mm. It is mounted on two shafts having 1.38 m distance between them and inclined at an angle of 20^0 with the horizontal. The conveyor gets its power from rear shaft through two sprockets keyed on both ends. Two idler rollers support it at the front. The width of the conveyor is 65 cm. Upper portion of the web is supported over two agitator sprockets and the lower portion over idler sprockets to avoid any undue sagging of the chain.

Power transmission system: Power from the tractor PTO shaft is taken to the gear box for driving the elevator conveyor through a telescoping shaft. Tractor PTO speed is reduced about 4.5 times by the hypoid gears driving the main shaft. Mounted on the main drive shaft is set of double groove V-pulley of 162 mm diameter which transmits power to a set of sprockets keyed to the same shaft. An idler is provided to adjust the tension in the V-belts.

Blades: The blade is essentially a scooping shovel of width 55 cm. The cutting edge nearly 100 mm is made of high carbon steel duly tempered and the rest of it is of case hardened mild steel. It is fixed directly to the slanting side members of the frame with a provision to adjust the angle of digging.

Windrow attachment: The soil and tubers are separated from each other over the elevator web due to the shaking provided by the agitator sprockets. The tubers dropped at the rear of the conveyor are spread over a strip of 50 cm width.

Gauge wheels: Two gauge wheels are provided on either sides which move in the furrow while operating the machine. They can be adjusted to maintain the desired depth of penetration.

Fenders: In order to prevent escaping of the potatoes from either sides while being carried over the web two fenders are provided. A single row elevator digger requires a tractor of 25 to 30 hp for its operation. The digging shovel may be either of the pointed type or conveying cutting edge type. The width of the digging shovel ranges from 50-60 cm.

Comparative performance of various types of diggers are given in Table 7.3.

Table 7.3 : Comparative performance of different methods of potato harvesting

S. No.	Particulars	Hand digging	Bullock-drawn digger	Tractor-drawn digging blade	Tractor-drawn potato spinner digger	Tractor mounted one-row potato elevator digger	Tractor mounted two-row potato elevator digger
1.	Output, ha/day	0.01	0.8	2.5	1.6	1.4	2.8
2.	Digging						
	Man-h/ha	650	6	4	5	5	3
	Bullock-h/ha	-	20	-	-	-	-
	Tractor-h/ha	-	-	4	5	6	3
3.	Picking exposed tubers, Man-h/ha	-	360	360	288	80	80
4.	Picking tubers in racks, Man-h/ha	40	40	40	40	40	40
5.	Planting of covered tubers, Man-h/ha	-	16	16	16	16	16
6.	Total labour requirement, Man-h/ha	690	422	420	349	142	139
7.	Bruising of tubers (%)	2.0	1.5	1.5	3.0	1.5	1.5
8.	Cutting of tuber (%)	5.0	0.5	0.5	1.5	1.5	1.5

Oscillating potato diggers

Oscillating or vibratory diggers are among the recent developments in the potato harvesting machinery (Verma, 2001; Verma *et al.*, 1992; Anonymous, 2012). Considerable research has been carried out in overseas countries to utilize the mechanical vibrations to reduce the implement draft and improve the soil-potato separation. It has been reported that draft reduction of the order of 40-50 percent can be achieved with an oscillating digger. The soil-separation efficiency is also significantly improves as compared to a fixed-blade digger. Oscillating potato digger works best when the frequency of oscillations is between 8 and 10 cps and the amplitude of oscillations leads to better soil breakup and soil-potato separation but balancing of inertial forces due to rotating and reciprocating masses is the major problem. Depending upon the forward speed, frequency, amplitude and direction of oscillations, 95-98% tubers can be exposed with the working of an oscillating digger. Relatively simple configuration, light weight, low-cost and compactness are among the main attributes of oscillating digger. However, such a digger requires careful design and balancing of the inertia forces to prevent transmission of vibrations from the digger blade to the digger frame and tractor links. Field capacity of a single-row-oscillating digger is found to be compatible with the single-row elevator digger with a passive blade. Tuber injuries are, however, much less.

Tractor drawn turmeric digger

The existing tractor drawn groundnut blade 'guntaka' is being used for harvesting turmeric crop (Anonymous, 2010, 2010a and 2013). The 'guntaka' blade is made of high quality steel. It consists of main frame made of M.S. pipe of 178 mm diameter, 5.5 mm thickness and 1800 mm length, two side supporting iron rods for digger blade of 700 mm lengths, three point linkage and cutting blade of 1.45 m length with 5 mm thickness (Fig. 7.95). Two holes are made on the circular main frame for fixing the side supporting frames for cutting blade. At the end of the side supporting frames a flexible sheet frame is attached for fixing and removing the cutting blade. It dugs four rows at a time at 300-350 mm depth with a field capacity of 0.36 ha/h.

Fig. 7.95 : Tractor operated turmeric digger

Potato combine harvester

A potato combine harvester dismantles the potato ridge, picks up the soil-potato mass, shifts out the loose soil and places the material on a slow-moving sorting conveyor to manually separate the un-separated clods and other foreign matter (Anonymous, 2012). Essentially, a potato combine comprises of a digging shovel with one or two pieces, or two inclined concave discs to dismantle the ridge and a primary conveyor to pick and elevate the soil-potato mass. Bulk of the loose soil is shifted out by the conveyor with the upper web supported over the agitator sprockets at one or two places. The material conveyed by the primary conveyor is fed on to a cross conveyor or a rotary cage-screen before delivering it onto a sorting conveyor belt.

The purpose of the intermediate conveying system is to effect further separation of the soil and feed the material onto a slow moving conveyor to enable manual picking of the remaining foreign matter. In some designs, the sorting belts have provision to alter the slope with the horizontal to separate the tubers from the clods and other foreign matter by using their differential repose angles. Potato tubers have smaller angles of repose than the clods. Electronic sensors have been used in potato combines for separation of tubers from the soil. Such machines are, however, quite expensive. Potato combines used in overseas countries are of single or two-row types. A potato combine works best on potato varieties with large tubers planted in relatively bigger fields with proper field layout.

For efficient operation of potato harvesting equipment following points should be kept in mind:

- Plant the crop in straight rows at 60 cm or wider spacing by means of a potato planter,
- Plant the crop length-wise to minimize the number of turning at the ends,
- Use a ridger for second earthing up if necessary,
- Provide adequate headland for turning of the tractor,
- Destroy the vines (haulms) about 2 weeks prior to harvesting,
- Add sufficient soil amending substances to develop a proper texture in the soil of the potato fields to eliminate/minimize clod formation,
- Ensure adequate moisture in the soil at the harvest time to avoid clod formation,
- Use right type of agitator sprockets to effect maximum soil separation without causing excessive injuries to the tubers,
- Ensure that the elevator conveyor web is neither too tight nor too loose. Links of a loose conveyor override the teeth of the driving-sprockets and that of extra tight conveyor cause undue wear of conveyor links,
- Maintain correct tension to minimize belt/chain slippage in the power transmission system,
- Lubricate the shaft bearings every week and those of the telescoping drive shaft daily,

- Check oil level in the gearbox daily before the start of the operation and top up when necessary,
- Alight the digger with the ridger(s) to keep the digging blade in the centre of the ridge(s),
- Avoid excessive depth of digging. The shovel should run 20-25 mm below the tuber zone. Adjust the depth with the help of the tractor top link and the gauge wheels. Digger frame should remain horizontal when the blade is operating at the desired depth,
- Lift the digger at the turns,
- Use correct forward speed. Never run the digger too fast otherwise percent exposed tubers will be reduced and the labour requirements will increase,
- The conveyor speed index (CSI) should be kept between 1.2 to 1.3. In other words, the speed of the elevator conveyor should be 20-25 percent higher than the forward travel of the digger,
- Never postpone the repair/replacement of the links, sprockets and the digging blade to ensure optimum results,
- Keep critical and fast-wearing components like conveyor links, agitator sprockets, and nuts and bolts handy to be used in the hour of need,
- Harvest the crop in alternate rows if adequate labour for picking the tubers is not available, and
- Store the digger under a shed during slack season. Apply mobil oil/grease on the blade edge and other parts to avoid corrosion.

7.5 Harvesting equipment for sugarcane

Mechanization of sugarcane cultivation in general and harvesting in particulars has been rather very slow all over the world. Most of the world's sugarcane is still cut by hand which is costly and time consuming operation (Srivastava, 2001; Yadav, 2003). It requires large number of labourers at the peak time of harvesting and shortage of labourers at that time delays harvesting operations. Labour requirement ranges between 850-1000 man-h/ha. With the increasing wages, unless we mechanize sugarcane cultivation, the cost of production of sugarcane will be going higher and higher.

Sugarcane harvesting in India and in most of the other developing countries is done manually. In northern India, during winter season, when there is not much agricultural activities and demand of labour is less, cane harvesting at most of the places is done free of cost in exchange of green tops which are used as cattle feed. Villagers in large number collect in one field, harvest the cane, clean them and bundle them in exchange of green tops. However, after the month of February when harvesting of Bengal gram and other crops starts, free labour is not available and harvesting is to be done by paid laborers. In southern India it is done by paid labourers only. In north India, during cold season, harvested cane kept in shade or covered with dry leaves to maintain its juice quality for 2-3 days but looses its weight unless it is kept moist. In the late season and in tropical areas quick disposal of harvested cane is most essential.

Harvesting knives used in India

In India 100% cane cutting is done by manual labourers (Fig. 7.96) using conventional types of cane knives that differ in size, shape & weight from one place to another (Fig. 7.97). These knives cut sugar cane much above soil level, thus leaving a portion of the cane as stubble which results not only in reduced sugarcane yield per ha but also in loss of valuable sucrose as maximum sucrose is concentrated at the base of sugarcane stalk. The operator requires bending his trunk too much and stubble shaving operation becomes essential for ratoon crop (Fig. 7.98). De-husking, cleaning, bundling of cut cane and loading for transportation are done manually. Indian Institute of Sugarcane Research (IISR), Lucknow has developed a sugarcane stripper (Fig. 7.99) which does the job of detopping and dressing of the cane. Sugarcane stripper, designed and developed at the IISR, is a simple device by which all the dried as well as green leaves can be removed in one stroke and green top portion can simultaneously be cut easily. The stripper consists of a pair of tongs, the jaws of which close to form a square and then extend beyond the square to form a V-shape in the front. A light tension spring holds the jaws closed. A knife is welded on the stem of the stripper for cutting the green tops. The cane stalk is gripped between the jaws of the tool and drawn downward. In one or two strokes, dry leaves are separated and pushing away from the stalk. Green top is then cut with the knife provided for the purpose. During trials, it has been found to increase the efficiency of human labour by about 15-20%. The hand tool alone, however, is not sufficient to handle the bulk of cane.

Fig. 7.96 : Manual harvesting and cleaning

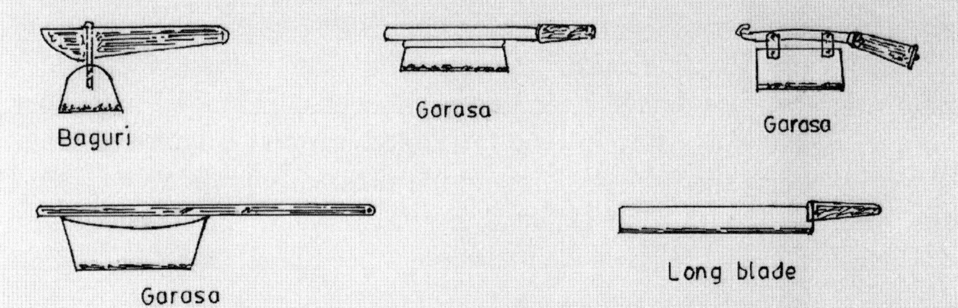

Fig. 7.97 : Varying shape and size of cutting knives

Fig. 7.98 : Cutting of cane by conventional knife & V.S.I knife

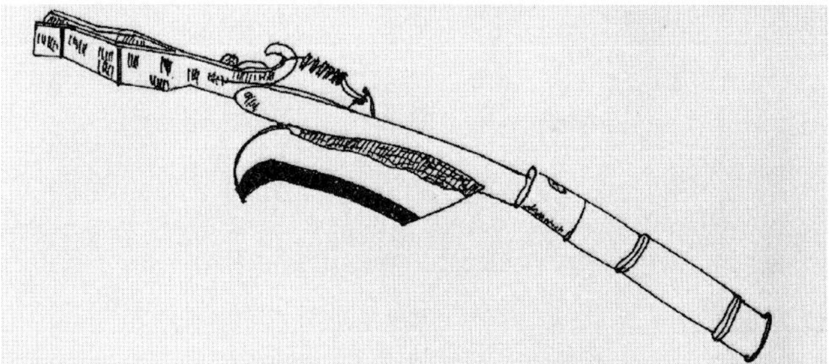

Fig. 7.99 : Sugarcane stripper

Sugarcane harvesting involves base cutting of the crop, detopping, detrashing, bundling, loading and transportation. Detopping and detrashing of crop itself takes about two-third of manpower required for harvesting. Several types of sugarcane combines and harvesters are used world over. They are normally used for crops, which are burnt in the field prior to harvesting for trash removal. Some harvesters are used in green crop and cane is burnt in windrow after harvesting. Some machines have been developed which can be used in the cane field without burning. This is particularly done where environmentalists object burning of cane. The sugarcane combine is a one-pass machine, which cuts the cane, detops, cuts in billets, cleans and conveys to transport cart/trolley. In case of whole cane sugarcane harvester, it cuts the crop, detop and put on the ground in windrow, which are loaded in trolleys by mechanical loader or grabber. Combine harvested cane must be processed within 16 hours to avoid deterioration and sucrose loss. Most of the sugarcane combine harvesters is self-propelled machine. However, some tractor-drawn machines are also available. Commercial sugarcane harvesters used in other countries may be classified into five groups viz. i) Push rake and grab type; ii) Tractor operated whole stalk harvester; iii) Self-propelled whole stalk harvester with linear or transverse windrowing; iv) Self-propelled billet type harvester; and v) Manually guided walk behind self propelled harvester.

Whole stalk sugarcane harvesting systems

In the beginning all harvesters were aiming at toping, base cutting and bundling whole stalks of sugarcane. These bundles were left in the field for picking up by grab type of front end loaders. Popular whole stalk sugarcane harvesters and systems are described below:

Push-rake system: Push rake operation, essentially involves pushing and piling of sugarcane. These piles of cane are grabbed and loaded into transport trucks and delivered to the mill. It is made of very sturdy tines welded in a frame. The push rake is mounted in front of the tractor. When pushed into standing cane, it broke the stalks off at the surface of the ground and left them in piles. Push-rake system is one of the oldest and crudest systems of mechanical harvesting of sugarcane. This method of harvesting is not very satisfactory. Push rake harvesting uproots many plants, which results in poor ratoon crops. It requires substantial replanting. The overall system is not very good and has very little hope in future, except in special circumstances where other systems cannot be used.

Windrow harvest system: In this system, sugarcane is cut and windrowed. Depending on the machine, pre-harvest or post-harvest burning is practiced. Sugarcane in windrow is mechanically loaded in trucks or field carts and transported to the mill. There are many machines that harvest the cane and place it in windrow for mechanical loading. Some of the most popular windrow harvesters are described below:

V-cuter harvest system: The cutters on the machine are popularly known as V-cutter. It is called V-cutter because its front end is shaped like a V. It is a two row harvester. There are two circular horizontal blades spaced 152.4 cm apart for cutting two rows of the cane at a time and a vertical parting blade in the centre. The unit is mounted on a track type tractor. Front end can be raised or lowered hydraulically for transport and field requirements. It is sturdy machine with few moving parts and very little down time. The machine is most suited for high tonnage cane. Wherever burning is permitted, the cane is burnt prior to harvesting. V-cuter opens the field and makes a windrow of two rows. So far, most of the machines are not equipped with a topper. There is no mechanism for positioning the cane prior to harvesting either. It only cuts and windrows the cane. The windrow is poorly laid which result in poor loading. V-cutter harvesting has similar problems as push rake system. V-cutter harvested cane delivered to the mill has a very high percentage of foreign matter. Under poor conditions it might reach around thirty per cent. Many times a quarter of the cane is left behind the loader, requiring another operation of manual cane picking. Under right field conditions its effective field capacity is about 0.8 ha/h. Its use is now being discouraged because of high percentage of foreign matter delivered to the mill, cane left in the field, ratton damage, and other problems.

Soldier type harvester: Soldier type harvester positions the cane for base cutting, tops it, base cuts it and places in windrow for mechanical loading. It cuts one row of green cane at a time. A topper with a gathering chain and disc removes the top from standing cane and drops to the right of one being harvested. Two sets of pick-up chains arranged in a V-are use for picking and feeding the cane to the base cutter. The harvested cane is conveyed through a cane conveying system to the windrow. The cane is burnt after to 12 to 16 hours of harvesting to eliminate the top and trash. After burning, cane loaders load the cane in to transport carts for the mill. Under the desired field and crop conditions, it average effective field capacity is about 0.4 ha/h. It does not uproot the plant and very little replanting is required. Foreign matter is within 10 per cent. Can left in the field is not very significant. It has many moving parts which break down frequently in heavier tonnage.

McConnell harvest system: The natural weak points exit in all most all sugarcane varieties. One at the extreme base and other all the junction between the mature and immature cane. The McConnell harvester is unique in the sense that it exploits these natural weaknesses of the cane for base cutting and topping. McConnell Harvest system uses the machines that can be mounted on some standard tractor, it tops, base cuts and partially cleans the cane. It leaves the cane in a windrow. The cane is further cleaned by labour and loaded manually or mechanically. It harvests one row at a time. When above harvester is used as a stage I machine of the two stage McConnell harvest system it does not have a topper. The first pass over the crop is made by the harvester which rolls the cane forward (and partially snaps it) and then base cuts it, leaving the cane in a cut orderly and continuous windrow with all the tops still attached to the cane and lying in the same direction. As the topper is not used, the machine travels fast. It can be used satisfactorily in yields up to 112 tons/ha. It is designed for row spacing of 152 cm or more. It can be with most tractors in the 75-90 hp class with an external hydraulic output of 18.16 liters/min or more. The second pass is made by the stage II machine traveling in the opposite direction to the stage I. If the cane is fed in the machine it tops first and picks the cane and feed it over a gap on to the conveyor. The gap prevents stones and soil entering the machine. The cane passes up the conveyor at the end of which the cane tops are gripped between a pair of rollers which accelerate it over a gap and off the conveyor. Trash is stripped off the cane by this action and allows to fall through the gap to the ground. The top leaves are encouraged to warp and is snapped off as the cane is accelerated rearwards into the bin. The tops are then blown clear by fans of special design for the purpose. The bin is capable of dropping piles of cane of any size up to ½ ton. Trash from under the pile is cleaned by tines fixed to the underside of the bin which engaged the ground as the bin opens.

Cut-crop-harvest or combine harvest system : All harvest system described so far namely push, pile and grab system and windrow harvest system have one operation in common. It is the cane being placed on ground for loading after cutting. This operation is partly responsible for cane left in the field and soil and rocks delivered to the mill. A combine harvest system eliminates this operation. The basic components of a sugarcane combine are gathering mechanism, topping mechanism, base cutter, feed conveyor, chopper, and elevator and cleaning by air blast.

Gathering mechanism: Its function is to separate cane and align the row to be harvested. They are made of revolving scrolls fitted on gathering walls. It consists of two triangular walls, approximately 140 cm apart at the tips and converging to the throat width of the machine just forward of the base cutters. The tips of each wall are fitted with ground engaging points to get under and lift stalks that are lying on the ground.

Topping mechanism: Its job is to gather, cut and discard non productive tops. The gathering operation is performed by gathering chains while cutting is done by a horizontally rotating disc fitted with mower blades which cut against a fixed anvil.

Base cutter: Its function is to cut the stalk at or just below ground level. At least one manufacture uses twin contra-rotating discs fitted with a number of replaceable knife blades. Some use single diameter blades for base cutting the cane. Recommended tip speeds is 1524 to 1828 m/min.

Feed conveyor: Its function is to convey the whole stalks of cane from the bass cutter to the choppers. In some machines it is made of endless chain slat conveyor. In others a series of rollers are used. In some machines augers are used for feeding cane to the chopper.

Choppers: Their function is to receive the whole stalks from the feed conveyor and chop them into short uniform billets. The design used is a pair of parallel shafts each with paddle shaped blades, which as they rotated came together in the plane containing the shafts and so gives a flying or traveling cut. There is other mechanism used for chopping. However, flying cut mechanism has the advantage of being aggressively self feeding, and that once the swath is engaged by the chopper; it will be pulled in continuously until broken or forcibly interrupted.

Elevator: Its function is to receive the billets from the choppers and convey them into a receptacle for transporting to the mill. An inclined chain and salt conveyor is used. The elevator can be rotated 180 degrees in most machines.

Air-blast cleaning: One of the biggest problems with mechanically harvested cane is the foreign matter in cane delivered to the mill. Foreign matter consists of leaves, tops, dirt, stones and many other materials picked from the field. In some machines one while in others two fans are being used for extracting leaves and dirt from the cane. The basic components combined in a frame and provided with a power unit and vehicle, constitute a cane combine. It is generally powered with an engine of about 150 hp.

The Self-propelled whole stalk harvesters with linear or transverse windrowing are either single row or double row units and cut canes from bottom and top and windrows either under the machine linear to direction of travel or to one side from where they can be collected and loaded into wagons.

Self-propelled billet type harvesters are combine chopper harvesters where cane is lifted, topped and cut into small billets and loaded in specially design bins. Primary and secondary extractors removed most of the extraneous matter. The harvesting rate is very high, need specially design containers as bins and efficient management of the harvested crop.

Manually guided walk behind self-propelled harvester are suitable for small fields with low tonnage (Fig. 7.100). These are suitable for flat cultivation and cannot handle recumbent cane. They are available in single row with best cutter and chain driven windrowing systems.

Fig. 7.100 : Cutter bar type manually guided walk behind self-propelled harvester

Problems in adoption of imported harvesters in India

i) Small & scattered fields without any provision of headland for turning,

ii) Narrow row to row spacing of the crop (60-100 cm) as compared to 150-180 cm used in other countries where mechanized harvesting is done,

iii) Non-erect and lodging characteristic of the crop due to varieties and cultural practices,

iv) Combine harvested crop needs to be processed in about 16 h of harvest,

v) Modified transport system and higher level of management is required,

vi) Higher initial investment,

vii) Amount of foreign matter delivered to the mill with cane is much higher in comparison to accepted norms in Indian mills,

viii) Sugarcane purchase system prevalent in Indian mills,

ix) Sugarcane tops used as valuable animal feed in most parts of the country. Loss of green tops is considered negative factor by most prospective users of the machine, and

x) High cost of spares and services for repairs and maintenance.

Requirements of mechanical cane harvester in India

The mechanical harvesters used in other countries cannot suitably be used under Indian conditions because of the following reasons:

i) Many of these harvesters are side mounted and take more space. They can be worked only from one side (one end) of the rows.

ii) Due to their wider width they can be used only in fields where row to row spacing is between 1.5 - 1.8 metres.

iii) De-trashing in most of the cases is done by burning the cane which in India will mean tremendous loss of sugar by way of cane drayage due to rapid desiccation during transportation to mills.

iv) Most of the harvesters are two row harvesters and require larger power for their operation and propulsion. They are suitable either for large plantations or to be used on custom basis on bigger size forms.

There is, therefore, a need for designing of suitable mechanical cane harvesters for Indian conditions. There are certain technical problems that must be solved in order to design a suitable mechanical cane harvester for Indian condition. The fact that most of the cane stalks are in various stages of re-cumbency and not all standing erect should be considered one of the big problems in developing a harvester. Removal of leaves and tops are equally important problems. Following are the main requirements of a mechanical cane harvester:

i) It should cut cane just at or near the ground level,

ii) It should cut the tops at proper level,

iii) It should preferably strip the cane mechanically,

iv) It should windrow the cut canes to one side or should deliver into trailers,

v) It should be able to handle both erect and lodged canes,

vi) The cutting mechanism of the harvester should be in front of the harvester so that it could be worked from both ends of the rows, and

vii) The machine should preferably be a trailed one and should be such that it could be worked with the tractors of 35 hp range.

Under the present circumstances the harvester for Indian conditions should be developed in stages. In the first stage it should be able to cut and windrow the canes. One row or paired row at a time. In the 2nd stage it should be able to cut, top, detrash and windrow, one row or paired row at a time. In the third stage it should be able cut, top, detrash and convey clean cane sticks into trolleys. Only one row or paired row at a time. Assessing the performance of a single row harvester of type (3) above a two row harvester may be designed on the same principle. Since the row to row spacing of cane varies greatly in our conditions and since majority of fields are comparatively of smaller size a single row or paired row harvester may prove more versatile than a two row harvester.

At IISR Lucknow, two types of sugarcane harvesters have been developed, one a tractor rear mounted PTO shaft operated single row windrower harvester and other front mounted (Fig. 7.101). It is operated by a 35 hp tractor and can be used for cutting single row flush with the ground. Cut cane stalks are windrowed along the standing cane row to be harvested next. Windrowed cane stalks are collected manually for removing dry leaves and green tops by conventional hand tools. The unit consists of (i) framework of mild steel angle iron; (ii) standard tractor pulleys deriving power from tractor PTO shaft through universal cross joints and a telescopic shaft; (iii) a separate bevel gear unit to give drive to the base cutter unit provided as an offset to the main frame; (iv) base cutter disc; (v) pusher guards to take care of lodged crop; and (vi) gauge wheels for adjusting the height of cut and balancing the equipment during operation. The equipment is attached with three points linkage of the tractor. Power is transmitted from tractor PTO to the standard pulley and then to the separate bevel gear unit through coupling. The cutting blades operate at about 800 rpm in anti-clockwise direction. The base cutter unit cuts the stalks flush with the ground and throws them to one side. Pusher guard takes care of the lodged crop. It is essential to lift harvested cane stalks in a simultaneous action so as to facilitate harvesting of the next row. Although the equipment performed well and quality of cutting is satisfactory, simultaneous collection of cane stalks causes problems and the output is quite low. Lodged canes are the main hurdles and the pusher guard provided for the purpose does not work well. Moreover, 3-4 corner rows has to be harvested manually so as to facilitate entry of the machine in the field. These observations led to modifications and development of a tractor front mounted PTO operated double row harvester.

Fig. 7.101 : Tractor front mounted sugarcane harvester

Double row front mounted harvester

The basic design features of the rear mounted harvester have been utilized to develop double row tractor front mounted harvesting system. Major components of this harvester are framework and its attachment system with the tractor, drive system from tractor PTO shaft to the main gear unit, base cutters, front reel and the rotating crop dividers with spirals (Fig. 7.101). The various components such as main gear unit for operating base cutter, crop dividers and front reel have been provided on the main angle iron frame. Main frame is connected to the tractor with the help of an auxiliary frame. Two hydraulic cylinders have been provided and connected to the tractor hydraulic system for lowering or raising the equipment during operation or transportation. Drive from tractor PTO shaft is transmitted to the main gear unit through chains and sprockets and a well supported telescopic shaft and cross joints. Drive to the other base cutter unit fixed at 750 or 900 mm is given with the help of a pair of bevel gears. The base cutters are 500 mm is diameter and rotate at a speed of 800 rpm. Crop dividers get drive from the main shaft through a worm and disc gear and take care of the lodged canes. The equipment gives an output of 0.25 ha/h.

Tractor mounted detrasher

Presently sugarcane is harvested manually in our country. The harvesting of sugarcane involve base cutting of standing sugarcane stalks, detopping of green top and dry trash removal, bundle making of cleaned cane stalks and loading to transport vehicles. All these operations are done manually. It has been observed that detopping and trash removal consume about 50-60 per cent of the total time consumed in performing all the unit operations involved in sugarcane harvesting. IISR Lucknow has designed and developed a tractor PTO shaft operated detrasher (Singh, 2014). Its design is similar to the engine operated detrasher developed at the Punjab Agricultural University, Ludhiana. The IISR detrasher contains two rollers moving in opposite directions, one cane feeding beater, a cane take off beater and a blower unit (Fig. 7.102). The machine receives power from the PTO shaft of a tractor and is hitched through standard three point linkage system for transportation. The drive to various parts of the system is transmitted through sprocket chain drive, gear drive and through V-pulley. Cane feeding mechanism consists of a feeding chute and a roller. Provision is there to adjust the angle of inclination of feeding chute with the horizontal. The mild steel feeding roller is used. Rubber-canvass belt (5 ply- 10 mm thick) loops are at the periphery of the roller. The effective outer diameter of roller with loops is 300 mm. The looped roller provided better feeding of tops as well as cane stalks by providing better grip. Mechanism for detrashing has two counter rotating rollers (detrashing and lower roller), air blower and a comb type attachment. Detrashing and lower rollers are mild steel rollers, cushioned with rubber-canvass belt. The effective outer diameter of cushioned roller is 300 mm. Air blower is in such a way that air discharge forced the cane top and leaves into the pocket (formed between detrashing rollers) that in turn are pulled down through the counter rotating detrashing rollers. Since cane stalk is continuously moving forward through the feeding roller and detrashing rollers pull down cane top leaves, the cane top is removed from the weakest point i.e. joint of immature top from the cane. A comb type attachment (rubber-canvass fingers mounted on mild steel flat) is 5 mm ahead of air discharge pocket. Cane delivery mechanism consists of a pair of rollers (upper flap roller moving counter clockwise and the lower roller free to move). These rollers are so designed that it not only guides the cleaned cane but also brush the cane to remove left over trash sticking on the cane. A comb type attachment consists of rubber-convass belt attached on a mild steel flat is fitted at the outlet which tends to comb the detrashed cane coming out from the detrasher. The average trash percentage left in the cleaned cane is 5.5 percent on cleaned cane weight basis. For feeding of two to three canes at a time the output is 0.8-1.0 t/h.

Fig. 7.102 : IISR sugarcane detrasher

Self-propelled sugarcane harvester

Self-propelled billet type cane harvester consists of a diesel engine of 175 hp or more, base cutting unit, green top gathering and cutting unit, cleaning unit, billet cutting unit and conveying unit (Fig. 7.103); Singh (2014); Singh (2007); Yadav (2003). Disc-O-Gatherer topper provides highly efficient green top cutting (Fig. 7.104). Gathering facilitates better capturing of green tops for excellent top cutting making less green material in harvest. Lift-O-matic crop divider and base cutting unit works well even in lodged crop (Fig.

7.105). Crop divider prevents uprooting of adjacent row cane by better separation of the cane from adjacent row in case of entanglement giving less roots and soil in the harvest. Fine chopping of green leaves and dry trash results in better removal of trash by the extractor fan. It gives optimum size of the billets that result in better capacity, road transport with reduced cost of transportation. A cabin is provided on the top of the machine with all controls that gives complete operational controls and comfort to the driver (Fig. 7.106). Manufacturing of sugarcane harvesters has started in India by Tirth Agro Technology Pvt. Ltd. Rajkot (Gujarat). It is a billet type harvester (Fig. 7.107). It has features like chopping, cane straightening, leaf cutting, cane cutting and conveyer attachment to transport chopped cane directly to tractor trolley, hydraulic transmission and powered by 173 hp diesel engine. A fully loaded cabin is provided on the top of the machine with all controls for complete operational control and comfort of driver.

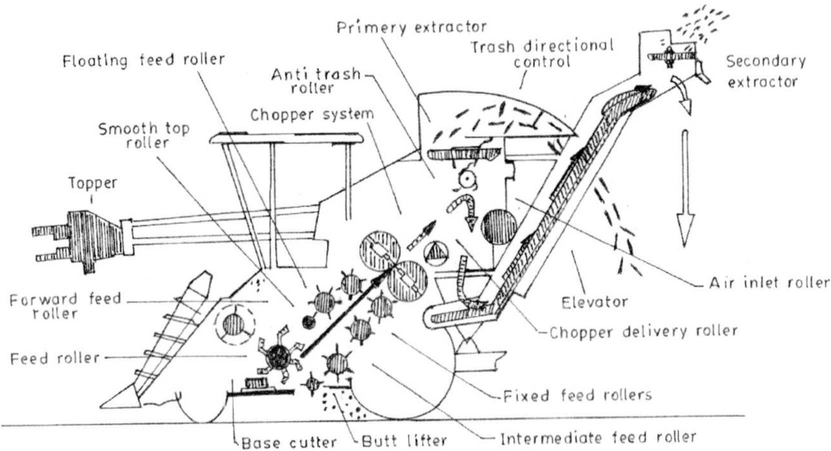

Fig. 7.103 : Details of self-propelled sugarcane harvester

Fig. 104 : Gathering and green top cutting unit of self-propelled sugarcane harvester

Courtesy: New Holland Fiat (India) Pvt. Ltd.

Fig. 7.105 : Twin crop divider and base cutting unit in self-propelled billet type sugarcane harvester
Courtesy: New Holland Fiat (India) Pvt. Ltd.

Fig. 7.106 : A view of self-propelled billet type sugarcane harvester
Courtesy: New Holland Fiat (India) Pvt. Ltd.

Fig. 7.107 : Self-propelled billet type sugarcane harvester
Courtesy: Tirth Agro Technology Pvt. Ltd. Rajkot (Gujarat)

7.6 Harvesting/Picking Equipment for Cotton

Manual harvesting/picking of cotton is quite a labour intensive operation (Fig. 7.108). Mechanical harvesting of cotton is widely used in USA, USSR, Egypt etc (Sandhar and Goyal, 2001). Machines used for cotton harvesting are basically of two types. These are either of stripper or picker type. The variety requirement for these two types of machine is somewhat different. Both these types of machines require the cotton variety with compact sympodial /semi-sympodial plants with synchronized boll opening. Cotton is a perennial, indeterminate plant, and unlike most commercial crops produces flowers and fruit continuously during the growing season. Flowers and new bolls may be formed during the harvest phase, even through the bolls may never mature. Crop maturity extends over a considerable period of time. For best use of the mechanical picker, cotton plants should be erect, sturdy and of growth habits that permit penetration of insecticides and defoliants. Inner surfaces of the dry burs should be relatively smooth; burs should be straight and approximately equidistant from each other. These are inherent conditions that are often influenced by adverse environment.

Fig. 7.108 : Manual picking of cotton

Where machine picking is contemplated, insect control must be timely and effective to insure good distribution of the crop over the plants. With well-balanced crops several known varieties stand erect, but if out of balance may lodge badly with much lighter yields. Lack of proper insect control is conductive to excess plant growth, severe lodging and other conditions which interfere with machine picking.

Application of defoliants: For dependable defoliation, fields should be evenly grown and all conditions, such as un-levelled land and salinity, should be corrected before the crop is planted. Fertilizers should be judiciously used and nitrogen in particular should be fairly well depleted before maturity. The harvest should follow close behind abscission of the leaves. Defoliants reduce the harvest season and increase the percentage of leaf drop depending upon the mode of defoliant application. Early defoliation increases the quantity of high-grade cotton which may be taken in first pick and late defoliation increases losses in the first pick over those from green picking.

Cotton picking machines: The two types of machines used for cotton picking are commonly known as cotton picker and cotton strippers.

The picker type: These machines are designed to pick the open cotton from the bolls by means of spindles, fingers or frongs at any time during the season without material injuries to the foliage and unopened bolls. Mechanical pickers are selective, in that seed cotton is removed from the open bolls, whereas green uneven boll is left on the plant to mature for later picking. In high yielding areas where serious weather hazards make it important to start harvesting as early as possible, it is common practice to go over the field twice.

The stripper type: This type of cotton harvester removes the cotton from the plant by combing the plant either with teeth or by drawing it between stationary slot and revolving rolls. Strippers are once groove machines and are being used successfully. All bolls whether open or closed are removed from the plant in a single pass. Harvesting with strippers is, therefore, usually delayed until the plants shed their leaves following the first frost. Chemical defoliants and desiccants are some time applied to permit earlier stripping.

The thresher type: This type of cotton harvester severs the stalk near the ground surface and takes entire plant together with the cotton into the machine where vegetative material and cotton are separated.

Electrical type: Electrical type of cotton pickers are being used to pick the cotton from the bolls but they have not been successful because the cotton fiber is not sufficiently attracted to the electrically charged belt or finer to be pulled from the boll. Electrical cotton harvesters depend on attracting cotton fiber to a statically charged belt or finger to remove the cotton from the boll.

Pneumatic type: The pneumatic cotton harvester removes the cotton from the bolls either by suction or by blasts of air. This type of cotton harvester works on the suction or vacuum principle for harvesting cotton (Fig. 7.109). The tight fitting of locks of cotton in bur makes it difficult to suck out the cotton. Several long hose having picking nozzle at their ends are attached to large tanks from which the air is pumped to create a vacuum. Cotton is picked by placing the nozzle close to the cotton bolls and opening a valve, the intrusting air draws the cotton out of the bur, through the hose and into the tank. It is rather difficult in more storm proof varieties, by air suction, to draw cotton out of the bolls. The blast type of picker consists of gathering cotton by use of blast which is directed against the bolls in such a manner as to separate the cotton from the stalk and propel them into receptacle. A blast of air is strong enough to blow cotton out of bolls and carries a high percentage of trash.

Fig. 7.109 : Cotton picking on suction principle

Knapsack type suction cotton picker

Pneumatic cotton picker (Fig. 7.110 and Fig. 7.111) has been developed to pick the cotton bolls and store in a drum made of a translucent polypropylene container of 50 lit capacity fixed on the frame (Anonymous, 2015). The machine picks about 9 kg/h cotton which is thrice the capacity of manual picking, the average picking efficiency is 95% and it can be increased with skilled labour with proper training. The average trash content of the plucked cotton varies from 5-6%. The machine saves about 35% in cost of operation and 57% in time as compared to manual picking. With multiple suction pipes the machine capacity can be increased (Fig. 7.111).

Fig. 7.110 : Pneumatic cotton picker with single suction pipe

Fig. 7.111 : Pneumatic cotton picker with multiple suction pipe placed on trolley

Battery operated manual cotton picker

The machine consists of a picking device, battery, and cotton collecting bag (Fig. 7.112). Picker is operated manually with collecting bag attached to picking device. It has picking efficiency of about 70% where as manual picking has about 98% efficiency. However, picking capacity is 2-3 times more with this machine as compared to manual picking. It has trash content of about 2%. It is very light in weight and compact in size. Along with this machine, a battery charger, battery holding bag are also provided. This machine being battery operated helps in preventing air and noise pollution.

Fig. 7.112 : Battery operated manual cotton picker
Courtesy: Almighty Agrotech Pvt. Ltd. Rajkot (Gujarat)

Mechanical pickers

The basic components of a mechanical picker are; i) an arrangement for guiding the plants into the picking zone and proving necessary support while the seed cotton is being removed; ii) device to remove the cotton from open bolls; iii) a conveying system for picked cotton; and iv) a storage basket or a container in which picked cotton is stored temporarily. Pickers are available as one row or 2–row machines. These are either self-propelled or tractor mounted. A mechanical picker must be capable of gathering mature seed cotton with a minimum of waste and without causing serious damage to the fiber plant or unopened boll to ensure the highest possible grade of ginned cotton. The harvested material should have a minimum of leave, stems, hulls, weed and other foreign material. Secondly stream lining of the plant passage way while it is passing through the machine, by way of use of suitable limb lifters and synchronization of moving picker parts with the forward travel of the machine, all tend to minimize disturbance to the plant.

The cotton picker performs the work of hand picker in that only the locks of seed cotton are removed from the plant. There are four ways of classifying cotton pickers (Anonymous, 2010). They are by method of mounting, by number of rows harvested, by height of picking drums and by type of spindle used. It can be tractor mounted machine or self-propelled of one or two-rows (Fig. 7.113 and 7.114). In this, the cottonseed is removed from open bolls; whereas, green and unopened bolls are left on the plant to mature for later picking. A mechanical picker consists of a device to guide cotton plants to come into picker (Fig. 7.115) device to remove cottonseed from open bolls; a conveying system for picked cotton and a storage basket. These machines are capable of gathering mature cotton with a minimum of waste and without causing serious damage to the fibre plant and unopened bolls. The high yielding, long fibres and open-boll varieties of cotton are defoliated before the first picking.

The mechanical cotton pickers have either tapered spindles or small-diameter straight spindles. Spindles are carried either on bars arranged in vertical drums or on vertical slats attached to endless chain. Tapered spindles, commonly employed on drum-type pickers, have 3 or 4 longitudinal rows of sharp barbs for engaging the cotton bolls. Tapered, barbed spindles enter the plant perpendicular to row and wound the exposed lint on the berbs. As the spindles passes slowly, faster rotating rubber-faced buffers remove lint. Speed of spindle varies from 1850 rpm at forward speed of 2.9 km/h to 3250 rpm at 5.0 km/h. Speed of spindle influences the picking efficiency for fluffy bolls and increases from 80% at 700 rpm to 95% at 2300 rpm. The loss at higher speed is mainly due to cotton thrown by spindles and loss at lower speed is mainly due cotton left in burs. Straight spindles are longer than tapered type but smaller in diameter. They may be round or square and may have smooth or rough surface. In general, picking ability of pickers depends upon the spindles being wet when they come in contact with cotton. A stationary cam and followers on the bar achieve the proper orientation of spindle bars in relation to crop row. There are about 15 or 16 spindle bars on the front drum and 13 or 12 on the rear drum. Each spindle bar has 20 spindles on high-drum pickers and 14 spindles per bar on low-drum pickers.

Fig. 7.113 : Self-propelled two row cotton picker
Courtesy: New Holland Fiat (India) Pvt. Ltd

Fig. 7.114 : Self-propelled two rows cotton picker
Courtesy: John Deere India Pvt. Ltd.

Fig. 7.115 : Cotton plants being trap into picker
Courtesy: New Holland Fiat (India) Pvt. Ltd

Spindles: The basic principle of a revolving spindle penetrating the cotton plant, winding the seed cotton from the open boll and retreating to a doffing zone is employed by all commercial pickers (Fig. 7.115). The rearward movement of the spindles while in the picking zone is substantially the same as the forward movement of the machine so that the spindles do not move forward or backward with respect to the cotton plant. Each rotating spindle merely probes straight into the cotton plant from the side of the row, works on an open boll and then withdraws straight into the side with a minimum disturbance and damage to the remaining plant. The spacing of spindles is such that they can slip past unopened boll and leave them onto the plant to mature for a later picking.

Drum type spindle arrangement: The two drum picks up the bolls from the two sides of the row towards the spindle in the picking zone

(Fig. 7.116). In current high drum machines the front drum has 15 or 16 spindles bar and rear drum has 13 or 12 bars, with 20 spindle per bar. This gives a total of 560 spindles per row of cotton, each spindle requiring a precession fit sleeve bearing and being driven through the bevel gear by a shaft inside the spindle bar. They are suitable for low growing or medium height cotton. The proper orientation on the spindle bar in relation to row is obtained by means of a stationary cam and a follower in the bar.

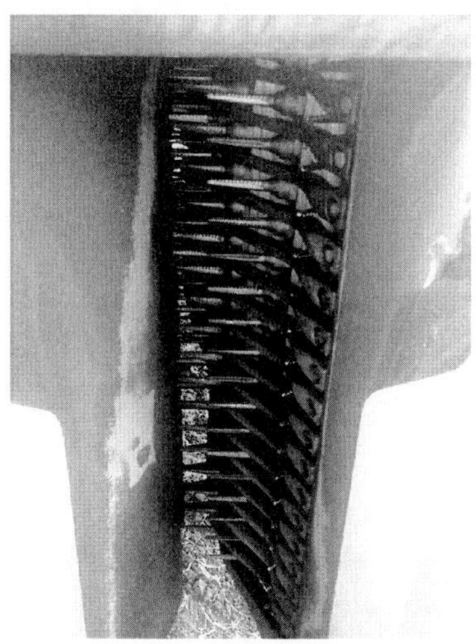

Fig. 7.116 : A view of drum type spindle
Courtesy: New Holland Fiat (India) Pvt. Ltd

Chain belt spindle arrangement: The picking process with a chain belt unit is essentially the same as with the drum type picker although the chain belt principle permits the spindle to remain in picking zone for a longer time. The spindle is normally straight. Each spindle is rotated by means of a roller in contact with a stationary, rubber drive rail, but only while on the picking side of unit. Guide strip holds the chain in position between the main sprockets and provide a curvature for moving the spindles laterally into and out of row. Each slat is pivoted between the upper and lower chains.

Spindle moistening: Spindle of either type are moistened with water for two reasons: i) as an aid in picking because of cotton adheres better to a wet steel surface; and to keep spindle clean as some gummy substances stick while these are in picking unit. Some wetting agents are also used which reduces the amount of water required for moistening and at the same time makes it more effective. A spindle moistening system is provided for each picking unit, water being metered in equal amount to each spindle. Application is made to each spindle just before it enters the picking zone, by means of a specially designed rubber wiping pad (Fig. 7.117). On machine with tapered spindle, the seed cotton is removed from the spindle by means of rotating doffer plates. The cotton is forced off as the doffer lug moves over the spindle surface toward the tip. With small diameter straight spindles stripping is accomplished by moving the spindles axially through the space between the closely fitted stripper shoes. Tapered spindles are rotating when doffed, whereas small diameter straight spindles are not.

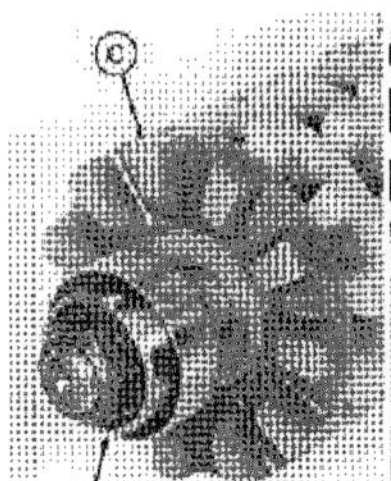

Doffer

Doffers and Moistening Pads in relation to spindles

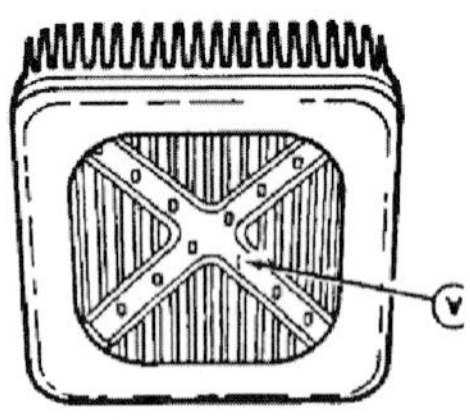

Moistening Pads

Fig. 7.117 : Spindle moistening system
Courtesy : New Holland Fiat (India) Pvt. Ltd.

Conveying and carrying: A pneumatic conveying system is used to move the cotton from the doffing area to storage container. The cotton is blown through the discharge duct against cleaning grate in the storage basket lid. This action removes some of the trash from the seed cotton. Machine with dual picking unit has 2 separate elevating systems to provide more uniform and positive conveying from each unit. Some machine use off set arrangement to eliminate contact between the cotton and the fan blades. Basket is emptied by with hydraulic cylinder.

Working of cotton picker: The plants are folded into the throat of the machine (Fig. 7.118) by rounded members and two limbs and bolls are lifted up by limb lifting fingers. The revolving drum or belt of rotating spindles passes moistening pads and spindles are thoroughly wetted. The spindles then are projected in among the limbs and bolls to emerge the cotton. The plants are passed against around the spindles by spring load pressure plates. The spindles rotate and come in contact with rubber doffers or stripper bar doffers. The cotton drops into the air conveying system, which conveys the cotton to the basket. Air, dust and trash are exhausted throughout grates, which deflect the cotton into the basket. Hydraulic cylinders tilt the basket and dump the cotton into a trailer (Fig. 7.119).

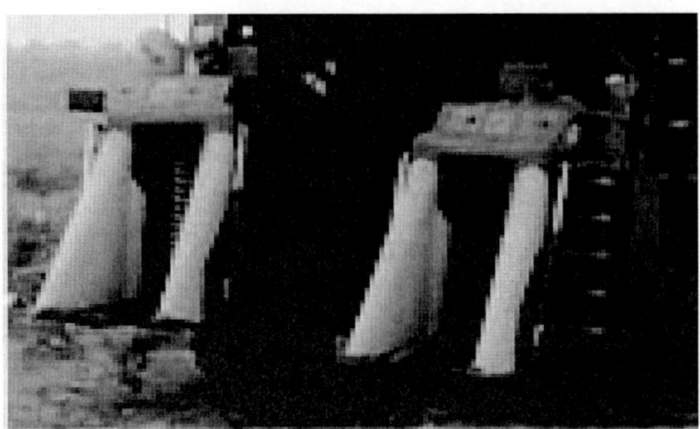

Fig. 7.118 : Picker units in two row self-propelled cotton picker
Courtesy: New Holland Fiat (India) Pvt. Ltd

Fig. 7.119 : Unloading of mechanical picker
Courtesy: John Deere India Pvt. Ltd.

Drum type cotton picker: Tapered spindles are used in drum type cotton picker. The spindles are mounted on a bar, the top end of which has a crank arm and a bearing that travels in a cam track. The cam actuated picker bars swing the bars around so that the rows of spindles are about 40 mm apart as they enter the cotton plants. As the spindles are spaced 40 mm on the bars the spindles are spaced 40 mm both vertically and horizontally. The proper orientation of the spindle bars in relation to the row or to the doffers is obtained by means of a stationery cam and followers on the bars. All linear motion of spindles while in the picking zone is at right angle to the row. The spindles enter the cotton plant through a grid section, which subsequently prevents the plants from being pulled into the doffing area as the loaded spring withdraws. The spindle drums are operated in pairs, one drum on each side of row, but not directly opposite. The front and rear drums have different speeds 60 rpm and 79 rpm. The spindles also rotate at different speeds (2200 rpm and 2700 rpm) at the same forward speed (generally 3.2 to 5.6 km/h) as the machine moves along the row. High drum has 20 spindles per bar and low drum model has 14 spindles per bar.

There are many factors that affect the performance of cotton pickers. Cotton pickers perform best when cotton plants are of medium size. Medium-sized plants flow through the machine and permit the spindles to engage the cotton better than large plants with many long limbs. The machine requires a well-opened boll with locks that are fluffy and fiber that is long enough to wrap around the spindle. Chemical defoliation is generally done before the picking operation, which helps the simultaneous opening of most of the cotton bolls. A delay in picking and early opening of cotton boll result in atmospheric damage to the exposed cotton fiber. A slight elevation of the soil at the base of plant and weed-free fields are essential for better performance of cotton pickers. Fairly thick and uniform spaced plants aid the performance of mechanical cotton picker.

Tractor operated two row cotton pickers

The tractor operated two row cotton picker is operated by 75 hp tractor (Anonymous, 2010). This picker is having two picking units and each unit has two picking drums (Fig. 7.120). Each drum has the revolving spindles to remove the seed cotton from the plants. The picker is having other important units such as doffer unit for removing seed cotton from the spindles, seed cotton conveying unit, storage basket and spindle moistening unit. The crop is sown with 100 cm row to row spacing. The plant to plant spacing can be 20 cm, 30 cm and 60 cm. As the mechanical picking is to be done the crop is defoliated by application of defoliant spray. The mechanization of cotton harvesting demands compact and synchronous maturity of the plant. The height and spread of the crop should be within 100 cm and 30-35 cm. The height of the lower bolls should not be less than 15-20 cm. It has field capacity of 0.5 ha/h and picking efficiency 85%.

Fig. 7.120 : Tractor operated two rows cotton picker
Courtesy: John Deere India Pvt. Ltd.

Mechanical strippers: There are several factors that have contributed to popularity of strippers in preference to pickers. These are: i) lower initial investment & maintenance cost; ii) better adapted to improved cotton varieties; iii) improved ginning equipment for separating trash; iv) trend toward closer row spacing; v) good recovery of cotton in the field; and vi) higher harvesting speeds. Cotton bolls exhibiting too much of storm resistant characteristic although well adapted to striping are difficult to pick mechanically. The size of plant, the type of growth and the nature of the boll all have more influence on the efficiency of the mechanical picker than does yield. Where the plant characteristic are suitable, a machine will pick up high yielding cotton just as efficiently as it will low yield cotton.

Cotton stripping machines are "once over" machines. In the stripper, all bolls whether opened or closed are removed from the plant in a single pass. Harvesting with a stripper is, therefore, usually delayed until the plants shed their leaves. Chemical defoliants are also some times applied to permit earlier stripping. An ideal variety for this type of machine is one with semi-dwarf plants with relatively short fruiting, short-nodded branches, storm-resistant bolls borne singly but having fairly fluffy locks for good extraction (stripping) and medium sized boll-stem. Stripping a variety that produces a widespread plant with numerous vegetative and fruiting branch results in low recovery of cotton and excessive field losses.

Cotton stripping machines are of single steel roller, double-steel roller or finger type. Double-brush nylon rollers used in place of steel rollers gives better performance. A plant population of 75-125 thousands per hectare in 1.2 m rows is commonly recommended for stripping harvesting. The double-roll cotton stripper may be centrally mounted on the tractor or it may be self-propelled. There are three different methods of conveying cotton from stripping unit viz. finger-beater rolls, augers and air. The finger-beater rolls are used with finger-type strippers. Auger-type of conveyer is suitable for roller-type strippers. Much dirt and trash can be screened out of the cotton through openings in the housing under the conveyers particularly where revolving beater conveyers are used.

There are many factors such as plant characteristics, cultural practices etc affects the performance of all types of mechanical cotton strippers. The desirable plant type for cotton stripper is one which has relatively short-noded fruiting branches 20-25 cm in length, less in height and has a stormy-resistant boll. The locks of story-resistant-type cotton are usually not very fluffy and are held tightly in the soil. Fluffy and loosely attached locks are easily caught and held between limbs and thus are pulled through the stripping space and lost. Every effort should be made to keep the field free of weeds, grass and vines. Pieces of grass collected with cotton are difficult to remove and if present in excess reduce the quality of cotton lint. The design and type of stripping unit also affects the performance of strippers.

Principle of cotton stripping: Mechanical stripping of cotton is accomplished by forcing the plants through an area too small for the bolls to pass. When plant is passed through the machine head, forward motion of machine applies a force in a direction perpendicular to ground, which separates the cotton bolls from plant. These snapped bolls are collected in the machine while plant remains in the row. The roots of plant should be strong enough to withstand the force applied by machine. Type of cotton stripper is finger type and brush type. The finger on the comb type stripper is 61.0 to 91.4 cm in length and set at angle of about 10 degree to 15 degree to the horizontal. Spacing between fingers is kept 16 mm. It consist of stripping fingers, stalk walker kicker, auger, cleaning grids, kicker wheels, green boll separator pneumatic air conveyor system, grate and auger to pneumatic conveyor system.

Working of finger type strippers: As the harvester moves through the field, the forward motion of the inclined teeth snapped the cotton bolls from the plant. Stripped cotton is moved up the inclined teeth by the next group of stalks being harvested. A reel mounted over the rear of the teeth kickes the cotton into a cross auger. The cotton passes through the cleaner which utilizes the air velocity method and combined cleaning techniques of combing and sling off type to remove the sticks, burs, crushed leaves and soil from the mature open bolls. The clean seed cotton is then moved pneumatically up into the basket. Limitation of finger type stripper's is that it works well only under certain conditions, such as plant should be small having low vegetative growth, properly defoliated and atmosphere should not be damp (humid). Operating speed usually does not exceed 3.2 to 4.0 km/h and sometimes must be considerably lower. To counter the limitation of finger type stripper, brush type stripper has been introduced. Main advantage is that the finger type stripper can be used for harvesting of any row width of cotton.

Brush type stripper: Brush type cotton harvester is classified according to number of rows it can harvest and type of brush. Brush type stripper is used primarily on standard single row planting, employ a pair of rolls that are approximately 1020 mm or 1170 mm long and 152 to 165 mm in diameter and are mounted at an angle of about 30 degree above horizontal. Their surfaces are fitted with longitudinal brushes alternated with rubber paddles at 45-degree interval around steel tube (behind). The rolls are driven at about 600 rpm with their adjacent surfaces moving upward next to the plants. As the bolls are snapped off, they are moved away from the plants by the surface of rolls and delivered on to the adjacent conveyors. These conveyors convey seed cotton to cleaner, which cleans the trash content from seed cotton, and then clean seed cotton is conveyed into basket. The clearance between rolls is adjustable and usually set at 3 to 6 mm. Centre to centre spacing of two rolls is 1.02 m. Advantage is that the flexibility of the roll surfaces permits a degree of automatic adjustment for light and heavy growth of plant. The spacing between the rolls can be adjusted manually to meet extreme conditions. Limitation is that at high rates of travel, it is difficult to steer the tractor so that narrow space between the stripping rolls is always in line with the row of plants. Two row brush type stripper has been developed to meet the demand of growing cotton in narrow rows having spacing 30 to 40 cm apart.

Two roll units have a centre to centre spacing of 356 mm instead of 1020 mm in single row stripper. This is the minimum practical spacing with the 165 mm diameter rolls. With careful driving double row spacing of 30 cm to 40 cm on each bed can be accommodated satisfactorily. Part of the stripped cotton from both rows passes down through the 25 mm space between the two centre rolls into an auger conveyer that required a tapered front section to fit into the limited space available between the rolls and top of bed. Cotton moved outward by the rolls is collected by two straights - auger located below and outside of the outer rolls and conveyed rearward to the cross auger. The rolls and the row unit augers are operated at 600 rpm.

References

Ahuja S S; Sharma V K. 2001. Development of Axial Flow Threshers for Sunflower & Paddy. Proceedings of Summer School on 'Advances in Seeding and Harvesting Machinery'. Dept. Farm Power & Machinery, PAU Ludhiana. May 21 to June 19. Pages 132-139.

Anonymous. 1984. Consolidated Report 1965-79. AICRP on Farm Implements & Machinery (ICAR). Dept. of Farm Power & Machinery, P.A.U. Ludhiana.

Anonymous. 2008. Research Highlight. AICRP on Farm Implements and Machinery, CIAE Bhopal. Technical Bulletin No.: CIAE/2008/141.

Anonymous. 2008a. Annual Report (2006-08). AICRP on Farm Implements and Machinery, OUAT Bhubneshwar.

Anonymous. 2008b. Annual Report (2006-08). AICRP on Farm Implements and Machinery, KAU Tavanur.

Anonymous. 2008c. Success Stories. AICRP on Farm Implements and Machinery, CIAE Bhopal. Extension Bulletin No.: CIAE/FIM/2008/80.

Anonymous. 2008d. Annual Report (2006-08). AICRP on Farm Implements and Machinery, MPUAT Udaipur.

Anonymous. 2008e. Annual Report (2006-08). AICRP on Farm Implements and Machinery, TNAU Coimbatore.

Anonymous. 2010. Research Highlight. AICRP on Farm Implements and Machinery, CIAE Bhopal. Technical Bulletin No.: CIAE/2010/151.

Anonymous. 2010a. Annual Reports (2008-10). AICRP on Farm Implements and Machinery, ANGRAU Hyderabad.

Anonymous. 2010b. Annual Reports (2008-10). AICRP on Farm Implements and Machinery, PAU Ludhiana.

Anonymous. 2010c. Success Stories. AICRP on Farm Implements and Machinery, CIAE Bhopal. Extension Bulletin No.: CIAE/FIM/2010/83.

Anonymous. 2010d. Annual Report (2008-10). AICRP on Farm Implements and Machinery, MPUAT Udaipur.

Anonymous. 2010e. Annual Report (2008-10). AICRP on Farm Implements and Machinery, MPKV Rahuri.

Anonymous. 2012. Research work on Potato cultivation. Department of Farm Machinery & Power Engg. PAU, Ludhiana

Anonymous. 2013. Research Highlight. AICRP on Farm Implements and Machinery, CIAE Bhopal. Technical Bulletin No.: CIAE/2013/158.

Anonymous. 2013a. Success Stories. AICRP on Farm Implements and Machinery, CIAE Bhopal. Extension Bulletin No.: CIAE/FIM/2013/40.

Anonymous. 2015. Research Highlight. AICRP on Farm Implements and Machinery, CIAE Bhopal. Technical Bulletin No.: CIAE/2015/179.

Garg I K; Singh Surendra. 2002. Farm equipment for Punjab agriculture. Department of Farm Power & Machinery, Punjab Agricultural University, Ludhiana.

Garg I K; Singh Surendra. 2003. Comparative performance of different harvesting and threshing systems for cereal crops. Paper presented at XXXVII Annual Convention of ISAE, held at MPUAT, Udaipur, Jan 29-31, 2003.

Garg I K; Singh Surendra; Sharda Ajay. 2003. Advances in Combine Harvesters for Cereal Crops" Paper presented at National Seminar on 'Recent Advances in Harvesting & Threshing of Cereal crops', Organized by ISAE & Department of Farm Power & Machinery at Dept of FPM, PAU, Ludhiana.

Gupta P K. 2001. Design and development of vertical conveyer reaper windrower. Proceedings of Summer School on 'Advances in Seeding and Harvesting Machinery'. Dept. Farm Power & Machinery, PAU Ludhiana. May 21 to June 19. Pages 158-161.

Mittal V K. 2001. Advances in Maize Harvesting and Shelling Equipment. Proceedings of Summer School on 'Advances in Seeding and Harvesting Machinery'. Dept. Farm Power & Machinery, PAU Ludhiana. May 21 to June 19. Pages 80-84.

Pandey M M; Ganesan S. 2005. Farm Mechanization Package for Dryland Agriculture, Technical Bulletin No: CIAE/FIM/2005/117. Central Institute of Agricultural Engineering, Bhopal.

Pandey M M; Majumdar K L; Singh Gyanendra; Singh Gajendra. 1997. Farm Machinery Research Digest. Technical Bulletin No. CIAE/97/69, Central Institute of Agricultural Engineering, Bhopal, 328 p.

Pandey M M; S Ganesan; R K Tiwari. 2006. Improved Farm Tools and Equipment for North Eastern Hills Region, Technical Bulletin No. CIAE/2006/121. Central Institute of Agricultural Engineering, Bhopal.

Saijpaul K K. 2001. Recent advances in grain harvesting equipment. Proceedings of Training Course on Advances in Agricultural Machinery Design under Centre of Advanced Studies. Department of Farm power & machinery, PAU Ludhiana. March 6-26. Pages 118-126.

Sandhar Nirmal Singh; Goyal R. 2001. Working Principle, Constructional Details and Operation of Cotton Harvesting Equipment. Proceedings of Training Course on Advances in Agricultural Machinery Design under Centre of Advanced Studies. March 6-26. Pages 191-204.

Sharma V K. 2001. Advances in threshing and straw bruising systems. Proceedings of Training Course on Advances in Agricultural Machinery Design under Centre of Advanced Studies. March 6-26. Pages 29-45.

Sharma V K; Gupta P K; Singh Surendra; Singh C P. 1986. Power requirements of different systems of spike tooth wheat threshers. *J. of Institution of Engineers* (Agril. Engg.). 66(1): 17-20.

Sharma V K; Singh Surendra; Gupta P K; Garg I K; Singh C P. 1986a. Development of multi crop threshers in India. *Agril. Engg. Today* (ISAE). 10(2-3): 11-18.

Singh P R. 2014. Souvenir, Tractor and Agricultural Machinery Manufacturers' Meet (TAMM 2014). Held at IISR Lucknow during February 8-9.

Singh S P. 2010. Ergonomical Interventions in Developing Hand Operated maize dehusker sheller for Farm Women. Ph. D Thesis submitted. MPUAT, Udaipur.

Singh S P; Singh Surendra; Singh Pratap. 2011. Hand operated maize dehusker-sheller: an option for small & hill farmers. Agrovet Buzz. Vol. IV, Issue 1: 15 January-15 March: 20-21.

Singh S P; Singh Surendra; Singh Pratap. 2012. Ergonomics in developing hand operated maize dehusker – sheller for farm women. *Applied Ergonomics. 43: 792-798.*

Singh Surendra. 2007. Farm Machinery – Principles and Applications. Directorate of Information & Publication of Agriculture, Indian Council of Agricultural Research, Krishi Anusandhan Bhawan-I, Pusa Campus, New Delhi.

Singh Surendra. 2009. Harvesting of Fruits and Vegetables. Lecture delivered during Winter School on 'Novel Technique in Packaging, Storage, Processing and quality Control of Fruits and Vegetables' organized by CIAE Bhopal from Feb. 24-March 16.

Singh Surendra; Pandey M M. 2008. X Plan Achievements (2002-2007). AICRP on Farm Implements and Machinery, CIAE Bhopal. Technical Bulletin No.: CIAE/2008/137.

Singh Surendra; Sharma V K; Garg I K; Gupta P K. 1986. Studies on harvesting & threshing systems for paddy in Punjab. *J Res Punjab agric Univ.* 23(2): 281-88.

Singh Surendra; Sharma V K; Gupta P K; Chauhan A M. 1985. A ridge planter for winter maize. *J. of Agril. Engg.* 22(2): 115-18.

Singh Surendra; Sharma V K; Gupta P K; Chauhan A M. 1988. Design, development and evaluation of two-row tractor mounted ridge planter for winter maize. *Indian J. of Agricultural Sciences*. 58(1): 45-47.

Singh Surendra; Verma S R. 2009. Farm Machinery Maintenance & Management. Directorate of Information & Publication of Agriculture, Indian Council of Agricultural Research, Krishi Anusandhan Bhawan-I, Pusa Campus, New Delhi.

Srivastava NSL. 2001. Advances in sugarcane harvesting equipment. Proceedings of Summer School on 'Advances in Seeding and Harvesting Machinery'. Dept. Farm Power & Machinery, PAU Ludhiana. May 21 to June 19. Pages 101-113.

Verma S R. 2001. Advances in potato and groundnut harvesting machinery. Proceedings of Summer School on 'Advances in Seeding and Harvesting Machinery'. Dept. Farm Power & Machinery, PAU Ludhiana. May 21 to June 19. Pages 140-157.

Verma S R; Sharma V K; Singh Surendra; Ahuja S S; Singh Santokh (eds.). 1992. Proceedings of National Colloquium on Potato Mechanization in India. *Punjab Agricultural University, Ludhiana.*

Verma, S R, Rawal, G S and Bhatia, B S. 1978. A study on human accidents in wheat threshers. *J. Agric. Engng.* ISAE Vol. 15(1): 19-23.

Yadav R N S. 2003. Sugarcane Production Machinery, CIAE, Bhopal. Bulletin No. CIAE/NATP/-SM/2002/95.

8

Forage Harvesting Equipment

Forage harvesting is more complex as compared to grain crops. Forage crop is of great bulk and mass, containing 70-80% water when first harvested. For storage it must be dried, either naturally or artificially, to a safe moisture content of about 20 to 25%. Due to low product value it limits the economic feasibility of mechanization of harvesting operation. The mowers are used to cut the crop and windrow in the field, which is manually collected for further chopping. The forage harvester consists of a combination of plant-cutting unit and a chopping unit and is called field forage harvester. The field forage harvester performs the functions of both row binder and silage cutter, as it severs the standing stalks from the ground and chops them into silage in one continuous operation in the field. It can be grouped into two according to the mechanism used to cut the crop viz. field choppers and flail harvesters.

A variety of fodder crops like barseem, corn, sorghum, bajra are grown throughout the year on most of the Indian farms. Fodder harvesting practices followed in India are far different from that accepted in developed countries (Singh, 2001). Regular harvesting of fodder with traditional or improved sickles is continued throughout the year, which requires labour of the range of 160 man-h/ha. In advanced countries where dairy is well-developed most of the operations are fully mechanized. The machines used include self-propelled and/or pulled type windrower, shredders, hay conditioners, field choppers, harvesters and choppers, balers (along with bales handling machinery) and even machines to make wafers, raking machines are useful machines and so on. Balers and cube making machines are useful when low moisture storage of fodder is required or when transporting distances are large. These machines are not of much relevance to our conditions where fresh harvested fodder is locally fed. This limits our interest to cutting and chopping machinery. Also average size of land holding and other socio-economic factors is also a hurdle for the large size bailing and handling machines. Taking such factors into consideration, the focus is limited to cutting and chipping machinery. Fodder harvesting includes operations of cutting the standing crop; its conditioning, transport and chopping. Therefore, in India simple and efficient forage harvesters are required that can harvest the crop economically and without causing much loss to the crop.

Shear cutting: To cut a material, force must be applied on it in shear at certain plane. Failing of the material in shear is desirable but generally shearing takes place along with certain deformation and bending of stems. This deformation and bending is undesirable which is wastage of energy. A common way of applying cutting forces is to press the material between two shearing elements, which move against each other without or with very small clearance. A pair of scissors is a good example. Either one or both of the elements may be moving. The motion between two elements may be linear, rotary or reciprocating.

Impact cutting: A single element shearing is also possible if the surface holding the plant or inertia of the stem provides sufficient resistance against the cutting element. If cutting is done close to the ground, ground itself serves as one cutting element. A high-speed cutting blade does not give much time to the plant stem to bend or deform before it is sheared off. This type of cutting is commonly known as impact cutting.

Impact type cutters

Rotary cutters: Rotary cutters have blades rotating in a horizontal plane, installed on periphery of discs or drums rotating on their vertical axes. Rotary cutters used for forage harvesting usually have counter rotating pair of drums or two pairs of discs (Fig. 8.1). Each drum or disc usually has two, sometimes 4 knives. However, machines having more sets of cutting drums may have width of cut up to 6 m. Peripheral speed vary from 50 to 80 m/s depending upon type of crop, its maturity etc. Blades are sometimes hinged through vertical pins on periphery of discs to provide a back swing if some hard object is encountered. The counter rotating drums have a tendency to make a windrow in between. Nice windrows can be obtained by using shields depending upon type of crop. The drums get their power through a common shaft rotated by PTO through V-belts. From common shaft drive is given through chains or bevel gears to drums so that relative rotating of drums is positively fixed. This avoids fouling of blades of two adjacent discs or drums, which have slight overlap. These types of cutters differ from conventional cutter bars in many ways. Unlike cutter bars they can be operated at high forward speeds (up to 14 km/h). There is hardly any chance of choking of the machine if proper speed of rotation is maintained. They need comparatively low maintenance. But their initial prices and power requirements are quite high in comparison to simple cutter bars.

Fig. 8.1 : Vertical rotary blade type forage harvester

Mower

The mower is a machine mainly used for harvesting grasses and forage crops. It cuts the stems of standing vegetation to make hay out of them. The mower cutter bar is capable of cutting the stems at 3-10 cm above the ground. There are different types of mowers used for cutting grass and forage crop such as cylinder (Fig. 8.2), reciprocating (Fig. 8.3), horizontal rotary and flail type mowers (Fig. 8.4). According to the source of power, mowers can be classified as manually operated, animal-drawn, tractor-drawn and self-propelled. According to the mode of hitching, mowers can be classified as trailed type, semi-mounted and integral mounted type. Semi-mounted and integral-mounted mowers can be further classified as rear, mid and front-mounted. According to drive used, mowers can be classified as ground-driven, engine-driven and PTO driven. Cylinder type mower is generally used for lawn mowing. It includes lawn mowers and gang mowers. It could be hand-propelled and self-propelled lawn mowers and tractor-drawn and tractor-mounted gang mowers. It has rotating helical blades arranged in horizontal plane (Fig. 8.5). There are three gangs in this mower having 3-4 helical blades each supported over vertical discs. The rotors of this type of mowers are supported on ball bearings at each end. With the rotation of blades, forage or grasses are cut for example lawn mower. It can cover about 1.5-2.0 ha/h at a speed of 8-10 km/h.

The flail type forage harvesting is simple in design and construction. It can be used for cutting any type of forage crops and grasses including lodged crops. Overload encountered by the machine during operation is absorbed by freely hanging flails and thus machine components are protected from field damage (Singh, 2007). Flail type mowers have large number of blades staggered in 3-4 lines on a drum rotating on a horizontal axis perpendicular to the direction of motion (Fig. 8.2 and 8.4). Width of blades varies from 50 to 150 mm and also vary in shape. These blades or knives are attached to the drum through hinges or small ring links to avoid the damage if any hard object comes in contact. Effective width of such machines varies from 1.2 to 4.0 m. Initially flail shredders were used to cut weeds before plowing or mowing lawns. The peripheral speeds of the knives were 50-60 m/s. When such machines were modified for fodder; the peripheral speeds were reduced to 40 m/s to avoid multiple cutting and field losses of the crop. Even though the field losses are more than the conventional cutter bars. Flail mowers also help in conditioning of the hay as they spread the crop in uniform windrow. Power requirement of these machines is also very high, about 25 kW per meter width. It depends upon many factors like crop density, forward speed etc. Their high initial costs are justified for their performance under difficult conditions. This makes it a unique machine to be used in stony, stumpy and uneven land. Forage crops cut by flail mower do not require any additional chopping by chaff cutter. This is done in the process of harvesting in the field itself by repeated beating action of the flails on the plants. Tractor rear mounted hydraulically controlled PTO operated flail type forage harvesting machine is versatile for numerous field operations such as fodder maize cutting, grass topping and mowing, green manure topping, cutting shrubs on river banks and drainage channels etc. These mowers are extremely sturdy and as such can work under varying field conditions. When used for grasses, it causes excessive field loss as short pieces are hardly recovered in raking or during picking from the windrows. Losses can be reduced using lower peripheral speeds. A flail mower has a full width adjustable gauge roller located behind the rotor to provide accurate control of cutting height and prevent scalping of high spots.

Pull type flail mowers are generally attached at the back of tractor and kept offset so that the tractor wheel runs on the cut hay rather than on the standing crop. Some of these units are provided with augurs to convey the cut material into trailers. It uses free-swinging chains and knives to sever the plants by beating or cutting action (Fig. 8.6). The flails or knives travel in the same direction the machine moves. It does not have chopping knives on the blower fan to chop the material into desired length for silage. The beating by flails more or less conditions the hay.

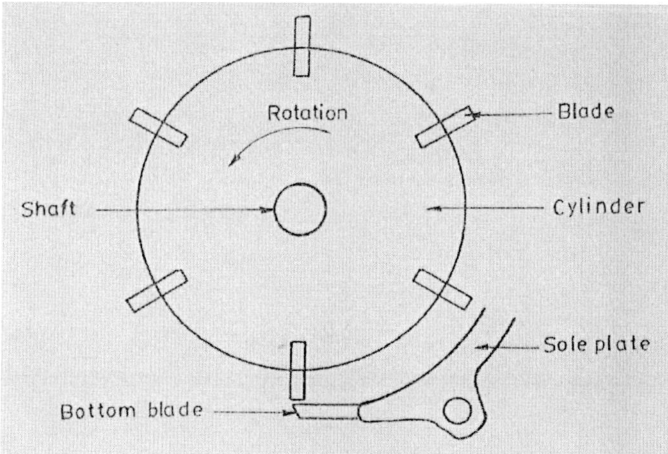

Fig. 8.2 : Cylinder mower

Fig. 8.3 : Reciprocating type mower

Fig. 8.4 : Flail type mower

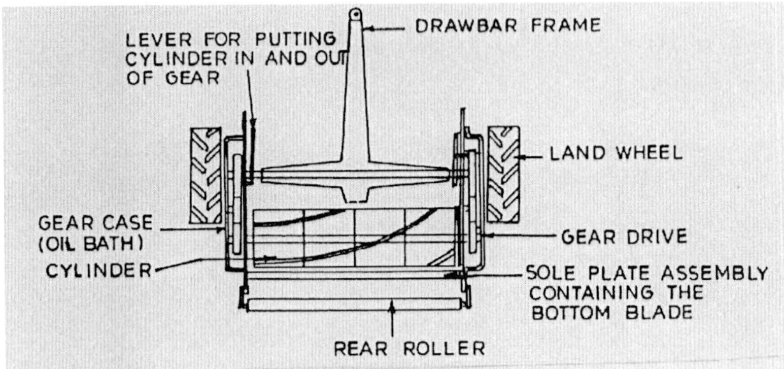

Fig. 8.5 : Gang mower

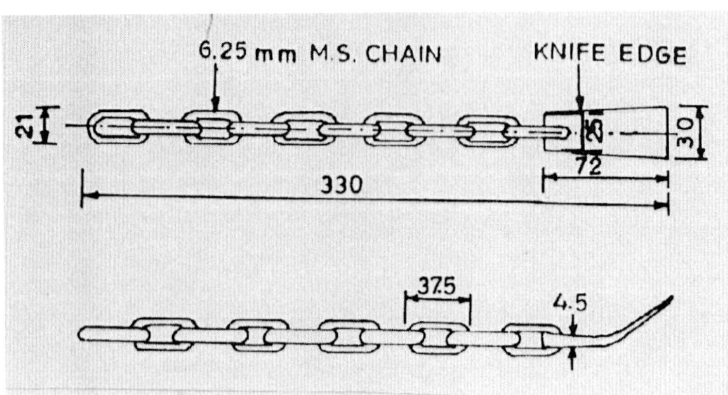

Fig. 8.6 : Free swinging chains

Reciprocating-knife type mower: It has knife that moves against stationary fingers. It is most common type of mower used for harvesting of forage crops. Horizontal reciprocating type mower has knifes which rotate at high speed in horizontal plane (Fig. 8.7). Due to reciprocating action of knife, grasses are cut uniformly. The conventional mower has the main parts: a cutter bar to cut the crop and separate it from uncut portion, power transmission unit to receive and transmit motion force, frame to support moving parts, wheels to transport and for operating the cutting mechanism, and auxiliary parts to lift and drop the cutter bar.

Fig. 8.7 : Reciprocating knife type mower
Courtesy: BCS India Pvt. Ltd., Ludhiana (Punjab)

Cutter bar: Cutter bar is designed to cut crop 25-50 mm above the ground level up to 100 mm or more. It must be able to cut a wide variety of crop materials from grasses to tough stocks without clogging. A number of attempts have been made to replace the reciprocating knife with continuous moving cutting units such as chain or band knives in helical roller cutters but none have performed so well that commercially feasible. It is steel flat made with high-grade steel. All other sub component included in the cutting mechanism is connected directly or indirectly to it.

Cutter bar assembly: It is heart of the mower. The knife is metal bar on which blades are mounted that are triangular shaped sections (Fig. 8.8). The cutting edges of these sections are mostly smooth edges. The knife sections move back and forth and cut plants in both directions. The section of the knife always stops at the center of the guard on each stroke. The length of the stroke is 7.5 cm. Guards are provided with ledger plates on which the knife sections move. Knife clips hold the sections down against ledger plates but allow it to move freely. Knife clips are placed together with the wearing plates to absorb the rearward thrust of the crop to the knife. A badly worn wearing plate or a loose knife clip may allow the knife to bend. This is a common cause of the knife head, breaking at the point where it is attached to the bar. There is a grass board on the outer end of the mower, which causes the cut plants to fall towards the cut material. The angle of the board can be changed for different crops. A shoe on each end of the cutter bar is provided to regulate the height of the cut above the ground. The inner shoe has a larger area of contact with the ground than the outer one. This results in smooth and easy sliding of the cutter bar on the ground.

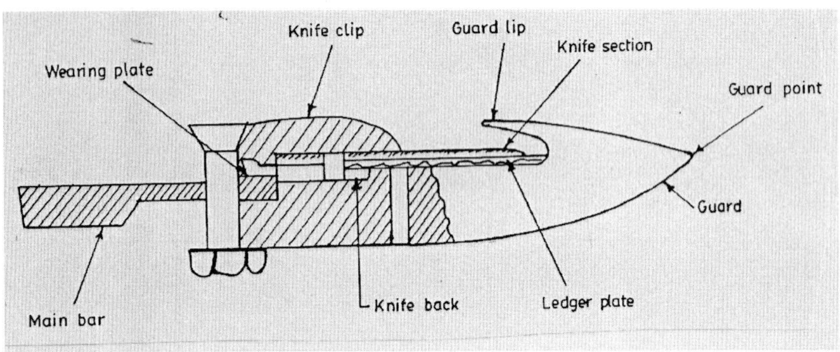

Fig. 8.8 : Cutter bar assembly

Guards: These are provided to hold the forage while it is being cut. These also serve to secure the cutting units and provide a place for the ledger plates. The guards are made with malleable casting or pressed steel, pointed at one end for parting and guiding the grass to the knife sections.

Ledger plate: The ledger plates are fixed over to the guards. They form one half of the cutting unit, the knife section acts as the other half. The edges of ledger plates are serrated to prevent stems of grass from slipping off the points of the shears.

Knife section: Triangular shaped knife sections are fixed over to the cutter bar. The rear part of knife section is kept square shaped for its easy fixing. These are perfectly smooth and work well in most of the grass and legume crops. Some sections are top serrated which work well in dry grass or straw.

Wearing plates: Wearing plates supports the knife from rear side. They provide vertical support to the rear of the knife sections and also absorb the forward thrust of the knife.

Knife-clips: These are provided to hold the knife section down close against the holder plates. They are made of malleable iron or steel. Knife clips are generally spaced three or four guards apart.

Inside shoe: The inner end of the cutter bar during the operation is supported with a shoe like runner. A replaceable sole is placed underneath the shoe, which is adjustable to regulate the height of cut.

Outside shoe: The outside shoe supports the outer end of cutter bar. It also has an adjustable sole to regulate the height of cut. The pointed front part of the other shoe acts as a divider. It separates the cut grass from standing uncut grass.

Power transmission unit: The knife of cutter bar is powered through a long pitman attached to the knife head. The power transmission consists of main axle, gears, crankshaft, crank wheel and pitman. The main axle receives power from one of the transport wheels. A spur gear mounted on the main axle drives the spur pinion on one end of the counter shaft in the gearbox. The bevel gear, which is provided to the other end of the counter shaft, engages with the bevel pinion on the crankshaft. The crank wheel and pitman are fixed on the outer end of crankshaft and the reciprocating motion is transmitted to the pitman, which in turn operates the knife in the cutter bar. The knife is connected to the pitman with a ball and socket joint. The tractor-drawn mowers are operated by PTO shaft. Most tractor-drawn mowers have a V-belt in the drive which provides overload protection and check high frequency peak torque and shock loads, sometimes dog clutch or a slip clutch is also provided.

Wheels: A pair of wheels made of cast iron with sufficient widths and number of lugs to develop better traction are provided in the mowers. Now, pneumatic wheels are also in use. The transport wheels transmit power to the knife. It is so arranged that when the wheel move forward the ratchet and pawl mechanism comes into play to drive the main axle. When the wheel is turned backward the pawl slips giving a clicking noise.

Frame: The main frame of the mower is made with heavy casting that supports other parts and provides openings for main axle, counter shaft and crank shaft. It provides space for gearbox, clutch and bearing. A triangular shaped welded frame is common in semi-mounted mowers over which power transmission unit is installed.

Pitman: The pitman is the connecting rod between the crank wheel and the knife. Pitman may be made of well-seasoned wood or of steel which supports weight and absorbs vibrations.

Grass board and stick: The grass board is provided to clear the cut material from narrow strip next to the standing crop and hence provide a place for the inner shoe for operation on the next turn. A stick is also provided to keep the forage falling correctly. It can be set high for tall grass and low for short material. The grass board is spring loaded and free to move to some extent when it meets an obstruction.

Operating speeds and adjustments: A system of force acts upon the material to cut it in such a manner as to cause it to fail in shear. A common way to apply the cutting force is by means of two opposed shearing elements, one or both may be moving. The motion may be linear with uniform velocity, reciprocating or rotary. Impact cutting principle is applied in rotary cutters and flail shredders. A continuous reciprocating knife type of cutter bar is commonly used to cut a wide variety of crop materials. Cutting in this case utilizes both impact and shear. There are two adjustments, which are very important for proper functioning of a mower. They are registration and alignment. Cutter bars are generally operated from the PTO through Pitman shaft. This is the cheapest mechanism but there are certain limitations. Pitman shaft and cutter bar should be ideally in line. Pitman drives can be operated at 850 to 1000 rpm. With dynamic balancing speed can be increased to 1200 rpm. A typical speed of 1000 rpm of Pitman at 76.2 mm stroke length gives maximum top speed to the knife about 4 m/s. Even this top speed of 4 m/s is not capable of cutting stems by impact as in rotary cutters (linear speed 50 m/s). Thus close clearance between knife sections and ledger plates needs to be maintained. Adjusting knife clips can reduce the clearance. Larger clearances result in clogging of knives and there can be considerable increase in power requirement due to deflection and deformation of stalks. The forward speed ratio to the strokes per unit time in PTO operated machines is adjustable. Thus cutter-bars can be operated at slow forward speed in dense crop and relatively high forward speeds if the crops are not that heavy. But deflection of stalks before cut and bunching of stalks before a cut in a particular stroke limits the top forward speeds which are in most of the cases below 6-7 km/h. Higher forward speed may cause frequent clogging of the machine. If crop is not adequately erect forward speed may further be reduced. Sharp and well-maintained knives with close adjustments help to maintain good forward speed.

Registration: A knife is in proper registration when midpoint of the knife section stops in the center of guard on each stroke (Fig. 8.9). It is very essential for an even job of cutting and unclogging of the cutter bar. If it is not properly registered then it can be adjusted by moving the entire cutter bar in or out with respect to the pitman crankshaft. The results of failing to register are an uneven cutting and an uneven loading of the entire mower. It also results in more draft and clogging of knife. There is provision to change the pitman length, which is provided in some of the mowers.

Alignment: The outer end of the cutter bar is fitted a little ahead of the inner end so that the outer end may align with the inner end, when the mower is pulled through crops. This is called lead and gives better cutting when the cutter bar is properly aligned i.e. the knife and the pitman run in a straight line. To allow for rear word deflection of the outer end of cutter bar during operation, it is customary to adjust the mower so that when not in operation, the outer end of cutter bar has a lead of about 2 cm per meter of cutter bar. To measure the lead of a mower, stretch a stirring parallel to the vertical plane of the pitman and determine the difference in the horizontal fore-and-aft distances from the string to the knife back at the inner and outer ends. Adjustment, if needed be done by rotating an eccentric bushing on one of the hinged pins.

Height of cut: Height of cut of crop can be adjusted by varying the adjustable shoes at the inner and outer ends of the cutter bar.

Energy and power requirement: Energy and power requirements in reciprocating cutter bar are quite low in comparison to rotary cutters, especially shredders, which need a lot of power to give acceleration to cut material. Since power requirements and thus energy requirements per unit mass of cut crop may vary widely depending upon type and condition of crop, condition of cutter bar, forward speed etc. On an average 0.91 kW per meter of cutter bar width may be required out of PTO at forward speed of 7 km/h and 100 rpm of the crankshaft. However, peak values (one peak per stroke) lie somewhere near 2.4 kW/m. Out of these loads two third loads is to cover frictional requirements of the moving parts and one third for cutting of the crop. Additional power is required to pull the machine forward.

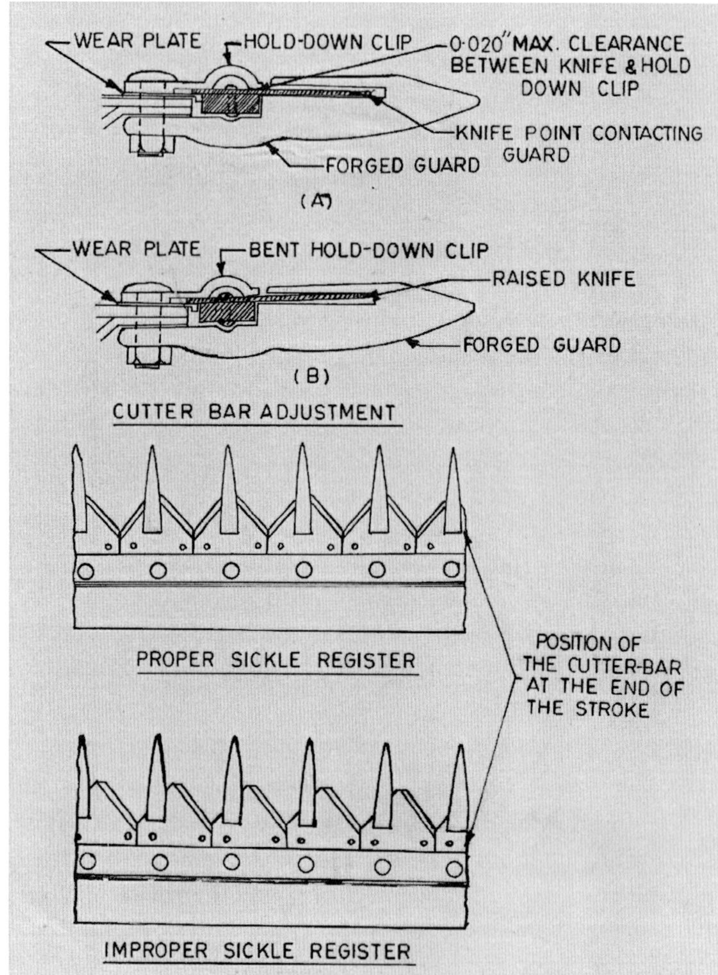

Fig. 8.9 : Adjustment in the cutter bar unit

Convention cutter bar mowers: It has a reciprocating type cutter bar, which is used to reap fodder as well as other crops. After harvesting it spreads the crop on the ground. Animal-drawn, tractor mounted as well as small engine operated cutter bars is available these days. Tractor operated machines are generally rear-side mounted (Fig. 8.10) or front mounted (Fig. 8.11). But front mounted machines provide better visibility to the operator and have easy control but additional mountings and drives are required to operate. Rear mounted cutter bars are most common. Effective width of cutter bars varies from 1.52 to 2.75 m. The offset drag force on the outer end of the cutter bar limits its width. Two shoes of the cutter bar help to maintain the uniform height of cut. During transport cutter bar can be folded to vertical position and when in operation it is spread to horizontal position.

Fig. 8.10 : Tractor rear mounted cutter bar type mower

Fig. 8.11 : Tractor front mounted cutter bar mower harvesting bajra for green fodder

Forage scythe: To reduce the labour required in harvesting fodder crop especially barseem which is very popular fodder, long handled scythe (Fig. 8.12) is being introduced. It is much faster than sickle and is operated in the standing posture conveniently. Labour required for sickle and scythe is measured to be 106 man-h/ha and 57 man-h/ha respectively.

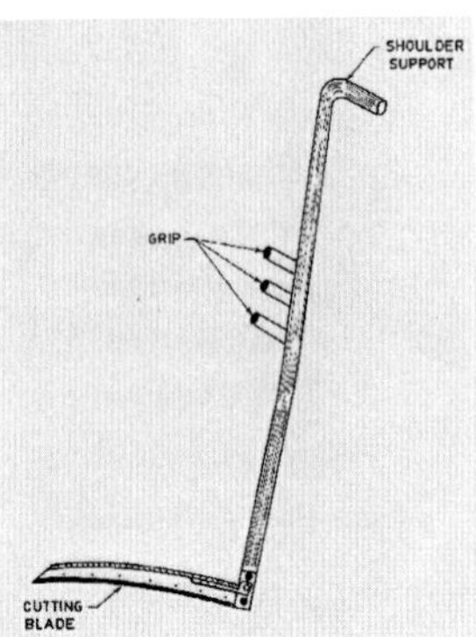

Fig. 8.12 : A view of **forage scythe**

Rotary disc type forage harvester: The machine is of tractor rear side mounted type (Fig. 8.13) and vertical rotary blade type (Fig. 8.14). Six swinging blades on each of the two rotary discs do the main function of cutting. Two conical drums are attached on the frame above each disc to give proper direction to the cut crop forming a windrow and prevent the crop from wrapping around the shaft (Fig. 8.13). Power from tractor PTO is transmitted to the discs through V-belts, chains, and sprocket and gear system. Height of cut can be adjusted by raising or lowering the skidding shoes, fitted on either side of the machine. An additional three-point hitch has been provided in the centre to reduce the transport width for ease during transport. Higher field capacity of 0.41 ha/h is obtained at higher forward speed of 5.4 km/h in case of rotary disc type forage harvester on barseem. Plant damage and uncut plants in general, decreased with increase in forward speed. Variation in the width of blade does not affect the plant damage and uncut plant in any appreciable manner. Windrow quality is satisfactory at all the forward speeds.

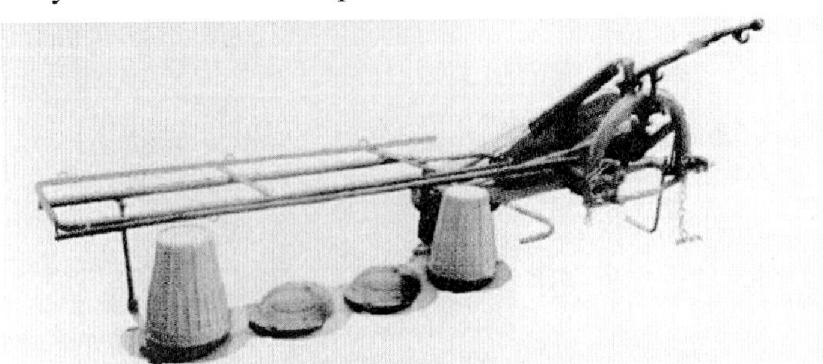

Fig. 8.13 : Tractor rear side mounted disc mower
Courtesy: BCS India Pvt. Ltd., Ludhiana (Punjab)

Fig. 8.14 : Tractor front mounted vertical rotary blade type mower

Tractor operated flail type forage harvester-cum-chopper

Tractor operated flail type forage harvester-cum-chopper is operated by a 26.1 kW tractor (Singh and Pandey, 2008; Anonymous, 2010, 2010a, 2013). This machine in a single operation can harvest, chop and load the chopped fodder like maize, *bajra*, and oats in the trailer attached to the machine. It consists of a rotary shaft on which flails are mounted to harvest the crop, auger for conveying the cut crop, cutters for chopping & conveying chopped fodder through outlet into the trailer (Fig. 8.15a). After the blades cut the crop, it comes in auger, which conveys it to the chopping mechanism. The chopping mechanism cuts the crop into pieces and chopped material is thrown out with a high speed and is filled into the trailer hitched to the machine (Fig. 8.15b). Working capacity of forage harvester is 0.2 ha/h. Weight of the machine is about 670.0 kg.

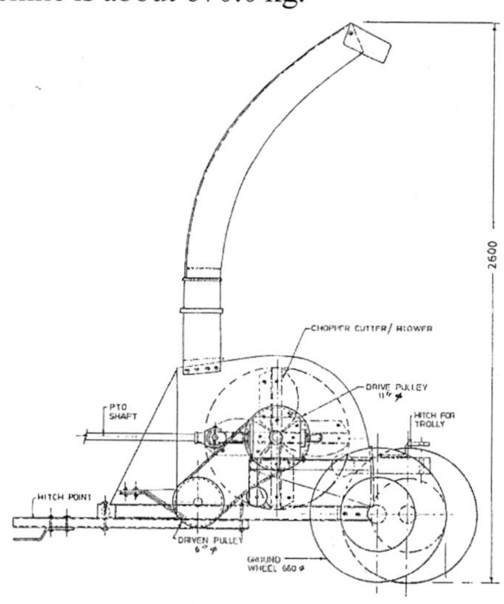

a) Details of forage harvester

b) A view of forage harvester in field

Fig. 8.15 : Tractor operated flail type forage harvester-cum-chopper

Tractor operated cutter bar type forage harvester and loader

The machine cuts the fodder crops and lays in the field which is manually collected and then fed into the loader where it is chopped in to small pieces and directly loaded into the trailer (Fig. 8.16). It is operated by a 26.1 kW tractor (Anonymous, 2008, 2010, 2010a). The system consists of a two separate units. First one consists of a mower (Fig. 8.16a) and the second one a chaff cutter-cum-loader (Fig. 8.16c). Both the machines are tractor operated. The mower has a working width of 196 cm and so designed that it can be folded easily during transportation. It has two powered feed roller and one compressing roller. It has two chaffing blades for chaffing of fodder and six thrower attachments being mounted on the periphery of the cutter. A reversing mechanism has also been provided for safety. The fodder is cut with mower (Fig. 8.16b) and is collected in the field manually. The chaff cutter-cum-loader chops the fodder at site and the throwers mounted on the periphery of the cutter throws the fodder directly into the trailer (Fig. 8.16d). Working capacity is 0.2 ha/h for cutting and 12-18 t/h for chaff cutter-cum-loader. The machine is designed to be used by low power tractor. The hydraulic lift, supplied as standard, allows the operater to direct the cutter bar while staying comfortably seated on the tractor. A spring linkage regulates the round pressure of the cutter bar. The four discs are suspended by eight small helical cutting blades. Two cones located on the first and the last disc act as conveyors of the forage facilitation formation of the pathway.

a) A view of tractor rear mounted cutter bar type forage harvester
Courtesy: Bharat Industrial Corporation, Moga (Punjab)

b) Tractor rear mounted cutter bar type forage harvester in field operation

c) A view of tractor operated chaff cutter-cum-loader
Courtesy: Bharat Industrial Corporation, Moga (Punjab)

d) Tractor operated chaff cutter-cum-loader chopping and loading in trolley
Courtesy: Bharat Industrial Corporation, Moga (Punjab)

Fig. 8.16 : Tractor operated cutter bar type forage harvester, chopper-cum-loader

Self-propelled fodder harvester (cutter bar type)

The self-propelled cutter bar type fodder harvester consists of 145 cm cutter bar (Fig. 8.17a); Anonymous (2008 and 2008a). It is mounted in the centre of the power transmission with the help of two side linkages made of high-pressure circular pipes to dampen the vibration of cutter bar. The unit is also provided with crop gathering drum-having flappers rotating inwards near the cutter bar. The vertical side covers around the transmission system are provided to avoid blockage of the fodder and extended vertical guides are provided for proper windrowing. The ground clearance of the machine is 480 mm to facilitate easy passage of fodder. Power to wheel is provided through extension with the help of chains and sprockets with the chassis suitably strengthened. The track width is also widened to 1040 mm and a caster wheel is provided at the rear to improve stability and maneuverability. The effective field capacity of machine is 0.1 ha/h at a forward speed of 1.5-2.0 km/h (Fig. 8.17b). The top and side views of machine are given in Figs. 8.17 c and 8.17d.

a) A view of self-propelled cutter bar type forage harvester

b) Self-propelled cutter bar type forage harvester working in field

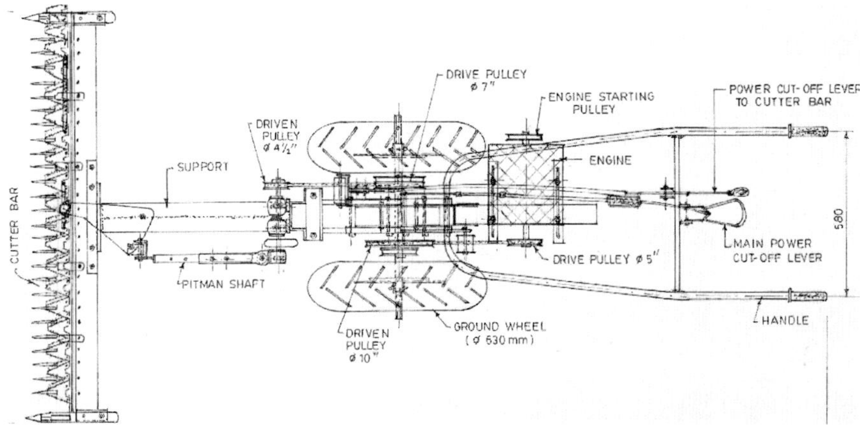

c) Top view of self-propelled cutter bar type forage harvester

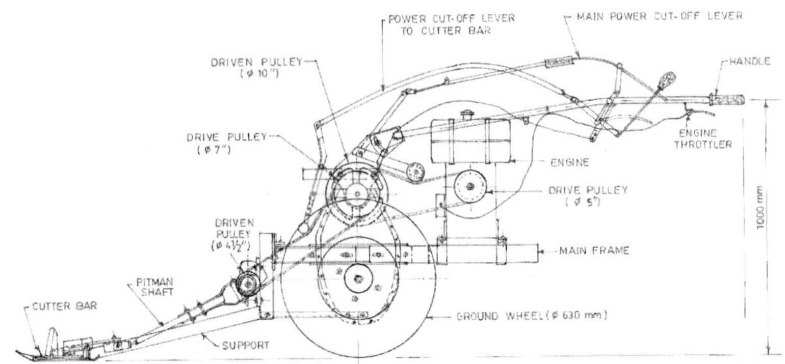

d) Side view of self-propelled cutter bar type forage harvester

Fig. 8.17 : Self-propelled cutter bar type fodder harvester

Self-propelled cutter bar type harvester for barseem, barley, sorghum & maize

A self-propelled machine has been developed for harvesting fodder crops like berseem, barley, sorghum, maize etc (Fig. 8.7); Anonymous (2010, 2010a). It cuts the fodder crop and makes a swath which can be collected easily (Fig. 8.18). The machine has cutter bar width of 130 cm. The power from engine has been provided to the traction wheels with the help of gear trains. The machine has four forward speeds and one reverse. The track width of machine is 1.23 m and the ground clearance of machine 41 cm. The steering of machine is done with the help of foot-operated pedal provided below the operator seat, which guides the wheel. The crop after harvesting falls behind the cutter bar and is guided with wooden sticks provided behind the cutter bar for small height/hard stem crops. But in case of tall and weak stem crops, binding attachment without twine is required to avoid wrapping of crop to cutter bar. The average field capacity of the machine for harvesting barseem and oat crop is 0.42 and 0.28 ha/h respectively.

Barseem fodder Bajra fodder

Fig. 8.18 : Self-propelled cutter bar type forage harvester in operation
Courtesy: BCS India Pvt. Ltd., Ludhiana (Punjab)

Self-propelled lucerne harvester

Lucerne is perennial leguminous plant growing to a height of 60 to 90 cm. It is a forage crop generally grown in basins having a very high yielding potential under appropriate fertilizer and irrigation management. Harvesting of crop is one of the important agricultural operations, which demands considerable amount of labour because it is done manually by sickles. The cutting and laying in the windrows consume 65-75% of labour and gathering, bundle making and transport in the field involve the rest of the labour requirement. The scarcity and high cost of labour and drudgery during harvesting are the serious problem faced by the farmers. It is, therefore, essential to adopt the mechanical methods so that the timeliness in harvesting operation could be ensured and field losses are minimized to increase the productivity and production on the farm (Anonymous, 2008, 2008b). The self-propelled lucerne harvester consists of a gearbox and cutter bar (Fig. 8.19). Type of cutter bar is bi-directional reciprocating type made from high carbon steel. Length of stroke for cutter bar is 25 mm and effective width of cutter bar is 860 mm. A man can walk behind the machine with an average speed of 2 km/h. The recommended speed ratio of the average cutter bar speed to the forward speed of machine is 1.3: 1.4. Two wheels are used for transportation purpose. The ground wheels drive the reel of the harvester. Ground drive provides the desirable feature of maintaining a constant speed ratio between peripheral speed and forward speed. The reel is made up from Ø 6 mm MS bar of diameter 70 cm and length 74 cm. The isometric view of machine is shown in Fig. 8.20. The speed ratio of the ground wheel to reel is 1:1. The effective field capacity is 0.113 ha/h and field efficiency 70-75%. There is saving of 52% in time and 90% in labour as against conventional method.

Fig. 8.19 : A self-propelled lucerne harvester in operation

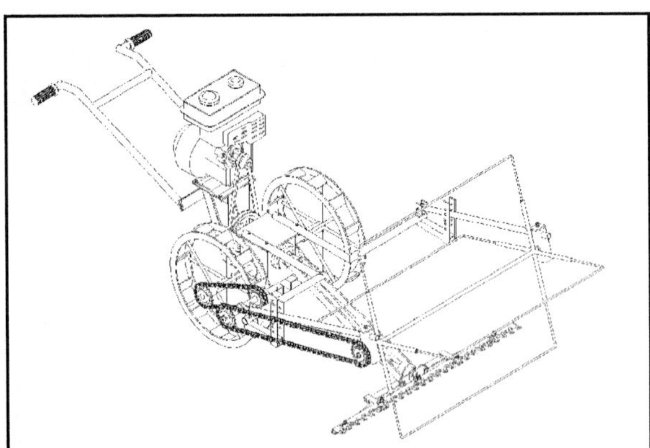

Fig. 8.20 : An isometric view of self-propelled lucerne harvester

Self-propelled flail type forage harvester for fodder crops

A self-propelled machine has been developed for harvesting fodder crops like maize, bajra and oat Anonymous, 2008, 2008a). The machine consists of a self-propelled forage harvester having a horizontal rotor on which nine L-shape flails are mounted in staggered position in three rows (Fig. 8.21). The maneuverability of the machine is controlled by a mechanism provided at foot of the operator. The cutting width of the machine is 60cm. The crop is cut by the flails by impact and is guided by the shield provided on the rotor. Harvesting efficiency and throughput capacity of the machine for harvesting maize, bajra and oat fodder increases with the increase in flail speed and it varies from 95 to 100 percent. With the increase in flail speeds, the height of cut from the ground decreases from 7.94 to 6.32 cm thus resulting in higher throughput capacity of the machine. The fuel consumption varies from 16.6 to 20.3l/ha. There is a saving of about 48 percent in labour cost. The machine has a very low capacity and the collection of fodder is difficult.

Fig. 8.21 : Self-propelled flail type forage harvester harvesting barseem crop

Forage chopping

Forage chopping involves reduction of size of cut stalks for direct green feeding or silage making. Mechanical handling of chopped fodder becomes easy through blowers and impellers. Chopping may be done along with field cutting with flail type field choppers or chopping may be a separate operation within a forage harvester after cutting. Field chopper can pick up windrower crop and blow the material into forage wagon. Stationery choppers are also available for silage making. Flail type field choppers differ from the flywheel and cylinder type choppers functionally.

Chaff cutters

A chaff cutter is a mechanical device for cutting straw or hay into small pieces before being mixed together with other forage and fed to cattle.

This aids the animal's digestion and prevents animals from rejecting any part of their food. Chaff and hay plays a vital role in most agricultural production as it is used for feeding cattle. Chaff cutters have evolved from the basic machines into commercial standard machines that can be driven at various speed and can achieved various lengths of cuts of chaff with respect to animal preference type. New chaff cutter machine include portable tractor driven chaff cutter where chaff cutter can cut the crop in the field and load in trolleys. Along with the field chopper its complementary equipment, the hay making or silage-making operation can be completely mechanized. The chopped material is blown directly into trailers, which are towed behind the chopper or pulled beside with a separate power source or driven behind or beside the chopper. The chopped material can be loaded mechanically at a controlled rate. Field choppers are used in several methods of harvesting forages.

Silage cutter: Silage cutter is a machine used for cutting up straw and green crop materials being ensiled in the preparation of feeds for livestock. Silage cutters may be motor-driven, horse driven, or manually operated. They are driven by an electric motor or a power take-off shaft from a tractor engine. A silage cutter consists of a cutting mechanism, a feed conveyor, and a drive mechanism; the cutting mechanism may be of the disk or drum type. Both types work on the principle of a scissors: rotating knives attached to a drum or disk serve as one of the blades, and the cutting edge of a fixed shear plate serve as the second blade. Drum-type silage cutters are usually stationary; disk-type cutters have wheels and are equipped with blower conveyors. Both types are used. Depending on the type of material being chopped and the purpose for which it is intended, the cutter may be positioned next to silos when green crop materials are being ensiled, in feed shops when livestock feeds are being prepared, and at locations where chopped straw is being prepared for livestock feed or bedding. Power-cum-hand operated chaff cutters is the power chaff cutters operate on 1.5-2 hp single phase electric motor and can be operated manually in absence of power. It is mounted on heavy steel legs and braces with forged feed which make the machine property bolt.

Shear bar type choppers: Shear bar type choppers cut the material with much uniformity in length as material is positively fed and held against the shearing tool. Two type of shear bar choppers are used i.e. flywheel type and cylinder type. Counter rotating feed rollers feed the material over the shear bar. Generally top rollers are spring loaded to accommodate varying thickness of the material being fed.

A flywheel type cutter head has 4 to 6 knives to cut the material. Chopping starts from one end of the Shear bar and proceeds to the other end to avoid peak cutting torque. The cut material is accelerated with 3 to 4 impellers fixed on periphery of the flywheel to the forage wagon or silage pit. A cylinder type cutter head bar generally has 6 knives, mostly cup shaped, to cut and simultaneously throw the material through the duct. In cylinder type cutter head, if shape of the knives does not support to accelerate the material a separate impeller may be used. Peripheral speeds vary from 18 to 30 m/s depending upon whether separate impeller is used or cylinder itself acts as impeller. Flywheel type choppers are very popular in India and are widely used.

Chopping lengths: Theoretical length of cut is the linear advance of material by feed mechanism between two successive cuts. To adjust the length of cut speed of feed mechanism can be adjusted. Increase in number of knives on flywheel will also reduce the length of cut. All the chopped forage does not exactly have same length. A little variation is acceptable. When materials are non directional, like barseem, variation in lengths is over wide range as compared to materials which are fed in one direction like bajra, sorghum, and corn. Specific energy required to cut material is inversely proportional to the length of cut. Thus cut lengths should be decided as large as possible until it affects storage characteristics like bulk density, handling by blowers or acceptability for direct feeding.

Power and energy requirements: Power required in operating a field chopper could be divided into the power requirements for; i) gathering and conveying of crop, ii) shearing/chopping of material, iii) accelerating the cut-material, iv) pumping air along with accelerated material, v) overcoming friction to move chopped material and vi) overcoming internal mechanical friction within the machine. Actual power requirements depend upon the type and condition of crop, type and condition of machine, its adjustments and length of cut. Under some conditions, the power requirements will increase as the feed rate or machine output per unit time increases. For a typical field chopper, power requirements under test conditions (13 mm length of cut, 35 m/s peripheral speed, crop moisture 74%) are 1.12 kWh/ton. Out of the total energy requirements about 35% power is required for cutting, 50% for accelerating the cut material (including its friction) and blowing air and the remaining 15% is used in gathering and conveying the crop.

Blade sharpness, clearance and velocity: These three factors have an important role in performance and power requirements of a chopping machine. Power requirement increases as sharpness of knives reduces. An edge width of 0.05-0.08 mm gives best results but it may be practically difficult to maintain this sharpness. Specific energy required for cutting may increase by 4-5 times if knives become very blunt. But clearance has very close interaction in blade sharpness. If blade is very sharp, increase in blade clearance has little effect. In average working conditions increase in clearance from 0.5-2.0 mm may increase the power requirements by 30%. Large clearances of blade from shear bar may result in poor cutting or rather may fail to cut the crop clearly. Increase in speed of blade may reduce the specific energy used in cutting. But since higher blade velocities are provided by high-speed cutter heads, there is an overall increase in energy consumption by the cutter head as speed is increased. This is chiefly because higher speeds pump more air and hence more energy is consumed per cut. Blade velocities of 30-33 m/s are common. Higher blade speeds can show good performance even when blunt or when blade clearances are large.

Hand chaff cutter

Green fodder crops are essential food requirement for on-farm animals. These fodder crops are harvested from the field daily either by sickles or reapers. Then crop is collected from the field and carried to the farmhouse where it is cut into small pieces. This is done to save storage, to aid in curing and to make it more palatable. The cutting of fodder into small pieces is done either manually using 'Gandasa' or by using manually or power operated chaff cutters. Depending on the type of cutting head, chaff cutters may be classified as cylinder cutting head machine and flywheel type cutting head machine. According to power used chaff cutters are classified as hand chaff cutters, animal operated chaff cutters and power chaff cutters or silage cutters. The hand chaff cutter consists of a feeding tray, cutting blades, flywheel, cover plate, feed rolls, shear plate, handle and stand (Fig. 8.22); Pandey *et al.* (1997). The blades are made of high carbon steel or alloy steel hardened and tempered to suitable hardness. The cutting edges are made sharp. These cutting blades are mounted on the flywheel. Two persons operate the machine, one feeds the forage or grass in the feeding trough and another rotates the flywheel with handle (Fig. 8.23). The material fed in the hopper is gripped between the feed rolls which pull it and the material get chopped between blades mounted on the flywheel and stationary shear plate. The length of chopped material can be changed. Dry or green fodder can easily be chopped with the machine. It is used to cut chaff, grass, green and dry fodder crops and paddy straw into bits for feeding to animals. The size of chaff or length of cut can be changed by changing the speed of feed rollers. Hand operated chaff cutters are provided with a two speed worm for adjusting the length of cut. Reducing or increasing the number of knives on the flywheel can also changes the length of cut of silage. This can be done easily on the hand operated chaff cutters. The length of cut of silage varies from 20-40 mm and for green fodder 25-50 mm or sometimes more.

Fig. 8.22 : Chaff cutter – hand operated
Courtesy: Bharat Industries, Varanasi (U.P.)

Fig. 8.23 : Two different types of hand operated chaff cutter used in country

Care and adjustments of hand operated chaff cutter: Since the hand operated chaff cutter is a simple and useful machine, it is very popular among Indian farmers. In order to get the most efficient operation for a long time; it must be properly installed on a rigid foundation preferably under shed. Many farmers have converted hand chaff cutters into power chaff cutters by using 1 hp electric motor or small engine. The following general procedure of maintenance and lubrication should be followed:

i) Before the machine is put into operation, its gears and bearings should be lubricated.

ii) The knives should be sharpened by hand filing. If the cutting edge has become too blunt, it should be sharpened on a grinder to the proper bevel only on one side.

iii) The clearance between the knives and the shear plate should be adjusted for the effective cutting.

iv) Loose nuts, bolts, and screws should be tightened to avoid accidents.

v) After the day's work is over, dirt from the gears and bearing parts and moisture from the knives should be wiped off.

vi) While the machine is idle, its flywheel should be kept locked so that the children do not operate it.

vii) Most of the cast iron parts and particularly the flywheel of the machine are quite susceptible to breakage by hammering, which should always be avoided.

viii) The moving parts should be properly lubricated.

Power operated chaff cutter

The feeding mechanism of a power operated chaff cutter consists of an apron and two feed rollers. Aprons are generally provided with metal or wooden slates to give a positive feed. The length of the table on which the apron moves is about 2 m so as to accommodate tall grasses or fodder crops. Spring-loaded feed rollers are provided to compress and feed the uncut material to the cutting head. Generally, corrugated or toothed rolls are preferred. Hand operated machines are not provided with aprons. A feed-table or merely a feed box is considered to be enough for this machine and the feeding is done by hand. Feed rollers of the hand-operated machine have variable spacing and the worm mounted on the flywheel shaft operates them. The bullock operated chaff cutters is similar to hand operated one but it is provided with a smaller diameter flywheel on which two or three knives are mounted. By means of a set of gears and universal joints, the slow speed of bullocks is utilized to operate the machine at about 250 rpm and thus its capacity becomes 4 to 6 times more than that of the hand operated machine. There are many types of power operated chaff cutters manufactured and used in different parts of the country (Fig. 8.24). These power operated chaff cutters of various sizes can be operated by 1 hp, 2 hp (Fig. 8.25), 3 hp, 5 hp motor/ engine and even by tractors' PTO (Fig. 8.26). The capacity of these chaff cutters vary from 400 kg/h for dry fodder to 900 kg/h for wet fodder or even more depending upon the size of chaff cutter, and condition of fodder. These chaff cutters have three blades, two rotating and third fixed.

a) Power operated chaff cutter

Courtesy : Bharat Industries, Varanasi (U.P.)

b) Another type of power operated chaff cutter

Fig. 8.24 : Power operated chaff cutter

Fig. 8.25 : Power operated chaff cutter
Courtesy: Bhagvati Krushi Udhyog, Ahmedabad (Gujarat)

Fig. 8.26 : Tractor PTO operated flywheel type chaff cutter
Courtesy: Bhagvati Krushi Udhyog, Ahmedabad (Gujarat)

Multi crop chopper

It is used to chop fodder crops such as maize, jower, bajra etc into small pieces to be used as cattle feed. It has a disc type cutting unit, chopping and conveying units (Fig. 8.27). The crop is cut, chopped in to small pieces and conveyed to container mounted on the tractor. It is operated by 70-80 hp tractors. It operates at 540 rpm and chopping size of the fodder is 12 to 35 mm. It has the capacity of 3 ton/h for green fodder and 1 ton/h for dry fodder. The container can be unloaded hydraulically to tractor trolley for transportation.

Fig. 8.27 : Tractor operated multi crop chopper
Courtesy: Santhosh Agri Machinery, Attur (Tamil Nadu)

References

Anonymous. 2008. Research Highlight. AICRP on Farm Implements and Machinery, CIAE Bhopal. Technical Bulletin No.: CIAE/2008/141.

Anonymous. 2008a. Annual Report (2006-08). AICRP on Farm Implements and Machinery, PAU Ludhiana.

Anonymous. 2008b. Annual Report (2006-08). AICRP on Farm Implements and Machinery, MPKV Rahuri.

Anonymous. 2010. Research Highlight. AICRP on Farm Implements and Machinery, CIAE Bhopal. Technical Bulletin No.: CIAE/2010/151.

Anonymous. 2010a. Annual Reports (2008-10). AICRP on Farm Implements and Machinery, PAU Ludhiana.

Anonymous. 2013. Research Highlight. AICRP on Farm Implements and Machinery, CIAE Bhopal. Technical Bulletin No.: CIAE/2013/158.

Pandey M M; Majumdar K L; Singh Gyanendra; Singh Gajendra. 1997. Farm Machinery Research Digest. Technical Bulletin No. CIAE/97/69, Central Institute of Agricultural Engineering, Bhopal, 328 p.

Singh Jaskaran. 2001. Design & development of forage harvesting machinery. Proceedings of Summer School on 'Advances in Seeding and Harvesting Machinery'. Dept. Farm Power & Machinery, PAU Ludhiana. May 21 to June 19. Pages 23-33.

Singh Surendra. 2007. Farm Machinery – Principles and Applications. Directorate of Information & Publication of Agriculture, Indian Council of Agricultural Research, Krishi Anusandhan Bhawan-I, Pusa Campus, New Delhi.

Singh Surendra; Pandey M M. 2008. X Plan Achievements (2002-2007). AICRP on Farm Implements and Machinery, CIAE Bhopal. Technical Bulletin No.: CIAE/2008/137.

❑❑❑

9

Farm Tools and Equipment for Horticulture, Plantation and Hill Agriculture

9.1 Farm Tools and Equipment for Horticulture and Plantation Crops

Agriculture in India is one of the most prominent sectors in its economy. Agriculture and allied sectors like forestry, logging and fishing accounts for about 14% of the GDP and employed 60% of the country's population. About 43% of India's geographical area is used for agricultural activity. Despite a steady decline of its share in the GDP, agriculture is still the largest economic sector and plays a significant role in the overall socio-economic development of India. Nearly 80 percent of the farm population operates smallholdings, the average size of holding being 1.16 ha. Out of a total geographical area of 329 m ha, about 142 m ha constitute the net sown area. Nearly 63 percent of this area is rainfed and its contribution to the overall production is 44%.

India is the second largest producer of fruits and vegetables in the world, the production of which has tripled during the last 50 years (Singh, 2011). Other sectors in horticulture like plantation crops, floriculture, root and tuber crops, mushroom and medicinal and aromatic plants have also progressed. Horticultural crops, which cover about 15 m ha, contribute about 150 million tonne, which account for 24.5% of the gross value of agriculture output. Fruits account for about 30.2% of the total horticultural production, while vegetables account for 60% and nuts, spices, flowers, mushroom and honey about 9.8%. Today horticulture crops contribute 25 per cent of the total agricultural exports of the country and during the last few years corporate have shown increasing interest in horticulture. In India, very little efforts have been made to mechanize farm operations for fruit and vegetables crops. It is mainly due to availability of labour and lack of organized large-scale fruit or vegetable farming. Now the trend is emerging towards the organized fruit and vegetable farming.

India has a large part of area under plantation crops. Tea, coffee and coconut palm trees are prominent among them. They employ a large amount of work force as most of the operations are being performed manually under very hard conditions. Not many equipment are available for doing various jobs because of adverse land topography, terraces, soil type and conditions and agro-climatic conditions. Some of the equipment already discussed in various chapters such as Tillage Equipment, Intercultural Tools, Fertilizer Application Equipment, Plant Protection Equipment and Horticultural Equipment can be used successfully, if suitable soil and environmental conditions persists. However, some equipment have been developed by various institutions for different plantation crops.

Mechanization of horticultural and plantation crops

Horticulture is the key area for diversification of agriculture in the country. Equipment for mechanization of orchard crops for pit making, transplanting of saplings, spraying, pruning, harvesting of fruits etc need to be identified/imported/designed, manufactured, introduced and popularized. Vegetable crop production needs to be mechanized. There is very little mechanization except for the potato crop. Equipment for seedbed preparation is available. However, equipment for planting, transplanting of seedlings, row cultivation, irrigation, spraying, harvesting, picking/digging need to be introduced by importing, modifying, designing and manufacturing. Use of plastic mulch reduces water requirement and checks weed growth. Equipment for laying plastic mulch, low plastic tunnels for cultivation of vegetables, cut flowers etc are required to be introduced and popularized. Green house technology has good scope in India for growing seedlings, flowers, high value off-season vegetables and some fruit crops. This technology needs to be promoted as a part of diversification efforts. Equipment for mechanization of cultivation in green houses needs to be introduced and popularized. R&D efforts need to be intensified. The futuristic farm equipment e.g. vegetable precision drills and seedling transplanters, vegetable harvesters and diggers for root crops are needed for the country. Some of the equipment developed in the country and suitable for fruits and vegetables crops is discussed here.

The fruit seeds are sown manually for raising the nursery/ seedlings for grafting or direct planting. For planting bold seeds, manual dibblers have been developed. The inclined plate and pneumatic planters have been successfully used for direct sowing of small vegetable seeds. Vegetable transplanters are undergoing extensive feasibility evaluation for planting vegetable seedlings. Hand tools such as budding and grafting knife, pruning knife and secateurs can be used for *plant propagation and pruning*. Telescopic handles can be used for the pruning of the twigs and branches, which are not in the reach of human hand. Hand tools or small power tillers with rotavators can be used for *weeding and interculture.* Different types of sprayers and dusters are available from manually to power operated such as knapsack sprayers, foot operated sprayers, power operated mist blowers and dusters for *plant protection.* Fogging machines and mist blowers are available which can be used in the green houses and for covered crop cultivation. Different types of pumps which includes centrifugal, turbine, submersible, axial flow, mixed flow pumps etc are available for lifting of the water for *irrigation,* including sprinkler and drip irrigation systems.

Harvesting is the most labour intensive operation due to delicate nature of fruits. For table purpose, harvesting would continue to be done manually. *Mechanical harvesters* for fruits can be used where fruits are sent immediately to processing industries for extraction of juices or making other products. *Grading* of the fruits is done according to the size and colour, which is presently being done manually. Attempts have been made to develop fruit graders for apple and oranges, which are yet to be commercialised. The need for special types of power units and/or cultivating implements arises when it is necessary to work between fruit trees, where planting distance has already been decided and equipment has to confirm it. Narrow rows of fruits sometimes necessitate the use of small tractors with matching equipment. There are special ploughs called hop ploughs suitable for very narrow work and turns two furrows to right and two to left simultaneously. Another type of implement desired is which can work as close to trees/ plants on one side as possible without destroying the roots.

Horticultural hand tools

Hand tools are widely used in horticulture because many of the areas cultivated are too small or it is almost impossible for other machines to do the job near plants/trees efficiently (Bhardwaj *et al.*, 2004; Agrawal, *et al.*, 2003; Manes *et al.*, 2007). One such hand tool is **spade** (Fig. 9.1), which can be used for digging, trenching and removing soil. Spades are available in normally different sizes in different parts of the country. The selection of spade depends on the type of soil and work to be done. The blade is made from tempered steel; however, some spades are also made of stainless steel. The large spade, usually known as a digging spade, has a blade approximately 280 mm long and 190 mm wide and weighs between 2 and 2.5 kg. A smaller border spade has a blade 220 mm long and weighs between 1.5 and 2 kg. Spade handles are made out of ash which is strong but light and the end may be 'D' or 'T' shaped. The 'D' shaped handle is more comfortable to use. Some modern spades have plastic grips and the shaft is covered with a plastic skin to protect it from the weather. The blade is made from tempered steel and the top edge or tread should be wide so that it does not cut the worker's boot. Stainless steel blades are used on some spades. This makes them more expensive but easier working. The blade is kept vertical during digging, as it requires minimum efforts.

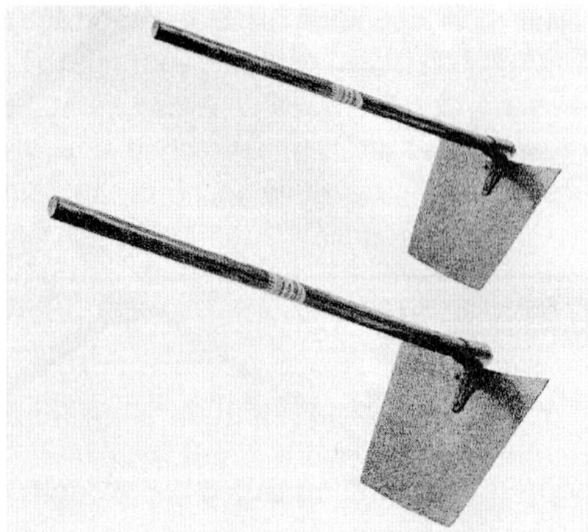

Fig. 9.1 : Spade

Courtesy: Falcon Garden Tools (P) Ltd., Ludhiana (Punjab)

The forks are also commonly used as horticultural tools. They are available in different types depending upon the type of job to be done. Digging forks (Fig. 9.2) are used for digging in the soil already turned by spade. They normally have four prongs, which can be either round or square in section. Border forks are used as weeders in the border area of field. They are narrower and lighter than digging forks and normally have 3 or 4 prongs. Potato forks are specially made for lifting the potatoes in the field. While lifting the potatoes in the field soil automatically drops through and not much damage is made to tubers. In this tool, the prongs are much broader than digging fork and usually have 4 or 5 in numbers. The forks vary in their type and use. Like the spade, the fork can vary in size and weight. The border fork is intended for weeding in borders, and two forks can be used back to back for root division. Compost and rubbish can easily be cleared with a potato fork.

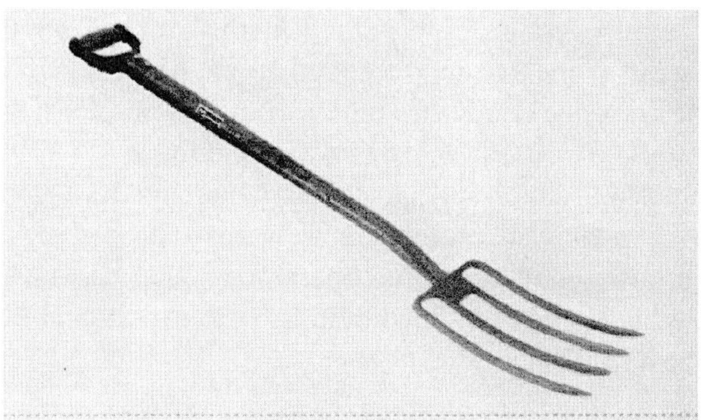

Fig. 9.2 : Digging fork

Courtesy: Falcon Garden Tools (P) Ltd., Ludhiana (Punjab)

Garden rakes (Fig. 9.3) are commonly used as a horticultural tool to break the soil to fine tilth, and to gather the grass, stone and hedge clippings in the field. It has about 10 rigid teeth. Lawn rakes and hay rakes are also used for collecting grasses and tree leaves in the orchard, which are necessary to keep the area clean. They have normally 12-20 teeth depending upon the size. Rakes like forks and spades also vary in their size and shape. The garden rake usually has ten rigid teeth and is used after the fork to break up the soil, level it and break it to a fine tilth. It can also be used for gathering stones, weeds and hedge clippings. The rake consists of a series of spring teeth arranged in twenty or more in number. The rake is made out of wood and has about twelve pegs for teeth. When not in use, the rake should never be laid down with the teeth upwards there is a danger of someone stepping on them receiving a blow from the handle.

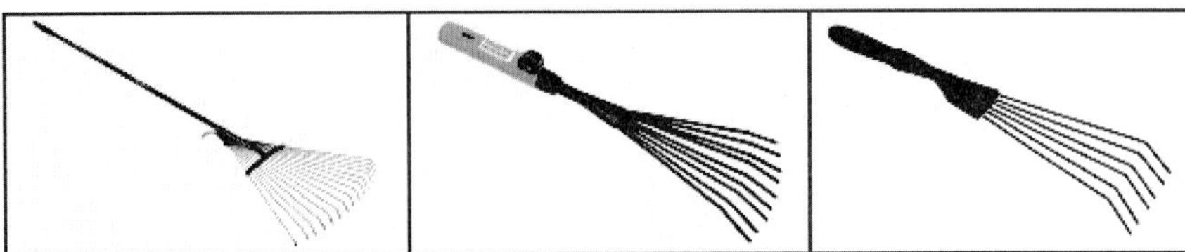

Fig. 9.3 : Garden rakes

Courtesy: Falcon Garden Tools (P) Ltd., Ludhiana (Punjab)

Hoes (Fig. 9.4) are used to destroy the weeds and loosen the soil around the trees. There are two types of hoes; draw-hoe being pulled by the worker and push-hoe. Triangular headed hoes are also used as horticultural tool for making shallow drill usually in conjunction with a garden line. Garden line is generally used when seed sowing, trenching, lawn edging and transplanting operations is performed in the garden. It has 6-20 m nylon twine with a pin at one end and reel at other end where it can be wound for storage. The draw hoe chops through the roots of the weeds, especially tap roots, and can also be used for seed sowing and for singling. It is used with a pulling or pushing action and the worker, who stands beside the drill being hoed, may move either backwards or forwards. The push or Dutch hoe is used with a pushing action as the operator walks backwards. The angle between the blade and the soil should be between 20 and 30 degrees to obtain the best results.

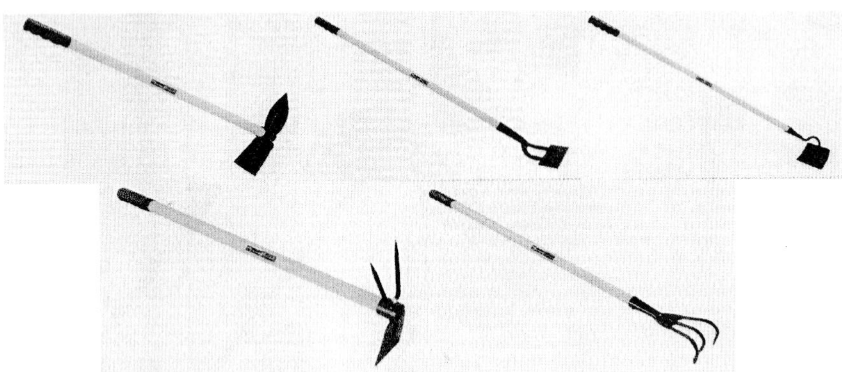

Fig. 9.4 : Different types of hoes

Courtesy: Falcon Garden Tools (P) Ltd., Ludhiana (Punjab)

Shears are used for cutting grasses and hedges. They are of two types viz. hand shears and edging shears. Hand shears have blade 200-300 mm long with wooden or plastic handles. The cutting action takes place between the two blades, which are pivoted at their centres, and material to be cut is sheared between two blades. The blade edges are ground at an angle just less than 90 degrees. Some shears have a pruning notch near the pivot of the blades so that thick twig can easily be cut. Edging shears are used for cutting lawn edges because they have handle about one meter in length and attached at an angle to the blades. Secateurs are also small hand tools used for pruning bushes and shrubs. Turfing iron and turf lifters are used for edging lawns.

Pruning secateurs

Pruning secateurs also known as pruning shears resembles a multipurpose combination pliers used in a workshop. The need of secateur arose to cut the branches or twigs, which are difficult to cut by pruning knives. Being handy and easy to operate, it is considered to be an essential tool of the gardener in plant propagation. Various types of pruning secateurs are fabricated for removing or cutting of unwanted branches or twigs, cutting of scion sticks, defoliation of leaves from the sticks and topping of small trees. These are single cut, double cut, parrot nose cut, roll cut, base cut, supa cut, replaceable blade type, easy cut, kiln cut etc. The pruning secateur consist of two cutting blades or one cutting blade and an anvil, handle, volute spring to keep the blade and handle in open position and a locking device for keeping the secateur in closed position. The blade is important part of the tool and is made from high carbon steel, tool steel or alloy steel. The blades are forged to shape, ground sharp at the cutting edge and hardened to 460-510 HB. Handles are made from aluminium or mild steel and in some cases a cover of plastic is provided on the arms of the handle. Usually the arms of the handle follow a fixed path during cutting operation but in some secateurs one of the arm of the handle is made rotating type for easy operation. For operation the branch or the twig is held in between the blades and handles pressed together which produces shearing action and cutting of the material. The secateur is selected according to the operation and size of the twig or branch.

The secateurs are small hand cutters used for pruning bushed and shrubs (Fig. 9.5). There are two common types: Parrot Bill Secateurs and Fixed Blade Secateurs. Parrot bill secateurs have two blades with curved which pivot and overlap as cutting takes place between them. Sharpening the parrot bill type is very difficult and needs extreme care. Fixed blade secateurs have one blade which cuts against a soft copper plate forming the second blade. The blade is much sharper than those of the parrot bill type and is usually replaceable. These are mostly used by professional gardeners and horticulturists. It has hardened steel blade with rust preventive coating. Some of them have solid steel handles with soft PVC grip, stainless steel coil spring and thumb operated safety lock. Total length is 225 mm with cutting capacity 15 mm. The unique rotating handle revolves on its axis of rotation allowing fingers to move naturally with maximum cutting power and fewer efforts. Some has solid forged lightweight aluminum alloy handles. A rubber cushion and shock absorber prevents wrist fatigue. The unique aluminum handle makes the pruner light in weight. Some have steel handle with soft grip, hardened steel blades with rust preventive coating, and leaf spring system with total length 200 mm with cutting capacity 12 mm.

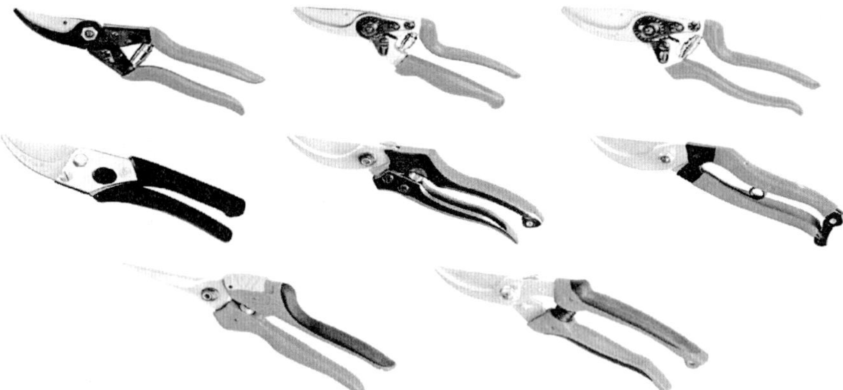

Fig. 9.5 : Different types of Pruning secateurs

Courtesy: Falcon Garden Tools (P) Ltd., Ludhiana (Punjab)

Pruning secateurs (Anvil type): Powerful anvil secateurs mostly used by professional gardeners and horticulturists. It has brass anvil, hardened steel blades, solid steel handle with soft grip; leaf spring system and safety lock (Fig. 9.6). It cuts stems up to 12 mm. Total length is 175 mm.

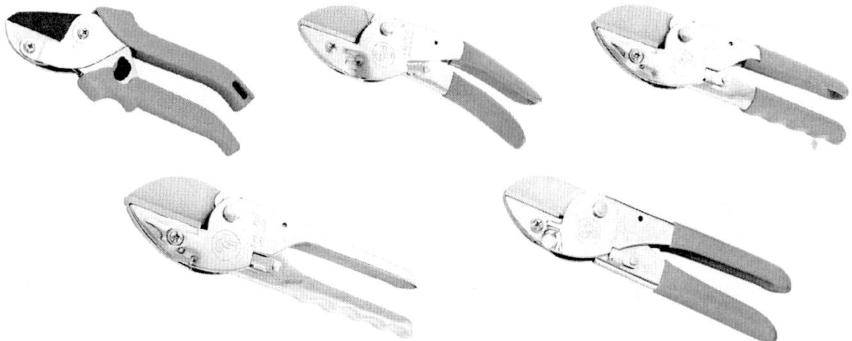

Fig. 9.6 : Different types of Anvil type pruning secateurs
Courtesy: Falcon Garden Tools (P) Ltd., Ludhiana (Punjab)

Pneumatic secateurs

The pneumatic secateurs, also known as pneumatic pruning shears, are used for pruning vines using pneumatic power (Pandey *et al.*, 1997). Gripping blade of the shear is stationary and shearing action is imparted by the other blade through the movement of piston, at the end of which it is fixed, with high-pressure air carried in a portable cylinder (Fig. 9.7). The device offers effortless, accurate and swift cutting, at the same time ensuring the quality of vines. The double acting piston facilitates easy pruning of even large branches. The extension member helps access to branches inside canopy. The cutting head of the shear can be adjusted as needed across 360°.

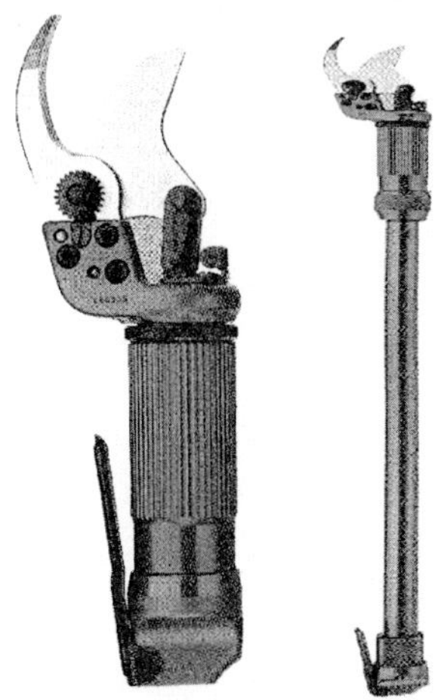

Fig. 9.7 : Pneumatic secateurs

Pruning and slashing knives

Pruning is a process of removing unwanted branches or twigs of a plant or tree for providing aeration, lighting and frame work which help in obtaining higher yields (Pandey *et al.*, 1997). Pruning and slashing knives are hand tools, which consists of a blade and tang joined rigidly to the handle (Fig. 9.8). The tip of the blade is either hooked or curved in order to cut or slash the small branches or twigs of plant or tree by pulling action. The tang of the knife is inserted in the handle and joined rigidly by riveting. The blade and tang are made from the single piece. The blade is made from high carbon steel, tool steel or alloy steel and hardened to 45-55 HRC. The handle is made from good quality wood or plastic. For operation the blade is engaged with the thin branch or twigs and pulled towards the operator to accomplish cutting. The blade is forged to shape and cutting edge is sharpened for cutting and slashing of thin branches and twigs of plantation crops and orchards.

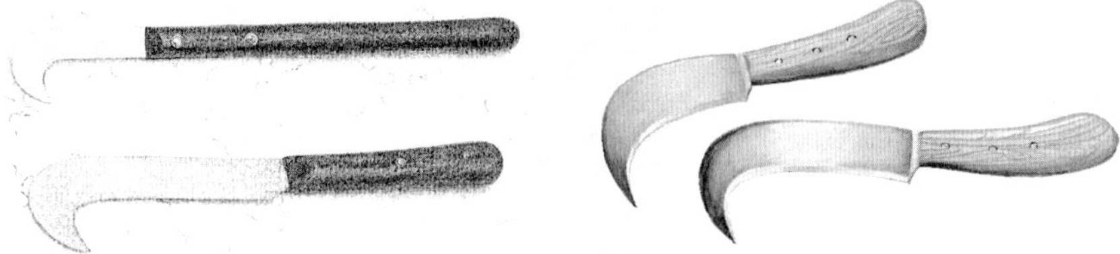

Fig. 9.8 : Pruning and slashing knives

Tree pruner

The tree pruner is a manually operated pruning tool for cutting of the branches or twigs of the orchard trees or plants in standing positions, which are beyond the reach of human hands for aeration and giving a shape for facilitating harvesting and adequate lighting (Pandey *et al.*, 1997). It essentially consists of a hooked anvil, spring actuated cutting blade, links for actuating the blades and a socket (Fig. 9.9). A long wooden handle is inserted in the socket in order to access the high branches or twigs for cutting. A rope is attached to the link for actuating the cutting blade. The blade is important part of the tree pruner and is made from high carbon steel, tool steel or alloy steel. The blade is forged

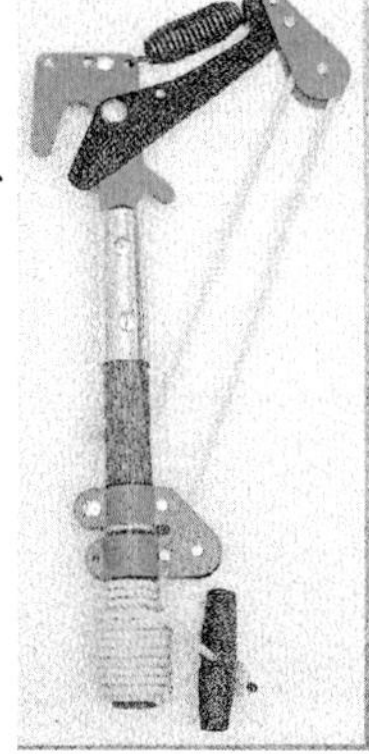

Fig. 9.9 : Tree pruner

to shape and the cutting edge grounded sharp. The blade is hardened to 425-450 HB. For cutting operation the branch or twig is brought under the hook. The blade is actuated by pulling the rope, which cuts the twig due to shearing action. When the rope is held loose the blade returns to its original position due to spring action. The tree pruners are available in various sizes. The size of the tree pruner is known by its overall length. One of the typical sizes is 360 mm including the length of the socket. The cutting capacity of the tree pruner is about 20 mm diameter branch or twig.

Pruning saw

The pruning saw is manually operated hand tool for cutting or trimming of the branches, which are beyond the capacity of the secateurs or tree pruner (Pandey *et al.*, 1997). Like carpenter saw it essentially consists of a serrated blade and a handle (Fig. 9.10). The blade has a longer pitch to avoid clogging during operation in cutting of green branches. The blade comes in straight and curved design and is made from tool steel having a carbon content of more than 0.7%. The cutting teeth are made sharp and hardened to 45-48 HRC. For efficient cutting the blade is provided with adequate rake and gullet angles. The handle is made from good quality wood and riveting joins the blade. For cutting the blade it is repeatedly moved over the branch and cutting is done in forward stroke.

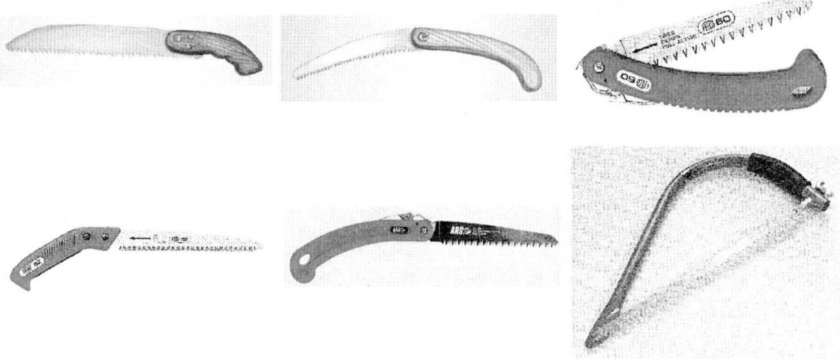

Fig. 9.10 : Pruning saw

Chain saw

It is also called power saw and is a light and portable machine normally operated by one person (Pandey *et al.*, 1997). Cutting is done by an endless chain fitted with cutters, which runs around a flat piece called the bar (Fig. 9.11). The drive link of the chain runs in a groove, machined in the edge of the bar and is pulled along by the teeth of a sprocket, which engage them. The sprocket in turn is driven at full speed either by small two- stroke petrol engine or electric motor. The power to the chain is transmitted through a centrifugal clutch mounted on crankshaft of the engine. The chain is of roller type and has left and right hand cutters spaced alternately along its length. In front of each of the cutters is a small projection called a depth gauge whose purpose is to control the depth of cut made by the cutter. The chain saw is used to trim dead or diseased wood from trees, to remove inconveniently placed branches or fell trees.

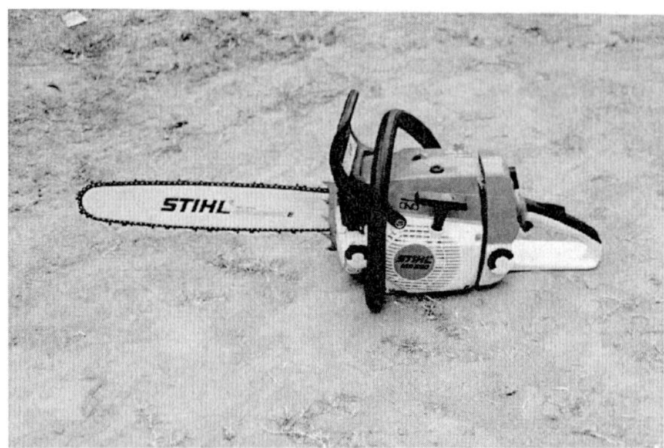

Fig. 9.11 : Chain saw

Budding grafting knife

This is used for accurate cut when budding & grafting (Pandey *et al.*, 1997). The budding knife is an important hand tool of a gardener, which consists of a folding blade and a handle. The blade has two edges. One of the edges is sharpened all along its length; whereas the blunt or the other edge is sharpened on the tip and is slightly curved. This sharpened curved portion is used to create a 'T' opening or slot on the bark of the mother branch or twig for the insertion of the bud. The edge sharpened all along its length is used for cutting of scion stick or defoliation of leaves from the scion and slashing of bud from the stick. Some budding knives have a short and round plastic blade at the end of handle called budder, which is used for raising of the bark of the slot for insertion of the bud. The blade when not in use is folded into the handle. The blade is made from high carbon steel, tool steel or alloy steel and hardened to 460-510 HB. The outer part of the handle is made from the horn, plastic or fine quality wood and the internal fittings from brass or aluminium alloy and a spring steel strip is provided to lock the blade in operating position. For operation the sharp edge of the blade is held against the scion stick and force is applied at an angle, which causes cutting of the stick.

The grafting knife is another important plant propagation hand tool, which resembles a household knife. The principal parts of the knife are blade and the handle. The cutting edge of the blade is sharpened all along its length and the other edge is blunt. The blade of the knife can be folded into the handle when not in use. A nail mark is provided in the blade to pull the blade from the handle. The blade is made from high carbon steel, tool steel of alloy steel and hardened to 460-510 HB. The operation of the knife is similar to that of budding knife and it is mainly used to cut the scion sticks for veneer grafting, cleft and stone grafting and inarching. Defoliation of the leaves of the scion stick, making 'V' groove for grafting and making of chisel point of the scion for insertion in the 'V' groove are the functions performed with the grafting knife. The outer portion of the handle is made from horn, plastic or good quality wood and the inner portion is fitted with aluminium or brass strips and a spring steel strip for locking of the blade in working position.

The budding and grafting knife is a multipurpose knife to accomplish both the budding and grafting jobs. It has stainless steel two blades & one brass budder with nylon comfortable grip. It consists two blades each for budding and grafting, which are either joined to a common hinge or are fixed to the ends of the handle (Fig. 9.12). A plastic budder is provided to the other end of the knife in which both the blades are joined to a common hinge or end of the handle. When not in use, both the blades can be folded into the handle. The blades are made from high carbon steel, tool steel or alloy steel and hardened to 460-510 HB. With both the blades it is a versatile knife of the gardener and is extensively used in orchards, vegetable gardens and plantations for budding and grafting purposes in order to evolve new varieties. The handle of the knife is thicker to accommodate two blades. The outer portion of the blade is made from horn, plastic or good quality wood and the inner part has brass or aluminium strips with spring steel strips for locking of the blade in working position.

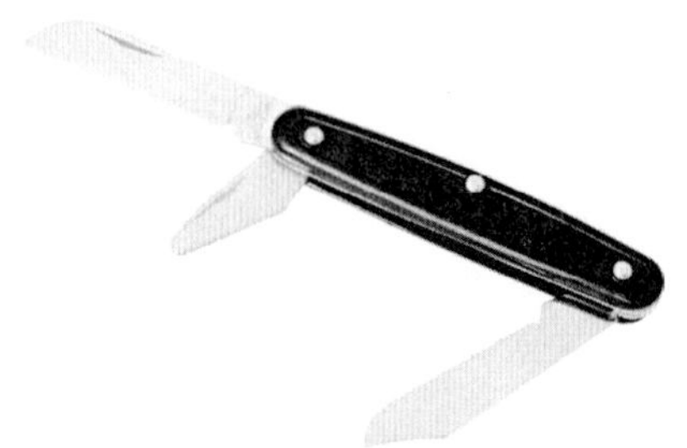

Fig. 9.12 : Budding grafting knife

Courtesy: Falcon Garden Tools (P) Ltd., Ludhiana (Punjab)

Knifes

Pruning knife: This is used for plants pruning (Fig. 9.13). It has stainless steel hardened blades, all corrosion resistance fitting, and nylon comfortable grip. **Patching knife** is used for plants Patching. It has stainless steel hardened blades, two knives and nylon comfortable grip (Fig. 9.13). **Mushroom knife used f**or mushroom cutting & sorting out. It has stainless steel hardened rust free blades and nylon comfortable plastic grip (Fig. 9.13).

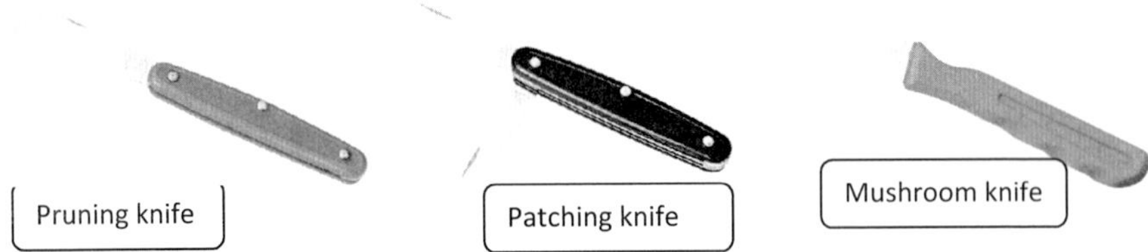

Fig. 9.13 : Different types of knifes
Courtesy: Falcon Garden Tools (P) Ltd., Ludhiana (Punjab)

Hedge shear: The hedge shear is manually operated hand tool for pruning, trimming and cutting of hedges and shrubs. The tool essentially consists of two blades with tangs (Fig. 9.14). The tangs are inserted in the wooden handle and secured by ferrule. The cutting action takes place between two blades, which are pivoted, and the material to be cut is sheared between these blades. The blades are forged to shape and edges are ground to obtain a bevel angle just less than 90 degrees. It is important to maintain the desired cutting while sharpening these blades to obtain clean cut. The blade and tang are made in single piece from high carbon steel, tool steel or alloy steel and hardened to 420-470 HB. The handles are made from high quality wood. For operation the handles are pulled apart to open the blades. The material or hedge twigs to be cut are brought in between these blades and moving the handles inward shears the twigs. This action is repeated fast for trimming of the hedges and shrubs. Some of the models are provided with pruning notch near the pivot of blades for cutting of thick twigs. The hedge shear is used for pruning and trimming of hedge and giving it desired shape. It is also used for cutting of shrubs and removing of haphazard growth in gardens and lawns. It is useful for trimming hedges when neat wall of foliage is required. It has blade size of 250 mm. I has hardened steel blades with rust preventive coating. It has plastic handle with innovative design for extra comfort cutting. Total length of hedge shear is 535 mm.

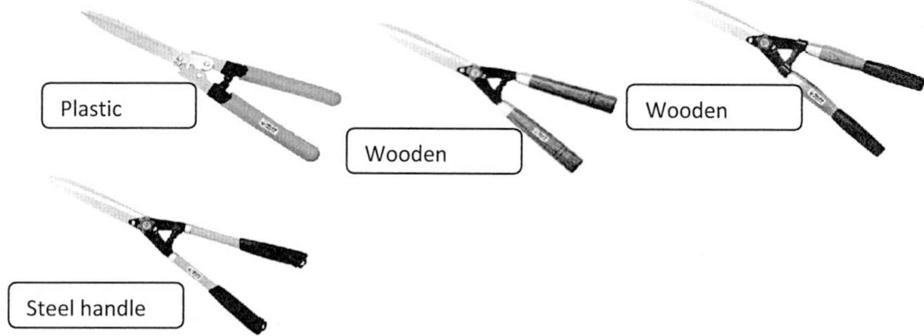

Fig. 9.14 : Different types of hedge shear
Courtesy: Falcon Garden Tools (P) Ltd., Ludhiana (Punjab)

Hedge trimmer

Hedge trimmer consists of a cutter bar having two sets of reciprocating blades (Fig. 9.15). The teeth along the top blade are diamond round and double edged to stay sharp for long (Pandey *et al*., 1997). It can cut even branches of up to 16 mm in diameter. The cutter bar is driven either by engine or motor. The unit can be moved in various directions- to the left, right, upwards or downwards. A baffle guard is provided to protect the user from flying leaves, stems or branches. The motor power unit is provided with flexible chord, which permits the movement of the trimmer to all places in the garden. An extra trigger switch is integrated in the handle for quick, error free operation. Hedge trimmer is used for trimming hedges, shrubs and brambles. It is also used for contouring plants in desired shapes and sizes for enhancing the aesthetics of the garden.

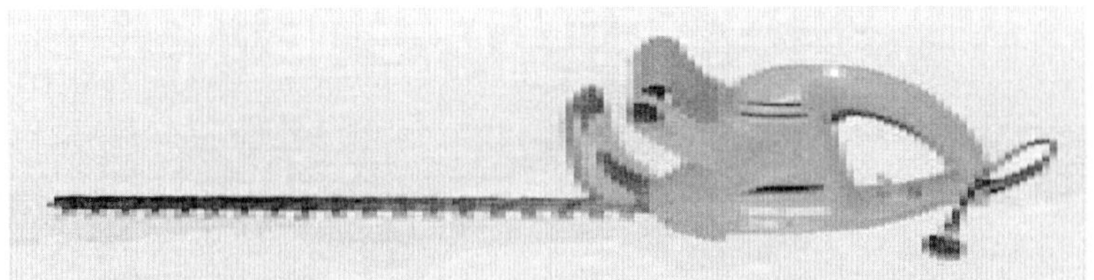

Fig. 9.15 : Hedge trimmer

Tea leaf plucking shear: Useful for tea gardening with reduced labour plucking and gathering together without the wastage of tea leaves. It covers more area with fewer efforts. It has sharp & tempered blades of high efficiency and reduces labour cost (Fig. 9.16). It has steel handle with soft grip. A bag is attached for collecting tea leaves.

Fig. 9.16 : Tea leaf plucking shear

Courtesy: Falcon Garden Tools (P) Ltd., Ludhiana (Punjab)

Loaper

The lopping shear is manually operated hand tool with long handles. It is used for pruning and cutting of branches and twigs of the orchard trees in standing position, which are beyond the reach, and cannot be cut with pruning secateurs. The shape of the lopping shear is similar to pruning secateur or hedge shear depending upon the manufacturers. The shear consists of two shearing blades joined to the sockets to which wooden handles are inserted (Fig. 9.17). The blades are fabricated from high carbon steel, tool steel or alloy steel, forged to shape and the cutting edges are hardened to 425-450 HB. The sockets are made from mild steel. Both the blades are pivoted at the common point, which allows them to open or close. For operation the handles are pulled apart which open the blades and the branch or twig to be cut is brought in between the blades. The blades are closed to put cutting pressure on the branch, which thus get sheared. Due to long handle thick branches can also be cut with the lopping shear. There are three types of loapers. First one gives finished cut of green limbs for the protection of the growth of the trees. It is light in weight, easy to use, has plastic handles and rust resistance. Overall length is 56 cm and can cut 30 mm thick branch. Second one is also light in weight & requires less effort for cutting thick and hard branches. It has PVC grip handle and is rust resistance. Overall length is 74 cm and can cut 36 mm thick branch. Third one is shearing type ideal for live and soft branches, producing a clean cut. It has tubular steel handles PVC grips. Overall length is 72 cm and can cut 20 mm thick branch.

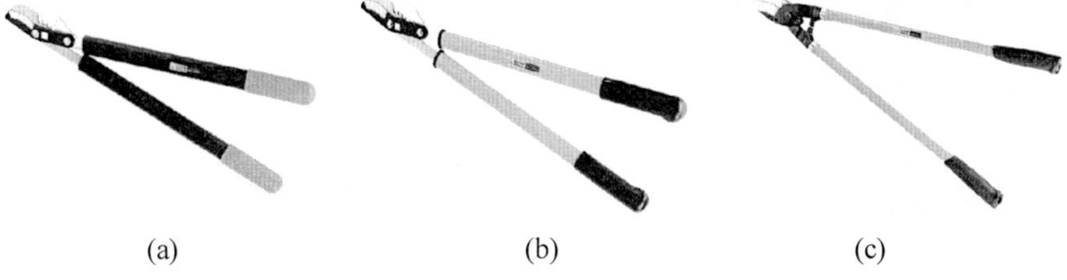

(a) (b) (c)

Fig. 9.17 : Loapers

Courtesy: Falcon Garden Tools (P) Ltd., Ludhiana (Punjab)

Forester shear

The forester shear is similar to the lopping shear for cutting the branches, twigs or bushes in standing position. The tool is provided with long handle, which allow the blades to exert more cutting pressure on the stock. The forester shear consists of a cutting blade and an anvil joined to the sockets (Fig. 9.18). Long wooden handles are inserted in the sockets, which help in cutting the branches or twigs beyond the reach of human hand. The blade and anvil which are important parts of the shear are made from high carbon steel, tool steel or alloy steel, forged to shape and the cutting edges hardened to 420-425 HB. For operation the branch/ twig is held between the blade and anvil and the blade and anvil are moved inward, which causes the blade to penetrate in the branch or twig where the anvil of the shear provide the support. The forester shear is used for pruning or cutting the unwanted thick branches or twigs up to 60 mm thick in orchard or forest. It is also used for cutting the shrubs or bushes and clearance of unwanted growth.

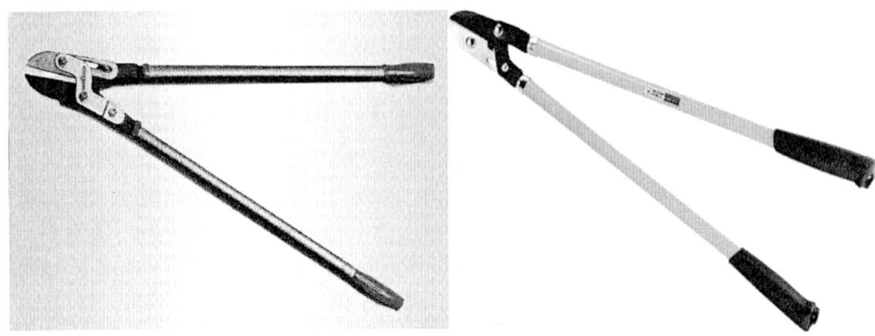

Fig. 9.18 : Forester shear

Courtesy: Falcon Garden Tools (P) Ltd., Ludhiana (Punjab)

Angular long reach pruner

Multipower long reach pruner used for clean cutting of green limbs from trees with less effort without using ladder (Fig. 9.19). It is light in weight and easy to use. It can cut at different angles. Cutting capacity is 22 mm thick branch. Extension pipe length can be increased to 2.3 m.

Fig. 9.19 : Angular long reach pruner

Courtesy: Falcon Garden Tools (P) Ltd., Ludhiana (Punjab)

Tree pruner with telescopic handle: Designed to cut hard branches and eliminates the need for ladder. It is easy & convenient to use with telescopic handle (Fig. 9.20). It is light in weight with fibre glass handle and has extra strong nylon rope.

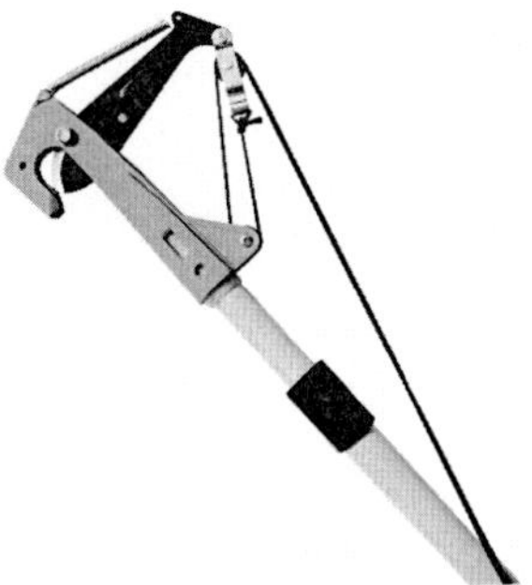

Fig. 9.20 : Tree pruner with telescopic handle
Courtesy: Falcon Garden Tools (P) Ltd., Ludhiana (Punjab)

Fig. 9.21 : Tree pruner with pruning saw &handle
Courtesy: Falcon Garden Tools (P) Ltd., Ludhiana (Punjab)

Tree pruner with pruning saw & handle: It is designed to cut hard branches and eliminates the need for ladder (Fig. 9.21). It is easy & convenient to use, light in weight and can cut 20 mm thick branches. It has extra strong nylon rope and extension pipe of aluminium alloy can be up to 2 m long.

Long reach cut & hold pruner: It is ideal for hand to reach branches and has long reach (Fig. 9.22). It is light in weight and easy to use. It can cut stems up to 10 mm. It has aluminium alloy holder.

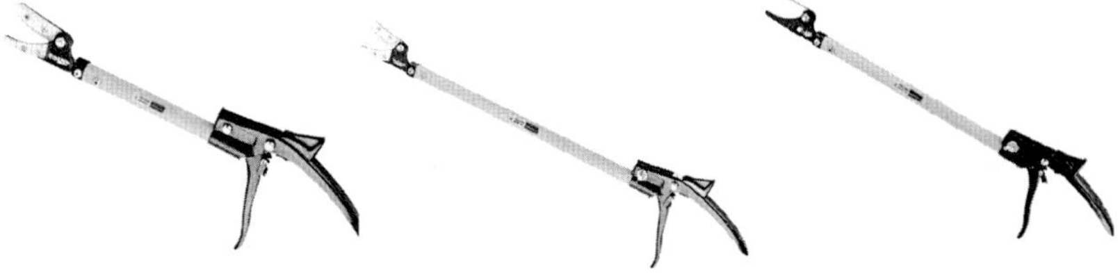

Fig. 9.22 : Different types and sizes of long reach cut & hold pruner
Courtesy: Falcon Garden Tools (P) Ltd., Ludhiana (Punjab)

Fig. 9.23 : Fruit catcher with extension pipe

Courtesy: Falcon Garden Tools (P) Ltd., Ludhiana (Punjab)

Fruit catcher with extension pipe: Fruit catcher makes fruit plucking & picking easy by adding the cutting blade (Fig. 9.23). It is light in weight and easy & convenient to use. It is used for plucking and picking fruits. Extension pipe is D type of aluminium alloy 2 m long.

Fruit gatherer with extension pipe: Useful for collecting/plucking fruits from tree (Fig. 9.24). It is light in weight and easy & convenient to use. It has extension pipe of aluminium alloy 2 m long.

Fig. 9.24 : Fruit gatherer with extension pipe

Courtesy: Falcon Garden Tools (P) Ltd., Ludhiana (Punjab)

Grafting tool: It is useful for easy grafting when slot cut is required, easy to use and light in weight (Fig. 9.25). It makes accurate and neat grafts. It is useful for easy grafting when omega cut and bud graft is required.

Fig. 9.25 : Grafting tool

Courtesy: Falcon Garden Tools (P) Ltd., Ludhiana (Punjab)

Digger: It is used to make holes for planting and transplanting seedlings or insert cuttings in the soil without damaging roots (Fig. 9.26). It has comfortable grip, rust preventing coating and easy to use.

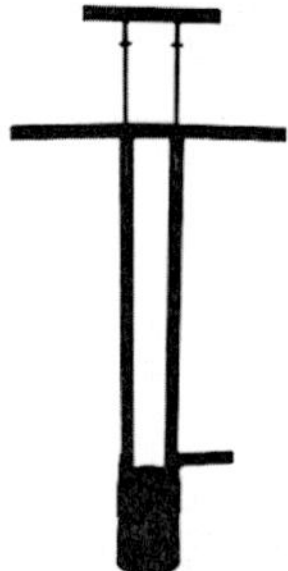

Fig. 9.26 : Digger

Courtesy: Falcon Garden Tools (P) Ltd., Ludhiana (Punjab)

Bulb planter: It is designed for digging right size holes for bulb plantation (Fig. 9.27). It has chrome finish, plastic comfort grip and can dig holes at 5 cm and 10 cm spacing.

Fig. 9.27 : Bulb planter

Courtesy: Falcon Garden Tools (P) Ltd., Ludhiana (Punjab)

Axe

The axe is a simple hand tool, which consists of cutting edge and an eye for fixing of a handle (Pandey *et al*, 1997; Bhardwaj *et al.*, 2004). It is forged to shape from a single piece. Axes are available in various sizes and shapes (Fig. 9.28). The common types are hand felling, felling estate pattern and felling trade pattern. For operation, the operator holds the handle with both hands at convenient position and the tool is raised to suitable position and struck with force against the work. The penetration is caused through impact action, which shears the slice of wood. The axes are made from high carbon steel and the cutting edge is hardened to 550-650 HB. Axe is multipurpose cutting tool used for felling and delimbing of trees, splitting of logs for firewood and dressing of logs for timber conversion. Small axes are also used for clearing of bushes.

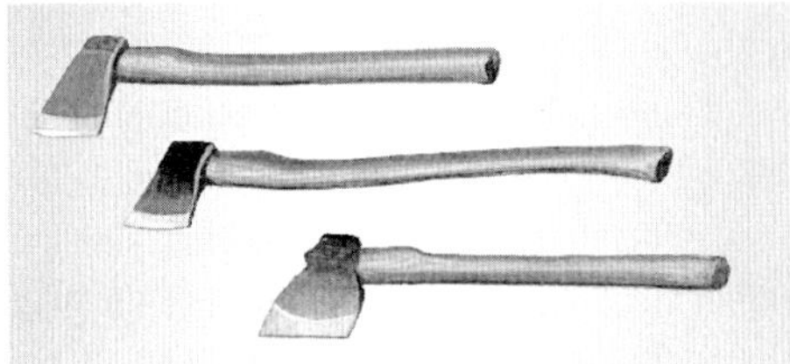
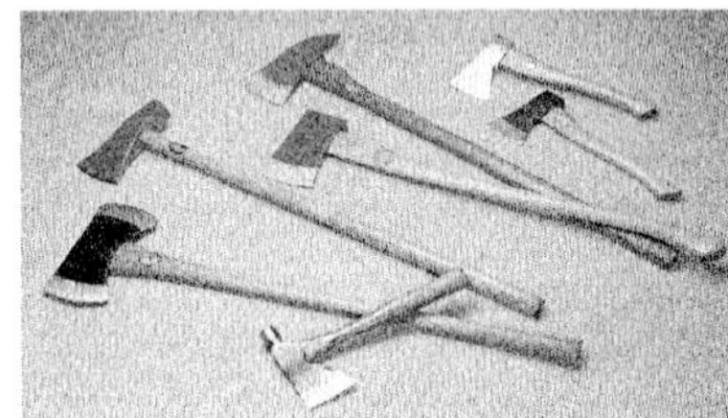

Fig. 9.28 : Different types and sizes of axe

Garden tools

Turf aerators: The purpose of turf aerator is to allow air to pass to the grass roots. It improves the surface drainage and enables top dressing to be easily absorbed. The hand aerator consists of a metal frame, with a wooden hand grip, onto which four or five tines are fitted. The aerator is pushed into the ground with the foot and then pulled out so that the tines leave deep holes. For large areas self-propelled aerators can be used. These consist of round cage onto which are fastened several rows of tines. The bars onto which the tines are mounted can pivot and are spring loaded so that they can enter and leave the soil vertically. As a result they perforate the turf cleanly and do not tear it. The cage is driven by a small petrol engine and carried on wheels for transport. Large tractor – operated aerators are in common use and can be either trailed or mounted. They consist of a series of four or six sided plates, equally placed along a shaft. At each corner of every plate fixed or free swinging tines are pulled forward, the tines entre the ground. If they are swinging they enter and leave the ground vertically so reducing turf damage. There are 4 common types of tine available viz. hollow tine, solid round tine, flat deep piercing tine and flat root action tine. The hollow tine takes out a core of soil and is used for turf renovation. The core of soil removed can be replaced by fresh soil brushed into the holes. The tines are self-cleaning as they are open at each end and the cores pass right through them. The solid round tine spikes the ground and helps in the removal of surface water. The soil around the hole is compressed. The flat deep piercing tine helps in soil aeration and cultivation. The flat root action tine is used to prune the grass roots and encourage healthier growth.

Rollers: Rollers are frequently used for consolidation on cricket wickets, bowling and golf greens. They are also used to press down small stones on areas of newly seeded grass before mowing. Rollers used for this purpose are of the flat type and are made of either steel or cast iron. They are available in a wide variety of widths and diameters. A good roller should consist of at least two sections, so that each section can rotate at different speeds when turning corners, to prevent damage to the grass. Rollers can be either hand –pushed, self-propelled or tractor-drawn. Most rollers are provided with some form of ballasting so that extra weight can be obtained. This may consist of tray on top of the roller so that extra weight in the form of concrete blocks, bags of sand or cast iron can be added. Other rollers are provided with closed ends so that the ballast in the form of water, sand or concrete can be put inside the roller. Water provides a good form of ballast, as it can be quickly removed and the roller weight can be changed very easily. Very heavy rollers are used on cricket wickets, lighter ones on bowling and golf greens. For large areas rollers can be pulled behind a tractor in gangs. Scrapers should be provided to prevent the buildup of soil on the roller, especially if the ground is damp.

Brushes: Brushes are used in turf maintenance for brushing the grass to encourage aeration, scatter worm casts and work in top dressing. The brushing are made from whale bone, plastic or hair. The working width of the brush can vary from 300 to 1800 mm. They can be attached to small hand pushed machines which are also adaptable for scarifying, piercing or aeration. Larger brushed are attached to the rear of tractor-draw machines.

Edge trimmers: With the large increase of ornamental lawns, greater lengths of lawn edges have to be trimmed. Powered edge trimmers have greatly reduced the time necessary for this from that taken by hand edging shears. The power supply for these machines is either a small petrol engine or a battery-operated motor. Edge trimmers are not usually self propelled and have to be hand pushed, the power being supplied to the blade only. The trimmer has a vertical blade which may be either mounted onto shaft of the motor or driven by a small V-belt. The blade has overhangs the edge of the lawn and rotates at high speed. The blade has a sharp edge and cut the grass on contact in the same way as a rotary mower. Good lawn edges are essential if a neat job to be made. The blade on some machines can also be used on the horizontal, so that the trimmer can be used as a small rotary mower for trimming in tight corners. The battery-operated edger will trim about 1.6 kilometers of lawn edge per charge, but this can be increased if large batteries are fitted. Included with the battery edges, but not necessary built in, is main charger, which will ensure that the battery is fully charged ready for use. Depth of cut is controlled by the front roller or wheel and raising the roller the depth of cut is increased.

Turf cutters: A small turf cutter consists of two L-shaped blades which are drawn under the turf at a depth which is adjustable. The depth of cut can usually up to 75 mm depth and also the pitch of the blade can be altered to suit the depth. The turf width is determined by the distance between the vertical parts of the blades and is about 300 mm on a small machine but can vary up to 600 mm on large machines. Turf cutters are driven by a cylinder petrol engine and the drive to the rubber tired roller is by V-belt and chain. The rear of the machine is supported by one or two rubber tired wheels. Some machines have an additional vertical blade which acts like a guillotine, working at right angles to the direction of travel and cutting the turf up to convenient lengths. After cutting the turf can be lifted and rolled. There are some machines designed to follow the turf cutter which lift, roll, and deposit the rolls of turf. A small machine cutting a 300 mm turf can cut up to 10 square meters of turf per minute.

Levelling devices: A small hand-operated leveler consists of a frame made out of angle iron with a handle hinged to it. The frame is drawn across the ground and smoothes out any unevenness in the surface. The frame can be used either way up. With the flat of the angle iron on the ground it is useful for spreading top dressing, leveling running tracks or leveling soil surfaces prior to seeding or turfing. It is used with the narrow side of the angle iron down. It is useful for top dressing into the turf.

The trowel: Trowels are small scoop shaped tools used for making small holes for plants. Their curved blades, usually 12 cm to 20 cm long, cut the soil round the plant leaving a ball of soil round the roots. Cheap trowels have steel handles which tend to buckle easily while better quality trowels have wooden handles. Smooth ends to wooden handles are essential.

The garden line: The garden line is a very important horticultural tool used for setting out straight lines. It is used when seed sowing, trenching, lawn edging, transplanting and for many other garden operations. It consists of between twenty and sixty feet of twine can be wound for storage. Polythene or nylon twine is best as it is rot proof and does not shrink if it gets wet. The pin and reel may be made out of wood or steel.

Rotating disc mower

In a rotating disc mower, a knife-edge moving at a high speed cuts the grass through impact. In the cutting unit, a circular disc in horizontal plane has two to four triangular or rectangular blades (Fig. 9.29). The blades are replaceable type. The disc rotates at about 3000 rpm. The blades are made from high carbon steel and they have to be maintained sharp, otherwise uprooting of grass will take place instead of cutting. In some designs, the blades are swinging type and move out of the way when they hit an obstruction. The surface finish obtained with a rotating disc mower is rough as compared to that obtained with a cylinder mower. Therefore, this is normally used for rough areas such as orchards where the finish is not critical. Rotating disc mowers are available with electrical motors or driven by engine. The mower is used for cutting grass in lawns and fields.

Fig. 9.29 : Self-propelled rotating roller lawn mower

Courtesy: Falcon Garden Tools (P) Ltd., Ludhiana (Punjab)

Cylindrical lawn mower

It consists of cylindrical reel on the surface of which cutting blades are mounted in spiral fashion, anvil stationery blade, handle, drive wheel, drive pawl, grass gatherer (bucket) and roller assembly (Fig. 9.30). The cutting blades are made of carbon steel and fastened to flanges mounted on a centre shaft. The cutting blades rotate, and because of their spiral mounting, cause progressive cutting action across the anvil blade. The anvil blade is a flat blade sharpened at the edges and can be adjusted. The grass is trapped between the rotating and anvil blades and cutting takes place due to shearing action. The machine has also front roller for adjustment of height of cut and a grass box at the rear to collect the cut grass while the machine is in operation. The mower is used for cutting grass in lawns and fields. It can be electric/engine operated machine (Fig. 9.31).

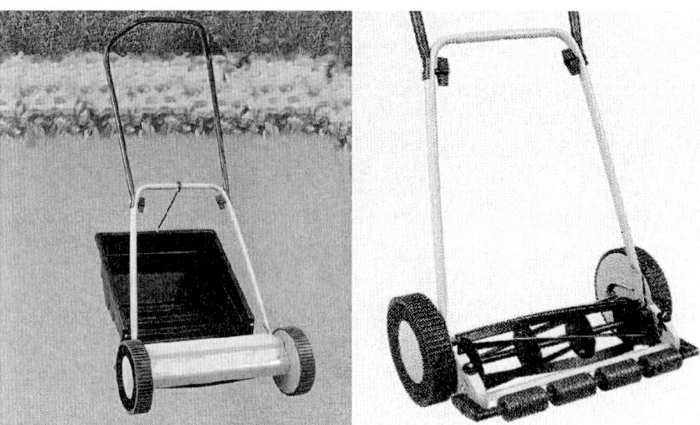

Fig. 9.30 : Cylindrical lawn mower

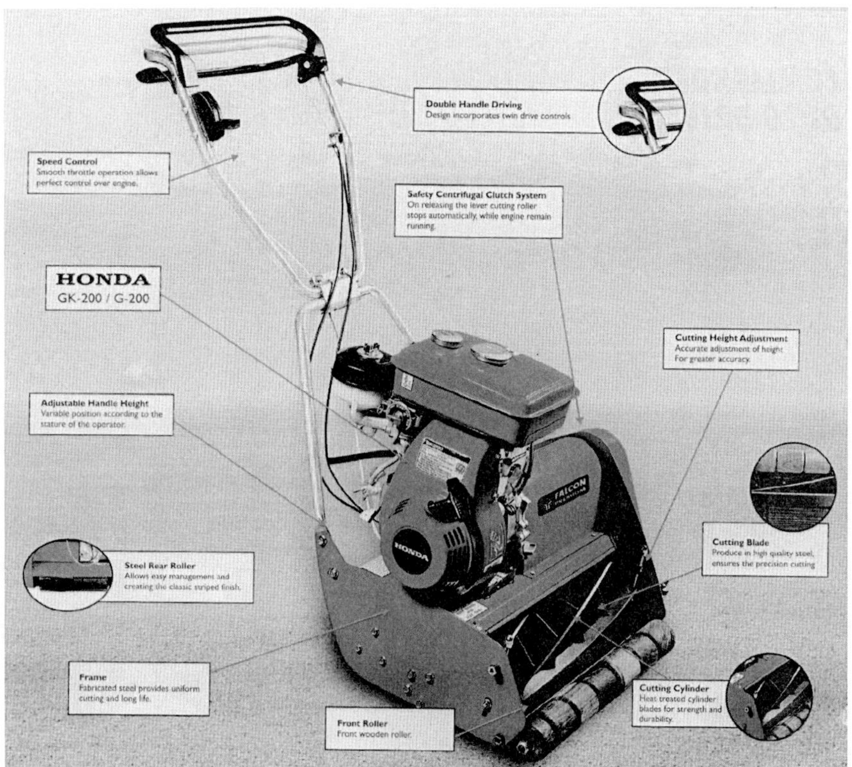

Fig. 9.31 : Power driven cylindrical lawn mower

Courtesy: Falcon Garden Tools (P) Ltd., Ludhiana (Punjab)

Post-hole digger

The post-hole digger is an implement that drills/digs a hole of varied sizes and depths, which are required for plantation of trees and samplings, fencing, erection of marking stones etc. The perennial crops require specialized operations like digging and pitting at the time of planting. Pitting is necessary to provide favourable conditions for early establishment and growth of young plants. In the conventional pits, the roots of tree seedlings are unable to penetrate deep into the hard soil and results in uprooting during high-speed winds. In the conventional method, the pits are dug with spade, pick axe etc which are laborious and time consuming. It can be attached to a 3-point hitch of a tractor, manually operated or power tiller operated. It saves a lot of time as compared to manual digging. It is safer and most economical method of drilling/digging the holes for plantations, nurseries etc.

Auger digger

It is a high speed machine fitted with diesel/petrol engine. It consists of auger, engine and handle (Fig. 9.32). The machine is easy to handle (Pandey *et al.*, 1997). For hill region it is efficient equipment for horticulture crops. The auger has length of 680 mm and diameter of 60 mm. It can dig 30 pits per hour.

a) Manual digger engine operated

b) Engine operated manually controlled by two person auger digger

Fig. 9.32 : Engine operated auger digger

Power tiller operated post hole digger

Power tiller operated post hole digger (also called pit digger or auger digger) consists of a small frame with the provision to lower and raise the soil-working element (Fig. 9.33a). Drive is provided to the unit with the help of a set of bevel gears and belt pulleys (Pandey *et al.*, 1996). Lowering and raising is accomplished by means of a rack and pinion arrangement which is operated by a hand wheel (Fig. 9.33b). It has two depth adjustment wheels, which support the weight of the implement, and provides stability. Pits can be dug up to a depth of 45 to 60 cm with up to 30 cm diameter. For operation the auger is mounted on a power tiller and is lowered with the help of a steering (Fig. 9.33b). The auger is lifted when the desired depth of the hole is achieved. It is suitable for digging circular pits for planting saplings. It is also suitable for use in orchards and forests due to its manoeuvrability.

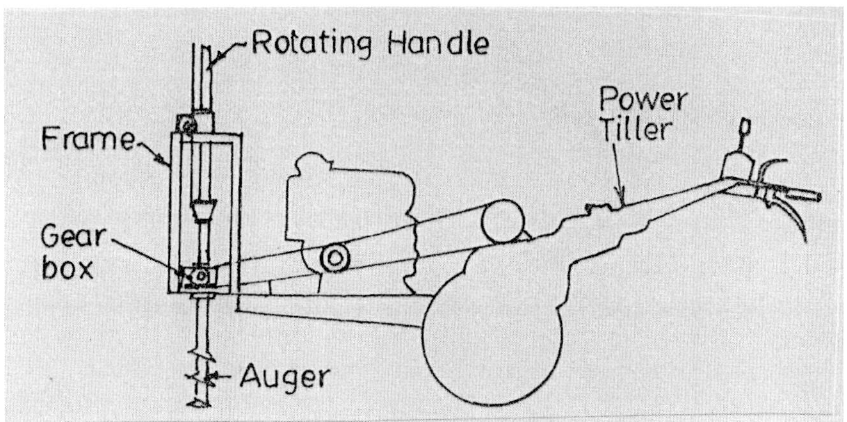

a) Components of power tiller operated post hole digger

b) Operating power tiller operated post hole digger

Fig. 9.33 : Power tiller operated post hole digger (pit digger)

Tractor operated post hole digger (pit digger or auger digger)

Manual digging of pits/holes is very expensive, laborious, and time consuming particularly when pits are required to be made in mass numbers. Posthole digger is an attachment to the three-point linkage of tractor (Pandey *et al*., 1997). It consists of an auger, which is driven through bevel gears (Fig. 9.34a). The auger gets drive from the tractor PTO through a propeller shaft and bevel gear box. The perpendicularity of digging auger is maintained with four-bar linkage formed by hitching system and the tie rod provided at the top. The tip of the auger is either diamond shaped or pointed with wings to suit to different soil conditions. The diameter and depth of hole can be changed by changing the auger assembly (Fig. 9.34b). Auger points and leading blades are replaceable. The soil counter-sinking attachment is fitted on co-operating auger of the post-hole digger. It includes a sleeve to which three ties of equal lengths are attached. The inner wall portion of sleeve is so designed that it just engages the outer wall portion of co-operating auger of the post-hole digger. Three soil cutting blades extending in the upward direction are welded with three equally spaced ties. The struts made of MS angle hold these ties. The whole assembly is then coupled to the propelling shaft of post-hole digger. The post-hole digger fitted with counter-sinking attachment is placed in a vertical position in the soil and drive is given through PTO shaft. The auger portion makes a cylindrical hole and counter-sinking attachment enlarges the hole in the form of an inverted truncated cone. The pits so made are suitable for planting and growing coconut palms and rubber. Tractor operated post hole digger is used to dig holes of different sizes (20 cm to 90 cm drill diameter) to a maximum depth of up to 90 cm for plantation of samplings like mango, coconut, pomegranate, teak, lemon etc (Fig. 9.34c). This is also used for

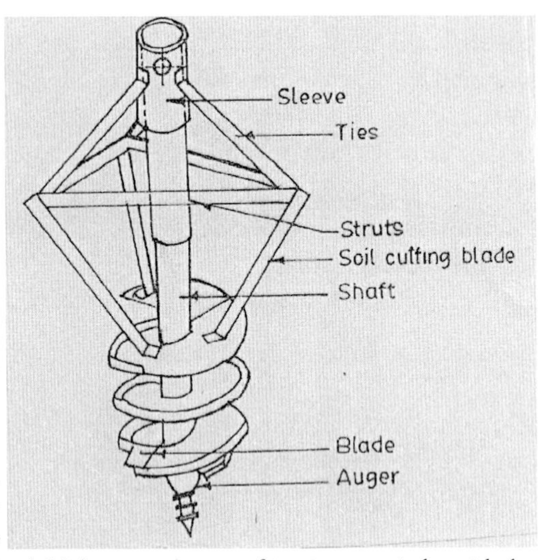

a) Major components of tractor operated post hole digger

erecting fencing poles, electric poles, road signs, marking stones etc. It takes less than a minute to dig a hole with well defined circumference and neater pit. Removed soil is placed near the pit which is used to fill the pit after plantation or erection work.

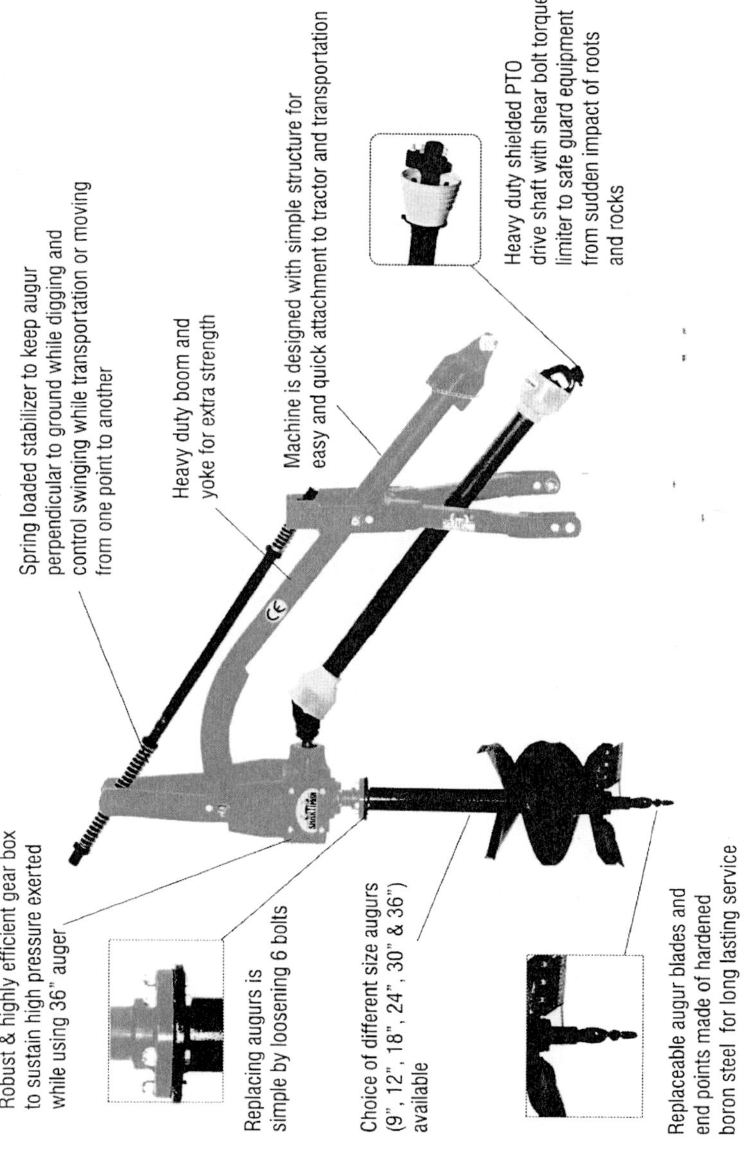

b) Details of tractor operated post hole digger
Courtesy: Tirth Agro Technology Pvt. Ltd., Rajkot (Gujarat)

c) A view of tractor operated post hole digger

Fig. 9.34 : Tractor operated post hole digger

Harvesting of fruits and vegetables

Harvesting of fruits and vegetables crop is an important agricultural operation, which demand considerable amount of labour. The availability and cost of labour during harvesting season are the serious problem. It is estimated that harvesting of crops consume about one-third of the total requirement of the production system. Timely harvesting of the crop is vital to achieve better quality of fruits, vegetables and higher yield of the crop. The shortage of labour during harvesting season and vagaries of the weather cause greater loss to the farmer. Some of the fruits and vegetables are highly perishable products and required to be harvested in a very narrow range of time. These may be processed, graded, stored or consumed fresh, soon after harvesting. Because of these reasons, it is very important to mechanize the harvesting of fruit and vegetables. High yielding and uniformly maturing varieties need to be harvested mechanically. Mechanical harvesting can result in an extremely high rate of product output in which case material handling methods are of major importance. Handling of fruits and vegetables should be such that, there should be minimum damage to the product. In general, mechanically harvested fruits or vegetables contain considerable quantities of trash, immature or damaged fruit, which has to be removed either during harvesting itself or separately and later on manually.

There are varieties of fruits available, which need to be harvested at their maturity. Many fruits do not mature uniformly, which needs several pickings to obtain maximum yields. In general fruits can be categorized into three types, namely, tree fruits, vine fruits and bush fruits. Tree fruits are mangoes, apples, papayas; vine fruits watermelon, muskmelons etc and bush fruits are raspberries, blueberries, cranberries etc. Various principles and devices have been tried for harvesting fruits.

Hold on and twist type

It is a manual-harvesting tool with which individual fruit is first held between two jaws of the tool and then twisted to shear off the stock (Pandey *et al.*, 1997: Bhardwaj *et al.*, 2004). The jaws are made of 14 gauge mild steel sheet. These are held together by a tension spring on a pivot fitted on 10 mm mild steel rod (Fig. 9.35). A handle is fitted to the tool. One of the jaws has a lever bracket and rope arrangement for operating the jaw. Three mm thick rubber sheet padding is provided on inside of the jaws to avoid any skin damage while holding the fruits. After its detachment, fruit is released by pulling the cord in to a ring. A cloth conveyor or net is provided below the jaws for collection of harvested fruits at ground level without any damage. The tool is suitable for harvesting peach, pear and orange. Its field capacity is 250-300 fruits/man-h. Another manual-harvesting device consists of an oval shaped rings. The bottom ring is meant for fastening nylon net. A cutting mechanism is provided at the top of the ring and it consists of double bladed triangular plate together with toothed wheel (Fig. 9.35). This toothed wheel is riveted at the centre of the two fixed cutting blades. The wheel rotates freely about its central rivet and acts as a conveyor of a mango stock. For fixing a bamboo handle of desired length, a holder is provided to the harvester opposite to the cutting mechanism. A plastic divider rod bisecting the cutting mechanism is provided in the ring to guide the stalk of fruits either to the left or to the right side of the cutting blade. For harvesting mango, the harvester is raised and fruit is taken in the ring by pulling the harvester. The pedicels of the fruit are taken in between the toothed wheel and blade. On rotation of a toothed wheel, the pedicle is guided over the sharp edge of the blade where it is sheared. Field capacity of the device is 140 fruits/h.

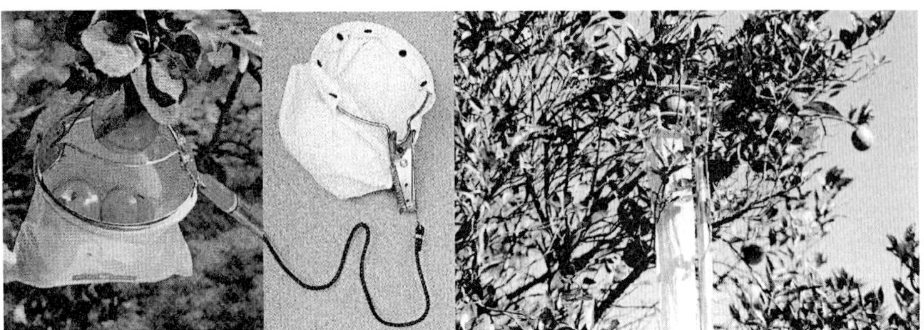

Fig. 9.35 : Hold on and twist type fruit harvester

Manually operated Sapota harvester

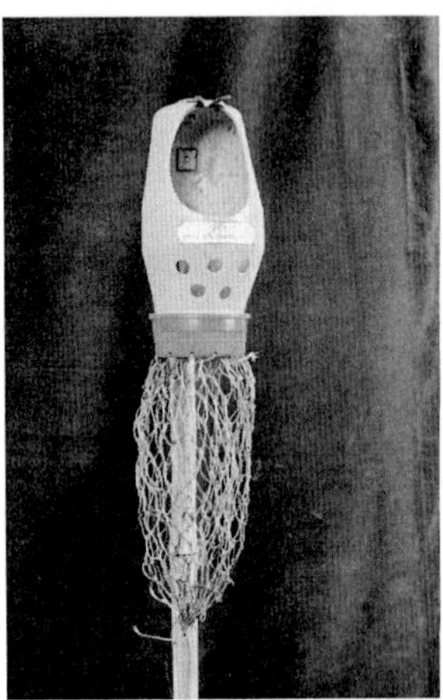

Fig. 9.36 : Manually operated Sapota harvester

It consists of main body of PVC having cylindrical shape (Pandey *et al.*, 1997). The upper end of the body is closed while bottom end is open to which nylon net for collecting the fruits are tied (Fig. 9.36). A stretched string closes the other end of the net. A gate is made on the body for entry of the fruits to be harvested. On the lower surface of the body a metal holder is fixed to hold the bamboo of required length. Two fingers cut in V-shape and with small sharp blades provided at the closed end of the body of the harvester. The fingers help to select and hold the fruit to be harvested from the bunch. By pulling the harvester, fruit is detached from the bunch, which falls in the body and rolls into the net. To unload the harvested fruits in the net a stretched string at the closed end of the net is loosened. The harvester is used for harvesting of small fruits like lemon, sapota etc.

Bhindi (Ladies finger) plucker

The plucker consists of two arms hinged together, cutting blades joined to open ends of arms and two rings joined to the arms (Pandey *et al.*, 1997); Fig. 9.37. The blades are made of medium carbon steel or low alloy steel, hardened and tempered to suitable hardness. Panicles are cut individually using this tool. The operator is spared of drudgery, discomfort and itching to skin of his hands, which are associated with conventional method of manual plucking without any aid. It fits in to the hand properly with the help of two rings, one over thumb and another over index finger. Force to cut the pedicle is exerted by pressing these two fingers against each other. Pedicle is sheared between two straight blades, one of which is notched for better grip. It is used for plucking of bhindi (ladies finger) from plant.

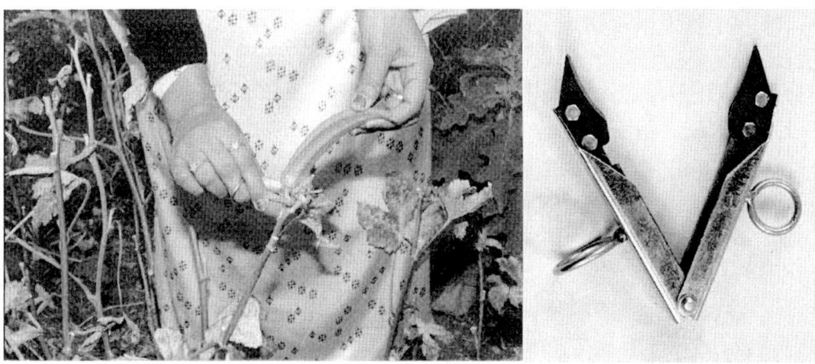

Fig. 9.37 : Bhindi (Ladies finger) plucker

Leafy vegetable harvester

It is a modification of hedge shear to which a gathering mechanism provided (Pandey *et al.*, 1997). The mesh is made of mild steel. The blades of the shear are made from high carbon steel, alloy steel or tool steel (Fig. 9.38). The edges are hardened and tempered to suitable hardness. The blades of the shear are joined to the wooden handle by tang. The handles are shaped for a comfortable grip. The harvester is operated by closing and opening of the blades with both hands. In the open position the stems of the vegetable are placed between the blades and during closing the vegetable is cut. The cutting takes place due to shearing action. The cut crop is collected in the mesh welded to the blades. The collected vegetable is placed over a twine for binding. It is used for harvesting of leafy vegetables such as spinach and fenugreek.

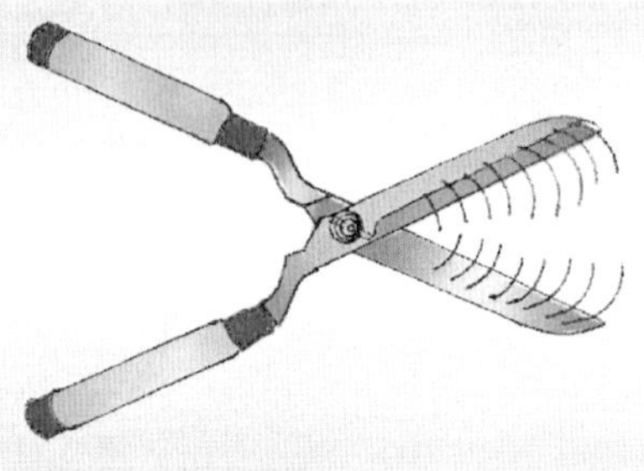

Fig. 9.38 : Leafy vegetable harvester

Tree fruit harvesting using shaking principle

This type of machine works on the principle of accelerating each fruit so that inertia force developed is greater than bonding force between the fruit and the tree (Singh, 2007). The shaking machines based on this principle are already in use for fruits like walnut, almonds etc. Tractor mounted cable shakers, fixed stroke boom shakers and boom type impact knockers are basically used for nuts. Impact knockers are preferred for almonds, because they are large and rigid trees. An impact knocker makes impulses with the help of mechanical, hydraulic or pneumatic means. An electric wheel or crank on the tractor drives fixed stroke boom shakers clamped to the limb of tree. It may also be powered through self-propelled unit.

The other type of manual fruit harvester is based on the principle of shaking the branches or trunk of the tree. It consists of two tongs, adjustable to different heights by means of extension bars. A crank arrangement, to be operated manually and provided with flywheel to shake the tree at proper frequency, is connected to tongs and placed on the ground. The tongs are fixed with main branch of the tree and shaken to detach fruit. In order to save fruit from damaging, a net around tree can be stretched to collect falling fruits. Fruit like apple, mangoes and pears can be harvested by shaking and catch method. The tree trunk or limb is shaken with a vibratory member and fruits are caught on canvas aprons. In some of the machines fruits may be dropped right on the ground. The methods for fruit removal from tree are limb shaking boom and inertia type, trunk shaking, persuading air and vibration. Inertia type shakers are preferred over the stroke shakers. In an inertia shaker, the exciting force is derived from acceleration of a reciprocating mass or two opposite rotating eccentric weight. Both arrangements are so designed that it provides sinusoidal or near to sinusoidal force vibration. The shaker is generally mounted on the catching frame itself and therefore needs other vehicle. In the trunk shaker type machines, the fruit removal occurs simultaneously over the entire tree and the falling fruits are distributed over the entire catching space. It requires a considerable amount of power and is not suitable for very large trees.

Limb shaker

Limb shakers are specifically designed for harvesting fruits and similar crops. The best timing for start harvesting fruits is from "mid to ripen stage". It is powered by a portable small engine with a pre-set RPM limit. The tool is provided with a shock dumping & anti-vibration system for the operator. It is the only tool of the kind that can be comfortably held leaning against the operator's body and effectively eliminating working toil and main causes of stress for operators. The hook is provided of a strong protective revolving rubber rolls which can easily be replaced reduces the wear & tear (Fig. 9.39). Before using hook the fruit tree limb (branch) is as close as possible at 90° angle and make sure the one hand hold the handle, never the hooking pole and allow the tool weight to hang, making sure the shock absorbing system has a free stroke. Trigger the gas for a few seconds and a powerful shaking force is transmitted to the branch felling fruits over netting usually placed over the ground, under the tree.

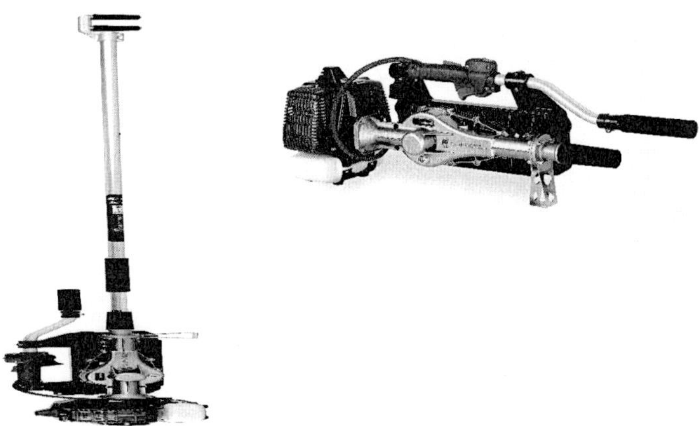

Fig. 9.39 : Limb shaker (hand held power operated)
Courtesy: GSM Co.

Harvesting of tree fruits with man positioners

Fresh fruits for marketing in the fruit markets are picked up by using a device called man position. In this device, a self-propelled machine is used which has an arrangement to position the worker's platform at three-dimensional direction. Picking platforms and other types of man positioners may reduce harvesting cost. This type of machine can have multilevel picking platforms, which can move along continuously and worker can pick the fruits. The fruits can be put in bin or conveyors.

Platform type fruit harvester

The fruit harvester consists of a hydraulically operated platform mounted tractor trolley (Fig. 9.40). The maximum height of reach of operator's platform is 6 m, and the sideways movement is restricted to about 2 m only. Close aerial access to the fruits provides better control on the harvesting operation thereby reducing the damage to fruits and tree branches (Anonymous, 2013; Agrawal *et al.*, 2003)). The tractor operated elevator attachment is a versatile and reliable worker-positioning platform from which the delicate fruits can be picked up or harvested very safely and efficiently from the trees. The positioning is adjustable by the operator himself both vertically and horizontally (Fig. 9.41). Hence selective picking of superior quality fruits with negligible fruit drop and higher capacity of harvesting fruits is achieved as compared to the traditional methods. It can harvest 70-150 kg of fruit per hour up to to the height of 7.6 m. It is used for safe and efficient harvesting of fruits. It can also be used for efficient spraying, pruning, etc.

Fig. 9.40 : Platform type fruit harvester

Fig. 9.41 : Modified version of platform type fruit harvester

Self-propelled platform type fruit harvester

Manually climbing on the tree for carrying out various operations such as fruit harvesting, coconut harvesting, pruning, lopping and spraying, are very risky, tedious and time consuming and more over requires trained man power. Self-propelled platform type fruit harvester can be used for all these operations safely and quickly (Fig. 9.42a); Anonymous, 2015. This machine is equipped with hydraulic transmission power by diesel engine of about 15 hp or so. Platform is provided with all controls for going up and down, for moving laterally for doing above mentioned operations (Fig. 9.42b). It increases productivity as compared to manual climbing and avoids risk and drudgery. Spraying becomes more easy and effective when does from the top and so other operations. It can work up to 8 m height with single operator.

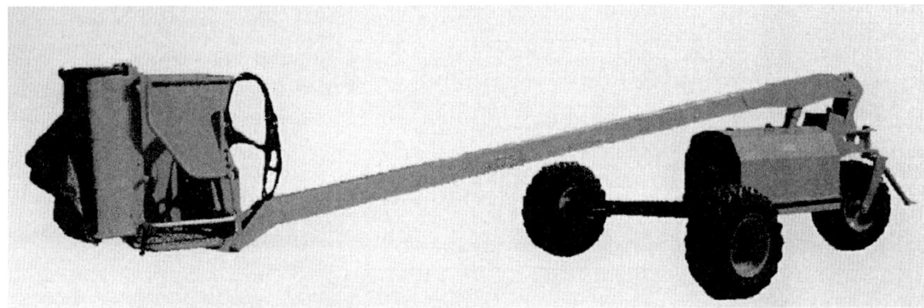

a) A view of self-propelled platform type fruit harvester

b) Self-propelled platform type fruit harvester under working condition

Fig. 9.42 : Self-propelled platform type fruit harvester

Courtesy: Tirth Agro Technology Pvt. Ltd. Rajkot (Gujarat)

Aerial access hoist for coconut and tall tree crop management

All existing tools and devices involve the operator to climb up the tree for harvesting and carrying out other management practices at the crown. Farmers who own large areas are interested in having a system which can elevate a person up to the tree crown by a portable aerial access platform. The existing aerial access platforms are having the limitations such as the access is vertically upward and not in the side wards, the machines are designed to operate by resting on firm surface, most machines have very wide stabilizing legs which cannot be operated under field conditions, and require long time for setting up and operating (Fig. 9.43); Anonymous, 2010, 2013. The machine was developed by Tamil Nadu Agricultural University, Coimbatore with support from M/s Vanjax, Chennai. Suitable safety devices have been incorporated to ensure stability of the hoist (Fig. 9.44). The positioning of the operator platform can be done by the operator himself using electro hydraulic controls. The tractor mounted aerial access hoist can be maneuvered inside coconut grove and four trees can be accessed from a single position. The time required for Operation of stabilizer is 1 min. The time required for positioning against a tree of 10 m height is 1.5 min.

Fig. 9.43 : Aerial access hoist for coconut and tall tree crop management

Fig. 9.44 : Another model of aerial access hoist for coconut and tall tree crop management

Khus root digger

Khus (Vetiver) is a very important and useful aromatic crop having its multiple uses. Generally Vetiver grows wild in some states but in some states like Andhra Pradesh, Uttar Pradesh, Tamil Nadu and Kerala its cultivation is practiced on limited scales also. The Khus roots have been traditionally used in India for its uses in soft drinks during summer and its valuable essential oil is used in various industries like, pharmaceuticals, cosmetics, perfumery and soap. The essential oil from Khus roots is extracted using distillation process. Realizing its commercial value, now the farmers of North Indian states have started commercial cultivation of vetiver being one of the most potential, commercially viable profitable aromatic crops. To promote its cultivation, the CSIR- CIMAP Lucknow has developed several high oil yielding varieties of Khus which are being widely used by farmers. Most of these varieties are ready for root harvesting within 12 months with better oil recovery (approx. 15 -20 l/ha); Tiwari (2011). Khus digger (Fig. 9.45a) contains only single detachable tyne and wider cutting blade/share made up of high carbon steel. The robust mounting frame is made up of heavy section of 10 mm MS pipe with rear mounting attachment system capable of sustaining the thrust of traction force involved in digging and throwing/turning of the harvested roots for continuous field operation. The shovel blade is quite sustainable and wear resistance made of high carbon steel provided with robust mould board curvature plate of 10 mm thick M.S iron and tool frame adds for its sturdy design to handle the heavy tractive force exerted during vetiver root digging. Disc attachment has been provided at a desired angle for minimising the load on the tractor as it makes a smooth pre-cut line of furrow for easy penetration of shovel and simultaneously for providing smooth turning of harvested Vetiver root. The weight of the modified Khus digger is 102 kg and the digger could effectively be operated with more than 25 hp tractor, which is easily available with farmers in the villages (Fig. 9.45b). The field capacity of digger is 0.4 ha/day as compared to 0.003 ha/day manually.

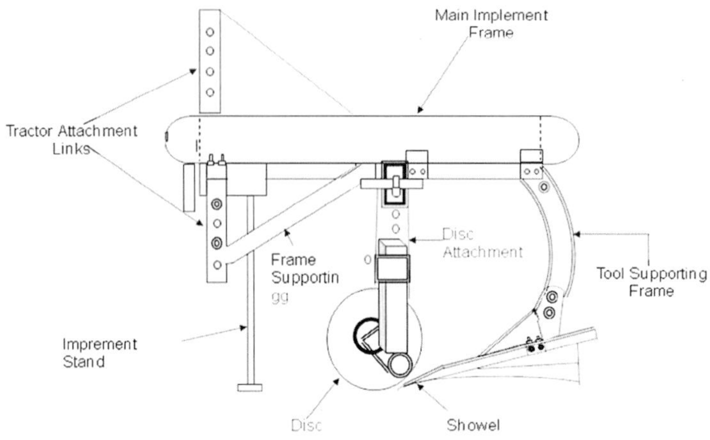

a) Details of khus root digger

b) A view of single row khus root digger

Fig. 9.45 : Tractor operated single row khus digger

Compost shredders and mixers

Vermi compost is a major ingredient in organic cultivation prepared from animal dung or biogas slurry. Though cattle population is increasing still there is shortage of dung as it is mainly used by farmers for cooking stove thereby reducing its availability for vermi compost. There is several biomass weeds material grown in nature rich in NPK content but not being used for productive purposes. A plant shredder is an effective machine for shredding of biomass weeds into small pieces which can supplement dung to prepare vermi compost (Fig. 9.46). The weed biomass is shredded by shredder in 2 to 4 cm pieces with output of 175 kg/h. It has an inclined hopper lower side of which consists of a high speed combing belt. The endless rubber belt is covered with steel teeth and driven by a built-in electric motor or petrol engine. As the belt moves upward through the hopper, its surface combs and shreds the material until it is fine enough to pass out in a thin layer through the opening between the belt and upper end of hopper. The shredded material is thrown out from the machine by the speed of belt and may be delivered into a large heap. Mixing may be achieved by putting the various constituents through the machine together. When the manure compost is ready for application, it can be spread in the field using farmyard spreader.

Fig. 9.46 : Plant shredder in working

Grading machinery

Grading of size and quality is an essential for the marketing of fruits and vegetables. Sizing machines may be grouped broadly into two based on diameter and based on weight. Machines, which grade by diameter, vary in design and size. It gives higher output than the machine grade fruits by weight. Roller feed conveyor is used for quality sorting of fruits. A good weight grader has many advantages. It can be used for any shape, easily adjustable and can be used for crops that are easily blemished. Grading by weight can be more accurate than grading by diameter.

The fruit grader employs stepwise expanding pitch fruit grading mechanism based on the principle of changing the flap spacing along the length of movement of fruits (Fig. 9.47). The grader can be used to grade fruits such as apple, sweet lemon, oranges and larger size sapota (Agrawal *et al.*, 2003). It is provided with elevator feeder for constant feeding of fruits in grading mechanism. One inspection platform is provided to feed the material to elevator feeder and sort out the damaged fruits by visual inspection. The bags containing fruits can be put on inspection platform and an adjustable gate can regulate feeding. The grading system is operated by a 2.5 kW electric motor with the help of belts and pulleys. The grading mechanism consists of two tracks of conveyor chains, matching sprockets, stainless steel flaps, and conveyor supports, flap space adjusting mechanism, sidewalls and fruit collecting chutes. The grading mechanism has provision to separate fruits into four grades. The grading mechanism has provision to adjust flap spacing between 55 to 140 mm.

Fig. 9.47 : Fruit grader

Expanding pitch rubber spool vegetable grader

It comprises of a frame, an elevator feed conveyor, an intermediate receiving conveyor with rubber spools, two identical driving rollers with helical grooves of gradually increasing pitch, collecting platform with adjusting partitions and gates, fenders, transport wheels and a power transmission system (Fig. 9.48); Verma (2001), Verma *et al.*, (1992). The elevator conveyor is inclined at an angle of 38 degree with the horizontal and is made of 13 mm rods with a pitch of 25 mm. The rubber spools of the sizing conveyor are mounted on 13-mm mild steel rods and two endless link chains at a spacing of 102 mm. The rods carrying the rubber spools rest on the mild steel railings on both sides. Power to the sizer is provided by 2-hp electric motor mounted on a platform. Potato graders are used for grading potato tubers into different grades. The grader can grade the potatoes into four different grades and its capacity is 2.5 t/h. The performance of the grader in terms of sizing accuracy is observed to be better for potato tubers between 20 and 60 mm diameter.

Fig. 9.48 : Expanding pitch rubber spool vegetable grader

Tree lopping machine

Lopping of tree is highly labour-intensive and difficult task. It is necessary for having good growth of trees and also from an aesthetic point of view. There is high risks involve in doing this job because the man has to climb the large trees and branches. A power tiller mounted tree-lopping machine can alleviate the problems associated with lopping of tall branches of trees (Jayan, et al. 1998). The machine consists of a circular saw as a cutting unit, a flexible shaft as power transmission unit and a GI pipe frame to support the shaft (Fig. 9.49). Power is taken directly from the engine crankshaft. The machine with an optimum operating speed of 1800 rpm can cut branches up to 5 cm thick at a maximum height of 5.4 m from the ground. The average capacity of machine is 0.393 m^2/h, which is higher than manual labour (0.057 m^2/h). On an average, the machine can cut 312 numbers of branches in an hour if all the branches are of 40-mm diameter.

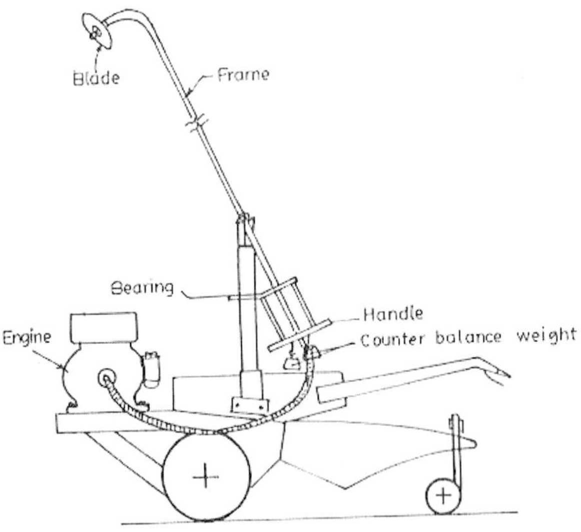

Fig. 9.49 : Power tiller operated tree lopping machine

Axial flow vegetable seed extracting machine

The extraction of seeds from different vegetables is commonly done manually, which is time and labour consuming and unhygienic. It often leads to injuries to the hand and seed productivity is low. An axial flow vegetable seed-extracting machine has been developed at Punjab Agricultural University, Ludhiana (Verma and Singh, 1988). The machine is quite simple and is capable of extracting seeds from a large number of vegetables such as tomato, brinjal, chillies, cucumber, summer squash, watermelon etc. It is based on wet extraction principle. It is operated by 26.11 kW tractor's PTO. It consists of frame, feeding chute, a primary chopping chamber, a crushing chamber, seed collecting chamber, rotor, concave screen a seed out let and a waste (pulp) outlet, water sprinkler and power transmission system (Fig. 9.50a). The fruits are cut into small pieces in the primary chamber. Thereafter, these are cut and crushed by means of axially arranged blades attached to a rotor shaft. The shaft is rotated at a speed between 250-300 rpm. Apart from the cutting blades, conveying rakes have also been provided on the shaft to move the pulp and course material along the axis to eject it out. The waste materially is finally ejected from the waste outlet. Concave screens have been provided under the rotor for separating fine, medium and large seeds. Three 25mm diameter pipes with holes, two on sides and one on top have been provided along the length of the rotor. Water is supplied to these pipes by a small centrifugal pump to wash seeds and finally crushed materials and collect separately (Fig. 9.50b). The machine can extract 1.25-3.00 kg of different vegetable seeds per man-hour. Weight of the machine is about 164 kg. Working capacity is 5.5 kg/h for brinjal; 3.78 kg/h for tomato; 9.42 kg/h for chilli; 4.68 kg/h for summer squash; 3.60 kg/h for water melon; 6.60 kg/h for squash melon; and 1.42 kg/h for cucumber. Savings are 60 - 85% in labour requirement and 50 - 65% cost of operation as compared to manual threshing.

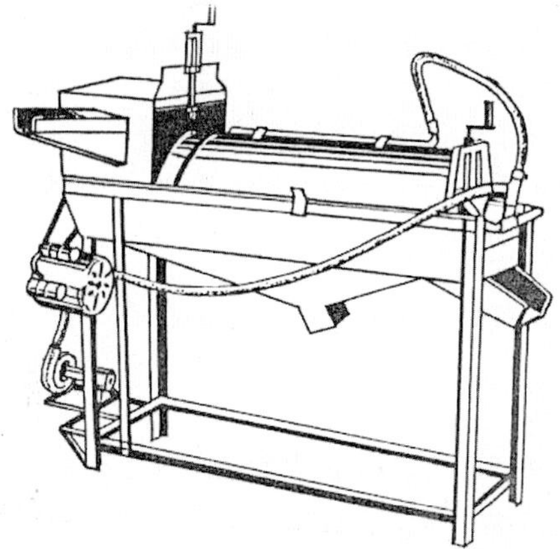

a) Isometric view of vegetable seed extracting machine

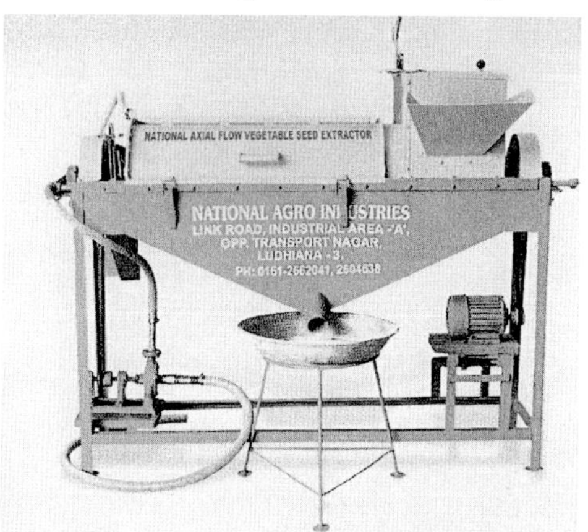

b) Vegetable seed extracting machine

Fig. 9.50 : Axial flow vegetable seed extracting machine

Courtesy: M/s National Agro Industries, Ludhiana (Punjab)

Areca nut dehusker

The areca nut palm occupies a prominent place among the cultivated cash crops in the country. It has a considerable socio-religious and medicinal value as well. It is used as medicine against leucoderma, leprosy, cough, fits, obesity, nasal ulcer etc (Bhardwaj *et al.*, 2004). Dehusking of areca nut takes about 35-40% of total cost of production and is mostly done by women folk. This method is highly labour intensive, time consuming and uneconomical. Dehusking of areca nut can be achieved by using impact type, scissors type and shearing type devices. The dehusker consists of a hopper, feeder, lead plate, shearing roller, friction plate and scraper (Fig. 9.51). The cutting blade is provided just below the lead plate. The angle of blade can be adjusted. The cutting blade and lead plate are integrated into a single unit with provision of adjusting the distance of this unit from the shearing roller in vertical as well as horizontal direction. The shearing roller is the heart of the machine and fixed on MS main shaft, which is driven by a belt drive. The feeder receives the dried and graded nut from the hopper and delivers it on the lead plate. The nut is then compressed between rotating shearing roller and lead plate. The teeth on roller peel off the husk and then kernel is ejected through the slot on the lead plate. The husk separated from the nut is caught in between sheering roller and friction plate and carried away by the shearing teeth. The scraper helps in removing the portion of husk sticking to teeth before teeth return for the next cycle of dehusking. A single-phase 0.5-hp motor operates the machine. It gives the output of about 9 kg/h and has 84.5% dehusking efficiency.

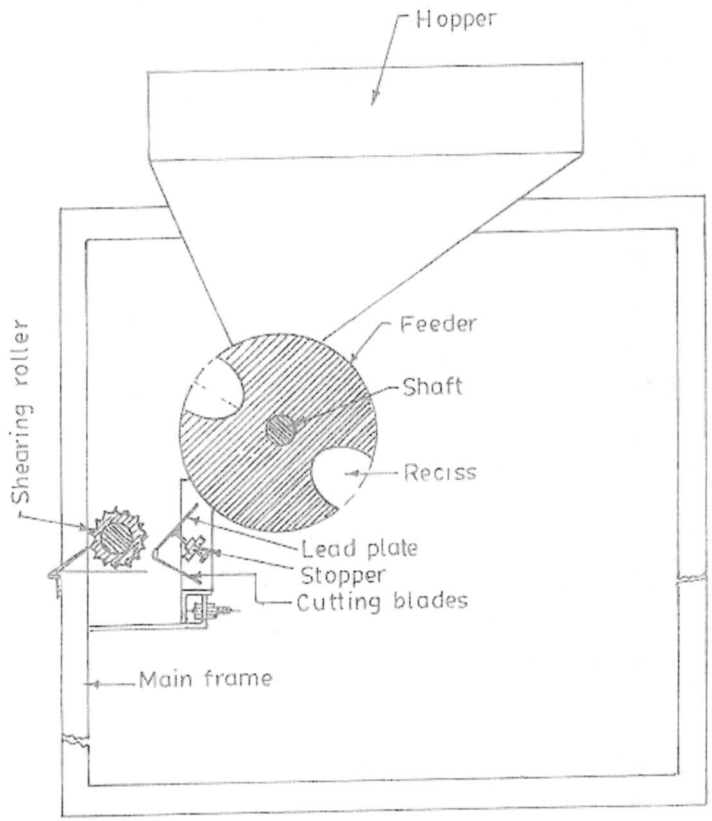

Fig. 9.51 : Sectional view of areca nut dehusker

Dehusker for fresh areca nut

Areca nut is being cultivated in large area in India. Most of the areca nut produced in Tamil Nadu is dehusked in green state. Labour requirement for dehusking is estimated at 7-8 kg of nuts per day per labour. This involves huge labour requirement and high cost. Existing models are expensive and also cause damage to the kernels (Bhardwaj *et al.*, 2004). Two different rotary dehusking mechanisms have been developed. The model-I has circumferential serrated blade and the model II had longitudinal profiled blades. The concave has two spring loaded rubber pads that press the fruit gently against the rotor. The fruits are fed at the top manually (Fig. 9.52). The fruits travel half the circumference and are dehusked in the process.

The dehusked kernels fall to the bottom along with the husk. The speed varies from 40 to 80 rpm. The results showes that the dehusking efficiency is lower in Model-I and higher in model-II. The best dehusking performance is observed at 60 rpm. The breakage is observed in the range of 8 to 10%.

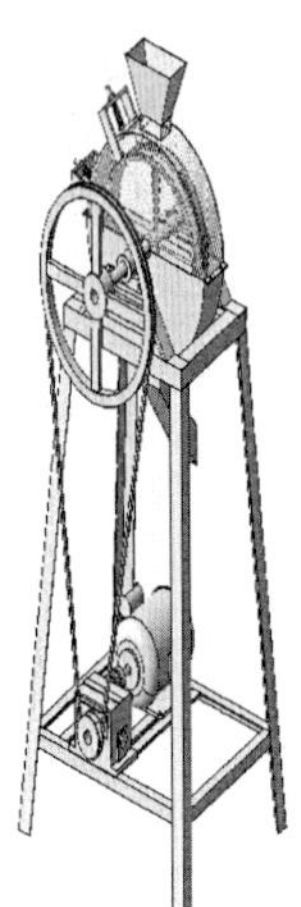

Fig. 9.52 : Dehusker for fresh arecanut

Arecanut stripper

Traditionally areca nut stripping demands enormous effort and energy from arecanut plantation workers. For minimizing drudgery of the workers with increased productivity at reduced expenditure levels, an impact type worker friendly areca nut stripper with safety features has been developed (Anonymous, 2010, 2010a; Bhardwaj *et al.*, 2004). The hold-on type areca nut stripper unit consists of a feed chute, a peg tooth cylinder, a stripping mechanism and an oscillating sieve (Fig. 9.53). The prime mover is a 2.2 kW, 3600 rpm petrol start kerosene run engine mounted on one side of the unit with supporting frame. The power from the prime mover is transmitted with a reduction ratio of 14:1. The stripping mechanism is a hollow peg tooth cylinder having 15 pegs equally spaced at 42 mm apart. The oscillating sieve is made of 7 mm mild steel rod spaced at 20 mm apart so as to retain the stripped areca nuts leaving the chaff to pass through the gap. A feed tray of 90 mm width is provided at a height of 1030 mm from the operator platform. The feed tray is inclined at 15° to facilitate easy handling of areca nut bunch while feeding. To restrict hand contact with the stripping cylinder, the feed chute length is adjustable form 380 to 550 mm (adjustable for female and male). The areca nuts are separated from bunches due to the impact force of pegs of rotating cylinder and the stripped areca nuts fall on the oscillating sieve. The oscillating motion of the sieve separates areca nut from the chaff and other impurities. The entire unit is mounted on wheels for easy transportation inside areca plantation. Protective guards have been provided on all moving parts to prevent accidental contact of persons or parts of clothing being caught in the transmission system. It eliminates the high work stress and back pain disorders caused due to lifting and forceful striking the areca nut bunch as followed in traditional method of stripping. It is suitable for stripping both green and ripen areca nut bunches. It can strip 650 - 950 kg of areca nut per hour. It reduces damage caused to the stripped areca nut. Striping efficiency of 99.5 per cent is achieved and it saves 66 and 77% cost and time as compared to conventional areca nut stripping.

Fig. 9.53 : A view of arecanut stripper

Arecanut peeling machine

A cost effective automatic arecanut peeler has been designed and fabricated by M/s. V- TECH Agri Engg. Thithahalli tq, Shimoga dist. (Karnataka). Its unique features are: it has the auto-feeding, separator, self-adjustment mode, and can peel various varieties of areca nut except "gotu" without causing any damage to the nuts (Fig. 9.54). The machine automatically separates the nuts from the peeled shells. The machine can peel about 45 kg of areca nuts in about an hour, which would have taken at least three to four hours if done manually. The capacity of the peeler can be increased up to 100 kg/h.

Fig. 9.54 : A view of arecanut peeling machine
Courtesy : M/s. V- TECH Agri Engg. Thithahalli tq, Shimoga dist. (Karnataka)

Tractor mounted banana stem shredder

Banana is a major cash crop of the country. Banana is mostly check row planted with the spacing varying from 125x125 cm to 150x150 cm, depending upon the variety. Plant population is about 4,500 per ha. After the harvest of the banana bunch, the stem is manually cut and left in the rows. After the harvest of the whole field, these are collected and left near the boundary for drying and subsequent burning. This process is tedious and time consuming. The banana stem shredder helps in disposing of the stem immediately after harvest. Shredded material is suitable for mulching in the banana garden and also for vermi compost. The average diameter of banana stem is 225 mm at the bottom and 100 mm at the top with the average height of 2.4 m. The banana stem consists of 95% of water and only 5% of fibre. Tractor mounted banana stem shredder has been developed (Anonymous, 2008, 2010, 2010a). The shredding unit consisted of 4 blades placed perpendicular to each other at 225 cm distance (Fig. 9.55). Additionally, 12 nos. of spikes with flat cutting edge are fitted with a gap of 120 mm between the rows. The whole device is mounted on a frame made of MS angle. The blades are driven by the PTO of the tractor with a bevel gear box and the hopper is trapezoidal in shape with a height of 800 mm. The stem is cut into small pieces and the water and fibre are separated. It takes 1.2 minutes to shred the stem having average height of 2.4 m. About 52 stems are required for shredding in one hour. The shredded material can be used for mulching in the banana garden. The shredded material takes 3-4 days to dry. The shredded fibre can be used for preparing vermi compost also.

Fig. 9.55 : Tractor mounted banana stem shredder

Tractor mounted banana clump remover

Banana crop is maintained for two years to get the benefit of two harvests. The crop needs removal of clumps (plants along with root portion) after two years. During the process of removal of the clump, the entire mother plant along with the rhyzome and side suckers as a whole mass has to be removed so as to prepare the land for the next crop. Manual labours using crowbars and spades do this operation. The labours have to dig to a depth of up to 45 cm to remove the clumps and hence the operation is cumbersome. In view of above, a tractor operated banana clump remover has been developed at TNAU, Coimbatore to mechanize the clump removing operation and to reduce human drudgery involved (Anonymous, 2008, 2010, 2010a). The nine-tyne cultivator frame has been adopted for the development of the equipment. Two numbers of 100 x 15 x 1000 mm sub-soiler shanks with shares of size 190 x 40 x 5 mm are fitted in the nine tyne cultivator frame at 225 mm spacing. These two sub soilers perform as a fork while removing the banana clump. A deflector has been provided to push the soil sideways. The equipment is attached to the 3-point linkage of a 26 kW tractor. For removal of banana clump the sub soiler shank is positioned behind it, and pressed into soil with the hydraulic system and the tractor is gently moved forward, simultaneously lifting the sub-soiler (Fig. 9.56). This combined action removes the entire clump along with its root portion as a whole mass. The field capacity of machine is 0.5 ha/h. The unit saves time and labour to a tune of 85 and 90 per cent respectively.

Fig. 9.56 : Tractor mounted banana clump remover (Stationery and in operation)

Tree climber

Skilled tree climbers are dwindling day by day because of the drudgery involved in this operation (Fig. 9.57). A coconut tree climbing device has been developed successfully which facilitates any person to safely climb the coconut trees and do the required operations (Fig. 9.58). The trunks of the coconut trees are uniform in cross section and there is not much variation in the girth and diameter from the base to the top since the tapering of tree trunk is very mild (Anonymous, 2008, 2010, 2010a). The climber made of M.S. square pipe consists of two components. Adjustable belts connect the components. The upper component is provided with a seating arrangement and lower component is having provision for holding the foot. The rubber cushioning is provided at the portion of frames, which comes in contact with tree to avoid any damage of tree. By standing on the lower component, the upper component can be moved up or down over the tree (Fig. 9.58). The operator can safely climb a tree of 10 m height in 1.5 min without any risk.

Fig. 9.57 : Manual method of climbing

Fig. 9.58 : Coconut tree climber

Manual climbing device for Palmyra

Unlike coconut trees, the variation in the girth of palmyra tree from base to top is considerable. The tree girth during drought is drastically reduced and become normal during normal rainy period. The traditional palmyra tree climbers use the rope type tree climbing device (Fig. 9.59) which could not be effectively used by the successors of the traditional tree climbers or unskilled persons (Anonymous, 2008, 2010, 2010a). Developing a palmyra tree climber device will be useful to the unskilled users in rural areas. The existing tree climber developed for the coconut trees have been re-designed to suit the conditions of the palmyra trees with provision for adjusting the unit to suit the girth variation in a particular tree so as to adjust the frame width on-the-go to correspond with the girth of the tree at each height. To accommodate a wider girth range, the mechanism has been improved by providing two number of sliding inclined bushes, which provide V shape to grip on the tree. Simple cranks and flexible shafts provided for easy adjustment (Fig. 9.60). Adjustment is now very easy. In addition provision is also made to carry the kits and pots for transporting the tapped neera. The unit consists of two components, the upper component operated by the hand and the lower component operated by the foot (Fig. 9.60). Adjustable belts connect the two components. The upper component is provided with a seating arrangement and the lower component is having provision for holding the foot. Now the power operated tree climbers have been developed which is remote controlled (Fig. 9.61); Anonymous, 2010b.

Fig. 9.59 : Palm climber

Fig. 9.60 : Multi tree climber

Fig. 9.61 : Different views of remote controlled palm climber

Shelling of palm fruitlets

There are three steps involved to accomplish the stripping of oil palm fruitlets in the field. The first step is to reduce the strength of fruit-stem joint by applying ethephon on fresh fruit bunch just after harvest. Then prepare threshed materials by producing spikelets from the bunch stalk. And finally detach loose fruitlets from cut spikelets using a drum thresher. The removing of woody shell from coconuts with reasonable protection of meat against damage is tedious job (Fig. 9.62). When meat is to be used for copra then the shell is broken by striking with a large knife either manually or mechanically. However, when meat is to be used for shredding, different kinds of knives or saws are used. Variations in size of coconuts, thickness of shells, curvature of the surface and method of holding the coconut are some of the parameters, which affects the peeling. A number of machines are available for peeling of coconuts. The first peeler developed consists of convex and concave cylinders and correspondingly concave-convex cutter and endless band bearing on the surface for paring coconut meats. The second machine developed consists of a suitable frame and table of revolving curved comb-toothed-radial cutters arranged around and at a distance from the central shaft for support and is driven to remove shells from coconuts. Another machine developed has blades connected with a rotating head for cutting the shell from a coconut. A desirable machine is one that (a) removes the shell from a coconut in a clean-cut and facile manner without injury to the nutmeat and to remove the skin with minimum of waste; and (b) has a duplex cutter mechanisms one for producing a saw-like cut on the coconut from end-to-end at a predetermined depth and other for paring the skin from the nut. The machine is motor-driven, compact, inexpensive and simple to use (Fig. 9.63). It has a provision of disposing of the shell-dust and skin fragments during the operation. The capacity of machine is 500 coconuts/h. In yet another machine, a knife is inserted into an eye of coconut and then it is engaged by a toothed endless chain, the teeth of which penetrates into the shell and move on the knife to remove the shell from meat.

The most of the machines for shelling coconuts involve the use of clamps and employ pressure against the surface to hold the nuts. Whether these clamps are smooth or pronged, many of them are ineffective in holding the nut adequately for smooth shelling. Many of the shelling methods involve complex and expensive equipment requiring constant maintenance. The manner of shelling the nuts includes parallel as well as helical cuts of hard shells, either of which results in inefficient separation of nutmeat and consequently high losses occurs. To minimize the losses, a machine has been developed for shelling of whole coconuts. The machine comprises of gripping mechanism to hold the nut rigidly at its ends with grippers that penetrate into shell. The grippers are locked into an axially fixed position and rotated with nuts. Rotary cutter cuts the shell layer-making spiral cut around the central portion of nut while it is in rotation.

Shredding: Large quantities of shredded coconuts are used throughout the world. To prepare it, the shell is clipped from the fresh nut with a sharp axe and the brown cover is shaved off the kernel. It is then washed in water several times, shredded and dried in ovens. It is graded on the basis of fineness of grain and packed for further use. The hand knives are generally used for cutting the kernel or meat from coconuts. The nuts are split in halves and then skilled man cuts the meat from the shell by holding the half in one hand and manipulating the knife with other. This is slow and extremely tiring work and sometimes it becomes necessary to cut the meat into small pieces. The commonly used mechanical device for shredding large quantities of coconut meat consists of a rotating disc having tangential arranged series of perpendicular comb teeth just above the plane of disc. As the disc rotates, chunks of fresh coconut meat are pressed against it and are squeezed between the comb teeth and the cutting knife to form the coconut shred. This causes a large degree of compression of nutmeat. The coconut shreds are then dried for further use by the consumers.

Fig. 9.62 : Manually operated dehusker

Fig. 9.63 : Power operated coconut dehusking machine

Mechanization of Lac cultivation

Lac cultivation is carried out mostly by the tribal in the states of Jharkhand, West Bengal, Chattishgarh, Madhya Pradesh, Maharashtra, Orissa and Gujarat, generating about 4 million man-days every year. About 55-60 per cent of the world demand for lac is met by India. Lac is an export-oriented commodity as 75-80% of lac produced in the country is exported mainly in refined/semi-refined form to over 100 countries all over the world (Prasad *et al.,* 2009). Lac is nature's boon to mankind, and a bio-resource unique in several respects. It yields basically three useful materials i.e. resin, wax and dye. These are natural, renewable, non-toxic and eco-friendly and can be put to an unbelievably wide range of applications. India possesses a huge untapped potential of lac production and it is possible to achieve a multifold increase in lac production. The mechanization of lac production can play an important role in increasing lac production and reducing the drudgery of lac growers.

Lac cultivation involves five major operations *viz.* pruning of lac host tree, broodlac placement on host tree, used up broodlac removal, pesticide application, lac crop harvesting and lac scraping. The operations like pruning, broodlac placement, used up broodlac removal and harvesting are carried out by men, where as selection of broodlac, bundle making and scraping are carried out by women. To carry out operations like pruning, broodlac placement used up broodlac removal and harvesting, farmers climb on the tree, which is time consuming, laborious and dangerous. Lac farmers use locally made pruning knife for host tree pruning, mature crop harvesting and lac scraping. Sometimes they also use tree pruner, secateur, pruning saw and axe for pruning and harvesting operations. In some cases, use of axe and sickle are also reported for pruning and harvesting operations.

Indian Institute of Natural Resins and Gums (IINRG), Ranchi has developed Hand Operated Roller type Lac Scraper, Pedal Operated Roller type Lac Scraper (Prasad et al., 2001c) and Power Operated Roller Type Lac Scraper (Anonymous, 2002-03). Birsa Agriculture University, Ranchi has developed Peg Type Lac Sheller (Pandey *et al.*, 1997). Central Institute of Post Harvest Engineering and Technology, Ludhiana has developed Lac Scraper-cum-Grader (Anonymous, 1998-99). Broodlac placement-cum-removal tool (Fig. 9.64) has been developed for placing broodlac on branches of lac host tree and its removal after larval emergence (Prasad *et al.*, 2001a). For small and medium size lac host tree broodlac can be placed from the ground level up to 6 m height and the farmers are not required to climb on the tree. Even female worker can place and remove broodlac using this tool. It is usually 1.5 kg in weight and a person alone can operate it.

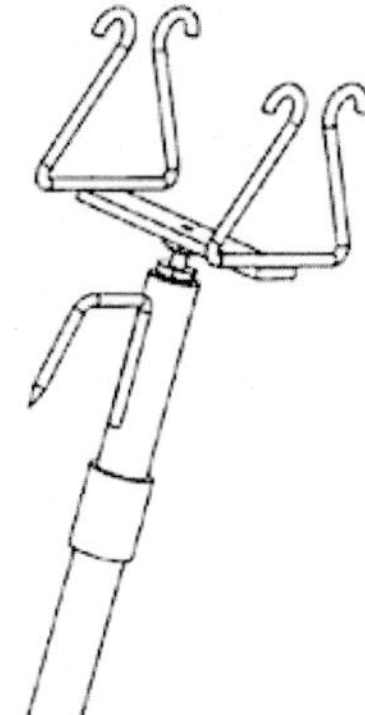

Fig. 9.64 : Broodlac placement-cum-removal tool

Broodlac removal hook (Fig. 9.65) is used to remove broodlac bundle from the branches of the host tree. In traditional method, farmers climb on the tree for removing used up broodlac. Sometimes they are required to climb even on thin branches. Thus, the operation is not only tough but dangerous too. For small and medium size lac host tree hook can be used for broodlac removal. The hook consists of a tubular body, a hook shaped cutting blade and a ring for bundle collection. For removing broodlac bundle, the string with which bundle is tied to the branch, is cut. The pointed hook, having sharp edge cuts the string, when it is engaged in the string and pulled down. Once, the string is cut, bundle falls down. To collect the bundle, a ring (Fig. 9.66) is provided under which a net is tied. By collecting the bundle in the net, shattering loss is minimized, as it prevents the bundle from falling on the ground. Generally used up bundle is removed after 3 weeks. By that time lac encrustation dries up and becomes susceptible to shattering with even minor impact. It is usually 1.5 kg in weight and a person alone can operate it.

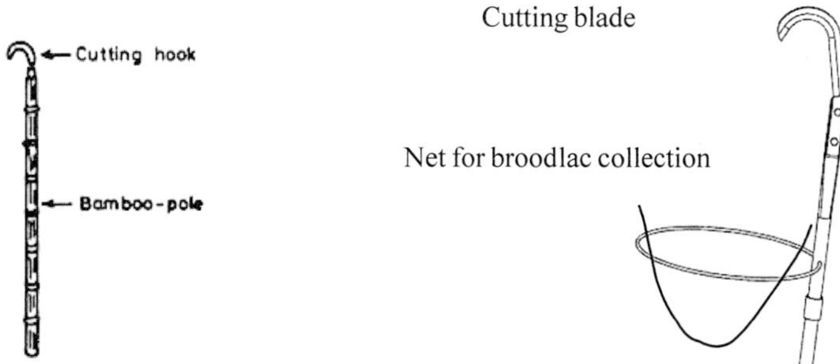

Fig. 9.65 : Broodlac removal hook

Fig. 9.66 : Broodlac removal cutting blade

The removal of shoots bearing mature lac encrustation by cutting is known as lac harvesting. The operation is similar to pruning, but requires more care than pruning, as shoots contain lac encrustation, which are either used as broodlac or scraped lac for sale. Any damage to the encrustation affects broodlac quality. Further, if harvesting operation gives a jerk to the lac bearing sticks, then there is a possibility of shattering loss. The farmers harvest crop with the help of sharp edge knife popularly known as pruning knife. The harvested crop is collected and the lac encrustation bearing twigs are selected and cut into pieces with help of a secateur, to be used as broodlac. Male labourers do the harvesting, while female labourers do collection of harvested material and broodlac selection. Lac scraping involves removal of lac encrustation either from used up broodlac (dried encrustation) or fresh lac encrustation from harvested lac stick. Female labourers mainly scrap lac from lac sticks using traditional tools *i.e.* pruning knife, scraping knife, sickle etc. A peg type lac sheller has been developed by the Birsa Agricultural University, Ranchi (Pandey *et al.*, 1997). It consists of two disks. One disk is spring-mounted and remains stationary while the other disk is with pegs on working surface and it is driven by hand cranking handle. There is a pair of gears mounted on rotary disk shaft and hand cranking shaft. The number of tooth on the two gears are in 6:35 ratio. The gap between two disks can be adjusted with the gap-adjusting lever. The machine scrapes 5 kg lac sticks in an hour and separates about 93.7 per cent lac from the lac stick. One person alone is adequate to operate the machine.

Indian Lac Research Institute, Ranchi has developed a hand operated roller type lac scraper (Fig. 9.67); Prasad *et al.* (2001c). The machine scrapes lac with the help of two scraping rollers, which move at differential speed in opposite direction. A handle is provided for operating the machine. The lac stick is fed in the machine through feeding hopper. After scraping, the material is discharged on an inclined sieve of 10 meshes. The material finer than 10-mesh size pass through the sieve and the remaining scraped lac along with stick slide down the sieve and come out of the machine. The capacity of the machine is 5.7 kg/h, which is more than any of the traditional equipment used by the farmer. The machine is 50 kg in weight and a person alone can operate it. Pedal Operated Lac scraper (Fig. 9.68) is similar to Hand Operated Lac scraper. But in pedal operated lac scraper power to the machine is supplied by operating the pedal of the machine. While operating the machine, the operator pedals and feeds the material in the machine. The capacity of the machine is 6 kg/h. The machine is 55 kg in weight and a person alone can operate it (Prasad *et al.*, 2001c).

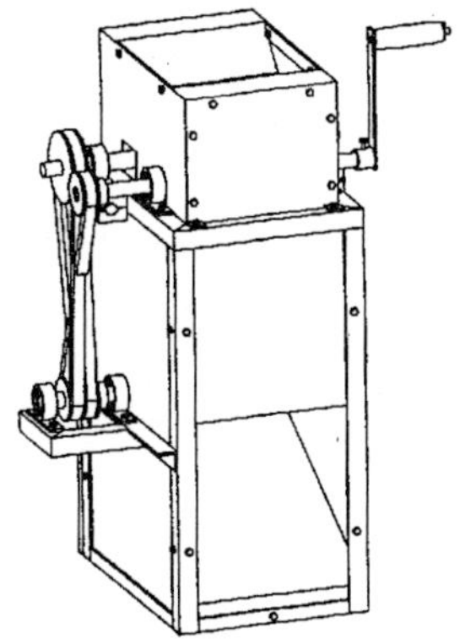

Fig. 9.67 : Hand operated roller type lac scraper

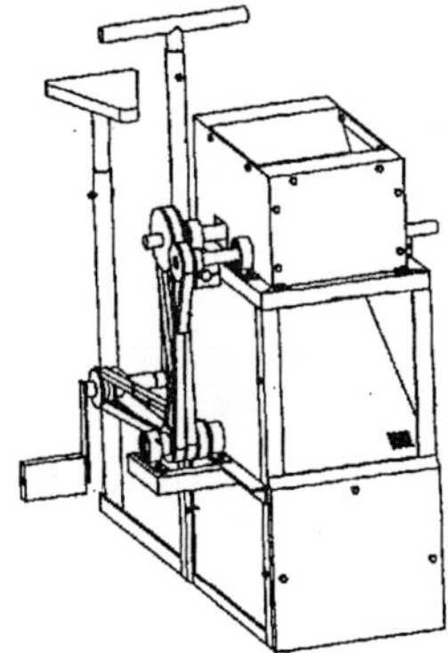

Fig. 9.68 : Pedal operated roller type lac scraper

Motor operated roller type Lac scraper has been developed to increase the output and to reduce the drudgery of the lac growers (Anonymous, 2002-03). It is driven by 0.5 hp single-phase electric motor. Only one person is required to operate the machine. The capacity of this machine is 10 kg/h, which is more than any of the traditional equipment used by farmer. The machine is 65 kg in weight. Power operated lac scraper-cum-grader does three operations at a time *i.e.* lac scraping, crushing scraped lac and grading crushed lac (Anonymous, 1998-99). Power to different parts of the machine is supplied by a 2 hp 3-phase electric motor through V type belt and pulley. The place where electricity is not available the machine can be operated by 3 hp small diesel engine. Two persons are required to operate the machine. The capacity of this machine is 20 kg/h. The machine is 150 kg in weight.

9.2 Farm Tools and Equipment for Hill Agriculture

Approximately, one third of the country's geographical area is hilly or undulating or has high altitude. High and sloping areas are the two main characteristics of hills, due to which the natural elements, like rain, sunshine, wind and gravity, exaggerate and aggravate the eroding forces. Unbearable human and live-stock pressures on many hilly area and indiscriminate felling of trees for commercial purpose have already led to rapid depletion and destruction of much forest cover. Water retention capacity and productivity of land have also seriously suffered in such areas and has impaired the ecology. Their unfavourable impact on the economic condition of the hill people is obvious. Such seemingly harmless activity as prolonged slope grazing by livestock, especially by goats and sheep, has exposed many hill areas to serious ecological degradation.

A large number of crops - cereals, pulse, oilseeds and others are cultivated in the region with an objective of attaining self-sufficiency in food grains production. Horticultural crops are well adapted to this region; these include fruits (apple, almond, walnut, plum and peach, cherry), vegetables (brinjal, chillies, potato, pea, cabbage, cauliflower, knol khol, turnips, onion, garlic etc), spices (saffron, kala zeera), and flowers (rose, gladiolus, marigold, tulip, carnation, chrysanthemums etc.); Agrawal et al., 2003; Singh, 2008; Singh and Mishra, 2009; Tandon and Singh, 2009. Forests occupy 60% of the total geographical area of the Himalayan ecosystem, though substantial area has been reduced to scrub forests. The agrarian economy of the hills is heavily dependent on forest and pasturelands, for energy supply, fodder, non-timber products and livestock rearing. This vital sector of hill economy, however, due to high biotic pressures and a biotic factor is degrading.

Valley land mechanization

Valley agricultural is a permanent type of cultivation practices throughout the hilly terrain both at low and high elevations. Under the system the land is used continuously year after year. Rice is the main crop, which is grown in the rainy season. The land utilization for double or multiple cropping is rather poor and mono cropping is practiced. Since the amount of rainfall is very little in winter, crops requiring less water, for example rape mustard, potato, tobacco, pulses and vegetables are grown in this season. Bench terraces are well adapted in some areas more particularly in mild slopes with irrigation facilities.

Systems of cultivation

Agriculture of North-Eastern hill region is unique in itself which is probably due to its agro-climatic conditions, type of tenancy of the farm holdings, steep slope, shifting cultivation, terrace and valley land cultivation (having narrow and smaller plots). Various cultivation practices are followed and some of them are discussed here.

Shifting cultivation: In shifting cultivation hill slope is selected and forest is cleared by cutting down trees/shrubs. Cut and dried material is burnt before the onset of monsoon, left out un-burnt materials are cleared by brooms made of bamboo sticks. No tillage is done and seeds are dibbled in zero tillage conditions. In the second year of jhum, light hoeing is done to loosen the soil before planting the crops. Raised soil beds are prepared on slopes at some places with the help of small hoe and spade for planting potato and other tuber crops. Shifting cultivation locally known as jhuming involves operations such as selecting forest land (virgin or secondary growth), cutting down the jungle during December–January and allowing to dry till February; burning it before onset of rains; planting an intimate mixture of seeds by broadcasted or dibbling; weeding, watching and protecting of crops; harvesting, threshing and storing. Weeding is the most strenuous and labour consuming part of the shifting cultivation; at least four weedings are necessary. The tools used for various operations are given in Table 9.1. The tools used for various operations are simple hand tools such as dao for cutting, spade for land preparation, wooden stick for dibbling, small spade and hand hoe for weeding etc. Absence of tilling or use of animal power, active participation of women, non-existence of artificial irrigation, little initial investments and minimum number of items are some other distinguishing features of this primitive cultivation. Under the pure jhum economy nobody offers labour for hire and they sell or purchase very little. The average size of jhum plot varies from 1.0 to 2.0 ha.

Terrace land cultivation: Irrigated and un-irrigated terrace cultivation is practiced on a limited scale. Soil manipulation is done by way of spading and hoeing. In Sikkim terraces are tilled with country plough using bullock power. Paddy is transplanted in terraces after puddling.

Valley land cultivation: Valley agriculture is a permanent type of cultivation practiced throughout the hilly terrain both at low and high elevations. Under the system the land is used continuously year after year. Valleys receive major share of runoff water and eroded silt from slopes. They form most fertile fields for growing paddy and jute crop. Rice is the main crop, which is grown in the rainy season. Paddy fields dug by spades and puddling is created by wooden planks or ladder pulled by two persons. Sometimes animals are also used for field preparation using country plough or wooden tillers. The land utilization for double or multiple cropping is rather poor and mono cropping is practiced. Since the amount of rainfall is very little in winter, crops requiring less water, for example rapeseeds & mustard, potato, tobacco pulses and vegetables are grown in this season. Bench terraces are well adapted in some areas more particularly in mild slopes with irrigation facilities.

Bun Mechanization: It is practiced by *khasi* and *jaintias* tribes for cultivation of vegetable crops such as potato and spices like ginger, turmeric on series of beds formed along the slopes of the hills. Bun making involves: cutting of shrubs and grasses, lining up rows of dry vegetation along slopes, overlaying with soil so collected from both sides so as to form "buns" and planting on the buns afterwards. Planting is done during mid March and harvesting is done during the month of December. Weeds are controlled either mechanically or by spraying weedicides. For controlling pests pesticides are sprayed. All operations are performed using locally made traditional hand tools and implements (Dao for jungle clearing and removing rhizomes from stalks, spade for bun making and harvesting, small hand hoe for weeding and conical bamboo baskets for transporting the farm produce), Table 9.1. No draft animal or mechanical power units are being used for bun cultivation. However, power tiller and bullocks with matching implements are used for ploughing the low land paddy fields. As the slope of the fields is more than 60% hence, it is difficult to use power tiller and tractors on the slope. It is necessary that bun method be replaced by terrace cultivation to promote use of power-operated equipment in this area.

Table 9.1: Level of Jhum and Bun cultivation

Operation	Activities	Level of mechanization	
		Traditional tools	Improved tools
Nursery raising	Raising of seedlings	Spade & Dao	Small engine operated cultivator
Land clearing	Cutting of trees bushes and grass	Dao and axe	Bush cutter
Pit digging	Making pits	Spade/pick axe	Post hole digger (engine operated) & power tiller operated
Land Preparation	Burning cleaning and soil working	Spade and hoe	Bush cutter with cultivator attachment
Planting or seeding	Seed distribution and dibbling	Wooden dibbling stick	Wooden stick with steel tip, improved dibbler, rotary dibbler
Crop husbandry	Weeding, spraying	Hoe and sweep (khurpi), manual sprayer	Wheel hand hoe,power sprayer & dusters

Contd...

Harvesting	Cutting or removing crop produce	Plain sickle, Dao and spade, manual fruit harvester	Serrated sickle, manual telescopic fruit harvester, reapers
Threshing	Removing ear heads, shelling etc.	Sticks and bamboo, bullock treading	Small 1 hp thresher, animal operated diggers
Handling	Carrying farm produce	Conical bamboo basket	Carts, power tiller operated trolleys

Indigenous farm tools and implements in Hilly region

Because of the dominant farming of the shifting cultivation, hand tools have remained very important agricultural equipment in the Hilly region. Conventional farming using bullock-drawn wooden plough, wooden harrow, and leveller wooden plank is prominent in this region (Table 9.2). In the absence of improved equipment, these have served the farmers of the region but these are of very low capacity, involve lot of drudgery and one of the causal factors for low productivity. Modern educated youth do not get a thrill and job satisfaction working with these outdated tools. Their shape, size, metallurgy, manufacturing process vary from craftsman to craftsman. The fact that these have remained in use so long established that these do have their strengths. Farm equipment has to meet agronomic specifications emerging from ongoing research and development, thus requiring constant efforts. Rationalized designs ought to be mass manufactured for reduced unit cost, quality, work efficiency and economy of effort and time.

Table 9.2 : Indigenous farm tools and equipment used in NEH region

S. No.	Name of implement	Purpose	Power source	Material make	Field capacity
1.	Deo	It is a multipurpose tool used for clearing jungles, cutting trees etc.	Human	Mild steel	-
2.	Dibbler	Dibbling of seeds under zero tillage conditions	-do-	Wood	0.002 ha/h
3.	Khasi spade	Used to prepare seed beds, for digging paddy fields, for preparing bunds, and terraces	-do-	Mild steel	0.004 ha/h
4.	Khasi spade (small size)	Weeding and interculture operations	-do-	-do-	0.002 ha/h

Contd...

5.	Local plough	Ploughing paddy fields	Bullock	Wood and steel	0.014 ha/h
6.	Leveller	For puddling and levelling	-do-	Bamboo	0.014 ha/h
7.	Peg tooth harrow	Puddling in wet conditions, weeding and interculture	manual	Wooden and bamboo	0.014 ha/h
8.	Knives, sickles	Harvesting crop	Human power	MS sheet	0.009 ha/h
9.	Conical baskets	Transport of farm inputs and farm products	-do-	Bamboo	-
10.	Maize sheller	Shelling	-do-	GI sheet	0.032 ha/h

Improved implements suitable for Hill region

Availability of improved agricultural tools and implements in the hilly region is limited to few farmers and commercial centres at some district head quarter. The state Department of Agriculture under various Central Sponsored Scheme has been providing subsidy towards purchase of improved agricultural implements and machinery like sprayers, dusters, threshers, winnowers, pump sets, bullock/power tiller /tractor drawn implements and machinery and power tillers and tractors. However, it is still beyond the reach of the common farmer in the hilly areas to purchase power tiller and tractor with the present subsidy provision. Some of the prominent and improved farm equipment developed and suitable for hill agriculture has been listed in various chapters (Table 9.3).

Table 9.3 : Traditional and improved agricultural implements being used by the farmers for different operations and improved agricultural machinery proposed to be introduced for popularization

Operation	Implement being used		Improved implements suggested for introduction
	Traditional	Improved	
Seed bed preparation			
Ploughing	Bullock drawn Desi Hal, local Tangroo (spade)	Shalimar plough (Senior and Junior),	Tractor drawn ploughs (in Terai), 9-tyne tillers, Tawi plough, soil stirring plough, power tillers, Tawi plough with Pora attachment, Tractor operated disc

Contd...

			plough and disc harrow, Rotavator.
Harrowing	Garden rake, wooden spike tooth harrow	Shalimar puddler; tractor mounted disc harrow	9-tyne tiller, Peg tooth harrow, disc harrow, tyne tiller, Bar harrow, paddy puddler-cum-leveller, tractor mounted disc harrow, harrow puddler, Rotavator
Clod Crushing	Kudali and wooden mallet	Plank	Tractor mounted clod crusher, harrow puddler
Land levelling	Wooden plank	-	Rake-cum-leveller, Hill puddler-cum- leveller, scrapper, floats, Singh patella
Seeding/Planting			
Sowing (name crops)	Wheat: Broadcasting, desi hal with seeding tube	Tractor mounted and animal drawn fertilizer drill	Zero-till seed-cum-fertilizer drills, Raised-bed-planter, Tractor drawn/Animal drawn inclined planter
Planting	Maize: dibbling Potato: dibbling	Tractor mounted fertilizer drill and planter	Multi crop seeder (cup metering), tractor mounted / Power tiller planter, Automatic/ semi-automatic planter in valley and plains
Transplanting	Rice: manual Vegetables: manual Saplings: manual	-	Manual/self-propelled rice transplanter, Semi-automatic vegetable transplanterTractor operated auger for pit making, plant replacer
Weeding & hoeing			
Weeding & interculture	Khurpi, Kudali	Grass cutting sword, weedicide, wheel hoe, desi hal, 3 tyne cultivator	Wheel hoe for upland and star weeder, cono weeder for wetland; broadcaster, wheel hoe,

Contd...

			cultivator, power weeder
Bund/furrow making	Spade, Karaha	Bullock drawn and tractor mounted bund former	Tractor mounted scrapper and disc type bund formers (in plains)
Spraying & dusting			
For field crops	-	Foot sprayer, knapsack manual sprayer	Power operated sprayer, knapsack power sprayer, knapsack rocking sprayer with long boom.
For tall crops	-	Foot sprayer	Power/Power tiller operated sprayer, Aero blast sprayer
For trees/shrubs	-	Foot sprayer	Orchard sprayer
Harvesting/ digging/ uprooting	Local sickles for field crops, manual plucking of fruits and vegetables, Kudali	Sickle	Potato digger, walking / riding and tractor mounted vertical conveyor reaper, mini combines
Threshing/ shelling	Rice: beating on a hard surface.	Pedal and power operated drum paddy thresher.	Power operated paddy thresher; sunflower power thresher, multi crop thresher.
	Wheat and barley: animal treading.	Power thresher for wheat, pulses and oilseeds	Multicrop thresher
	Maize: beating with flail	Tubular maize sheller	Tubular sheller, and power maize sheller, power dehusker- sheller
Winnowing/ Cleaning/ Grading	Winnowing in natural air	Manual and power operated winnowing fan	Power operated winnower, fruit grader, rotary/oscillating screen pre-cleaner
Straw manage-ment Straw reaping	Manual sickle harvesting close to the ground	-	Straw reaper (combine), Straw shaver

Farmers of hilly areas use a number of farm implements and tools for cultivation (Prasad *et al.*, 2012). Some of the important hand tools and implements are reported below. Merits and demerits of indigenous hand tools are given in Table 9.4.

***Chem*:** It is made up of high carbon steel blade fitted in bamboo root handle. To fabricate *Chem*, iron blade base is heated in fire till red hot and pushed inside the handle. Farmers believe that bamboo root handle is very strong, durable and holds the blade very strongly. Bamboo root does not crack or break during fabrication while pushing the red hot blade inside it. It is of two types viz. chem puri and chem kawm. Chem puri is mainly used for cutting of trees and bamboo, and clearing jungle. Its blade is very strong, straight and comparatively thicker than other types of *Chem* (Fig. 9.69). Chem kawm is mainly used for weeding and cleaning operations in the crop fields, therefore, its blade is comparatively sharp and thin than that of *Chem puri.* The characteristic feature of this tool is that its blade is curved at tip (Fig. 9.70). Farmers explained that curved tip of *Chem kawm* helps in cleaning weeds near the base of plants. Weeds at base of crop plants are taken into curved tip of the *Chem kawm* and then pulled towards the person using it. In this manner weeds are cleaned without damaging the crop plants.

Tuthlawh (Small spade): It is used for making shallow furrows for sowing of crops like ginger, turmeric, colocasia etc, digging of small pits and earthing-up operation. It is not exactly a spade but resembles to a spade. A triangular thick iron blade's base is pierced inside a curved bamboo root handle by heating the blade. It is also of two types i.e. small tuthlawh and big tuthlawh. Small *Tuthlawh* is used for sowing of vegetable and paddy seeds in the *jhum* fields (Fig. 9.72). Big *Tuthlawh* is used for sowing and harvesting of colocasia, turmeric, ginger, and yam in the *jhum* fields (Fig. 9.71).

***Chemkawi*:** It is used for making pot holes for sowing crops like paddy, maize, beans, okra etc and weeding and earthing-up in crops (Fig. 9.73). Generally it is fabricated using old worn out *chem.* The worn out *chems* are given shape of *chemkawi.*

Zam Pher: It is like a small drying platform which is used for drying of maize cobs, turmeric, ginger, etc (Fig. 9.74). Cane or bamboo battens of 2 to 3 cm width and 1 to 1.5 cm thickness are used for making the frame. Bamboo strips of 2 to 3 mm thickness and 5 to 8 mm width are woven in the form of mat and tied with the farme by using same bamboo strips or locally available grass called *Nag*. According to need and uses, the size and shape of holes in the mat varies. Mat of *pher* which are used for drying of ginger, turmeric, maize cobs, etc are having 1-2 cm size triangular /square /

rectengular holes. The mats of *pher* used for drying of rice, paddy and other small seeded crops are of very compact and even a single grain can not pass through the mat. According to the farmers these gaps or holes are being given for aeration so that produce can dry easily, quickly and to avoid rotting of the produce also. The size of *Zam pher* varies place to place and farmer to farmer, however, a common size is found to be 1.5-2.0 m × 1.25-1.5m.

***Thlangra*:** It is used for cleaning of paddy, maize and grains of many other crops. A compact bamboo mat is prepared from 2.3 mm thick, 1-2 cm wide bamboo strips and tied with a bamboo frame (Fig. 9.75). The frame is egg shaped, having a peripheri of 1-1.5 m with three distinct corners at the same distance intervals. Two corners are used for holding *thlangra* during cleaning operation and remaining one at the front for removing dirt, dust and other foreign materials from the produce. Produce to be cleaned is placed in *thlangra* and by the hands it is moved up and down. Produce is bounced over its surface due to up and down movements, and different materials get seperated owing to difference in their specific gravity. Generally the dirt, dust, and other foreign materials get seperated towards the front of *thlangra* and are removed from the produce. Up and down movements are repeated till produce is cleaned.

***Kho*:** It is used for storing maize cobs, vegetables, etc in open. *kho* is also used for drying freshely harvested maize cobs over fire oven which can be later used as seed material. *kho* full of cobs is placed in the first compartment constructed with wood over the fire oven at a height of 50 cm. This is made by tying bamboo mats over a frame of bamboo pieces (Fig. 9.76). *Kho* rim hight is 15-20 cm and peripheri varies from 50 -70 cm. At the base it is of oval shaped and at the top its shape is elliptical.

Local cart: These carts accommodate more load than the back-packs and could go a long distance without any discomfort to the carrier. Push-cart of the Mizo tribes deserves a special mention for its construction and control mechanism (Fig. 9.77 and Fig. 9.78). This shows inherent talent of the tribes for their scientific approach towards construction of push-cart. These push-carts are made of wood and bamboo. To reduce the friction, wood-bearings are used in wheels. To provide better traction and cushioning; rubber mounting is placed on the periphery of wheels. Frame of cart is constructed with bamboo or wood logs. Wheels are made of wood in the shape of a disc and lined with hard rubber which acts as tyre. Axle is made of wood of high tensile strength. Front axle is pivoted from the centre at front end of chassis. Steering is controlled by pulling/relaxing the two ropes attached to both ends of the front axle. Suppose the rider wants to turn the cart towards right side, he/she stretches rope of the right side and relaxes the rope of left side. At the front end of the cart, a vertical hand-brake lever is provided which is connected through a wire to a wooden log provided behind the rear wheels. When rider pulls the lever, the wooden log is pressed against rear wheels which ultimately controls speed or stop the cart as per need. Cart is operated by pushing it from the back on a steep ascent. Cart-rider sits on front seat of cart and places both legs on the front axle for ease in steering and holds the brake lever to control speed. These carts may be used on all types of road. When cart is loaded to its capacity then two people are required to push it on the steep ascent however it depends on size and capacity of cart.

***Em* (carrying baskets):** The *Em* is used for carrying vegetables, paddy, wood, etc on the back of a person (Fig. 9.79 and Fig. 9.80). *Ems* are available in different sizes and capacity like small, medium and large. The small one is used for carrying rice, vegetables, etc from market; the medium one is used for carrying paddy from the *jhum* and large one for carrying vegetables, maize cobs and other materials which are bulky but comparatively light in weight. Its frame is made up of bamboo batten of 8-10 mm square. Strips of 2-3 mm thickness and 6-8 mm width are woven in frame in the form of mat. To give strength to upper edges of *Em*, a ring of bamboo batten is woven with it. Base of the frame is of square shape and upper portion is either circular or semi-circular. Another type of *Em* is made by fastning bamboo pieces in the frame at 5- 10 cm interval which is particularly used for carrying wood from the forest (Fig. 9.80). To support basket while carrying, a structure made of bamboo strips is fastened with the basket which rests on head of the person carrying it. After fastening it with basket, both ends are tied together and widest portion of it placed over head of the carreir. The speciality of this rope like structure is that its width at the centre is 3-5 cm however width decreases gradually towards both the ends and at the ends its width varies from 1.5 -2 cm. Its length varies from 125 cm to 175 cm.

Table 9.4 : Merits, and demerits of indigenous hand tools

Tools	Merits	Demerits
Chem puri	Very suitable for pruning trees. Due to lighter weight and sharper blade small branches, bamboo, trees, etc can be cut with shorter strokes by single hand.	Due to shorter length of the handle drudgery is more.Grip of the handle is also poor.
Chem Kawm	It is very suitable for weeding operation in crop fields.No damage to the crop plants.	It cannot be used for pruning of trees.Grip of handle is poor.
Tuthlawh	Can be used in any kind of terrain and soil type.Suitable for all field operations in tuber crops	Causes fatigue as the worker has to bow down while working with this tool.
Chemkawi	Ease in operation, less energy requirement and no damage to the crop plant in weeding because it is done by scratching.	It cuts the aerial portion of the weeds leaving underground portion intact because of which weeds emerges quickly therefore frequent weeding is needed.Less field capacity because as the worker has to remain seated throughout the operation
Chemkawi	Ease in operation, less energy requirement and no damage to the crop plant in weeding because it is done by scratching.	It cuts the aerial portion of the weeds leaving underground portion intact because of which weeds emerges quickly therefore frequent weeding is needed.Less field capacity because as the worker has to remain seated throughout the operation

Fig. 9.69 : *Chem pui*

Fig. 9.70 : *Chem kawm*

Fig. 9.71 : Big *Tuthlawh*

Fig. 9.72 : Small *Tuthlawh*

Fig. 9.73 : *Chemkawi*

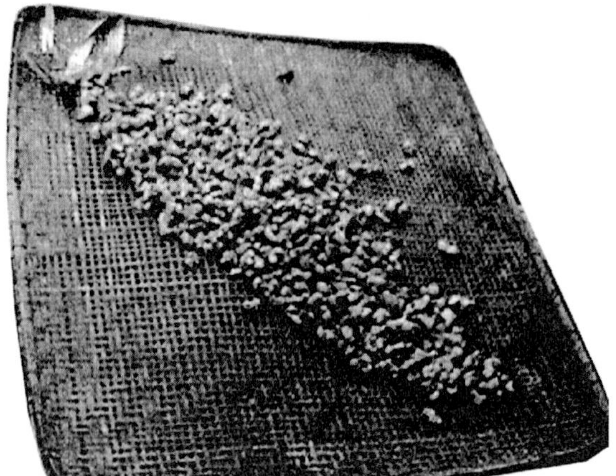

Fig. 9.74 : *Zam pher*

Fig. 9.75 : *Thlangra*

Fig. 9.76 : *Kho*

Fig. 9.77 : Mizo Local Cart (*Tawlailir*)

Fig. 9.78 : Mizo Local Cart (*Tawlailir*)

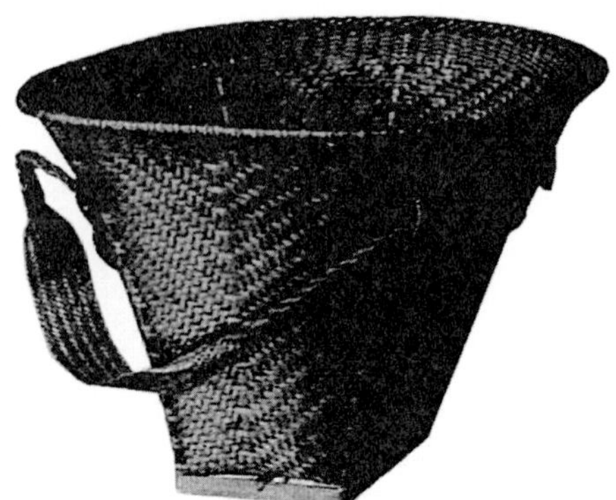

Fig. 9.79 : *Em*

Fig. 9.80 : *Em*

References

Agrawal K N; Ghadge S V; Singh R K P; Satapathy K K; Pandey M M. 2003. Directory of Horticultural Tools and Machinery for Hill Region. AICRP on FIM, ICAR Research Complex for NEH Region.

Anonymous. 1998-99. Annual Report, Indian Lac Research Institute, Namkum, Ranchi-834010 (India) pp 24-25.

Anonymous. 2001. Statistical Bulletin. Department of Agriculture, Govt. of Mizoram.

Anonymous. 2002-03. Annual Report, Indian Lac Research Institute, Namkum, Ranchi-834010 (India) pp 20-21.

Anonymous. 2008. Research Highlight. AICRP on Farm Implements and Machinery, CIAE Bhopal. Technical Bulletin No.: CIAE/FIM/2008/141.

Anonymous. 2010. Research Highlight. AICRP on Farm Implements and Machinery, CIAE Bhopal. Technical Bulletin No.: CIAE/FIM/2010/151.

Anonymous. 2010a. Annual Reports (2008-10). AICRP on Farm Implements and Machinery, TNAU Coimbatore.

Anonymous. 2010b. Annual Reports (2008-10). AICRP on Farm Implements and Machinery, KAU Tavanur.

Anonymous. 2013. Research Highlight. AICRP on Farm Implements and Machinery, CIAE Bhopal. Technical Bulletin No.: CIAE/FIM/2013/158.

Anonymous. 2013b. Annual Reports (2010-12). AICRP on Farm Implements and Machinery, TNAU Coimbatore.

Anonymous. 2015. Research Highlight. AICRP on Farm Implements and Machinery, CIAE Bhopal. Technical Bulletin No.: CIAE/FIM/2015/179.

Bhardwaj K C; Pandey M M; Singh Gyanendra. 2004. Horticultural Tools and Equipment. CIAE Bhopal. Technical Bulletin No.: CIAE/2004/110.

Jayan, P R, Sujith, G, Jyothiraj, C R, and Naushad, A. 1998. Development and testing of a tree lopping machine as an attachment to the power tiller. *J. of the Institution of Engineers* (India), Vol. 79 (September): 5-7.

Manes G S; Ahuja S S; Dixit Anoop; Singh Surendra. 2007. Horticultural Mechanization as a Tool for Diversification. Paper presented at 20th National Convention of Agricultural Engineers and National Seminar on "Farm Mechanization for Diversification of Agriculture" held at Punjab Agricultural University, Ludhiana from January 19 to 20.

Pandey M M; Ganesan S. 2005. Farm Mechanization Package for Dryland Agriculture, Technical Bulletin No: CIAE/FIM/2005/117. Central Institute of Agricultural Engineering, Bhopal.

Pandey M M; Majumdar K L; Singh Gyanendra; Singh Gajendra. 1997. Farm Machinery Research Digest. Technical Bulletin No. CIAE/97/69, Central Institute of Agricultural Engineering, Bhopal, 328 p.

Pandey M M; S Ganesan; R K Tiwari. 2006. Improved Farm Tools and Equipment for North Eastern Hills Region, Technical Bulletin No. CIAE/2006/121. Central Institute of Agricultural Engineering, Bhopal.

Pandey M M; Surendra Singh. 2009. Farm Tools and Equipment for Horticulture and Hilly Region. Paper published in proceedings of 9th Agricultural Science Congress held at Sher-e-Kashmir University of Agricultural Sciences and Technology of Kashmir, Shalimar, Srinagar – 191 121 (J&K) during June 22-24, 2009.

Prasad Kamta; Gupta Rajendra; Brajendra; R P V. 2012. Indigenous Farm Tools Used by the Tribal Farmers of Mizoram in *Zhum* Cultivation. *AET.* Vol. 36(2): 19-25.

Prasad N; Kumar K K; Pandey S K; Bhagat M L. 2001a. Design and development of hand operated roller type lac scraper. Paper presented in 35th Annual Convention of ISAE held at CAET, Bhubaneswar (22-24 January, 2001).

Prasad N; Pandey S K; Bhagat M L; Kumar K K. 2001c. Design and development of pedal operated roller type lac scraper. *Journal of Agricultural Engineering* (ISAE) 38(3):40-44.

Prasad N; Pandey S K; Kumar K K; Agarwal S C. 2001b. Scope of mechanization in lac production. *AMA* 32(2):65-67.

Prasad Niranjan; Pandey S K; Bhagat M L. 2009. Mechanization of Lac Cultivation. *AET* Vol. 33(2)

Singh Surendra. 2007. Farm Machinery – Principles and Applications. Directorate of Information & Publication of Agriculture, Indian Council of Agricultural Research, Krishi Anusandhan Bhawan-I, Pusa Campus, New Delhi.

Singh Surendra. 2008. Farm implements and machinery for hill agriculture. Paper presented at Conference on "Agricultural Development in NEH Region – Challenges and Opportunities" held at ICAR Research Complex for NEH Region, Barapani, Meghalaya from June 18-19, 2008.

Singh Surendra. 2011. Farm mechanization and impact on agricultural productivity. Paper presented at Interaction Meet on Agricultural Engineering Technologies with Farmers, Manufacturers of Farm Equipment and Scientists during March 12, 2011 at UAS Bangalore.

Singh Surendra; Mishra A K. 2009. Farm implements and machinery for hill agriculture in North Western Himalaya. Paper presented at 22nd National Convention of Agricultural Engineers and National Seminar on 'Emerging trends of Agricultural Engineering for farm mechanization of hilly regions' held at CSKHPKV Palampur from Jan. 20-21, 2009.

Singh Surendra; Pandey M M. 2008. X Plan Achievements (2002-2007). AICRP on Farm Implements and Machinery, CIAE Bhopal. Technical Bulletin No.: CIAE/2008/137.

Singh Surendra; Verma S R. 2009. Farm Machinery Maintenance & Management. Directorate of Information & Publication of Agriculture, Indian Council of Agricultural Research, Krishi Anusandhan Bhawan-I, Pusa Campus, New Delhi.

Tandon S K; Singh Surendra. 2009. Improved Agricultural Equipment for Introduction in Hilly Region. Paper presented at 9th Agricultural Science Congress at Sher-e-Kashmir University of Agricultural Sciences and Technology of Kashmir, Shalimar, Srinagar – 191 121 (J&K) from June 22-24, 2009 (Abstract published).

Tiwari J P. 2011. Low cost, high-tech implement for khus root digging. In: The Fifth International Conference on Vetiver, 28-30 Oct., 2011. P. 78-79, organized at CSIR-Central Institute of Medicinal and Aromatic Plants, Lucknow, India.

Verma S R. 2001. Advances in potato and groundnut harvesting machinery. Proceedings of Summer School on 'Advances in Seeding and Harvesting Machinery'. Dept. Farm Power & Machinery, PAU Ludhiana. May 21 to June 19. Pages 140-157.

Verma S R; Sharma V K; Singh Surendra; Ahuja S S; Singh Santokh (eds.). 1992. Proceedings of National Colloquium on Potato Mechanization in India. *Punjab Agricultural University, Ludhiana.*

Verma, S R; Singh, Hari. 1988. Development of an axial-flow vegetable seed extracting machine. *J. Agric. Engng.* ISAE Vol. 25(1): 98-104.

10

Farm Tools and Equipment for Dryland Agriculture

Indian agriculture is predominantly a rainfed agriculture under which both dry farming and dry land agriculture is included. Dry farming was the earlier concept for which amount of rainfall (less than 500 mm annually) remained the deciding factor for more than 50 years. In modern concept, dry land areas are those where the balance of moisture is always on the deficit side. In other words, annual evapo-transpiration exceeds precipitation. In dry land agriculture, there is no consideration of amount of rainfall. It may appear quiet strange to a layman that even those areas which receive 1100 mm or more rainfall annually fall in the category of dry land agriculture under this concept. The success of crop production in these areas depends on the amount and distribution of rainfall, as these influences the stored soil moisture and moisture used by crops. The amount of water used by the crop and stored in the soil is governed by the water balance equation: ET = P-(R+S). When the balance of the equation shifts towards right, precipitation (P) is higher than ET, so that there may be water logging or it may even lead to run off (R) and flooding (S). On the other hand, if the balance shifts to the left, ET becomes higher then the precipitation, resulting in drought in the various severity. Taking the country as a whole, as per meteorological report, severe drought in large area is experienced once in 50 years and partial drought in five years while floods are expected every year in one part of the country or the other, especially during rainy season. In fact the balance of the equation is controlled by the weather, season, crops and cropping pattern. Out of 142 million ha of net sown area in the country, rainfed agriculture is practiced in 95 million ha (67%). Nearly 67 m ha of rainfed area falls in the mean annual precipitation range of 500-1500 mm.

In arid and semi arid areas with high average soil temperature and dry spells, there is a need to break the soil, which becomes very hard. A pointed tool like chisel or bar point are used on country plough to break soil without inverting or disturbing crop residue, in order to collect and store rain water and reduce wind erosion and evaporation losses. Under such conditions lister plough, rigid tine cultivator, duck foot sweeps and other similar equipment are useful and can be operated for one or two passes. Under black soil regions (vertisols) of Madhya Pradesh, Maharashtra, Gujarat and Andhra Pradesh, soils dry up and develop deep cracks during hot summer weather, and hence ploughing is not very essential. Mould board ploughing may be done once in 3 to 4 years to destroy weeds. For such soils shallow cultivation by a blade harrow or sweep cultivator is sufficient to prepare a good seedbed, when weeds are under control. Continuous operation of mould board for few years may be required to control the weeds. In humid areas, it is desirable to have deep tillage accompanied by soil inversion and burying of crop residues. This helps in enhancing nitrogen fixation in soil and incorporation of biomass. In dry land areas tillage requirements are mainly linked with improved moisture intake and retention, reduced evaporation and checking of weed growth. Studies have indicated that increased infiltration rate and higher crop yields can be achieved under dry land conditions by performing deep tillage by mould board plough.

Rainfed farming comprises about 91% area of coarse cereals (sorghum, pearl millet, maize and finger millet), 91% pulses (chickpea and pigeon pea), 80% of oilseeds (groundnut, rape seed, mustard and soybeen), and 65% of cotton (Singh and Singh, 2014). Also, about 50% area under rice and 19% area under wheat is rainfed. During the past 25 years there occurred significant changes in the area and yield of imported crops of rainfed farming areas. The area under coarse cereals decreased by about 10.7 million ha and most of this has under sorghum. The area under oilseeds increased by 9.2 million ha and most of this increase is due to irrigated rapeseed and mustard and soybean. The total area under pulses and cotton remained constant but more of cotton area became irrigated and shifts in the area occurred from one agro-ecological region to others. Area under chickpea in the northern belt decreased but increase in the central belt. This change occurred due to increase in area under rice-wheat cropping system which displaced chickpea and also pearl millet to a great extent and maize to a small extent.

In dry land agriculture, scarcity of water is the main problem. Apart from the low and erratic behavior of rainfall, high evaporative demand and limited water holding capacity of the soil constitute the principle constraint in the crop production in dry land area. Yield fluctuations are high mainly due to vagaries of weather, often much behind the risk bearing capacity of the farmers. In dry land area deficiency and uncertainty in rainfall of high intensity causes excessive loss of soil through erosion which leaves the soil infertile. Owing to erratic behavior and improper distribution of rainfall, agriculture is risky, farmers lack resources, tools become inefficient and ultimately productivity is low. V*ertisoles* have high clay content and high moisture retention capacity. Owing to its swelling and shrinking characteristics, permeability is low and hence the rate of infilteration of water is minimum. This causes more surface and high soil loss from the top layer owing to surface erosion. It is estimated that 68.5 tones/ha per year soil is lost from *vertisoles*. Due to high clay content it develops cracks during Rabi season at flowering stage of crops. *Alfisols* are, by and large, light textured soils which have low moisture holding capacity but high water intake. The rain water falling in such areas gets soaked up and saturates the profile. The soil water percolation is more and therefore, is lost for crop use. Owing to faster intake of water in the profile the surface runoff is limited and soil loss from erosion is low (3.05 t/ha/year). Soil crusting is a common problem in low rainfall areas. *Entisols* are generally loamy sand or sandy loam. Depth in these soils is not a constraint. These soils have very low clay content and hold water up to 200 mm per meter of soil profile. Its nutrient holding capacity is poor. In low rainfall areas monsoon cropping is practiced and in high rainfall areas double cropping is possible. *Submontane soils* are medium in texture and depth is medium to deep as well as moderate in clay content. Moisture retention capacity is high (300 mm/m profile). Phosphorous may be limiting in high production system. Due to high rainfall double cropping is possible in these soils. *Sierozems* are extremely light soils, effectively depth being influenced by the $CaCO_3$ concentration in soil profile. Its moisture holding capacity is low (150 mm water.m). Sierozemic soils are low in nitrogen and sometimes inadequate in phosphorous. Subsoil salinity is common. These soils are mostly monsoon cropped, except in deep sandy loams where post-monsoon cropping is also possible.

Improved dryland technology

The improved techniques and practices, which have so far been generated and recommended for achieving the objective of increased and stable crop production in dryland areas, have been summarized in following lines (Pandey *et al*., 1997; Pandey and Ganesan, 2005).

In order to take full advantage of annual precipitation in dry land agriculture, higher doses of energy input is essential. Farmers in dry lands have been using traditional and outdated farm equipment which not only perform poorly but also demand a lot of energy and time and post-harvest operations. Farm implements can help to conserve as much rain water *in situ* as possible and to harvest rain water. Shallow off season tillage with pre-monsoon showers ensures better moisture conservation and lesser weed intensity. It has resulted in 20% yield increase in sorghum. Deep tillage helps in increasing water in soils having textural profiles and hard pan. This has resulted in 10% yield increase in sorghum and 9% yield increase in case of caster. For in-situ moisture conservation, land has to be opened so that it can cause hurdle to flow of rain water. Tillage machines of appropriate size and type matching the power sources need to be used. Location specific seeders have been developed for dry land areas and these have shown good prospects and promise. A feature of these machines is that the seeds and fertilizers are placed in the moist zone of the soil resulting in a high percentage of seed germination and good crop vigour. In deciding farm mechanization in dry land areas, where farmers are generally poor, and their socio-economic condition should always be kept in mind. The foregoing discussions show that technology of crop production in dry land areas have been generated to a great extent. What is important now is to view it in socio-economic context of the farmers. Once the technology is adopted by the farmers, the contribution of dry land areas to the total production can be sizably improved and the living standards of the farmers of these areas can be improved.

Characteristics and problems of drylands and technological intervention

The drylands are the areas of meagre or undependable rainfall, in which the average precipitation is deficient in relation to water requirements of crops during its growth period. These include the arid zones in which arable crop production is not possible without irrigation and also their semi-arid fringes, in which rainfall, though precious; crops are raised by special techniques. Dryland farming is defined as an efficient system of soil and crop management in regions of low and mal-distributed rainfall. It is the practice of crop production entirely with rain precipitation received during the crop season or on conserved soil moisture in low rainfall areas of arid and semi-arid climate and the crops may face mild to very severe moisture stress during their life cycle. The problems associated with dryland agriculture are:

i) Inadequacy and uncertainty of rainfall and its erratic distribution,

ii) Late onset and early cessation of rains,

iii) Prolonged dry spells during the crop period,

iv) Low moisture retention capacity,

v) Poor soil fertility conditions and erratic rainfall,

vi) Socio-economic constraints, particularly because of the predominance of small and marginal farmers,

vii) Lack of improved technology,

viii) Limited infrastructure development and improper and untimely availability of credits and agricultural inputs,

ix) Subsistence farming, poverty, illiteracy, and

x) Low productivity of cattle and lack of fodder.

Needs of dryland mechanization

Mechanization plays a very important role in enhancing the production and productivity on dryland areas. It could be at least double in *alfisols* by appropriate management of all resources through timely and precision application. Mechanization is defined as improving efficiency of doing work at low cost, faster speed, with high precision and more comfort. Dryland mechanization is a process of adoption of needs based, location specific, efficient and precision tools, devices, equipment and machines matching to available power source, suitable for local soil, crop and socio-economic conditions. It helps to lower the operation cost, improve resource use efficiency, timely operation, and enhance crop productivity and profitability. Drylands require a tillage practice which can conserve basic resources namely soil and water. Both prevention of soil erosion and *in situ* moisture conservation are of utmost importance. Tillage practices also aim at complementing basic soil and water conservation measures such as graded bunds, land levelling and/or smoothening of surface. Rapidity of tillage operations for keeping the field ready for seeding immediately after the rain sets in is the key to the success of crops. It gains further importance when a second crop is to be sown subsequent to the harvest of first crop. Mechanization of agriculture has shown benefits in terms of: i) increase in productivity to the extent of 12-34 per cent; ii) seed-cum-fertilizer drill facilitates 20 per cent saving in seeds and 15-20 per cent saving in fertilizer; iii) enhancement in cropping intensity is 5-22 per cent; and iv) increase in gross income and return of the farmers is 30-50 per cent.

Improved farm implements and machinery for dryland crops

Drylands need special skill and judicious manipulation of the soil to retain the available moisture for the crops to get established. As the soil moisture availability is limited to a very short period, timely sowing operation is crucial for optimum crop establishment. Farm implements for dryland agriculture and irrigated agriculture do not differ substantially but some are more suited to the dryland areas (Singh, 2007; Pandey and Ganesan, 2005). The details of these equipment in terms of working principle, and performance results are given in previous chapters. These have been presented in the order of farm operations, i.e., seed bed preparation, sowing and planting, weeding and interculture, plant protection, harvesting and threshing. However, millet crop which is very common for dryland agriculture as well as hill agriculture; threshers have been developed of various capacity and also multi crop to be used in both conditions.

Multi millet thresher

Millets are very important food crop of tribal people mostly grown in tribal and hilly areas of India and under dryland conditions. These crops are grown in rainfed areas and temperature more than 20°C where other crops yield are very poor and are less prone to disease and pests. In India, millets are major staple food in the state of Uttrakhand, Chhattishgarh, Tamil Nadu, Odisha and Karnataka where these are grown widely with yield as high as 3 t/ha. There are many types of millets such as Finger millet, Barnyard millet, Kodo millet, Little millet, Foxtail millet, and Proso millet. Millets are very nutritious and important crop for balanced diet, rich in vitamins, protein, carbohydrate, minerals, fibers, iron, amino acid, phosphorus, magnesium and potassium (Rao, 1986). The epidemiological evidences indicate that person on millet based diets have good resistance for degenerative diseases such as heart disease, diabetes, hypertension etc (Anonymous, 2001). Millets are also good sources of energy, sources of essential amino acids except lysine and threonine but are relatively high in methionine (Gopalan *et al.*, 1997). In spite of health benefit and good source of nutrition, the utilization of millet for food is mostly confined to the traditional consumers and population of lower socio economic condition (Maleshi and Desikhachar, 1985) may be due to the lack of appropriate technologies available for management of these crops.

Harvesting of some millets take place at 16 to 20% moisture content in month of October-November and kept under sun for reduction of moisture up to 10 to 12%, followed by staking the harvested crop for 30 to 45 days in a threshing yard for loosening the glumes of the grains. The major and important operation in millet production is threshing. Traditionally in tribal and hilly areas, threshing of millet crop is done either beating by sticks or by treading out the crop panicle under the feet of oxen. These threshing operations are most time consuming, and energy intensive, labour intensive, drudgery prone and uneconomical. These methods of threshing have low output with low quality. The mechanized threshing of millets can reduce the drudgery of farmers/labours, and improve the quality of product. With existing socio economic condition of millet growing farmers, the large capacity threshers are inappropriate and even the small size thresher with large scale sophistication are difficult to be adopted (Grame, 1998; Singh *et al.*, 2002). Therefore, development of thresher for all millet crops was found necessary which can do threshing of all millets.

Physical properties of crop are very important for the design and development of the machine (Baryeh, 2002; Subramanian and Viswanathan, 2007). The dehuller of 40–50 kg/h capacity thresher was developed at Vivekananda Institute of Hill Agriculture (ICAR), Almora, Uttarakhand for dehulling barnyard millet using a 3.7 hp electric power source (Singh *et al.*, 2002). The dehulling drum is fitted with a canvas strip as a cutting device, to provide gentle impact and shear on the grain. The dehulling chamber is fitted with a sliding sieve, which allowed repetitive impact and shear to detach the husk from the grain. The friction force between canvas strip and barnyard millet grain provided the required force on grain for detaching the hulls from the grains (Fig. 10.1). The throughput, dehulling efficiency and specific energy consumption are 46.5 kg/h, 98.8% and 0.054 kWh/kg respectively. Dehulling efficiency decreased with decrease in moisture content from 9 to 7.22%, and the rate of decrease was rapid thereafter. Low dehulling efficiency at higher moisture content might be due to excess deformation of seeds, which withstood the impact force given by canvas strip on it, without splitting the hulls.

Fig. 10.1 : A view of thresher for dehulling of barnyard millet grain

It was observed that threshing of millet is better in case of combination of impact and shear on the crop (Singh *et al.,* 2003). Therefore the machine for threshing of all six minor millets was developed on the principle of combination of shear and impact at Central Institute of Agricultural Engineering, Bhopal, India. Many researchers have worked on the optimization of process parameters like milling, threshing (Singh *et al.,* 2004; Tiwari *et al.*, 2007; Singh *et al.,* 2008; Mangaraj and Singh, 2011). Different millets like little millet, kodo, proso, foxtail, barnyard and finger millet of local varieties were collected from small village Patalkot/Dhindhori tribal areas of MP. Physical properties of all millets are suitable for the proper threshing of millet (Jain and Bal, 1997; Kamble *et al*., 2003). The moisture content of the crops is kept 5 to 11% for performance analysis.

A multi millet thresher has been developed for threshing of the millets based on the different properties of the minor millets (Singh, 2013; Singh *et al*., 2011). Developed machine works on the principle of impact and shear on the ear head of the crop for the purpose of threshing of millets. The threshing drum is fitted with three rows of canvas strips and three rows of cutting knives placed alternately as some of the millet crop requires cutting action and some requires shear for complete threshing (Fig. 10.2). An isometric view of machine is shown in Fig. 10.3. The knives provide impact cutting of crop stem during threshing and the canvas strip rows gives gentle abrasion and shear on the grain for removing the grains from the glumes. The threshing chamber is fitted with a sliding sieve which is present closely to allow repetitive impact and shear to complete detach of glumes from the grains which helps in complete threshing of millets. The length of the feeding chute is kept 900 mm as per IS: 9020"2002. Grip handles of threshing sieves were made as per inner grip diameter for better comfort of the worker. Shaker assembly for cleaning system is provided with packing for reduction of vibration and noise. Rubber transportation wheels instead of cast iron wheels are provided for easy transportation and for absorption of vibration during operation. The machine is attached with safety guards over power transmission system. A flapper is fitted in the feeding chute to arrest the dust which may create health problems of the worker. A two hp, single phase electric motor is used as power source and the power is transmitted to the threshing drum, aspirator by the help of belt drive. For variation of drum speed from 500 to 1000 rev/min different size of pullies are used according to the requirement.The air flow rate is maintained below the terminal velocity of the grains and above the terminal velocity of chaff. The speed of aspirator is maintained by the use of belt drive. The threshing efficiency is 95% and cleaning efficiency 94% at 7.8% moisture content, 105 kg/h feed rate, 625 rpm cylinder speed, and 5 mm threshing sieve size. The thresher is found suitable for threshing of all six minor millets.

Fig. 10.2 : Multi millet thresher

Fig. 10.3 : An isometric view of multi millet thresher

References

Anonymous. 2001. Grain of truth an analysis. *Science and Environmental fortnightly magazine*, 15 May, 31-41.

Baryeh E A. 2002. Physical properties of millet. *Journal of Food Engineering*, 51: 39-46.

Gopalan C; Ramasastri B V; Balasubramanian S C. 1997. Nutritive value of Indian Foods. National Institute of Nutrition. Indian Council of Medical Research, Hyderabad.

Grame R.Q. 1998. Global assessment of power threshers for rice. *Agricultural Mechanization in Asia, Africa and Latin America*, 29 (3): 47-54.

IS: 9020. 2002. Indian Standard. Test code for general and safety requirements for power threshers. Bureau of Indian standard (Indian Standard Institution), New Delhi.

Jain R K; Bal S. 1997. Physical properties of pearl millet. *Journal of Agricultural Engineering Research,* 66: 85-91.

Kamble H G; Srivastava A P; Panwar J S. 2003. Development and evaluation of a pearl millet thresher. *Journal of Agricultural Engineering*, Vol 40(1): 18-25.

Malleshi N G; Desikhachar H S R. 1985. Miling, poping and malting characteristics of some minor millets. *Journal of Food Science and Technology*, 22: 400.

Mangaraj S; Singh K P. 2011. Optimization of machine parameters for milling of pigeonpea using RSM. *Food bioprocess technology* (2011) 4:762-769.

Pandey M M; Ganesan S. 2005. Farm Mechanization Package for Dryland Agriculture, Technical Bulletin No: CIAE/FIM/2005/117. Central Institute of Agricultural Engineering, Bhopal.

Pandey M M; Majumdar K L; Singh Gyanendra; Singh Gajendra. 1997. Farm Machinery Research Digest. Technical Bulletin No. CIAE/97/69, Central Institute of Agricultural Engineering, Bhopal, 328 p.

Rao M A. 1986. Rheological properties of fluid foods. In M. A. Rao& S. S. H. Rizvi (Eds.), Engineering Properties of Foods (pp. 1-48), New York: Marcel Dekker, INC.

Singh K P. 2013. RPF-III. Design and development of multi millet thresher. Central Institute of Agricultural Engineering, Bhopal. CIAE/AMD/2012-16.

Singh K P; Mishra H N; Saha S. 2011. Design, development and evaluation of Barnyard millet dehuller.*Journal of Agricultural Engineering*, 48 (3): 17-25.

Singh Surendra. 2007. Farm Machinery – Principles and Applications. Directorate of Information & Publication of Agriculture, Indian Council of Agricultural Research, Krishi Anusandhan Bhawan-I, Pusa Campus, New Delhi.

Singh Surendra; Singh R S. 2014. Energy for Production Agriculture. Directorate of Knowledge Management in Agriculture, Indian Council of Agricultural Research, Krishi Anusandhan Bhawan-I, Pusa Campus, New Delhi - 110012.

Singh K P; Kundu S; Gupta H S. 2002. Vivek thresher for madua and madira.*Anonymous, Annual report, VPKAS (ICAR), Almora,*88-89.

Singh K P; Kundu S; Gupta H S. 2003. Development of higher capacity thresher for ragi/ kodo.*Recent trend in millet processing and utilization*, CCS, HAU, Hissar, India,109-116.

Singh K P; Pardeshi I L; Kumar M; Srinivas K; Srivastva A K. 2008. Optimisation of machine parameters of a pedal operated paddy thresher using RSM. *Biosystems Engineering,*100: 591-600.

Singh S K; Agarwal U S; Saxena R P. 2004.Optimization of process parameters for milling of green gram (Phaseolusaures*).Journal of Food Science and Technology,* 41(2): 124-30.

Subramanian S; Viswanathan R. 2007. Bulk density and friction coefficients of selected minor millets grains and flours. *Journal of food engineering*, 81: 118-126.

Tiwari B K; Jagan Mohan R; Bhasan B S. 2007. Effect of heat processing on milling of black gram and its end product quality. *Journal of Food Engineering*, 78: 356-60.

11

Straw Management Equipment

There are many environmental risks associated with stubble burning. Thus the adoption of alternative straw management strategies is the best interest of all concerned. Key points when considering harvesting surplus cereal straw are: i) maintain sufficient crop residue on the land to protect the soil from erosion. This can be accomplished by keeping the stubble standing particularly after seeding; ii) harvest surplus crop residue with an appropriate frequency so as not to lower soil organic matter, soil fertility and crop productivity; and iii) fertilize crops according to soil test recommendations. The decision regarding the amount of straw to be removed should be based on the inherent value of the straw for maintaining the viability of the cropping system and protecting the soil resource versus the value of the straw for other uses (e.g., feed and bedding for livestock, industrial strawboard and feedstock for alternative energy forms). The value of retaining the straw on the land is difficult to determine but should be based on a number of factors such as: value of straw for soil erosion control; equivalent fertilizer value of the nutrients contained within the straw; value of the straw for building soil organic matter, soil quality, and soil tilth; and value of the straw for soil moisture conservation (Singh *et al.*, 2003; Singh *et al.*, 2009). At harvest, it is best to chop the straw as fine as possible and spread both the straw and chaff across in the field. A number of equipment have been developed for managing the straw in the field and discussed here.

Stubble shavers (Rotary slasher)

It is a tractor rear mounted PTO operated machine. The machine consists of 2-3 blades mounted on the ends of a rotating arm hinged with pivot pins (Fig. 11.1a). The power to the blades is supplied from PTO of a tractor through a speed reduction gearbox and belt and pulley drive giving a speed ratio of 1.8:1 (Garg and Singh, 2002; Anonymous, 1999 & 2002). The blades rotate in a horizontal plane near the ground and cut or save the stubble left after combine operation in paddy fields. The pulley and tip diameter of blades is 250 and 1500 mm respectively. The tip speed of blade is 76 m/s at PTO rated speed. The overall assembly is covered with an MS sheet to protect the machine from dust and stubble. The height-adjusting wheel has been provided to adjust the height of cut. The machine has been provided with fenders in front and two sides as safety measures for diverting the trash on the rear. It can cover about 0.2-0.3 ha/h. The blades need sharpening after every 10-12 ha of use and need replacement after 100 ha. The use of stubble shaver saves about two disking operations, which saves time, energy and money. It requires about 2.5-4.0 lit/h of fuel. It is a versatile machine for cutting any crop residues such as wheat, paddy and grasses from the ground level. It is operated by 26.11 kW tractor's PTO. It is available in various sizes of width 1.2 m, 1.5 m, 1.8 m. This can easily cut thick and tall weed plants and grass and slash plant residue as tall as 2.5 cm (Fig. 11.1b). It offers adjustable cutting height up to 2.5 cm. It is very useful for cutting rough and tough shrubs.

a) A view of rotary slasher

b) Tractor operated rotary slasher in operation

Courtesy : Tirth Agro Technology Pvt. Ltd. Rajkot (Gujarat)

Fig. 11.1 : Tractor operated stubble shaver/rotary slasher

Power tiller operated shredder/ slicer cum *in-situ* applicator incorporator

One of the major difficulties in vegetable production is to clear the field of old vegetable plants after harvesting. At present, population of power tillers is increasing rapidly due to its suitability to small and medium farmers. Hence a roto shredder-cum-*in-situ* applicator as an attachment to power tiller has been developed for increasing the versatility and annual usage of the power tiller (Anonymous, 2008b). It has a shredder assembly, power transmission system, hitch frame and rotary tiller attachment (Fig. 11.2). The shredding unit consists of main shaft and shredding blades. Two number of swing back type rotary blades are hinged to the main drive shaft, which receives power through a bevel gear box from the engine pulley of the power tiller. The shredder speed is 900 rpm. An additional 50 kg of counter weight is added to the rear of the power tiller for balancing the system. The rotary tiller of the power tiller provides fine degree of pulverization enabling the necessary incorporation. The field capacity is 0.57 ha/day with shredding efficiency 90%.

Fig. 11.2 : Power tiller operated stubble shaver

Chopper type tynes for power tiller rotavator for sugarcane trash shredding

India is one of the major sugarcane growing countries in the world. For sugarcane crop cultivation, sugarcane trash management is of vital importance since it is usually burnt in the fields because the cost of labour required in collection and transportation of the trash is prohibitive. The sugarcane trash can be used for making organic matter. Power tillers are used for different operations in sugarcane crop viz., stubble uprooting, interculturing, earthing up, transportation etc. Power tiller rotavator is equipped with L-shaped hatchet type tynes for trashy conditions for weed control under wetland condition (Anonymous, 2008). If these tynes are used for sugarcane trash shredding, entanglement of trash around the rotavator shaft is observed. The normal angle for L shaped tynes is 118^0 + /-2^0 which if changed to 160^0 to 180^0 makes it chopper type tynes (Fig. 11.3). Effective field capacity of power tiller rotavator equipped with chopper type tynes is observed to be 0.065 to 0.085 ha/h with 66% to 82% field efficiency.

Fig. 11.3 : Power tiller rotavator with chopper type tynes for sugarcane trash shredding

Straw baler

The rice-wheat fields harvested by combine are generally left with long loose straw/ stubble in the field. Nearly 75% of rice-wheat straw goes as waste besides causing environmental pollution due to burning by farmers prior to tillage for subsequent sowings. The primary reason for burning is due to lack of suitable straw management practices for incorporation or retrieval of straw from combine-harvested rice-wheat fields. Rice-wheat straw, being burnt, are rich renewable soil enriching resource and nutrient building materials when incorporated properly into soil and if retrieved, it is useful as animal feed and for industrial purposes. Baler is a useful machine for retrieval of straw which picks up loose hay or straw, compresses it into bales of even size and weight, ties them with twine or wire and discharges the completed bale out of the machine. Baling of straw following combining involves only that part of the straw, which passes through the combine. Loose straw that is thrown by the combine contains more leaves and is of better nutritive value but the amount of this straw is hardly one-third of the total obtained in hand harvesting at ground level. Although, combining is the cheapest harvesting practice yet it sacrifices 40-60% of recoverable straw. Thus, it is important to introduce suitable reaping mechanism in existing baling machine for simultaneous cutting of left out stubbles and baling complete straw.

Manually gathering of loose straw after combine harvesting is very laborious, tedious, expensive and difficult to handle transport and storage. A straw baler collects the loose straw such as paddy/wheat straw; sugarcane trash left in the field after harvesting and makes the bales. The bales have immense commercial value such as cattle feed, bio mass power generation, Ethanol production, paper production, packaging material, mushroom cultivation, particle board manufacturing, organic manure and many more (Singh *et al.*, 2007; Singh, 2007; Singh *et al.*, 2003; Anonymous, 2002). It consist of a three major operating units like picking unit, compression unit where straw is compacted and knotter unit for binding the bales tightly (Fig. 11.4a 11.4b). Plunger works at a speed of 93 strokes per minute. It consists of reel type straw pick up assembly, and straw compaction and tying units. It automatically picks up the residue straw from field with the help of reel which is transferred into bale chamber with the help of feeder and then straw is compressed with the reciprocating ram into a compact variable length size. It also automatically ties the knots using metal wire or nylon rope. The capacity of the machine is up to 18 tonnes per hour. The size of rectangular bales can be varied from 80 x 45 x 45 cm to 150x 45 x 45 cm and bale weight can be varied from 15 to 45 kg. The number of bales formed varies according to the extent of loose straw in the field. It has been observed that for better machine performance and full recovery of straw, stubble shaver should be operated in combine harvested field. A straw baler can be operated in the combine harvested field where loose straw is picked and there after baled, standing stubble remains untouched (Fig. 11.4c). However for higher number of bales per unit area stubble shaver can be operated after combine harvesting in paddy. It is used for baling of straw into bales of rectangular cross section. It is operated by 26.1 kW tractors PTO. Working capacity of baler is 0.25-0.35 ha/h.

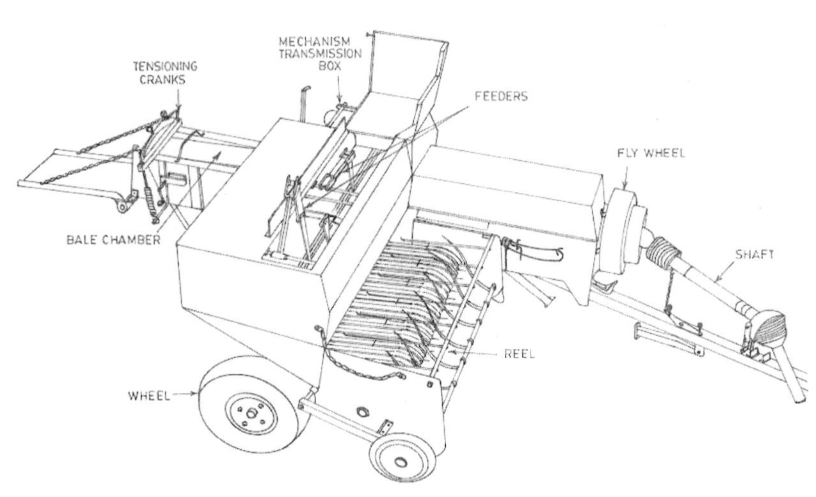

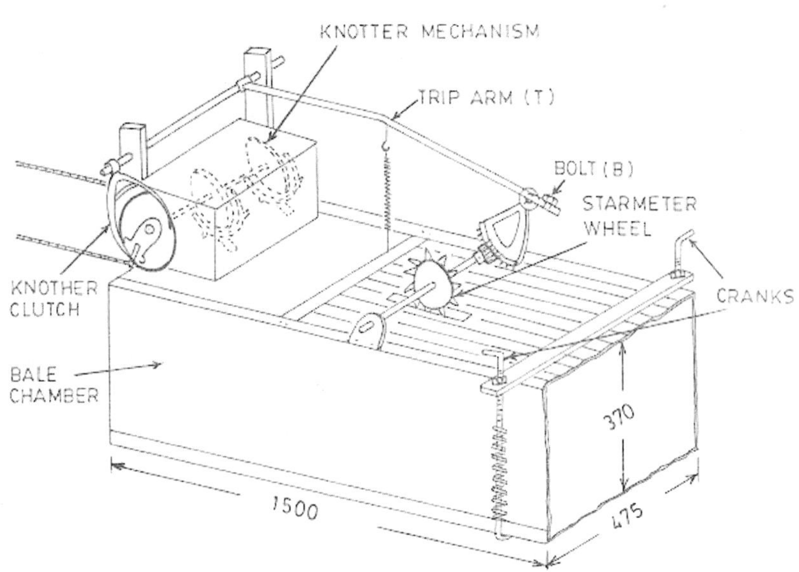

a) Details of baler

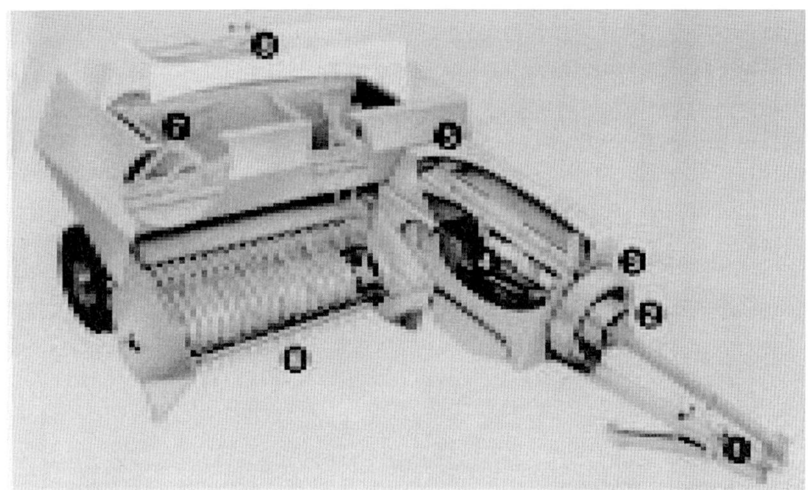

1) Adjustable drawbar to ensure correct hitching; 2) slip clutch and shear bolt to secure the baler transmission; 3) Heavy flywheel ensures smooth transmission and maximum power transfer; 4) Hypoid gearbox for smoth drive and resistance to wear; 5) shaft drive for feeders and knotter; 6) positive crop pick up with tines; 7) each feeder tine protected by a shear bolt; and 8) shaft driven knitters.

b) Components of baler

c) A view of baler in field
Courtesy: CLAAS India Pvt. Ltd.
Fig. 11.4 : Straw baler

Round baler

It is used to cut and compress crops such as rice, cotton, hay etc in to compact round bales (Anonymous, 2015). It is operated by 45 hp tractors. It has a pick up reel and compacting unit (Fig. 11.5 and Fig. 11.6). Pick up reel width is 92.7 cm. It can make bales of size 95 x 60 cm with weight up to 300 kg. The capacity of machine is 40 bales/h. Bale loader has also been developed to load bales dropped in the field by balers to tractor trolleys for transporting them to end use. It picks up the bales from the fields and elevates them high enough to load a high trailer. It has three hooks in front which enters in the round bales and lift it up (Fig. 11.5). It is operated by 55 hp tractors. It can lift up to 900 kg.

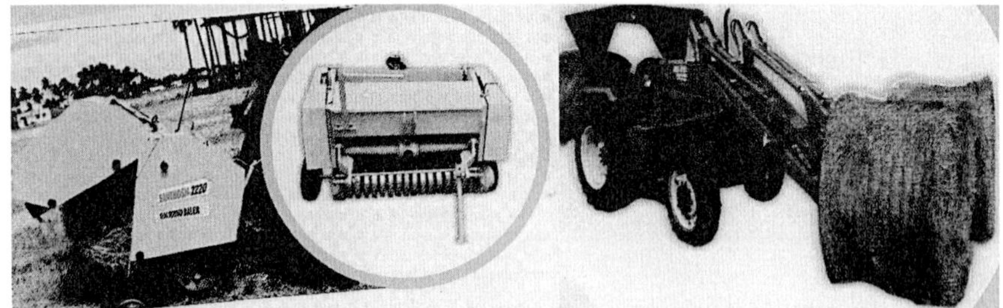

Fig. 11.5 : Tractor operated mini round baler along with loader
Courtesy: Santhosh Agri Machinery Attur, Salem (Tamil Nadu)

Fig. 11.6 : Tractor operated round baler
Courtesy: Tirth Agro Technology Pvt. Ltd. Rajkot (Gujarat)

Tractor operated baler with cutter bar

The cutter bar baler has been developed at Central Institute of Agricultural Engineering, Bhopal with the support from M/s. Kartar Agro Industries Pvt. Ltd., Bhadson, District Patiala (Punjab) for retrieval of paddy straw after combine harvesting (Fig. 11.7). The average field capacity is found as 0.34 ha/h at 2.75 km/h forward speed. The range of cut stubble length by the cutter bar of the baler is 70-155 mm. It is possible to recover 55.5% more stubbles using the baler with cutting bar. However, this recovery could be increased further, if the combine is operated with greater height of cut, so that longer cut stubbles could be picked up by the baler fingers.

Fig. 11.7 : Tractor operated baler fitted with cutter bar
Courtesy: M/s. Kartar Agro Industries Pvt. Ltd., Bhadson, District Patiala (Punjab)

Straw combine

Wheat straw is extensively used as a cattle feed for milch animal in India and other Southeast Asian countries. Due to ever-increasing population, area under fodder is likely to shrink making wheat straw more and more important ingredient of the animal feed. Combine harvesters for paddy and wheat were introduced in early seventies. However, these combine waste the wheat straw in the field. Therefore, this important cattle feed is left in the field for burning, as mixing requires additional energy. This intern causes intense heat and smoke and along with threshing floor dust pollutes the environment to a very high degree, raises the surrounding temperature.

Straw combine is a threshing machine which cuts, threshes and cleans the straw in one operation (Ahuja, 2001; Ahuja *et al.*, 1993; Anonymous, 2008a, 2010 & 2013, 2010b; Garg *et al.*, 2002; Garg and Singh, 2002; Singh and Pandey, 2008; Verma *et al.*, 1992). The wheat stalks left after combine harvest are cut by an oscillating blades while revolving reel pushes them back towards the auger. The stalks are conveyed into the machine by the auger and reach the threshing cylinder which cut the stalks into small pieces against concave. It gives superior separating performance. The short fragments fall through the bars of the concave. After cutting the wheat by combine harvester, it accumulates the wheat stalks and after threshing it, put into the drum. Different type of straw can be threshed to change over the sieves according to our need. Stone Trap Tray (S.T.T.) is used to avoid the machine accident due to stone or any hard product. Machine is fully belt operated. Cutting height is adjustable.

Straw combine recovers the straw left behind by combine harvesters and coverts it to finely bruised straw (bhusa). It can be tractor operated (45-50 hp) or self propelled (Ahuja, 2001). It consists of stubbles cutting unit, straw collecting unit, feeding unit, straw brushing unit and bhusa blowing unit (Fig. 11.8). The cutting unit consisting of cutter bar of length 2.1 m, crop reel, and platform type auger. The straw bruising system has a chain type feeding conveyor fixed on frame directly behind the auger to carry the cut materials in to the serrated tooth type bruising cylinder. Straw sieving unit consists of a rectangular sieve shaker which is suspended by hanger having reciprocating motion. The machine works well at concave bar spacing of 14 mm, feed rate of 1.4 t/h and cylinder peripheral speed of 32.25 m/s. An enclosed trailer is required to collect bhusa (Fig. 11.9). Brushing mechanism may be spike tooth cylinder type, chaff cutter type or serrated circular saw type. Serrated-tooth system has been incorporated recently in straw combines in view of their claimed lower power requirements. Material being fed to the bruising unit also contains few grains mostly on account of cutter bar losses. Material passing through the concave falls on a reciprocating sieve. Chaff and lighter materials are sucked by the aspiration fan and are delivered to the attached trolley through an adjustable blower duct. Aspirator normally has 3 to 6 blades with outer diameter ranging from 50 to 80 cm. Fan tip speed varies from 30-50 m/s (800-1000 rpm). Conventional tractor trolley is used as a transient storage during harvesting operation. This trolley is covered with a fine net, so that dust is blown off and bruised chaff is stored in this. In order to maintain a continuous field operation of straw bruising unit, two to three trolleys have to be netted (or canvassed) and employed in this operation. This is essential because while the first trolley is being unloaded the second is used for straw combining. Number of trolleys to be used depends upon the transportation distance. One wheat straw combining unit and three transport trolleys and two tractors (one of 45 hp and another of 35 hp) are being used by the farmers. Average field capacity is 0.24 to 0.4 ha/h, straw output 15 to 25 q/ha, and height of cut 5 cm to 7 cm. It can recover 55-65 percent of straw successfully as compare to the traditional method of threshing. The fuel consumption varies from 3.5-4.0 l/h. There is also a recovery of 75-100 kg of grain per hectare. Thus there is an additional saving. For straw combine one person operates the machine and 3-4 persons are needed to unload the trolleys of straw. To avoid the hitching of additional trolley, a container has been designed and developed at Central Institute of Agricultural Engineering, Bhopal with support from Rattan Agro Industries, Moga (Punjab) which is mounted over the straw combine (Fig. 11.10a). Fig. 11.10b gives the view of commercial straw combine with mounted container.

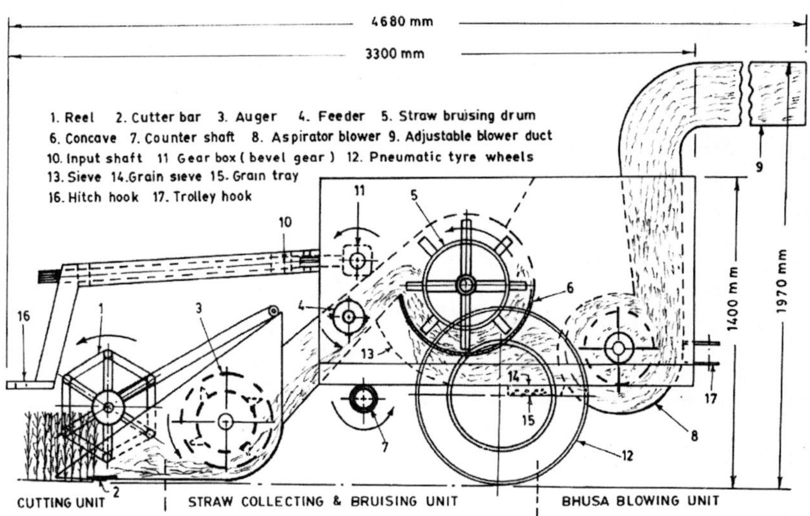

Fig. 11.8 : Material flow in straw combine

Fig. 11.9 : A view of Straw combine in field with trolley attached
Courtesy: Dasmesh Mechanical Works, Amargarh (Punjab)

a) Tractor operated straw combine with straw storage unit mounted and unloading of straw

b) Tractor operated container mounted straw combine
Courtesy: Bharath Agro Products, Paramakudi (Tamil Nadu)
Fig. 11.10 : Tractor mounted straw combine

Flail type straw chopper-cum-spreader

It is used for chopping of paddy straw, which helps in easy incorporation of straw in the soil in fewer operations by using traditional equipment. The machine in a single operation can harvest the straw left after combining, chop it into pieces and spread on to the field. It is operated by 26.1 kW tractors' PTO (Manes *et al.*, 2007; Manes *et al.*, 2010; Garg *et al.*, 2002; Anonymous, 2008, 2010). It consists of a rotary shaft mounted with blades named as flail to harvest the straw and chopping unit consisting of knives (Fig. 11.11). The harvesting unit has 52 flails mounted on 4 rows with width of cut 149 cm (Fig. 11.12). The straw after cutting by the flail, pass on to the chopping mechanism, which cuts the straw into pieces. The chopping mechanism has a 65 cm cylinder with 8 rows of serrated knives and three counter rows each having 18 knives fixed at the bottom. A reel is attached in front of the cutter bar to feed the straw to the cutter bar for proper cutting. The cut stubbles are conveyed to the chopping cylinder with the help of feeding cylinder attached between the cutter-bar and chopping mechanism. The machine is operated in combine harvested fields for chopping the stubbles (Fig. 11.13). The chopped and spread straw is easily incorporated in to the soil by the use of single operation of rotavator or disc harrow and decayed after irrigation. Thus instead of 7-8 operations which are required for straw incorporation in traditional way; by chopping straw it is incorporated in a much easier way and there is a saving of about 40-50 % time and cost. The speed of operation varies from 2.0-2.4 km/h when operated in first low and second low gear respectively. Height of stubbles after chopping varies from 5-14 cm. The field capacity and fuel consumption of the machine ranges from 0.4-0.5 ha/h and 4.0-4.5 litre/h respectively.

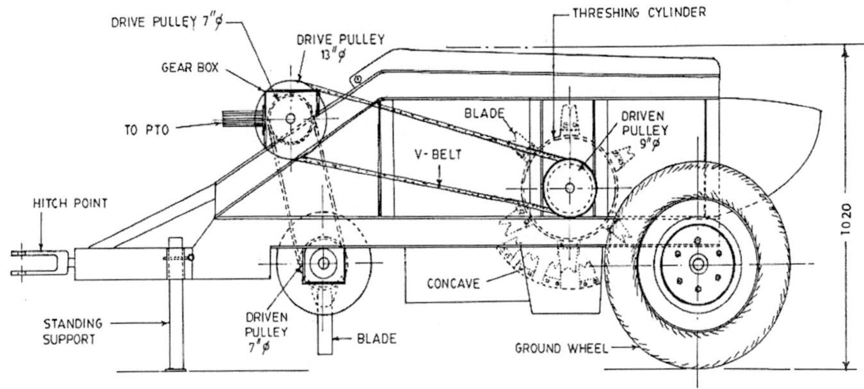

Fig. 11.11 : Details of flail type straw chopper cum spreader.

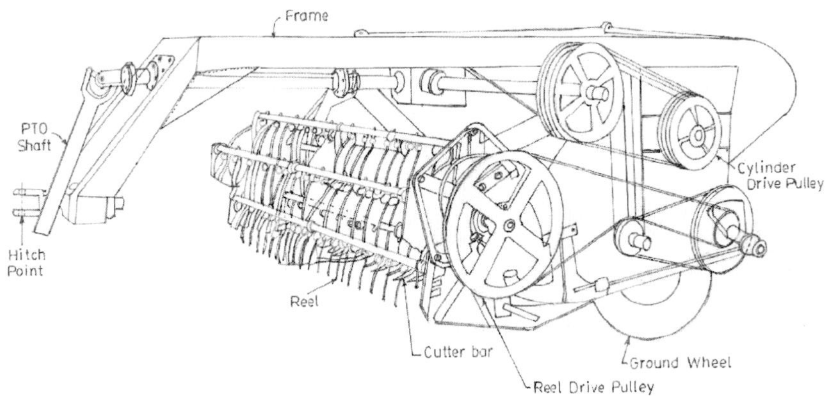

Fig. 11.12 : An isometric view of cutter bar type rice-straw chopper-cum-spreader

Fig. 11.13 : Tractor operated cutter bar type straw chopper-cum-spreader

Happy seeder

Rice wheat is the major cropping system in India. The increasing constraints of labour and time have led to the adoption of mechanized farming in highly intensive rice-wheat system. The area under combine harvested rice and wheat is about 80% and 70% of the total area under the two crops, respectively. While, at present more than 80% of wheat residue is collected by the farmers after combine harvesting using straw combine and often fed to animals, paddy straw is considered poor feed for animals due to its high silica content. Burning is the normal and easiest method of rice stubble management because residues interfere with tillage and seeding operations for the next crop. Despite the ban imposed by the Govt., burning of residues has been doing great damage to the environment. A Happy seeder has been developed at Punjab Agricultural University, Ludhiana for rice straw management (Singh *et al.* 2003; Singh *et al.*, 2009). The machine consists of a rotor for managing the paddy residues and a zero till drill for sowing of wheat (Fig. 11.14). Flail type (Gamma) straight blades are mounted on the straw management rotor which cuts (hits/shear) the standing stubbles/ loose straw coming in front of the sowing tine and clean each tine twice in one rotation of rotor for proper placement of seed in soil. The rotor blades/flails guide/push the residues as surface mulch between the seeded rows. This PTO driven machine can be operated with 45 hp tractors and can cover 0.3-0.4 ha/h. Adoption of happy seed drill technology holds great promise in retaining precious nutrients in the soil and minimizing the harmful effects due to burning of crop residues. Mulched rice residue is likely to result in N immobilization than incorporated residue and can also provide non-N benefits such as conservation of soil water and control of weeds. It is broadly accepted that residues decompose faster when incorporated than placed at the surface.

Fig. 11.14 : Tractor operated happy seed drill being operated by 45 hp tractor

Majority of the farmers own 35 hp tractors which is a main limitation in the adoption of 45 hp driven happy seed drill. A new prototype of happy seed drill has been developed which works efficiently in heavy straw load with 35 hp tractors mostly available with farmers in the region (Singh *et al.*, 2009). The new developments include modification of row spacing, blade geometry (J and gamma type), blade tip speed, and machine weight and rotor size/curvature (Fig. 11.15). The new machine requires less power than the 45 hp tractors but maintains the latter's performance. The new prototype having 25.7 cm row to row spacing is 19% lighter in weight with 30% more tip speed of modified rotor blades, 40% more windows opening for easy loose straw movement compared with 45 hp happy seed drill. It is also observed that the new Gamma type blades consumes 34% less fuel as compared to the J-type blades.

The strip of stubble in front of the sowing tynes is cut, picked up and placed on the side of the drilled seed as mulch. Machine covers 0.2-0.3 ha/h. Earlier model of happy seed drill called Çambo happy seeder' throws the chopped straw behind the machine after seeding (Fig. 11.16). Mulched crops residue improves the soil hearth and added organic matter to the soil. Substantial water saving has also been recorded due to avoidance of first irrigation and mulching.

Fig. 11.15 : Tractor operated happy seed drill being operated by 35 hp tractor

Fig. 11.16 : Tractor operated Combo happy seeder (Stationery view and working in field)

Specially modified No-till drill for straw management

This drill can be used for sowing in under chopped & spread paddy straw conditions (Singh *et al.*, 2009). Specially modified no-till drill consists of furrow openers mounted on 3 members instead of 2 members in the conventional no-till drills thus staggering furrow openers to provide more lateral clearance for passing of straw (Fig. 11.17). Vertical clearance of the frame from the ground is increased from 30 cm to 60 cm by using longer shank of furrow opener. This drill can be used after operation of straw chopper.

Fig. 11.17 : Specially modified No-till drill for straw management working in field

Conveyor seeder

A nine row conveyor seeder has been developed to sow wheat in combine harvested paddy fields (Anonymous, 2015). It consists of an elevator-cum-pickup reel, a seed-cum-fertilizer drilling unit with 9 inverted T type furrow openers and power transmission system (Fig. 11.18). Elevator-cum-pickup reel is fitted with spikes to lift the loose paddy straw. Wheat is drilled in the standing stubbles and loose stubbles settle as mulch on the drilled seeds. The power to the pickup conveyor-cum-elevator is provided by tractor PTO through gear box. The field capacity of machine is 0.35 ha/h. There is saving of 24% in cost of operation as compared to conventional method.

Fig. 11.18 : Tractor operated conveyer seeder working in field

Paddy straw chopper-cum-loader

A paddy straw chopper-cum-loader has been developed at Punjab Agricultural University, Ludhiana, which chops the straw into small pieces and loads on the trailer, attached rear side of the implement (Anonymous, 1999 and 2002; Singh *et al.*, 2003). Two types of paddy straw chopper have been developed in collaboration with manufacturers (Fig. 11.19 and Fig. 11.20). These machines are operated by 45 hp tractor and can chop the paddy straw into small pieces and load in the trolley attached behind. The machine covers 0.4 ha/h. The chopped straw could be mixed with other fodder crops and used as feed to the animals.

Fig. 11.19 : Tractor operated paddy straw chopper-cum-loader working in field

Fig. 11.20 : Tractor operated paddy straw chopper-cum-loader working in field

Paddy straw spreader

The average length of loose straw and standing stubbles in rice field after combine harvesting is about 54 and 27 cm respectively and its distribution is 44 and 56% by weight. The yield of paddy straw is about 100 to 125 q/ha. A paddy straw spreader, an attachment to combine harvester has been developed to spread the paddy straw coming out of combine in the field (Fig. 11.21), which provides uniform field condition for subsequent field operations such as stubble shaving, paddy straw chopping and tillage (Anonymons, 2002). It has four blades mounted on spreader. The length of blades is 30 cm. Diameter of spreader is 115 cm. Rotational speed of spreader is 300-400 rpm.

Fig. 11.21 : Tractor operated paddy straw spreader working in field

Cotton plant puller or uprooter

Cotton is one of the important crops grown in the country. After the picking is over, the stalk of cotton needs to be removed from the field to prepare the seedbed for next crop. The usual practices followed are either to cut the stalk with an axe and remove the stumps with a digging tool or to pull the stalks manually by hand after irrigating the field. Considerable time and labour is thus involved in this operation. Apart from this manual pulling results in injury to the hand. A cotton puller can pull the stalk in the irrigated field but there is not much labour saving. A manual stalk puller can be used to pull individual stalk from the field, but is very laborious and tedious operation. Tractor operated cotton uprooter has also been developed at Punjab Agricultural University, Ludhiana and it can uproot the cotton stalks (Fig. 11.22). A cotton stalk puller with components such as frame and supporting units, stalk pulling wheels, tynes, spring unit, transmission unit, ground wheel unit, and float have been designed and fabricated. The pull force required to uproot a plant is a function of the taproot depth and independent of stalk height and thickness/diameter. Pulling force is not affected by pull angle. Pull force increases with the decrease in soil moisture. Plant pulling efficiency of the machine is largely affected by pull angle and peripheral speed. With the increase in pull angle, plant pulling efficiency increases. The plant breakage decreases with increase in pull angle.

Fig. 11.22 : Tractor operated cotton stalk uprooter

Another type of cotton stalk puller has been developed at Punjab Agricultural University Ludhiana for pulling cotton stalks (Gupta, 2006). It consists of frame and supporting units, stalk pulling wheels, tynes, spring unit, transmission unit, ground wheel unit, and float (Fig. 11.23). It is observed that the pull force required to uproot a plant is a function of the tap root depth and independent of stalk height and thickness/diameter. Pulling force is not affected by pull angle. Pull force increases with the decrease in soil moisture. Plant pulling efficiency of the machine is largely affected by pull angle and peripheral speed. With the increase in pull angle, plant pulling efficiency increases.

Fig. 11.23 : Tractor operated cotton stalk puller

Tractor operated ratoon management device with discs

A tractor operated ratoon management device with discs has been developed at Indian Institute of Sugarcane Research, Lucknow for performing stubble shaving, off-barring and fertilizer application in sugarcane ratoon crop without disturbing the trash-mulch and is suitable for widely spaced single row or paired row crop (Anonymous, 2014). It is equipped with two discs for off-barring of both sides of row and opening of slit (without disturbing the trash mulch) for application of fertilizer near to root zone (Fig. 11.24). Provision has been made for adjustment of height of the discs as per the height of the ridges in the field. An innovative approach has been developed and incorporated in the equipment for application and regulation of fertilizer. The rate of fertilizer is varied by varying the rotational speed of the metering augers (specially designed mild steel augers for free flow of fertilizer). The rotary speed of metering augers is varied by varying the peripheral length of the mild steel ground wheel (used for transmission of power to metering augers). The ground wheel has been provided with spikes. The peripheral length of ground wheel is varied by varying the length of spikes. The greater the peripheral length of ground wheel lesser will be the rotational speed of the metering augers for a particular forward speed of the equipment. The equipment is designed to vary the fertilizer from 75 to 150 kg/ha. A serrated disc has been provided for shaving of stubbles close to the ground surface. The height of disc could also be varied according to the height of ridges in the field. The power to serrated disc is transmitted from tractor P.T.O. through a power train consists of pulley gear box, propeller shaft and universal joint crosses.

a) A tractor operated ratoon management device with discs developed at IISR Lucknow

b) A tractor operated ratoon management device
Courtesy: Deccan Farm equipment Pvt. Ltd. Kolhapur (Maharashtra)
Fig. 11.24 : Tractor operated ratoon management device with discs

Tractor operated plant residue shredder

After harvesting of sugarcane large quantity of sugarcane trash and stubbles are left in the field. In order to remove the trash from the field, farmers either burn the trash *in situ* or employ labour for collection and removal from the field. Now-a-days due to environmental concerns of burning and labour scarcity in collection and removal of trash, cane growers are opting for keeping the trash in the field for decomposition for improving the soil health by way of addition of organic matter or using it as trash mulch for moisture conservation (Anonymous, 2014). In view of the above, a tractor operated plant residue shredder has been developed at Indian Institute of Sugarcane Research, Lucknow for reduction of size of the trash *in situ* in the field and shaving of stubbles. It consists of a mild steel framework, pair of rotating blades in zig-zag profile, power transmission unit and chemical application unit (Fig. 11.25). Power to the blades is transmitted through tractor PTO. Power transmission unit consists of telescopic propeller shaft, universal joint crosses, and gear reduction box.

Fig. 11.25 : Tractor operated plant residue shredder (Stationery and field operation)

Tractor operated shredder

Manually clearing of crop stalks from the field requires 70-100 man-h/ha to pull out or cut stalks, gathering and removing. Tractor operated shredder is used to cut and shred the cotton stalk in the field (Fig. 11.26). Feeding system consists of two feeder drums with disc cutters for cutting stalks and two pressure roller drums one with spring loaded swing type for uniform feeding of material to shredding unit (Fig. 11.27). Shredding system consists of fly wheel with 6 blades rotating at 1600 rpm for more productive chopping. Paddle fitted on the periphery of fly wheel gives added lift to chopped material that spreads evenly in the field or can be collected into trolley. It can be used for row crops such as cotton, caster, maize, chilly and other similar crops. The machine can be operated by 40 hp tractor and above with dual clutch. It is attached to PTO of the tractor and operated at 540/1000 rpm. Shredded crop can be spread in the field or can be collected in the trolley which is rich bio mass for agricultural usage such as fodder and bio fuel; and industrial usage such as paper pulp, particle board, power generation etc.

Fig. 11.26 : Tractor operated shredder (stationer as well as in field)
Courtesy: Tirth Agro Technology Pvt. Ltd. Rajkot (Gujarat)

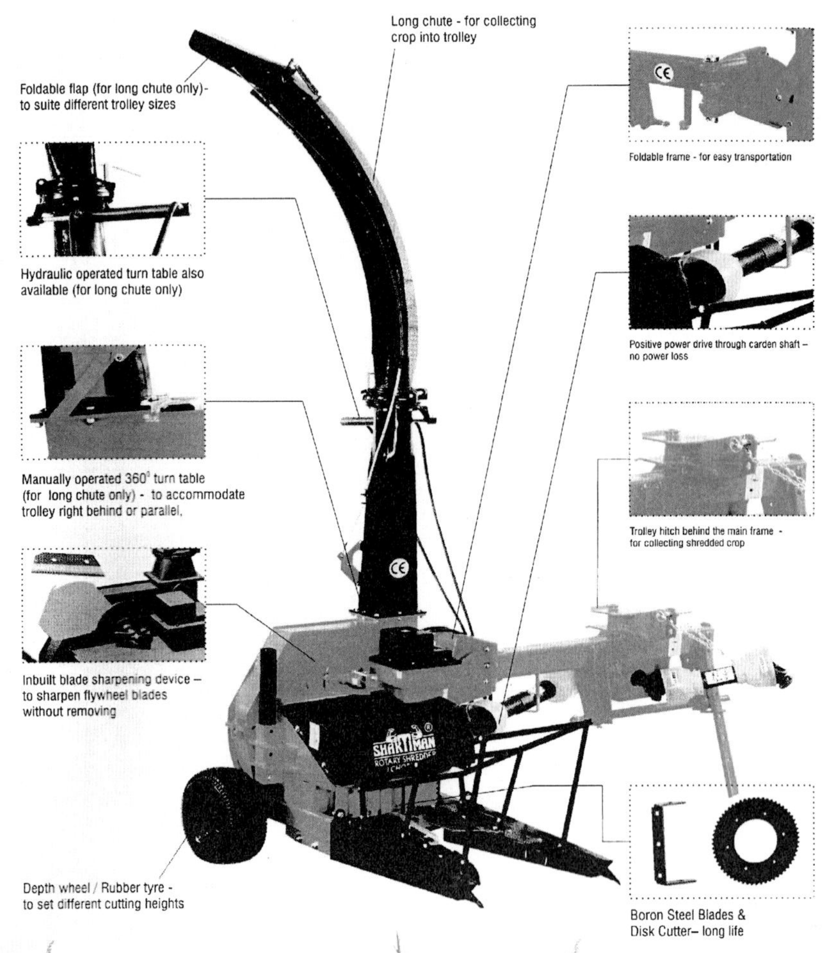

Fig. 11.27 : Details of tractor operated shredder
Courtesy: Tirth Agro Technology Pvt. Ltd. Rajkot (Gujarat)

Plant shredder for vermi compost

The organic cultivation of agriculture crops and horticultural crops is getting vide popularly in the country. The organic products are fetching good export value in the international market thus area is increasing under organic cultivation. The vermi compost is one of the major ingredients in organic cultivation prepared from animal dung or biogas slurry. Though cattle population is increasing still there is shortage of dung as it is mainly used by the farmers for home cooking thereby reducing its availability for vermi compost. There are several biomass material grown in the nature that are rich in NPK contents but not being used for any productive purposes. A plant shredder is a very effective device which can be used to cut the plants or weeds into very small pieces, which can be used as supplement raw material for vermi compost (Anonymous. 2010a). Plant shredder is having a rotating disc cutting unit at an rpm of 2150 operated by 3 to 5 hp motor. Machine has self feeding system. The rotating disc with mounted knives cut the plant in to very small pieces which can be mixed with cow dung as a raw feeding material for worms (Fig. 11.28). The weed is shredded at 120 kg/h, with length of cut 1.3-4.6 cm and average length of cut 2.8 cm at load conditions.

Fig. 11.28 : Electric operated plant shredder for vermi compost

Plant shredders are also called bio shredders. This is a machine available with a electrical single phase motor with direct transmission or with a petrol/ diesel engine with belt transmission or a tractor version with three point linkage. It is used to mulch garden waste, hedge clippings and small pruned branches. The material to be shredded is fed into loading hopper and is shredded by a fixed blade of tempered steel. The blades on the back of the shredding rotor eject the material and create a strong suction inside the hopper which facilitates descent of green material and leaves. They are available in different sizes as per power availability and can shred the materials from 50 to 90 mm diameter. The machine has multiple cutting systems formed by a special rotor with 16 mobile hardened steel hammers, four fixed steel blades placed on the edge of the rotor and four buckets inside the rotor for ejecting the cut material (Fig. 11.29). The direct ejection of the material through the rotor without the use of additional fan and of a related transmission system, reduces noise level and make maintenance operation faster. Inside the hopper there is a hydraulic feeding roller with adjustable speed. With four pneumatic tyres the machine can be taken to the place of shredding material.

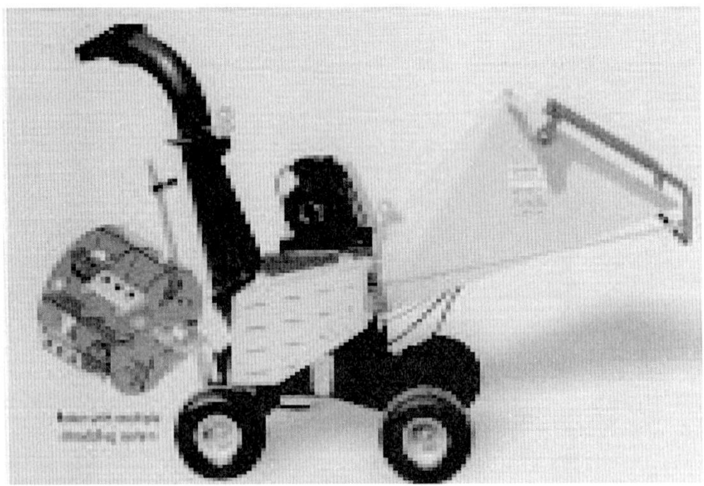

Fig. 11.29 : Bio shredder

Courtesy: Falcon Garden Tools (P) Ltd. Ludhiana (Punjab)

Ratoon management device (RMD)

Ratoon manager cuts the sugarcane stubble or any other stubble at ground level (Anonymous, 2014). It has a fertilizer box assembly applying fertilizer simultaneously (Fig. 11.30). It can be operated by 35 hp tractor and above (Fig. 11.31). Ridge cutting discs can also be attached to smoothen the ridges. It has ratoon manager cutting blade assembly (Fig. 11.32). It can cover 0.12 to 1.50 ha/h. It functions as stubble shavings, deep tilling, manure/ bio-fertilizer and liquid chemical application, and earthing up.

Fig. 11.30 : Tractor operated ratoon management device

Fig. 11.31 : Tractor operated ratoon management device working in field

Fig. 11.32 : Ratoon manager cutting blade assembly

Tractor operated rake

Tractor operated rake is a farm equipment used for collecting swath of bio mass spread in the field into a windrow (Singh, 2007). This helps enable the field bales to pick up the bio mass from the windrow made by rake. The working width is about 3.5 m. It has 9 arms and number of double tines per arm three (Fig. 11.33). Rotor diameter is 2.9 m. It has four tyres of tandem type to support the machine in the field as well as for transport purpose. Transport width is 1.5 m. It is operated by PTO of tractors of 35 hp and above and PTO rpm is maintained in the range of 350 to 450. It increases the efficiency of baler by 25-60% by reducing the number of passes. It can work in all types of terrain and slopes. Pivoted frame design and shock absorbers present in the machine gives minimum vibration and uniform and comfortable raking. This machine is used in biomass (such as cotton stalks, sugarcane trash, sorghum, bajra, maize, paddy straw and other similar bio mass) power generation and co-generation.

Fig. 11.33 : Tractor operated rake working in the field

Courtesy: New Holland Fiat (India) Pvt. Ltd.

References

Ahuja S S. 2001. Wheat Straw Combines: Field Performance and Economic Evaluation. Proceedings of Summer School on 'Advances in Seeding and Harvesting Machinery'. Dept. Farm Power & Machinery, PAU Ludhiana. May 21 to June 19. Pages 185-192.

Ahuja S S; Kalkat H S; Sharma V K. 1993. A perspective on wheat straw combine-its field performance and economic evaluation. Paper presented (FPM-93-2-96) at XXVIII Annual Convention of ISAE held at CIAE Bhopal, 2-4 March, 1993.

Anonymous. 1999. Souvenir on Paddy Straw Management at Annual Day of Punjab Chapter of ISAE, held at PAU, Ludhiana on February 26, 1999.

Anonymous. 2002. Annual report of Mechanization of rice-wheat cropping system for increasing the productivity (NATP-project), Department of Farm Power and Machinery, PAU, Ludhiana.

Anonymous. 2008. Research Highlight. AICRP on Farm Implements and Machinery, CIAE Bhopal. Technical Bulletin No.: CIAE/2008/141.

Anonymous. 2008a. Success Stories. AICRP on Farm Implements and Machinery, CIAE Bhopal. Extension Bulletin No.: CIAE/FIM/2008/80.

Anonymous. 2008b. QRT Report. 2002-07. AICRP on Farm Implements and Machinery, TNAU Coimbatore.

Anonymous. 2010. Research Highlight. AICRP on Farm Implements and Machinery, CIAE Bhopal. Technical Bulletin No.: CIAE/2010/151.

Anonymous. 2010a. Annual Report. 2008-10. AICRP on Farm Implements and Machinery, MPUAT Udaipur.

Anonymous. 2010b. Success Stories. AICRP on Farm Implements and Machinery, CIAE Bhopal. Extension Bulletin No.: CIAE/FIM/2010/83.

Anonymous. 2013. Research Highlight. AICRP on Farm Implements and Machinery, CIAE Bhopal. Technical Bulletin No.: CIAE/FIM/2013/158.

Anonymous. 2014. Annual report (2013-14). Division of Agricultural Engineering, Indian Institute of Sugarcane Research, Lucknow.

Anonymous. 2015. Research Highlight. AICRP on Farm Implements and Machinery, CIAE Bhopal. Technical Bulletin No.: CIAE/FIM/2015/179.

Garg I K; Singh Surendra. 2002. Farm equipment for Punjab agriculture. Department of Farm Power & Machinery, Punjab Agricultural University, Ludhiana.

Garg I K; Singh Surendra; Sharda Ajay. 2002. Paper presented at Group meeting on "Straw management in Combine Harvested rice wheat fields", May 6-7, 2002, held at Department of Farm Power & Machinery, PAU, Ludhiana.

Gupta Ram Autar. 2006. Design, Development and Evaluation of a Tractor Operated Cotton Stalk Puller. Unpublished Ph. D. thesis. PAU Ludhiana

Manes G S; Dixit A K; Singh Surendra. 2010. Comparative evaluation of flail type and cutter bar type straw chopper-cum-spreader for rice straw management. *Journal of Institution of Engineers (Agril. Engg).* Vol. 91 (December): 3-5.

Manes G S; Dixit Anoop; Singh Surendra; Sharda Ajay. 2007. Performance Studies on Rice Straw Chopper-cum-Spreader. Paper presented at 20th National Convention of Agricultural Engineers and National Seminar on "Farm Mechanization for Diversification of Agriculture" held at Punjab Agricultural University, Ludhiana from January 19 to 20.

Pandey M M; Majumdar K L; Singh Gyanendra; Singh Gajendra. 1997. Farm Machinery Research Digest. Technical Bulletin No. CIAE/97/69, Central Institute of Agricultural Engineering, Bhopal, 328 p.

Singh Sandhya; Singh Surendra; Dixit Anoop. 2007. Performance evaluation of field balers. *J. of Agril. Engg.* Vol. 44(1): 43-47.

Singh Surendra. 2007. Farm Machinery – Principles and Applications. Directorate of Information & Publication of Agriculture, Indian Council of Agricultural Research, Krishi Anusandhan Bhawan-I, Pusa Campus, New Delhi.

Singh Surendra; Garg I K; Sharda Ajay. 2003. Machines for Straw Management. Paper presented at National Seminar on 'Appropriate Mechanization Technologies for Energy Management in Agriculture' held at Northern Region Farm Machinery Training and Testing Institute, Hissar (Haryana) on March 26.

Singh Surendra; Mehta Vaishali; Sharda Ajay. 2005. Economics of using straw baler for paddy-straw management. *J Res Punjab agric Univ* 42(1): 78-82.

Singh Surendra; Pandey M M. 2008. X Plan Achievements (2002-2007). AICRP on Farm Implements and Machinery, CIAE Bhopal. Technical Bulletin No.: CIAE/2008/137.

Singh Yadvinder; Sidhu H S; Singh Manpreet; Dhaliwal H S; Blackwell John; Singh Rajinder Pal; Singla Neena. 2009. Happy Seeder - A Conservation agriculture technology for managing rice residues. Technical Bulletin, Department of Soils. PAU Ludhiana.

Verma S R; Kalkat H S; Singh Joginder. 1992. Straw combine-A new development in Agril. Machinery Agril. *Engineering Today*, ISAE 15-16 (1-6):24-30.

List of Manufacturers

Who have supported in writing of this book through photographs and drawings

S.No.	Name of Firm & Address	Contact Details
1.	A J Precision & Automation Pvt. Ltd. C-40, Sector-81, Noida Distt-G.B.Nagar-201305 (Uttar Pradesh)	Ph: 0120-4558773 Mob: 9818991973; 9210900192 Email: ajprecision@gmail.com
2.	Amar Agricultural Implements Works Amar Street, Janta Nagar, Gill Road Ludhiana-141003 (Punjab)	Ph: 0161-2491780, 2493128, 3299540 Fax: 0161-5019886, 2814036 Mob: 98720-18040 Email: amaragri@gmail.com
3.	Almighty Agrotech Pvt.Ltd. G/1934-35 Lodhika-G.I.D.C. Almighty Gate Kalawad Road, Metoda – 360021 Dist. Rajkot (Gujarat)	Ph: 02827-287307 Fax: 02827-287508 Mob: 9825612142 Email: aap@almightyagro.com Website: www.almightyagro.com.
4.	Bhagwati Krishi Udhyog A/6, First Floor, Satyamev – I Opp. New High Court, S.G. Highway Sola Ahmedabad – 380 060 (Gujarat)	Ph: 079-27665468 Fax: 079-27665216 Mob: 9825050051, 9377177203 Email: bhagwatikrushi.gokul@yahoo.co.in
5.	Bharat Industrial Corporation Akalsar Road, Moga – 142 001 (Punjab) Er. Baldev Singh Hunjan	Ph: 01636-224075, 222075Fax: 01636-224076 Mob: 98140-69075, 9878029075Email: baldev_moga @satyam net.in
6.	Birar Equipments B-21, Road A, Street No. 4 NICE M.I.D.C., Satpur Nashik – 422 007 (Maharashtra)	Mob: 9822258471, 9028318954 Ph: 0253-2350148 Email – anilnavyug@gmail.com Website: www.navyugagrotech.com

7.	Bull Agro Implements SF No. 200-1B6, Kannampalayam Post, Ravathur Pirivu, Coimbatore - 641402 (Tamil Nadu)	9842272522 Ph: 0422-2910080 Email: chandramohan@bullagro.com
8.	Bull Machines Pvt Ltd. Sf No:5/1A, L&T Bypass Jm., Trichy Road Chinthamanipudur Coimbatore - 641103 (Tamil Nadu)	Mob: 9842906634 Ph: 0422-2270859
9.	Bharat Industries C.K. 65/468, Piari Kala, Kabir Road Varanasi (Uttar Pradesh)	Mob: 9935413241, 9936521139
10.	Balaji Agricultural Ind. (P) Ltd., Mama Bhanja ka Talab, Reewa Road Allahabad (Uttar Pradesh)	Mob: 9838201474
11.	BCS India Pvt. Ltd. Vill. Maangarh, P.O. Kohara Kohara-Machhiwara Road P O Kohara, Dist. Ludhiana – 141 112	Ph: 0161-2848597 Fax: 0161-2848598 Mob: 9872874743 Web: www.bcs-ferrari.in Email: skbansal@bcs-ferrari.in
12.	Bharat Agro Products 1/145-C, Madurai-Mandapam Road Opp. Union Office Paramakudi-623707 (Tamil Nadu)	Ph: 04564-229230, 223244, Mob: 9345810555, 9443323185 Email: anupreethi@india.com, bharathagro@india.com
13.	Bihar Ma Durga Agro Industries Pvt. Ltd. Plot No. 61-62 P, Industrial Estate Biada, Pandual Dist. Madhubani (Bihar)	Mob: 9431860000, 9431870000 Email: mashyama2009@gmail.com
14.	CLAAS India Ltd. 15/3 Mathura Road Faridabad – 121 003	Ph.: 0129-4297000, 2270660, 2255977 Website: www.claas.com Email: jan.tobias@claas.com manish.pradhan@claas.com
15.	Dashmesh Mechanical Works Nabha-Malerkotla Road, Amargarh Distt. Sangrur-148022 (Punjab)	Ph: 01675-284221, Mob: 98151-74313 Fax: 01675-285999 Email: info@dasmesh.net

16.	Deccan Farm Equipment Pvt Ltd C- 35,M.I.D.C. Shiroli Kolhapur – 416122 (Maharashtra)	Mob: 9764999857 Email: accounts@deccanequipments.com bharat@ deccanequipments.com
17.	Droli Mechanical Works Near Majestic Cinema Majestic Road, Moga-142001 (Punjab)	Mob: 98413-60027 Ph: 01636-223487 Emaol: droli_dmw_moga@yahoo.co.in drolidmwmoga@yahoo.co.in
18.	Droli Industries (Regd) (BASANT) Near Dhaliwal Hospital, Majestic Road Moga-142001 (Punjab)	Ph: 01636-223557 Mob: 98140-29811 Email: basant_droliind@yahoo.co.in
19.	Essey Engineering Co. MIDC, Satpur, Nashik (Maharashtra)	Mob: 9422256754 Email – info@esseyloaders.com
20.	Falcon Industries 2524, Janta Nagar, St. No. 5,Gill Road, Ludhiana – 141 003 (Punjab)	Ph: 0161-2490877,2494840 Fax- 0161-2503156 Mob: 09872620996 Email: falcongardentools@rediffmail.com falcontools@rediffmail.com
21.	Farm Implements (India) Pvt. Ltd. 13, Kumarappa Street Nungambakkam Chennai- 600 034 (Tamil Nadu)	Ph: 044-28261676, 28273493, 42137084 Fax: 044-28265345 Email: rotavato@vsnl.com balachandra.babu@gmail.com
22.	Ganesh Agro Equipments At & PO: Vadpura Opp. Navajivan Hotel, Ta: Kadi Dist: Mehsana - 382706	Mob: 09925031817 Email: sanjay@ganeshraj.com
23.	Govind Industries Pvt. Ltd., Lucknow Road, Near Railway Crossing Barabanki – 225 001 (Uttar Pradesh)	Ph: 05248-223056, 227285, Mob: 9415048612 Email:Deepak_gobind@rediffmail.com
24.	Greenfield Equipments India Pvt Ltd. 237-A4, AVG Layout SIDCO Industrial Estate Coimbatore - 641021 (Tamil Nadu)	Mob: 9367676767 Ph: 0422-2676767 Email: visu@greenfield.in

25. Gursukh Agro Works
Vill. Jhakroudi, P.O. Samrala
Ludhiana – 141 114
Ph: 01628-262422
Mob: 9417100150
Email: ppsguron@rediffmail.com

26. John Deere India Pvt. Ltd.
Cybercity, Magar Patta city
Hadpsar, Pune – 411018 (Maharashtra)
Ph: 020 – 66481775,
Mob: 9158892273
Email: patilkrishant@johndeere.com
jainsachin2@johndeere.com

27. Kailash Krishi Yantra Udyog
NH-8, Ajmer Road Bagru
Jaipur – 303007 (Rajasthan)
Ph: 0141-2865189
Mob: 9829433761, 9829114604
Email: info@kkyujpr.com

28. Kartar Agro Industries Pvt. Ltd.
Amloh Road, Bhadson-147202
Patiala (Punjab)
Ph: 01765-260136
Mob: 9217100236
Email:
kartarcombine@kartarindia.com
kartar1234@rediffmail.com

29. Khalsa- Punjab Engineers (Khalsa)
Dashmesh Nagar, Bagpat Road, Meerut
(Uttar Pradesh)
Mob: 9837340086
Email: khalsa_agri@yahoo.co.uk

30. Khedut Agro Engineering
Plot No. 6, Survey No. 191
Shantidham Society Road
Near Orke Farm
Veraval (Shaper) - 360024
Ta. Kotada Sangani, Dist. Rajkot (Gujarat)
Mob: 9426206420
Telefax: 02827-253312
Email: dinesh_patel63@yahoo.com
Website: www.khedutagro.com

31. Kisan Engineering Works
C-57, MIDC, Awadhan
Dhule – 424 006 (Maharashtra)
Mob: 9422289812
Ph: 02562-246612
Email –
kisanengg23812@yahoo.co.in

32. Kisan Engineering Works
Bela Industrial Estate
P.O: R.K. Ashram (MIC Bela)
Muzaffarpur- 842005 (Bihar)
Mob: 09431238670, 09931436950
e-mail: vishnukec@yahoo.co.in

33. Lemken India Agro Equipment Pvt. Ltd.
D-59, MIDC Butibori
Nagpur - 441108 (Maharashtra)
Ph: 07104-305401
Mob: 9545022258
Email: s.chhabra@lemken.com

34.	Madho Agro Industries B-3, Industrial Focal Point Moga-142001 (Punjab)	Ph: 01636-226135, 329906, 220106 Mob: 98151-34421, 98155-33106 Email: 1mai@indiatimes.com Web: www.madhoagro.com
35.	Mecfa Enterprises C.F.C. Buildings, Big Industrial Estate, Chandpur Varanasi – 221 106 (Uttar Pradesh)	Mob: 9839056569; 9838656569 Email: nerajparikh@yahoo.com namitparikh@yahoo.com
36.	National Agro Industries Link Road, Industrial Area A Ludhiana-141003 (Punjab)	Ph: 0161-2222041, 5087853 Fax-0161-2220299 Mob: 9815043000, 9815064000 Email: sales@nationalagro.com
37.	New Holland Fiat (India) Pvt. Ltd. Plot No. 3, Udyog Kendra Greater NOIDA – 201 306 Distt. Gautam Budh Nagar (U.P.)	Ph.: 0120-2350401/02/03 Fax: 0120-2350424-25 Mob: 9811432965, 9313869529 E-mail:gaurav.sood@cnh.com, sanjay.vij@cnh.com
38.	Redlands Ashlyn Motors 2/575, Madukkai Road, Mulamachampatty Coimbatore – 641021 (Tamil Nadu)	Mob: 9387103530 Ph: 0487-2427392 Email: sales@redlandsmotors.com
39.	Renaaissance Power Products Pvt. Ltd. SF No. 92-A/2, Door No. 1/138, Karadivavipudur Karadivavi (PO), Palladam Taluk Coimbatore 641 658	Email: renaaissancepp@yahoo.in Mob: 9600930640
40.	RJ Sekar Industries Kuppayeevalasu Road, Porulur Post Oddanchatram Dindigul - 624616 (Tamil Nadu)	Mob: 9787242452 Ph: 04553 260749
41.	Redlands Ashlyn Motors (P) Ltd. Regency Complex Karikkath Lane M G Road, Thrissur (Kerala)	Mob: 9446568563 Email: ashlynchem@hotmail.com
42.	Santhosh Agri Machinery 282, Kartar Complex Salem-Cuddalore Main Road Narasingapuram Attu (Taluk) Salem (Dist) – 636108 (Tamil Nadu)	Mob :7373032407 Email: santhosh.machinery@gmail.com

43. South East Farm Equipments Pvt. Ltd.
214/94-C Trichy Main Road
Thammampatty Post
Salem District - 636 113
Ph: 04282 – 226638, 227326
Telefax: 04282-226638
Mob; 9244344326, 9244344328
Email: agrotop@india.com

44. Tractor & Farm Equipment Ltd. (TAFE)
Hazur Gardens, Sembian
Chennai - 600 011
Ph: 044-28279073
Fax: 044-28228902
Mob: 9841096923
Email: trkesavan@tafe.com
rammohan@tafe.com

45. Tirth Agro Technology Pvt. Ltd.
Near Hotel Krishna Park
NH-8B, Gondal Road, Vivdi
Rajkot – 360004 (Gujarat)
Ph.: 0281-2386047, 2377204
Mob: 9825072637
Fax: 0281-2380852
Email: info@shaktimanagro.com
Web: www.shaktimanagro.com

46. Umiya Agriculture Industries
52-53, G.I.D.C. Estate
B/h D. Patel Petrol Pump
Kheralu Road, Visnagar – 384315 (Gujarat)
Telefax: 02765-231593/220324
Mob: 9825055707, 9825231875
Email: uai_25@yahoo.co.in

47. V- TECH AGRI ENGG.
Kuntuvalli, Melige -577514
Thithahalli tq, Shimoga dist. (Karnataka)
Mob: 9448105006

48. V.S.T. Tillers Tractors Ltd.
P.B. No. 4801, Mahadevapura P.O.
Whitefield Road
Bangalore – 560 048 (Karnataka)
Ph.: 080-28510805/06/07,
28510275, 28510318
Fax: 080-28510221
Website: www.vsttillers.com
Email: vstgen@vttlhq.com

❑❑❑